mm-Wave Silicon Power Amplifiers and Transmitters

HOSSEIN HASHEMI
University of Southern California

SANJAY RAMAN
Virginia Tech

CAMBRIDGE
UNIVERSITY PRESS

CAMBRIDGE
UNIVERSITY PRESS

University Printing House, Cambridge CB2 8BS, United Kingdom

Cambridge University Press is part of the University of Cambridge.

It furthers the University's mission by disseminating knowledge in the pursuit of education, learning and research at the highest international levels of excellence.

www.cambridge.org
Information on this title: www.cambridge.org/9781107055865

© Cambridge University Press 2016

This publication is in copyright. Subject to statutory exception and to the provisions of relevant collective licensing agreements, no reproduction of any part may take place without the written permission of Cambridge University Press.

First published 2016

Printed in the United Kingdom by Clays, St Ives plc

A catalog record for this publication is available from the British Library

Library of Congress Cataloging in Publication data
mm-wave silicon power amplifiers and transmitters / edited by
Hossein Hashemi (University of Southern California), Sanjay Raman
(Virginia Tech).
    pages   cm – (The Cambridge RF and microwave engineering series)
Includes bibliographical references and index.
ISBN 978-1-107-05586-5
1. Millimeter wave devices – Design and construction. 2. Power amplifiers – Design and construction.   3. Metal oxide semiconductors, Complementary.   I. Hashemi, Hossein, editor.
II. Raman, Sanjay, editor.   III. Series: Cambridge RF and microwave engineering series.
TK7876.5.M588   2015
621.381'325–dc23   2015008270

ISBN 978-1-107-05586-5 Hardback

Cambridge University Press has no responsibility for the persistence or accuracy of URLs for external or third-party internet websites referred to in this publication, and does not guarantee that any content on such websites is, or will remain, accurate or appropriate.

Contents

List of Contributors		*page* ix
Preface		xi

1   **Introduction**   1
Hossein Hashemi and Sanjay Raman

1.1	Why mm-waves?	1
1.2	Why silicon?	3
1.3	Wireless communication basics	4
1.4	Wireless transmitter architectures	6
1.5	Power amplifier basics	7
1.6	Examples of commercial mm-wave applications	8
1.7	Examples of military mm-wave applications and initiatives	9
1.8	Conclusions	13
	Acknowledgments	13
	References	13

2   **Characteristics, performance, modeling, and reliability of SiGe HBT technologies for mm-wave power amplifiers**   17
David Harame, Vibhor Jain, and Renata Camillo Castillo

2.1	Introduction	17
2.2	Bipolar device physics	18
2.3	SiGe HBT processing and structures	27
2.4	Key circuit design metrics	37
2.5	BiCMOS passive devices and features	47
2.6	SiGe HBT modeling and characterization	52
2.7	Reliability	58
2.8	Performance limits of SiGe HBTs	64
2.9	Summary and conclusions	66
	References	68

3   **Characteristics, performance, modeling, and reliability of CMOS technologies for mm-wave power amplifiers**   77
Antonino Scuderi and Egidio Ragonese

3.1	Introduction	77
3.2	Materials for high frequency: CMOS and its evolution	78

	3.3	CMOS active devices	79
	3.4	Nonlinearities	91
	3.5	Noise	98
	3.6	Thermal effect	102
	3.7	Large-signal performance degradation and reliability	105
	3.8	CMOS passive devices	114
	3.9	Measurement and modeling issues	118
	3.10	CMOS trends: SOI	121
	3.11	Conclusions	132
	Acknowledgment		133
	References		133

4 Linear-mode mm-wave silicon power amplifiers — 139
James Buckwalter

4.1	Why linear?	139
4.2	Linear amplifier design: large-signal device characterization	142
4.3	Gain of mm-wave amplifiers	148
4.4	Linear classes of operation	155
4.5	Optimization of mm-wave amplifiers: why linear?	162
4.6	Case study: Q-band SiGe power amplifier	166
4.7	Doherty amplifiers	170
4.8	Case study: a Q-band Doherty power amplifier	175
4.9	Summary	177
References		178

5 Switch-mode mm-wave silicon power amplifiers — 180
Harish Krishnaswamy, Hossein Hashemi, Anandaroop Chakrabarti, and Kunal Datta

5.1	Introduction to switching power amplifiers	180
5.2	Design issues for CMOS mm-wave switching power amplifiers	181
5.3	Design issues for SiGe HBT mm-wave switching power amplifiers	187
5.4	Linearizing architectures for switch-mode power amplifiers	200
5.5	Conclusions	203
References		204

6 Stacked-transistor mm-wave power amplifiers — 207
Peter Asbeck and Harish Krishnaswamy

6.1	Introduction	207
6.2	Motivation for stacking	207
6.3	Principles of transistor stacking	211
6.4	Transistor stacking for switch-mode operation	216
6.5	Application of stacking at microwave frequencies	218
6.6	Si device technology for stacked designs	223
6.7	Stacked FET mm-wave design	226

	6.8	Design of mm-wave stacked-FET switching power amplifiers	232
	6.9	Stacking versus passive power-enhancement techniques	238
	6.10	Harmonic matching in stacked structures	240
	6.11	Active drive for stacked structures	240
	6.12	Case studies and experimental demonstrations	241
	6.13	Summary and conclusions	253
	6.14	Acknowledgments	254
	References		254

7 On-chip power-combining techniques for mm-wave silicon power amplifiers — 257
Tian-Wei Huang, Jeng-Han Tsai, and Jin-Fu Yeh

7.1	On-chip power-combining techniques	257
7.2	Direct-shunt power combining	262
7.3	2D power combining	265
7.4	3D power-combining technique	282
7.5	Conclusion	297
References		298

8 Outphasing mm-wave silicon transmitters — 302
Patrick Reynaert and Dixian Zhao

8.1	Introduction	302
8.2	Outphasing basics	304
8.3	Outphasing signal generation	307
8.4	Outphasing signal combining	313
8.5	Outphasing non-idealities	317
8.6	Case study: 60-GHz outphasing transmitter	321
8.7	Conclusions	329
References		331

9 Digital mm-wave silicon transmitters — 334
Ali M. Niknejad and Sorin P. Voinigescu

9.1	Motivation	334
9.2	Architectures for high efficiency/linearity	337
9.3	Digital mm-wave transmitter architectures with on-chip power combining	344
9.4	Digital antenna modulation	356
9.5	Conclusion	371
References		373

10 System-on-a-chip mm-wave silicon transmitters — 376
Brian Floyd and Arun Natarajan

10.1	Introduction	376
10.2	Multi-Gb/s wireless links at mm-wave frequencies	376

	10.3	On-chip mm-wave transmitter architectures	380
	10.4	Single-element transmitters	386
	10.5	Phased-array transmitters	387
	10.6	Millimeter-wave transmitter examples	398
	10.7	Conclusion	415
	References		416
11	**Self-healing for silicon-based mm-wave power amplifiers**		**419**
	Steven M. Bowers, Kaushik Sengupta, Kaushik Dasgupta, and Ali Hajimiri		
	11.1	Background	419
	11.2	Introduction to self-healing	421
	11.3	Sensing: detecting critical performance metrics	426
	11.4	Actuation: countering performance degradation	433
	11.5	Data converters: interfacing with the digital core	439
	11.6	Algorithms: setting the actuators based on sensor data	442
	11.7	System measurements of a fully integrated self-healing PA	445
	11.8	Conclusions	453
	Acknowledgment		453
	References		453
	Index		457

Contributors

Peter Asbeck
University of California, San Diego

Steven M. Bowers
University of Virginia

James Buckwalter
University of California, Santa Barbara

Renata Camillo Castillo
IBM

Anandaroop Chakrabarti
Columbia University

Kunal Datta
University of Southern California

Kaushik Dasgupta
Intel

Brian Floyd
North Carolina State University

David Harame
Formerly of IBM, now with Global Foundries

Ali Hajimiri
California Institute of Technology

Hossein Hashemi
University of Southern California

Tian-Wei Huang
National Taiwan University

Vibhor Jain
IBM

Harish Krishnaswamy
Columbia University

Arun Natarajan
Oregon State University

Ali M. Niknejad
University of California, Berkeley

Egidio Ragonese
ST-Microelectronics

Sanjay Raman
Virginia Tech

Patrick Reynaert
Katholieke Universiteit Leuven

Antonio Scuderi
Qualcomm

Kaushik Sengupta
Princeton University

Jeng-Han Tsai
National Taiwan Normal University

Sorin P. Voinigescu
University of Toronto

Jin-Fu Yeh
National Taiwan University

Dixian Zhao
Katholieke Universiteit Leuven

Preface

Silicon has become the uncontested technology of choice for commercial radio-frequency integrated systems such as those in smartphones, tablets, and televisions. Research over the past decade has demonstrated the feasibility of realizing complex silicon integrated systems at millimeter frequencies. There is little doubt that operation at millimeter waves not only offers advantages, but also is necessary in many commercial and noncommercial applications. Millimeter-wave integrated circuits for automotive radars and high-speed wireless connectivity are already in the market. The fifth-generation wireless standards will include millimeter-wave operation as an essential component to increase the overall capacity. The volume of millimeter-wave integrated systems may soon exceed billions of units per year.

Millimeter-wave operation has a long history. Sir Jagadish Chandra Bose demonstrated transmission and reception of 60 GHz electromagnetic waves over a distance of 23 m in 1895. The application of solid-state devices in the millimeter-wave range started in the second half of the twentieth century. In the 1970s, solid-state transceivers at 60 GHz were demonstrated primarily by using diodes for signal generation, frequency conversion, and amplification. Monolithic millimeter-wave receivers and transmitters were reported in the 1980s using III–V transistors capable of providing power gain well into the millimeter-wave region. Applicability of silicon technologies, including CMOS, for radio-frequency applications was established in the 1990s. Complex silicon integrated systems at millimeter waves were reported in the 2000s, and commercial products started entering the market shortly after.

Throughout history, technology has always been a limiting factor in the amount of radio-frequency signal power that can be generated. In the absence of devices that can provide power gain, generating electromagnetic signals will be power inefficient. Early demonstrations of electromagnetic signal generation at higher frequencies typically involve nonlinear processes such as harmonic signal generation. These approaches are gradually replaced with more linear amplification approaches once supporting technologies become available. In other words, efficient high-power electromagnetic signal generation and amplification oftentimes lags the demonstration and even deployment of wireless systems operating at those frequencies. It is hence natural to see that efficient high-power generation of millimeter-wave signals using silicon technologies is an ongoing research topic nearly a decade after the early demonstrations of complex silicon millimeter-wave integrated systems.

Not too long ago, silicon was considered to be incapable of serving as a proper technology for the realization of power amplifiers even at radio frequencies. In fact, the viability of CMOS in certain commercial RF wireless systems is still a debated topic. In 2009, the US Defense Advanced Research Projects Agency (DARPA) postulated that watt-level transmitter output power can be achieved in silicon technologies with efficiencies significantly beyond the state-of-the-art and these transmitters could be linearized on-chip to support high-order digitally modulated waveforms. This led to the launching of the Efficient Linearized All-Silicon Transmitters ICs (ELASTx) program, of which we were both key players (Sanjay as founding program manager, Hossein as leader of one of the key performer teams). In June 2012, we organized a workshop at the IEEE Radio Frequency Integrated Circuits (RFIC) Symposium with an ambitious title of "Towards Watt-Level mm-Wave Efficient Silicon Power Amplifiers." The workshop included talks by prominent individuals from academia and industry, including several ELASTx team members, covering challenges and research efforts around this topic. The enthusiasm from speakers and participants was accompanied by realistic skepticism about the viability of such an outrageous proposition in the near future. It is extremely gratifying to witness watt-level silicon power amplifiers and transmitters generating millimeter-wave signals efficiently from various research groups across the world a few short years after the workshop.

The seeds of this book were planted at the same IEEE RFIC Workshop. Cambridge University Press, led by Dr Julie Lancashire, concurred with our vision that a book on the topic of silicon millimeter-wave transmitters and power amplifiers is timely. We did not want the book to be a mere collection of research results that have appeared as papers over the past few years. The intent was to draft a book that includes technology, challenges, theory, and a systematic approach towards realization of silicon millimeter-wave power amplifiers and transmitters with research results offered as proof-of-concept case studies. Most chapters of the book are written in an advanced textbook style suitable for graduate students and practicing engineers. Many graphs and tables include comprehensive data about the relevant technologies, devices, and circuits, and serve as complete up-to-date references for researchers and developers. Maintaining consistency and flow across various chapters is a challenge in a multi-authored book. The authors have been very cooperative in drafting and revising their chapters in this spirit. It has been a pleasure to work with the world's top individuals in the area of silicon millimeter-wave integrated circuits for this project. We hope that all readers learn from reading this book as much as we did.

Editing a multi-authored book, especially when the authors are all prominent busy individuals, is not an easy task. It requires great patience and support. We have been lucky to work with the wonderful team at the Cambridge University Press on this project. Dr Julie Lancashire was nothing but graciously supportive and understanding over the past two years. Elizabeth Horne and Heather Brolly provided wonderful assistance. Thank you all!

1 Introduction

Hossein Hashemi and Sanjay Raman

Advancements in semiconductor technology have led to a steady increase in the unity power gain frequency (f_{max}) of silicon transistors, both in CMOS and SiGe BiCMOS technologies. This, in turn, enables realization of complex monolithic silicon integrated circuits operating at the millimeter-wave (mm-wave) frequency range (typically defined as 30–300 GHz). Prime target applications of silicon mm-wave integrated systems include high-speed wireless access, satellite communications, high-resolution automotive radars, and imagers for security, industrial control, healthcare, and other applications. However, scaling of silicon transistors for high f_{max} comes at the expense of reduced breakdown voltages, and hence limitations on output voltage swing and power. The link range and energy consumption of wireless systems are direct functions of the transmitter output power and efficiency, respectively. Efficient generation and amplification of radio-frequency (RF) modulated waveforms using silicon transistors is an ongoing challenge due to the reduced breakdown voltage of scaled silicon transistors, loss of passive components, and the conventional linearity–efficiency trade-off. This book covers the fundamentals, technology options, circuit architectures, and practical demonstrations of mm-wave wireless transmitters realized in silicon technologies.

1.1 Why mm-waves?

The main motivation to operate the wireless systems at higher carrier frequencies is the larger available bandwidth which translates to higher data rate in communication systems and higher resolution in ranging and imaging systems. Furthermore, the size of the antenna and circuitry, typically proportional to the wavelength, reduces with increasing carrier frequency. On the other hand, operating at higher frequencies poses two fundamental challenges. First, the loss of most materials increases with the frequency; therefore, compared with radio and microwave frequencies (below 30 GHz) the electromagnetic wave at mm-wave frequencies is attenuated more as it propagates in an environment (Fig. 1.1). It should be noted that over the mm-wave spectrum there are "windows" of relatively lower attenuation around 35 GHz, 90 GHz, 140 GHz, etc., and, consequently, these bands are often selected for mm-wave applications; on the other hand, high atmospheric attenuation levels around frequencies such as 60 GHz enable more aggressive frequency reuse, and are therefore often selected for small cell or secure communications applications. Second, the performance of semiconductor

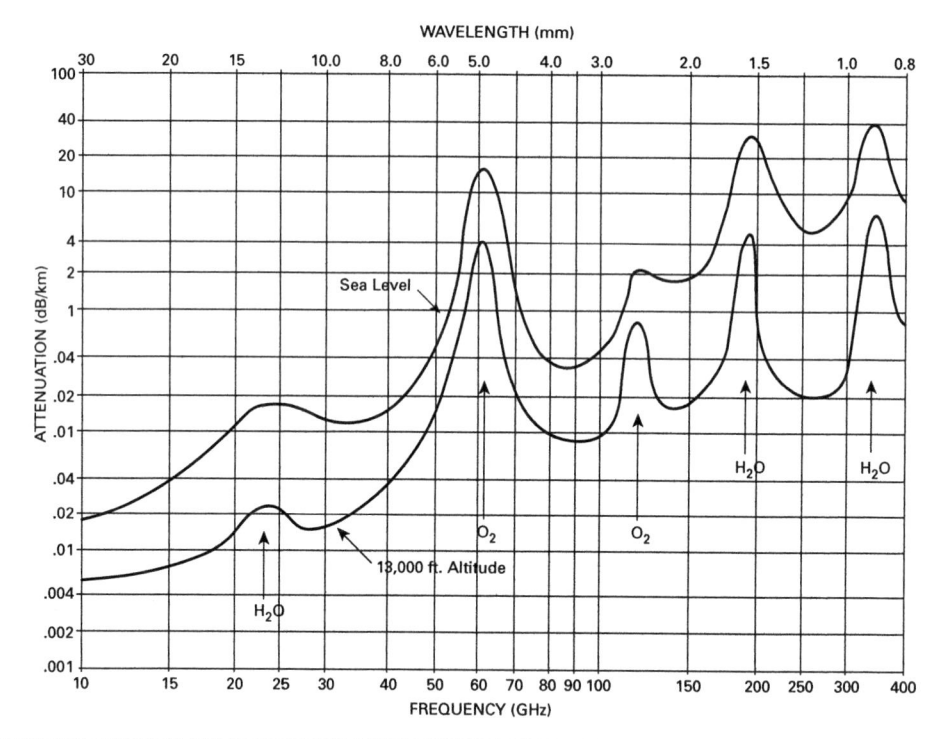

Fig. 1.1 Typical atmospheric attenuation in dB/km as a function of frequency.

devices worsens with frequency; this includes reduced gain, increased noise, and more nonlinearity at higher frequencies for a given technology.

Historically, mm-wave systems have been confined to defense, aerospace, and niche commercial applications due to the high cost of multi-chip-module (MCM) approaches and the need to use compound semiconductor processes to achieve required performance. However, over the past decade, there has been an explosion of research and development towards system-on-a-chip (SOC) realizations of complex wireless systems operating at mm-waves for high-volume commercial applications. Commercial complex mm-wave SOCs, such as 60 GHz phased array transceivers for high-speed wireless access, exist today [1–3]. Even certain defense applications, such as large-scale mm-wave phased arrays for helicopter operations, promise to be benefited by silicon implementations [4]. The main commercial applications being pursued at mm-waves include high-speed wireless connectivity with primary focus in the 60 GHz industrial, scientific, and medical (ISM) frequency band; high-resolution automotive radars with primary focus at the 77 GHz frequency band; mm-wave backhaul in the Ka and E bands; and active and passive RF imagers with primary focus at frequencies above 100 GHz. The fifth-generation commercial wireless standard (5G), targeting systems beyond 2020, is expected to include mm-wave operation for high-data-rate wireless access between small cells and mobile devices. High-resolution radar continues to be

IEEE Frequency Bands [GHz]	Frequency [GHz]	Applications	Applicable Standards
Ka (27-40)	28, 38	Mobile communications for 5G Cellular networks (emerging)	None yet available
	27.5-30	SATCOM uplinks (e.g., Inmarsat Global Xpress: 27.5-30) and downlinks (e.g., Iridium: 29.1-29.3)	MIL-STD-188-164, ITU-R S.524-9, FCC 25.138, ETSI EN 303 978
	24.25-30	L-3 ProVision imaging scanners at airports	IEEE C95.1
	35	Munitions and missiles seekers and sensors	Unknown
V (40-75)	43.5-45.5	U.S. AEHF military SATCOM system uplinks*	MIL-STD-3015
	57-66	"Last inch", short range wireless communications	IEEE 802.11ad, IEEE 802.11aj, IEEE 802.15c
	71-75	"Last mile", point-to-point backhaul wireless communications	ETSI TS 102 524
W (75-110)	75-76, 81-86, 92-95	"Last mile", point-to-point backhaul wireless communications	ETSI TS 102 524
	76-77	Autonomous cruise control (ACC) "long range" automotive radar	ETSI EN 301 091 parts 1 & 2
	77-81	Short range "stop & go" automotive radar	ETSI TR 102 263
	94	Missile seekers, collision avoidance radars	Unknown
	85-110	Imaging for medicine, biology, and security	IEEE C95.1
G (110-300)	110-120	Imaging for medicine, biology, and security	IEEE C95.1
	220-240	Long range wireless communications, atmospheric research radar	None yet available
	120, 183, 325	Short range wireless communications (emerging)	None yet available

* U.S. AEHF system downlink frequencies are located at 20.2 GHz – 21.2 GHz (IEEE K band).

Fig. 1.2 Summary of major mm-wave applications and applicable standards.

the primary application for mm-wave military systems. Figure 1.2 summarizes major mm-wave applications and their applicable standards.

1.2 Why silicon?

The advancement of silicon technologies, CMOS in particular, is motivated by performance gains of digital computation and signal-processing integrated circuits. Specifically, the computation speed and power consumption of digital circuits improve with technology scaling. The large investment required for advancing the silicon manufacturing technologies has been justified by the large demand due to the economy of scale. Thanks to groundbreaking research since the 1990s, today most of the RF functions of a wireless system are also realized in the same digital CMOS process leading to SOC realizations. Compared with the traditional multi-chip-module (MCM) approaches, SOCs reduce the cost, complexity, and power consumption, while enhancing robustness thanks to on-chip calibration, built-in self-test (BIST), and self-healing schemes. Furthermore, availability of "free" digital functions has enabled new system architectures with improved performance over conventional schemes.

The widespread usage of silicon technologies for complex mm-wave integrated systems is a result of large-scale research and development over the past two decades. Silicon technologies capable of operating at mm-waves were available in the 1990s [5, 6], followed by monolithic mm-wave circuit realizations shortly after [7]. Early efforts towards realization of complex mm-wave integrated circuits in standard silicon technologies were led by Caltech [8–11], IBM T. J. Watson Research Center [12, 13], UC Berkeley [14, 15], UCLA [16, 17], and the University of Toronto [18, 19] among others around the mid 2000s. Later, in addition to the aforementioned groups, several more research groups such as Georgia Tech [20], UCSD [21–24], National Taiwan University [25, 26], Intel [27], and Tokyo Institute of Technology [28] among others made significant contributions in the research and development of mm-wave silicon complex integrated systems.

While technology scaling provides transistors with higher transistor unity power-gain frequency, it also reduces the breakdown voltage and hence the maximum allowable voltage swing. In fact, there is an inverse relationship between maximum speed and breakdown voltage for a given semiconductor material. The lower allowable voltage swing degrades the signal-to-noise ratio (SNR) and linearity of many circuit building blocks, and also challenges efficient generation of high-power signals. Silicon technology does not lead to higher-performance circuit building blocks with a fixed topology when compared with a compound semiconductor realization. The main advantage of using a silicon technology for high-performance mm-wave systems, in addition to lower cost and footprint, is the higher performance of the entire system enabled by new integrated architectures that leverage combination of analog, mixed-signal, and digital designs.

Chapters 2 and 3 discuss the current state of the art in SiGe and CMOS technologies, respectively, for mm-wave transmitter applications.

Efficient, watt-level radio-frequency (typically <6 GHz) power amplifiers and transmitters now exist commercially. The choice of using silicon versus compound semiconductor technologies in an RF power amplifier depends on the specific application, market demand, and related economics. It is quite conceivable that the growth of wireless devices and connectivity at radio frequencies thanks to CMOS realizations will be repeated at mm-wave frequencies for a complementary set of applications.

1.3 Wireless communication basics

The general form of a modulated waveform, which can be the electric or magnetic field of a propagating electromagnetic wave, can be expressed in the so-called polar form as

$$x_{polar}(t) = a(t)\cos(\omega_c t + \phi(t)), \quad (1.1)$$

where ω_c is the carrier frequency, and $a(t)$ and $\phi(t)$ are the amplitude modulation (AM) and phase modulation (PM) portions of the waveform and contain the information. This expression can also be written in another form, commonly referred to as the Cartesian form, as

$$x_{Cartesian}(t) = I(t)\cos(\omega_c t) + Q(t)\sin(\omega t), \quad (1.2)$$

where $I(t)$ and $Q(t)$ contain the information and are referred to as the in-phase and quadrature-phase components, respectively. While the aforementioned two forms are mathematically interchangeable, they inspire different architectures for their generation and detection, in turn offering various implementation trade-offs. One notable difference between the aforementioned forms is that $a(t)$ and $\phi(t)$ occupy significantly larger bandwidth in the frequency domain as compared to $I(t)$ and $Q(t)$.

The main function of a wireless transmitter is the energy-efficient generation of a high-power modulated waveform as close to the aforementioned ideal forms as possible. Nonlinearity of the transmitter creates distortion in the generated waveform; moreover, noise can be added to amplitude and/or phase components as well. Distortion in the transmitted signal degrades the receiver detection, or, more specifically, the bit error rate (BER) in digital communication systems. Error vector magnitude (EVM), a metric that quantifies the deviation of constellation points in a digitally modulated signal from their ideal state, degrades with transmitter nonlinearity. Nonlinearity also causes the power of the bandlimited modulated signal to get spread to a larger bandwidth in a process commonly referred to as spectral regrowth. Spectral regrowth reduces the SNR of other users of the wireless network, and degrades the overall system capacity. Adjacent channel power ratio (ACPR) characterizes the spectral regrowth in wireless transmitters due to nonlinearity.

The maximum achievable data rate of a communication system, referred to as the channel capacity, is given by Shannon's capacity limit equation, under the assumption of additive white Gaussian noise (AWGN), as

$$C = BW \log(1 + SNR), \qquad (1.3)$$

where BW is the system bandwidth. The linear increase of capacity with bandwidth is the main motivation behind moving to mm-wave carrier frequencies where more bandwidth is available. On the other hand, increasing the wireless capacity through SNR requires an exponentially larger relative signal power, which often leads to an unrealistically large transmitter power requirement. In a line-of-sight (LOS) wireless link with distance d between the transmitter and receiver, the power of a received signal, P_r, is given by

$$P_r = P_t G_t G_r \left(\frac{\lambda}{4\pi d}\right)^2, \qquad (1.4)$$

where P_t is the power of the transmitted signal, G_t and G_r are the transmitter and receiver antenna gains, and λ is the wavelength. Therefore, for a system with equal transmit and receive antenna gains, the power of received signals with shorter wavelengths (higher frequencies) decreases. In order to maintain a sufficiently high received signal energy at mm-waves, the antenna gain is typically increased. This implies a more directional wireless link. Fortunately, the direction of the narrow electromagnetic beam can be steered electronically using phased-array beam-forming networks consisting of multiple antennas and the requisite control electronics. Fortunately, the smaller wavelength at mm-wave frequencies leads to a small footprint for the multi-antenna beam-forming scheme. The requirement for additional electronic circuitry in multi-antenna mm-wave transceivers necessitates a silicon SOC approach to reduce the cost, power consumption, and footprint. A major advantage of mm-wave beam-forming is that co-channel interference can be mitigated, leading to an increase in overall network capacity.

1.4 Wireless transmitter architectures

The modulated waveforms of Eqns (1.1) and (1.2) can be generated directly in the so-called direct I/Q (homodyne) and polar transmitter architectures. Direct I/Q modulators require accurate I/Q generation circuitry and linear amplification; as such, they often utilize additional feedback circuitry for I/Q correction and/or amplifier linearization. Furthermore, linear amplifiers are fundamentally inefficient. Polar modulators, on the other hand, can operate with switching amplifiers with a theoretical efficiency of 100%. However, they require accurate synchronization between the amplitude and phase signal paths that have a larger bandwidth compared with their I/Q counterparts (Fig. 1.3).

Modulation and frequency up-conversion can occur in multiple stages such as in the heterodyne architecture. Waveforms that do not carry amplitude-modulation information, also known as constant-envelope waveforms, can be created within phase-locked loops (PLL) or other forms of phase/frequency modulators, and consequently amplified with efficient switching class (nonlinear) power amplifiers (PA). Architectural trade-offs between various transmitters are well documented and will not be elaborated on here.

The increase of transistor speed has also led to digital generation and processing of RF waveforms. Modern RF transmitters, and transceivers in more general terms, use more digital building blocks for several reasons. The performance of digital circuits improves with technology scaling. Digital circuits consume less chip footprint compared with their analog counterparts. Multi-function and reconfigurable systems are easier to realize using digital circuits than with analog circuits. Digital circuits are more robust to device mismatches, as well as process, voltage, and temperature (PVT) variations. Finally, in principle, the design and verification of digital circuitry can be more straightforwardly automated compared with analog circuits. Assuming the continued advancement of transistor performance over time, it is conceivable that future high-performance mm-wave transmitters will also include significant digital circuitry at their core. Several of the chapters of this book cover digitally assisted or inspired power amplifier and transmitter concepts and circuits.

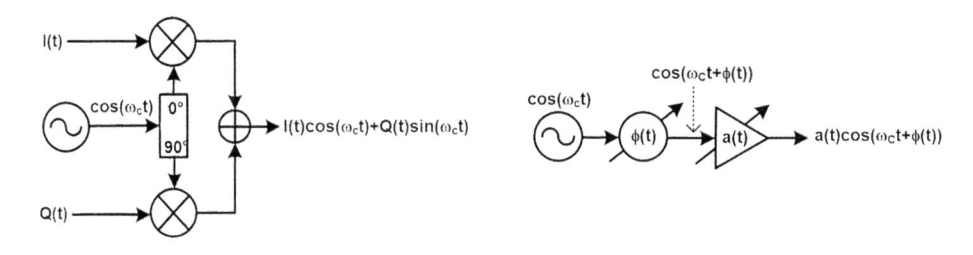

Fig. 1.3 Direct generation of modulated waveforms using direct I/Q and polar transmitter architectures.

1.5 Power amplifier basics

Historically, power amplifiers have been classified as either linear or switching amplifiers. This designation is somewhat misleading as several amplifier topologies that fall under the linear category are not strictly linear. In fact, with the exception of small-signal amplifiers (Class A topology with small input and output waveforms), all other amplifiers are nonlinear to some extent. The linearity versus efficiency trade-off of conventional (analog) power amplifiers is well documented. For instance, while the theoretical efficiency of Class A amplifiers is limited to 50%, switching amplifiers with Class D, E, F,... topologies can approach theoretical 100% power efficiency. Power added efficiency (PAE) measures the true efficiency of a power amplifier as it includes the AC power delivered at the input, P_{in}, as well as the power delivered through the DC supply, P_{DC}, as

$$PAE = \frac{P_{out} - P_{in}}{P_{DC}}. \quad (1.5)$$

At frequencies where the transistor can have significant power gain, the input RF power can sometimes be neglected – hence, instead of PAE, drain or collector efficiency, measuring the ratio of output power delivered to the load to the DC power drawn from the supply can be used. In fact, classical design criteria, optimization steps, and performance limits of power amplifiers do not consider the finite power gain of transistors. This assumption may not hold true at mm-wave frequencies where silicon transistors do not necessarily offer significant power gain. Several chapters in this book cover the effect of finite transistor power gain in the formulation and design steps of power amplifiers at mm-wave frequencies.

Power amplifiers can also be classified according to the main power transistor's operating mode. The output terminal of a transistor can be modeled as either a current source or a voltage switch. In single-ended amplifiers, the former applies to the forward active region of BJTs (bipolar junction transistors) and saturation region of FETs (field-effect transistors), while the latter applies to the saturation region of BJTs and triode region of FETs. Passive networks connected between the transistor terminals, and specifically at the output node, shape or scale the voltage and current waveforms. Waveform shaping is used primarily to reduce the overlap between the power transistor's voltage and current waveforms and hence increase the power efficiency. Waveform scaling is used to deliver the desired amount of power to the load given the limited breakdown limits of a transistor. Impedance transformation and power combining are the primary approaches used to scale the voltage or current waveforms.

Advancements in digital-signal-processing techniques, combined with the low overhead of complex digital circuitry, have led to widespread utilization of linearization techniques in power amplifiers and transmitters. Today, it is generally believed that, for the same specification, a transmitter can be designed to be more power efficient if linearization circuitry is combined with more efficient, nonlinear power amplifier cores, as opposed to relying on inefficient linear power amplifier cores. In fact, many modern transmitter architectures designed to support digital modulation formats now rely on switching power amplifiers. In envelope tracking approaches, the DC power delivered to the power amplifier output stage is modulated by the amplitude envelope of the waveform to be transmitted, thereby avoiding wasting DC power headroom while the signal amplitude is at lower swings.

It should be noted also that devices used in a power amplifier undergo large voltage and current swings, and oftentimes at elevated temperatures. This creates concerns over their short-term and long-term reliability. The concern is elevated at mm-wave frequencies as the transistors (with limited power gain) are typically pushed hard to deliver the desired amount of power. Fortunately, availability of "free" digital circuitry combined with advanced BIST and self-healing techniques mitigates some of the aforementioned risks. This will be elaborated on in a dedicated chapter. Power-combining strategies can also be utilized to increase the overall output power of the transmitter while maintaining more reasonable power dissipation levels for the individual transistor stages.

1.6 Examples of commercial mm-wave applications

1.6.1 60 GHz wireless communications: the IEEE 802.11ad standard

The frequency band around 60 GHz has been designated as an industrial, scientific and medical (ISM) band for unlicensed wireless communications. This band coincides with one of the resonance modes of the oxygen molecule, resulting in higher propagation loss. This higher propagation loss, while unfavorable for the usage of 60 GHz in long-range applications, enables tighter frequency reuse in a dense wireless network planning, resulting in higher aggregate capacity. The Wireless Gigabit Alliance (WiGig) was a trade association that developed and promoted the standardization of 60 GHz unlicensed wireless communications for creating interconnected home entertainment and office devices such as computers, displays, and mobile phones. The developed standard is called IEEE 802.11ad, and the WiGig alliance was subsumed by the Wi-Fi alliance. The physical layer (PHY) supports data rates up to 4.62 Gbps and 6.75 Gbps for single-carrier (SC) and orthogonal frequency division multiplexing (OFDM) modes, respectively, over 2.16 GHz channel bandwidths. The media access control (MAC) layer is compatible with the IEEE 802.11 standard.

Several research groups and industrial entities have produced chips or chipsets that are compatible with the IEEE 802.11ad standard. A few representative examples are here to provide a perspective of the technology evolution and state-of-the-art performance numbers. In 2011, SiBeam (later acquired by Silicon Image) reported 65 CMOS chips with up to 32-element beam-forming, capable of supporting the early drafts of 802.11ad standard with 3.8 Gbps data rate over 50 m (32TX/32RX) [1]. In 2013, Panasonic reported an 802.11ad CMOS chipset (90 nm RF and 40 nm BB), capable of supporting SC mode, achieving 1.5 Gb/s data rate over 1 m [2]. In 2014, Broadcom Inc. reported an 802.11ad 40 nm LP CMOS chipset with 16TX/16RX beam-forming, capable of supporting SC and OFDM modes, and achieving 4.6 Gbps data-rate (PHY rate) over 10 m [3].

1.6.2 77 GHz automotive radars

Various forms of sensors are used to enhance the safety and experience of driving. Millimeter-wave radars have been used in high-end cars for over a decade. The 76–77 GHz band has been the primary standardized frequency band for automotive radars throughout the world [29]. Current automotive long-range radars (LRRs) based on frequency-modulated continuous wave (FMCW) schemes are able to detect objects as far as 250 m away with fixed or steerable beams. There have been ongoing efforts towards utilizing more bandwidth within 77–81 GHz for automotive short-range radars (SRRs) with better range resolution for applications such as park assist.

While early realizations of automotive radars were based on discrete components, modern versions are highly integrated. A few representative examples are provided here. The advantage of silicon technologies towards realization of low-cost automotive radars has been recognized since the early 2000s [30, 31]. In 2008, a collaborative team led by Infineon Technologies reported a monolithic four-channel 77 GHz FMCW automotive radar chip capable of operating across $-40\,°C$ to $+125\,°C$ in a 200 GHz SiGe:C production technology [32]. In 2012, Freescale Semiconductor reported a transceiver chipset consisting of a four-channel receiver and a single-channel transmitter 200 GHz SiGe BiCMOS technology for 76–77 GHz LRR and 77–81 GHz SRR automotive applications [33]. More recently, UC San Diego and Toyota Research Institute have led the development of automotive phased array radars based on custom SiGe HBT chips [34, 35]. The increased interest and likelihood for commercial autonomous vehicles (self-driving cars) will undoubtedly result in further research and development towards low-cost high-performance mm-wave radars.

1.6.3 Future applications

The fifth-generation wireless communication standard (5G) is expected to include millimeter-wave for high-speed wireless access in mobile systems [36, 37]. Millimeter-waves continue to be among the legitimate candidates for high-speed connectivity in bulk-haul communications [38, 39], data centers, etc. Higher mm-wave frequencies (100–300 GHz) may be used for high-resolution imaging and sensing applications. In all of the aforementioned applications, economical as well as technical considerations will determine the advantages of a silicon solution.

1.7 Examples of military mm-wave applications and initiatives

1.7.1 Advanced EHF SATCOM

The Advanced Extremely High Frequency (AEHF) system is a US military satellite communications system developed to provide secure, jam-resistant global communications with data rates, overall throughput, and user coverage far exceeding those of the existing, legacy Milstar system [40, 41]. The planned system consists of four geosynchronous Earth orbit (GEO) satellites with inter-satellite links for global communications. The satellite downlink and uplink frequencies are 20.2–21.2 GHz and

43.5–45.5 GHz, respectively [42]. Data rates up to 8.2 Mb/s to future Army AEHF ground terminals are planned (the system will be able to support ~6000 terminals). Since the gain of an antenna is proportional to the antenna's physical area and inversely proportional to the square of the signal wavelength, high-gain antennas can be realized in relatively small physical sizes at 45 GHz. Currently low-cost terminal (LCT) technologies are under development by various companies for SATCOM-on-the-move (SOTM) and SATCOM-at-the-halt applications. It is thus envisioned that chip-scale, watt-level, highly linear, efficient silicon transmitters may be used in compact satellite terminals.

1.7.2 The DARPA ELASTx program

In 2009, the US Defense Advanced Research Projects Agency (DARPA) recognized that watt-level transmitter output power can be achieved in silicon technologies with efficiencies significantly beyond the state-of-the-art and with linearity sufficient to support high-order digitally modulated waveforms (e.g., 64QAM).

This led to the launching of the Efficient Linearized All-Silicon Transmitter ICs (ELASTx) program [43]. Figure 1.4 shows a conceptual block diagram of an envisioned efficient linearized all-silicon transmitter IC, based on extensive leveraging of integration available with today's silicon technologies. In the envisioned SOC, transmitter inputs are baseband I/Q symbol samples generated in a digital processor. The samples are converted to analog I/Q signals and passed to the I/Q modulator, where a modulated complex waveform is created and up-converted to the carrier frequency of interest. The output of the transmitter is linear RF power, where PA linearization can be accomplished by using a variety of techniques (e.g., feedforward, Cartesian feedback, pre-distortion, polar modulation, outphasing, etc.). Watt-level output powers at mm-wave frequencies can be accomplished through power-combining strategies such as planar parallel, spatial, and waveguide power combiners, as well as series device stacking to enable overall high output swings in low-voltage scaled silicon technologies. At the same time efficiencies >50% were targeted by the program, through various

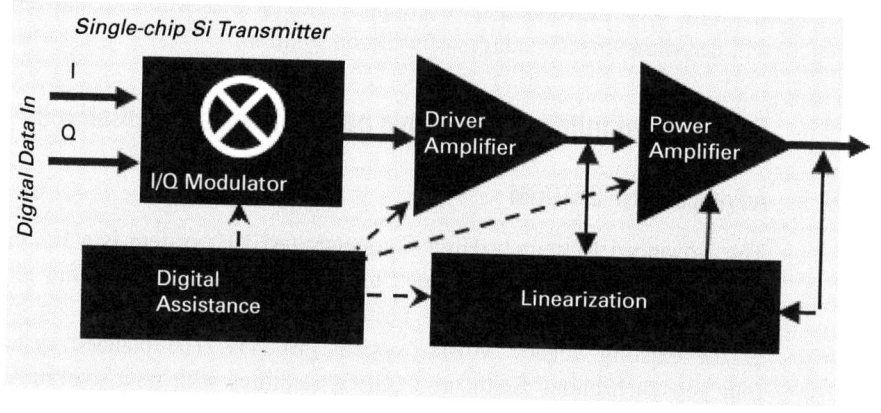

Fig. 1.4 Conceptual block diagram of an ELASTx silicon mm-wave transmitter SOC.

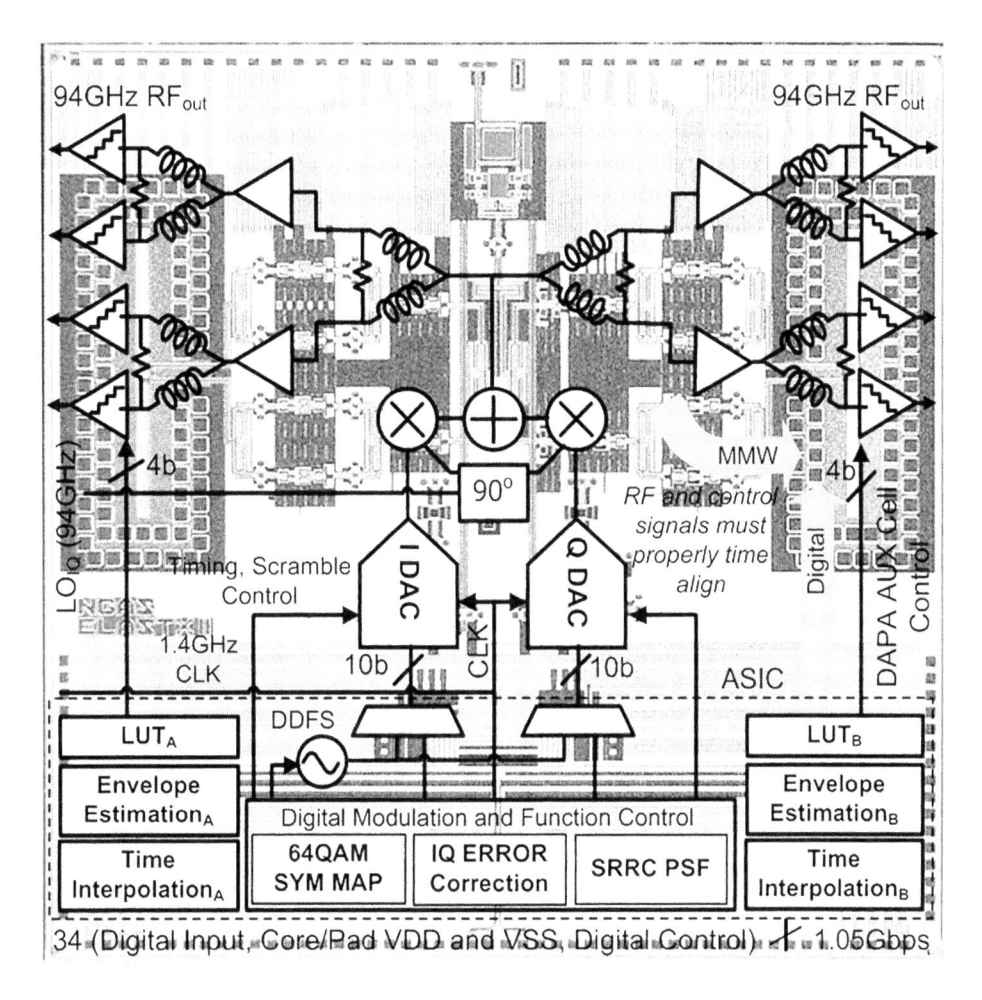

1.7 Examples of military mm-wave applications and initiatives

Fig. 1.5 Microphotograph of the W-band SOC from Northrop Grumman with functional blocks overlaid [45]. The chip size is 8 mm × 8 mm.

combinations of nonlinear/switching amplifier classes, outphasing, and Doherty-like schemes. As shown in Fig. 1.4, the envisioned role of the ultrahigh level of integration is to provide digital-computing capabilities needed to improve performance of all transmitter analog circuit blocks by means of digital assistance. Phase I of the program targeted 45 GHz carrier with 3.5 GHz bandwidth coverage; Phase II of the program targeted 94 GHz carrier with 5 GHz bandwidth coverage.

Outstanding demonstrations of the potential for silicon SOC integration of mm-wave transceiver functions under the ELASTx program were achieved by Northrop Grumman at 45 GHz [44], and more recently at 94 GHz [45]. Figure 1.5 shows a die photograph of Northrop's 94 GHz 64QAM 1 Gbps reconfigurable SOC CMOS transmitter with digitally assisted power amplifiers (DAPAs) and back-etched thru-silicon waveguide power combiners in 45 nm SOI CMOS technology. The SOC includes a 7-million-gate digital ASIC section with configurable digital modulation and transmit pre-coding. The digital section feeds two 10 b 1.4 GHz current-steering DACs followed by a direct conversion

I/Q modulator driving eight DAPAs. This SOC successfully transmitted a 1.05 Gbps 64QAM signal with 3.9% EVM and 50 dBc ACPR at 94 GHz.

Review article [46] provides an overview of the ELASTx program status circa 2014. Figures 1.6a and 1.6b show the state-of-the-art P_{out} and PAE performance versus RF

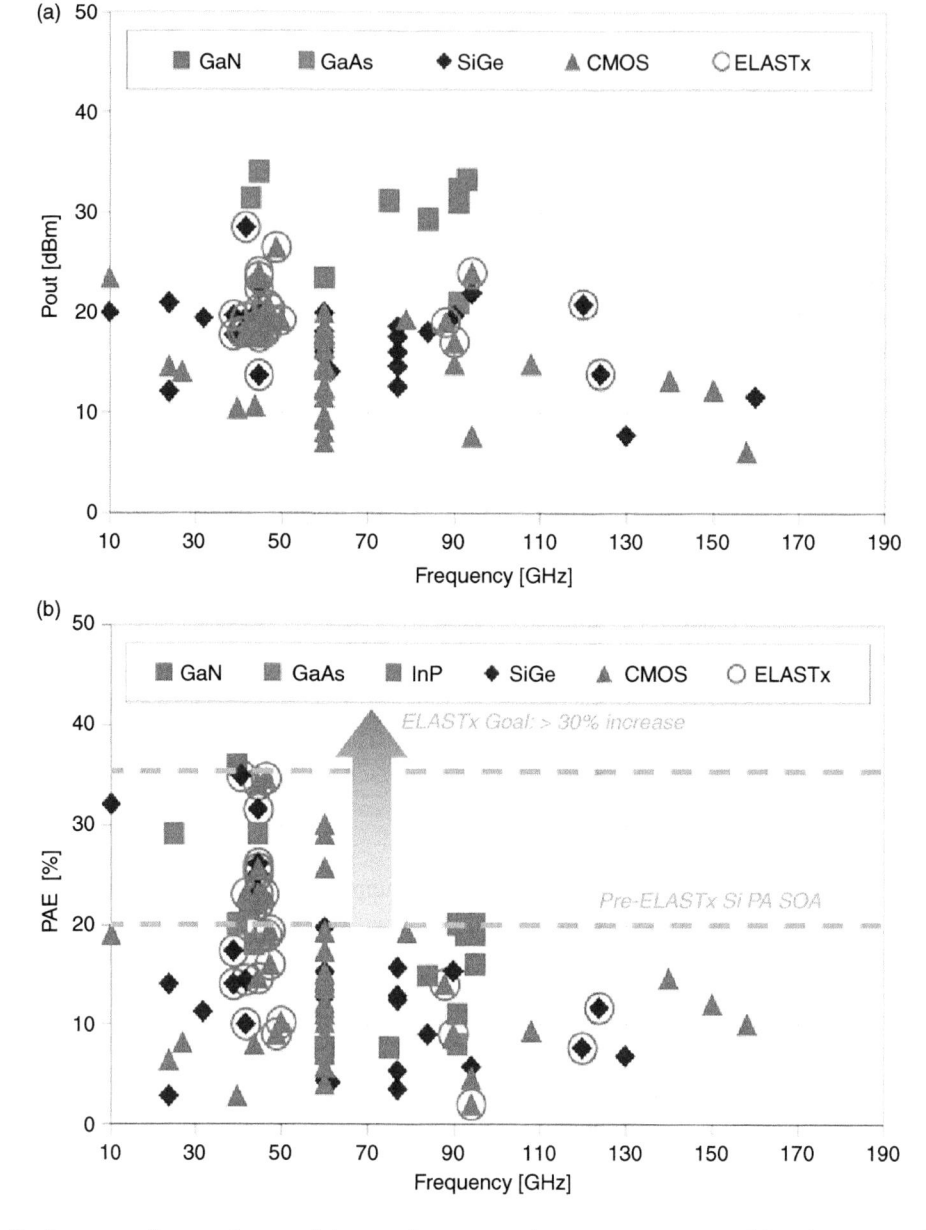

Fig. 1.6   Performance of reported state-of-the-art microwave and mm-wave power amplifiers in a range of various semiconductor technologies: (a) output power versus frequency; (b) power-added efficiency (PAE) versus frequency.

frequency. As can be seen, the ELASTx program has been responsible for driving significant advancements in silicon-based mm-wave PA output powers and efficiency at 45 GHz, while the commercial market pull has driven significant advancements at 60 GHz. More recent SOA results show advancements at 94 GHz and 140 GHz regimes, as well.

1.8 Conclusions

Silicon-based mm-wave integrated systems have already entered the commercial market for wireless communication and automotive-radar applications. Efficient high-performance power amplifiers and transmitters may be the last remaining frontier in this space. It is anticipated that mm-wave silicon power amplifiers and transmitters achieving and even surpassing the performance of alternative technologies will emerge. A holistic approach to designing a complete integrated system, leveraging digital signal processing, may be one key enabler towards achieving this goal.

Acknowledgments

The material of this chapter was inspired by the work of many researchers over several years. The contribution of all researchers in the general area of silicon mm-wave integrated systems is appreciated. The authors would like to thank the US Defense Advanced Research Projects Agency for their significant support of these advancements through the DARPA ELASTx program. The authors would also like to thank Iskren Abdomerovic of Booz Allen Hamilton, Arlington, VA, for his assistance with various figures in this chapter.

References

[1] S. Emami, *et al.*, "A 60 GHz CMOS phased-array transceiver pair for multi-Gb/s wireless communications," in *IEEE International Solid-State Circuits Conference Digest of Technical Papers (ISSCC)*, February 2011, pp. 164–166.

[2] T. Tsukizawa, *et al.*, "A fully integrated 60 GHz CMOS transceiver chipset based on WiGig/IEEE802.11ad with built-in self calibration for mobile applications," in *IEEE International Solid-State Circuits Conference Digest of Technical Papers (ISSCC)*, February 2013, pp. 230–231.

[3] M. Boers, *et al*, "A 16TX/16RX 60 GHz 802.11ad chipset with single coaxial interface and polarization diversity," in *IEEE International Solid-State Circuits Conference Digest of Technical Papers (ISSCC)*, February 2014, pp. 344–345.

[4] H. B. Wallace, "A W-band silicon based phased array for helicopter operations," in *IEEE International Symposium on Phased Array Systems & Technology*, October 2013, pp. 1–3.

[5] G. Patton, *et al.*, "75-GHz f_T SiGe-base heterojunction bipolar transistors," *IEEE Electron Device Letters*, vol. **11**, no. 4, pp. 171–173, April 1990.

[6] A. Schuppen, *et al*, "Multi emitter finger SiGe-HBTs with f$_{max}$ up to 120 GHz," in *International Electron Device Meeting (IEDM) Technical Digest*, December 1994, pp. 377–380.

[7] A. Gruhle, *et al*, "Monolithic 26 GHz and 40 GHz VCOs with SiGe heterojunction bipolar transistor," in *International Electron Device Meeting (IEDM) Technical Digest*, December 1995, pp. 725–728.

[8] H. Hashemi, *et al.*, "A fully integrated 24 GHz 8-path phased-array receiver in silicon," in *IEEE International Solid-State Circuits Conference Digest of Technical Papers (ISSCC)*, February 2004, pp. 390–391.

[9] A. Natarajan, *et al.*, "A 24 GHz phased-array transmitter in 0.18 μm CMOS," in *IEEE International Solid-State Circuits Conference Digest of Technical Papers (ISSCC)*, February 2005, pp. 212–213.

[10] A. Babakhani, *et al.*, "A 77 GHz 4-element phased array receiver with on-chip dipole antennas in silicon," in *IEEE International Solid-State Circuits Conference Digest of Technical Papers (ISSCC)*, February 2006, pp. 629–630.

[11] A. Natarajan, *et al.*, "A 77 GHz phased-array transmitter with local LO-path phase-shifting in silicon," in *IEEE International Solid-State Circuits Conference Digest of Technical Papers (ISSCC)*, February 2006, pp. 639–640.

[12] S. Reynolds, *et al.*, "60 GHz transceiver circuits in SiGe bipolar technology," in *IEEE International Solid-State Circuits Conference Digest of Technical Papers (ISSCC)*, February 2004, pp. 442–443.

[13] B. Floyd, *et al.*, "A silicon 60 GHz receiver and transmitter chipset for broadband communications," in *IEEE International Solid-State Circuits Conference Digest of Technical Papers (ISSCC)*, February 2006, pp. 649–650.

[14] C. Duan, *et al.*, "Design of CMOS for 60 GHz applications," in *IEEE International Solid-State Circuits Conference Digest of Technical Papers (ISSCC)*, February 2004, pp. 440–441.

[15] ———, "60 GHz CMOS radio for Gb/s wireless LAN," in *IEEE Radio Frequency Integrated Circuits (RFIC) Symposium Digest*, June 2004, pp. 225–228.

[16] B. Razavi, "A 60 GHz direct-conversion CMOS receiver," in *IEEE International Solid-State Circuits Conference Digest of Technical Papers (ISSCC)*, February 2005, pp. 400–401.

[17] ———, "CMOS transceivers for the 60-GHz band," in *IEEE Radio Frequency Integrated Circuits (RFIC) Symposium Digest*, June 2006, pp. 1–4.

[18] T. Yao, *et al.*, "65 GHz Doppler radar transceiver with on-chip antenna in 0.18 μm SiGe BiCMOS," in *IEEE International Microwave Symposium (IMS) Digest*, June 2006, pp. 1493–1496.

[19] S. Voinigescu, *et al.*, "SiGe BiCMOS for analog, high-speed digital and millimetre-wave applications beyond 50 GHz," in *Bipolar/BiCMOS Circuits and Technology Meeting*, June 2006, pp. 1–6.

[20] S. Pinel, *et al.*, "A 90 nm CMOS 60 GHz radio," in *IEEE International Solid-State Circuits Conference Digest of Technical Papers*, February 2008, pp. 130–131.

[21] Kwang-Jin Koh and G. M. Rebeiz, "An X- and Ku-band 8-element linear phased array receiver," in *Proceedings of the IEEE Custom Integrated Circuits Conference*, September 2007, pp. 761–764.

[22] Kwang-Jin Koh, J. W. May, and G. M. Rebeiz, "A Q-band (40–45 GHz) 16-element phased-array transmitter in 0.18 μm SiGe BiCMOS technology," in *IEEE Radio Frequency Integrated Circuits Symposium Digest*, June 2013, pp. 225–228.

[23] Sang Young Kim and G. M. Rebeiz, "A low-power BiCMOS 4-element phased array receiver for 7684 GHz radars and communication systems," *IEEE Journal of Solid-State Circuits*, vol. **47**, no. 2, pp. 359–367, February 2012.

[24] F. Golcuk, T. Kanar, and G. M. Rebeiz, "A 90–100-GHz 4 × 4 SiGe BiCMOS polarimetric transmit/receive phased array with simultaneous receive-beams capabilities," *IEEE Transactions on Microwave Theory and Techniques*, vol. **61**, no. 8, pp. 3099–3114, August 2012.

[25] Yi-An Li, et al., "A fully integrated 77 GHz FMCW radar system in 65 nm CMOS," in *IEEE International Solid-State Circuits Conference Digest of Technical Papers*, February 2010, pp. 216–217.

[26] Pang-Ning Chen, et al., "A 94 GHz 3D-image radar engine with 4TX/4RX beamforming scan technique in 65 nm CMOS," in *IEEE International Solid-State Circuits Conference Digest of Technical Papers*, February 2013, pp. 146–147.

[27] E. Cohen, et al., "A thirty-two element phased-array transceiver at 60 GHz with RF–IF conversion block in 90 nm flip chip CMOS process," in *IEEE Radio Frequency Integrated Circuits Symposium Digest*, June 2010, pp. 457–460.

[28] K. Okada, et al., "A 60 GHz 16QAM/8PSK/QPSK/BPSK direct-conversion transceiver for IEEE 802.15.3c," in *IEEE International Solid-State Circuits Conference Digest of Technical Papers*, February 2011, pp. 160–161.

[29] J. Hasch, et al., "Millimeter-wave technology for automotive radar sensors in the 77 GHz frequency band," *IEEE Transactions on Microwave Theory and Techniques*, vol. **60**, no. 3, pp. 845–860, March 2012.

[30] J. Bock, et al., "SiGe bipolar technology for automotive radar applications," in *Bipolar/BiCMOS Circuits and Technology Meeting*, June 2004, pp. 84–87.

[31] I. Gresham, et al., "Ultra-wideband radar sensors for short-range vehicular applications," *IEEE Transactions on Microwave Theory and Techniques*, vol. **52**, no. 9, pp. 2105–2122, September 2004.

[32] H. Forstner, et al, "A 16TX/16RX 60 GHz 802.11ad chipset with single coaxial interface and polarization diversity," in *IEEE Radio Frequency Integrated Circuits (RFIC) Digest*, June 2008, pp. 233–236.

[33] S. Trotta, et al., "An RCP packaged transceiver chipset for automotive LRR and SRR systems in SiGe BiCMOS technology," *IEEE Transactions on Microwave Theory and Techniques*, vol. **60**, no. 3, pp. 778–794, March 2012.

[34] P. Schmalenberg, et al., "A SiGe-based 16-channel phased array radar system at W-Band for automotive applications," in *European Radar Conference (EuRAD)*, June 2013, pp. 299–302.

[35] B. Ku, et al., "16-element phased-array receiver with 50° beam scanning for advanced automotive radars," *IEEE Transactions on Microwave Theory and Techniques*, vol. **62**, no. 11, pp. 2823–2832, November 2014.

[36] Zhouyue Pi and F. Khan, "An introduction to millimeter-wave mobile broadband systems," *IEEE Communications Magazine*, vol. **49**, no. 6, pp. 101–107, June 2011.

[37] S. Rangan, T. S. Rappaport, and E. Erkip, "Millimeter-wave cellular wireless networks: potentials and challenges," *Proceedings of the IEEE*, vol. **102**, no. 3, pp. 366–385, March 2014.

[38] D. Bojic, et al., "Advanced wireless and optical technologies for small-cell mobile backhaul with dynamic software-defined management," *IEEE Communications Magazine*, vol. **51**, no. 9, pp. 86–93, September 2013.

[39] R. Taori and A. Sridharan, "Point-to-multipoint in-band mm-wave backhaul for 5G networks," *IEEE Communications Magazine*, vol. **53**, no. 1, pp. 195–201, January 2015.

[40] U.S. Air Force. Advanced Extremely High Frequency (AEHF) Satellite System. [Online]. Available: http://www.losangeles.af.mil/library/factsheets/factsheet.asp?id=5319

[41] D. Brown and D. Schroeder, "Commercially hosted resilient communications," in *Military Communications Conference (MILCOM)*, November 2001, pp. 2227–2232.

[42] A. Einhorn and J. Miller, "Spectrum management issues related to the AEHF system," in *Military Communications Conference (MILCOM)*, October 2007, pp. 1–7.

[43] DARPA Microsystems Technology Office. DARPA-BAA-09-36, Efficient linearized all-silicon transmitter ICs (ELASTx). [Online]. Available: https://www.fbo.gov

[44] T. LaRocca, *et al.*, "45 GHz CMOS transmitter SoC with digitally-assisted power amplifiers for 64QAM efficiency improvement," in *Radio Frequency Integrated Circuits (RFIC) Symposium*, June 2013, pp. 359–362.

[45] ——, "A 64QAM 94 GHz CMOS transmitter SoC with digitally-assisted power amplifiers and thru-silicon waveguide power combiners," in *Radio Frequency Integrated Circuits (RFIC) Symposium*, June 2014, pp. 295–298.

[46] I. Abdomerovic, *et al.*, "Leveraging integration: towards efficient linearized all-silicon mm-wave transmitter ICs," *IEEE Microwave Magazine*, vol. **15**, no. 3, pp. 86–96, May 2014.

2 Characteristics, performance, modeling, and reliability of SiGe HBT technologies for mm-wave power amplifiers

David Harame, Vibhor Jain, and Renata Camillo Castillo

2.1 Introduction

SiGe BiCMOS technology is an excellent choice for RF (radio frequency) and mm-wave applications as it combines the best features of CMOS logic, high-performance SiGe heterojunction bipolar transistors (SiGe HBTs), and RF passives like transmission lines, capacitors, inductors, Schottky barrier diodes (SBDs), p-i-n diodes, and resistors. RF models for all of these devices and process design kits are readily available for the design and simulation of RF circuits. In addition, BiCMOS technologies generally support multiple HBTs having different breakdown voltage, noise, and performance specifications for varied applications.

The selection of any technology for mm-wave applications like power amplifiers (PAs) or transmitters is dependent not only on the performance of the technology but more so on the economics and total cost of the project integration. Compound semiconductor HBTs like InP and GaAs have better cut-off frequencies and breakdown voltage but lose to silicon technologies in terms of cost and integration. RFCMOS technologies may have similar performance specifications to the SiGe HBTs but BiCMOS is still preferred over RFCMOS for several reasons. At the same lithography node, SiGe HBT has superior performance and breakdown voltage to the CMOS devices. For equal performance CMOS requires much more aggressive lithography ($>n+2$ gen) and as such is significantly more expensive than SiGe BiCMOS. SiGe HBTs have a higher breakdown voltage and power handling capability, higher operating voltage, higher self-gain, better voltage swing, better impedance matching, and better linearity than CMOS. This chapter will explore these and other aspects of the SiGe BiCMOS technology that make it ideal for silicon mm-wave applications like power amplifiers and transmitters.

The chapter begins with a basic graded-base SiGe HBT device physics description followed by a discussion about the processing technology used to fabricate the device emphasizing aspects that impact high-performance mm-wave SiGe HBTs. A detailed comparison of performance trends across various technologies and industries is presented next. For mm-wave circuits there is an additional set of metrics such as breakdown voltage, noise figure, linearity, and thermal effects which are key to mm-wave applications. Several sections are devoted to discussion of breakdown voltage, thermal effects, and noise figure. In addition to the SiGe HBT transistors, silicon RF passives are also important and there is a discussion on Schottky diodes, PIN diodes, and other passives readily available in BiCMOS technologies. SiGe BiCMOS is a silicon-based technology which uses silicon reliability evaluation techniques for all of the devices and interconnects to guarantee robustness against failure mechanisms. There is a section on the reliability evaluation of SiGe HBTs. The chapter concludes with a last section on the ultimate limits of SiGe BiCMOS technology.

The objective is to give the reader a basic understanding of the SiGe BiCMOS technology and the trade-offs in manufacturing the technology for mm-wave applications.

2.2 Bipolar device physics

2.2.1 Graded-base SiGe heterojunction bipolar transistor

Bipolar junction transistors (BJTs) are typically utilized to achieve a current gain. Their operation relies on carrier injection from the emitter into the base and subsequently into the collector region, resulting in current flow from the emitter to the collector, as illustrated in Fig. 2.1.

To enable carrier injection from the emitter into the base, the emitter–base (EB) p-n junction is typically forward biased. The injected carriers are then transported across the base by diffusion. The base–collector (BC) p-n junction is typically reverse biased to create a large electric field in the BC depletion region that sweeps the injected carriers across the depletion region into the collector, resulting in the collector current (I_C). The magnitude of the electric current and the speed of the device are directly dependent on the width of the neutral base (W_B) and the total base doping level, known as the Gummel number (Eqn (2.1)). Neutral base is defined as the undepleted base region between the EB and BC depletion edges:

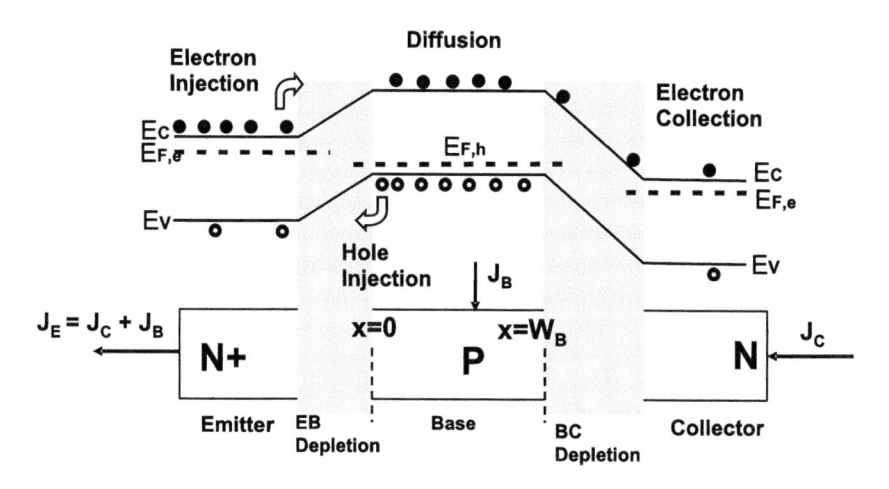

Fig. 2.1   Simplified band diagram of a silicon bipolar junction transistor.

2.2 Bipolar device physics

$$Q_B = \int_0^{W_B} p(x)dx, \tag{2.1}$$

where $x = 0$ and $x = W_B$ are neutral base boundaries on the EB and BC junctions of the base.

In the case of an NPN BJT, electrons are injected from the emitter into the base, whereas in a PNP BJT the injected carriers are holes. Hence for NPNs the forward bias current is essentially made up of electrons and can be modulated by minor changes in the forward bias voltage (V_{BE}), since the current varies exponentially with the voltage bias (Eqn (2.2)) [1]:

$$I_C = I_{C0}\left(e^{qV_{BE}/kT} - 1\right), \tag{2.2}$$

where I_{C0} is a constant dependent on the base material properties, base doping, grading in the base, and neutral base thickness.

During transistor operation, holes are simultaneously injected from the base into the emitter, resulting in a base current (I_B). In order to obtain a current gain ($\beta = I_C/I_B$) in BJTs, it is necessary to suppress the injection of holes from the base to the emitter. This is typically accomplished by the utilization of a base doping concentration level that is lower than that of the emitter and/or the use of an interfacial oxide; this will be discussed later on. Base current also arises due to recombination of majority charge carriers injected into the neutral base from the emitter (electrons for NPN) with the majority carriers in the base (holes for NPN). Another source is recombination of holes with traps present at an imperfect semiconductor–dielectric interface. Total I_B is the sum of these different components.

The speed of the BJT transistor can be increased by several means, but largely consists of reducing the width of the neutral base and the total charge in the base. However, these changes to the BJT result in increases in the base resistance, which negatively impact the RF performance, and in increases in the output conductance with a concomitant reduction in the device Early voltage. In the 1950s, Herbert Kroemer introduced the graded-base SiGe HBT in two seminal papers [2, 3]. Kroemer's graded-base SiGe HBT offered an immediate improvement to the BJT degradation associated with increasing the device speed. In the graded-base SiGe HBT a linear graded germanium concentration is introduced across the base with increasing concentration toward the collector. The germanium grade creates a quasi-electric field in the base that accelerates the minority carrier electrons across the base, significantly reducing the time it takes the electrons to transit the base. Additionally, there is an increase in the Early voltage, the current gain (β), and the current at a given bias condition. These changes to the device characteristics may be easily understood by examining the analysis of Cressler and Niu [4], shown in Eqns (2.3)–(2.5), in which a comparison of the SiGe HBT and the Si BJT of the same doping profiles is made.

Since germanium is a smaller bandgap material than silicon, a reduction in the electronic bandgap of silicon is realized when germanium in introduced into the lattice. It is well established that, for every 10% increase in the germanium concentration in the silicon lattice, an approximate 75 meV reduction in the silicon electronic bandgap is realized.

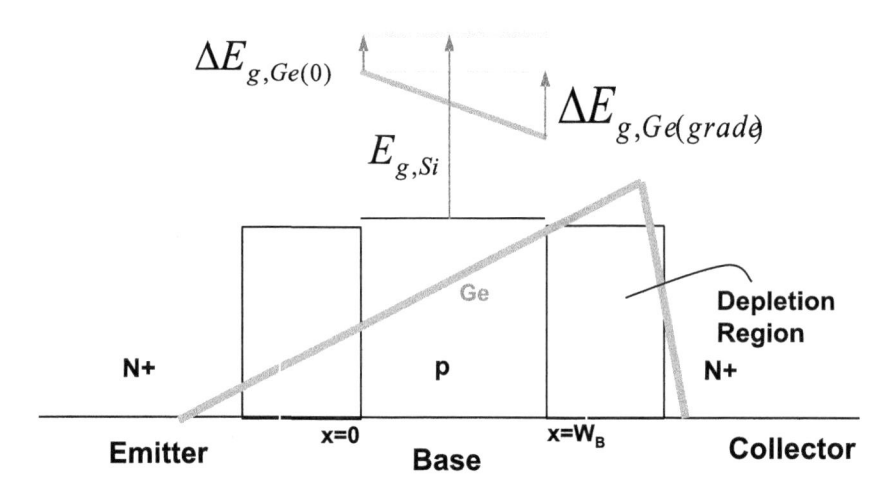

Fig. 2.2 Simplified profile depicting germanium grading in the base and corresponding bandgap grading of a graded-base SiGe HBT transistor.

This is captured by Cressler and Niu [4] in the term $\Delta E_{g,Ge}(0)$, which defines the reduction in base bandgap due to addition of germanium at the beginning of neutral base, and the term $\Delta E_{g,Ge(grade)}$, which represents the bandgap grade across the neutral base, illustrated in Fig. 2.2. By comparing the SiGe HBT and the Si BJT with the same doping profile, the enhancements from the graded-base germanium profile in the collector current density, J_c, the Early voltage, VA, and the base transit delay, τ_B, are as shown below:

$$\frac{J_{c,SiGe}}{J_{c,Si}} = \frac{\Delta E_{g,Ge(grade)}}{kT} \cdot \frac{e^{\Delta E_{g,Ge}(0)/kT}}{1 - e^{-\Delta E_{g,Ge(grade)}/kT}} \quad (2.3)$$

$$\frac{VA_{SiGe}}{VA_{Si}} = e^{\Delta E_{g,Ge(grade)}/kT} \left[\frac{1 - e^{-\Delta E_{g,Ge(grade)}/kT}}{\Delta E_{g,Ge(grade)}/kT} \right] \quad (2.4)$$

$$\frac{\tau_{B,SiGe}}{\tau_{B,Si}} = \frac{2}{\eta} \cdot \frac{kT}{\Delta E_{g,Ge(grade)}} \left\{ 1 - \frac{kT}{\Delta E_{g,Ge(grade)}} \left[1 - e^{\Delta E_{g,Ge(grade)}/kT} \right] \right\}. \quad (2.5)$$

Assuming the ratios $R_0 = \Delta E_{g,Ge}(0)/kT$ and $R_{GRADE} = \Delta E_{g,Ge(grade)}/kT$ are large, the equations are further simplified to Eqns (2.6)–(2.8):

$$\frac{J_{c,SiGe}}{J_{c,Si}} \approx \frac{\Delta E_{g,Ge(grade)}}{kT} \cdot e^{\Delta E_{g,Ge}(0)/kT} \approx R_{GRADE} \cdot e^{R_0} \quad (2.6)$$

$$\frac{VA_{SiGe}}{VA_{Si}} \approx e^{\Delta E_{g,Ge(grade)}/kT} \approx e^{R_{GRADE}} \quad (2.7)$$

$$\frac{\tau_{B,SiGe}}{\tau_{B,Si}} \approx \frac{kT}{\Delta E_{g,Ge(grade)}} \approx \frac{1}{R_{GRADE}}. \quad (2.8)$$

2.2 Bipolar device physics

In a typical graded-base SiGe HBT design R_0 may be relatively small, while R_{GRADE} may be more significant and is a function of the technology node and desired enhancement. A 10% increase in the Ge concentration across the graded base will produce an $R_{GRADE} \sim 3$, while a 20% increase in the Ge concentration across the graded base will correspond to an $R_{GRADE} \sim 6$. These simple calculations suggest the high germanium contents required in a typical design to obtain significant enhancements in J_c, V_A, and τ_B.

In the graded-base SiGe HBT, germanium is introduced into the base region of the device, such that the base sits on a linearly increasing germanium concentration region that extends from the emitter to the collector. Germanium is not present in a BJT and only silane gas is used for growing silicon. The germanium concentration grading is accomplished during the epitaxial growth of the base region by introducing germane gas into the system. The partial pressure of the germane gas in the growth chamber is increased relative to the silane flow as the base layer is grown for a linearly graded profile. The slope of germanium is controlled by the rate of change of germane partial pressure in the system. More details of the epitaxial growth process are in the next section.

The primary speed parameter of the SiGe HBT is the unity current gain cut-off frequency, f_τ, which is directly proportional to the inverse of the total emitter to collector carrier transit time. The expression for the total emitter to collector transit time, τ_{EC}, is given in Eqn (2.9) and is defined as the sum of carrier delay components associated with transit and RC charging delays:

$$\frac{1}{2\pi f_\tau} = \tau_{EC} = \tau_E + \tau_B + \tau_C + \tau_{CSCL}$$

$$= \frac{kT}{qI_c} C_{EB} + \left(\frac{kT}{qI_c} + R_E + R_C\right) C_{CB} + \frac{W_B^2}{\gamma D_n} + \frac{W_{CSCL}}{2v_{exit}}, \quad (2.9)$$

where

$$\gamma = \frac{kT}{\Delta E_{g,Ge(grade)}} \left\{1 - \frac{kT}{\Delta E_{g,Ge(grade)}} \left[1 - e^{\Delta E_{g,Ge(grade)}/kT}\right]\right\}. \quad (2.10)$$

The terms may be defined as follows: I_c is the collector current value where f_τ is calculated, W_B is the neutral base width, W_{CSCL} is the base–collector space charge region (SCR) width, v_{exit} is the electron velocity as it exits the base into the collector, γ is the field factor, which is a measure of the quasi-electric field established in the base, R_E and R_C are the emitter and collector resistance respectively, and C_{EB} and C_{CB} are the emitter–base and collector–base capacitances. Note that the impact of Ge is, at first glance, apparent only in the γ term multiplying the base transit time, but it also has an impact through the relationship of C_{CB}/C_{EB} and the transconductance g_m ($g_m = qI_c/kT$) since Ge enhances J_c at an applied emitter base bias and therefore effectively decreases the charging delay due to C_{CB} and C_{EB}. This relation illuminates what needs to occur to boost the device's performance, in order to reduce τ_{EC} and increase f_τ.

A simplified drawing of the vertical graded device profile is shown in Fig. 2.3, in which the changes required for device speed improvements are also indicated. A TEM

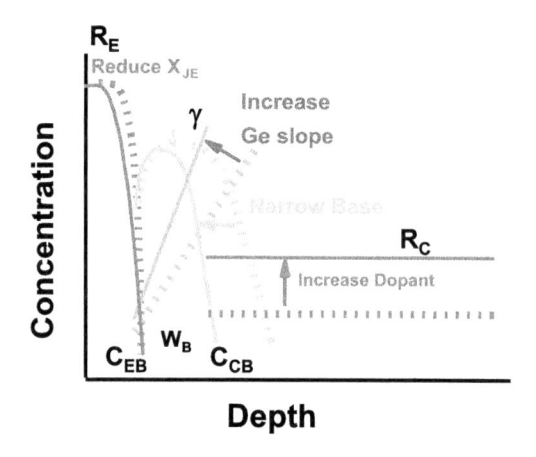

Fig. 2.3 Schematic dopant and germanium grade profile of SiGe HBT with the location of key parameters and scaling directions.

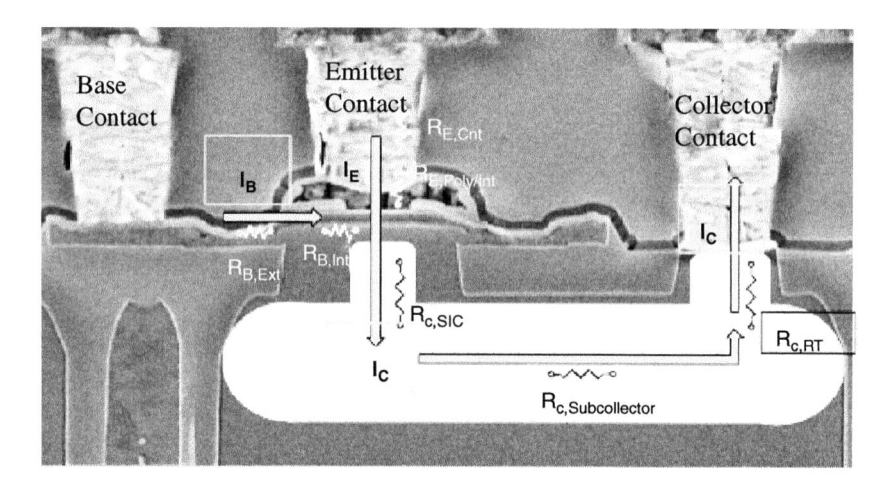

Fig. 2.4 Schematic cross section of SiGe HBT indicating key structural elements, capacitances, and resistances relevant to the delay components.

of a SiGe HBT transistor is shown in Fig. 2.4 with emitter, base, and collector regions marked. The current flow through the device and some of the key resistances and capacitances are also indicated in the figure. It is important to note that, in the SiGe HBT, the current flow from the emitter to the collector is vertical into the wafer, while the parasitic elements are dependent on the lateral device dimensions. For instance, the collector resistance, R_C, is a composition of several components: three major ones are the reach-thru resistance, sub-collector sheet resistance, and resistance of the silicon implanted collector under the intrinsic base.

The Kirk effect is a critical consideration in HBT device design that merits discussion. In order to achieve high-frequency operation, it is essential to operate the device at

2.2 Bipolar device physics

high currents, which reduces the capacitance charging delays associated with transconductance g_m. There is, however, a maximum current density at which the device can be operated, above which β and the frequency response of the HBT are drastically degraded. This maximum current density is called the Kirk threshold, J_{KIRK}, defined in Eqn (2.11), and the phenomenon is known as the Kirk effect:

$$J_{KIRK} = \frac{2\varepsilon_0 \varepsilon_r v_s}{W_{CSCL}^2} (\phi_{bi} + V_{CB}) + qN_c v_s. \qquad (2.11)$$

At this critical current limit, the effective neutral base width widens due to the space charge region injection into the collector. The Kirk limit exists due to the finite electron velocity in the collector. Under Shockley boundary conditions, it is generally assumed that the minority carrier charge at the base–collector junction at the base-edge is *zero*, requiring infinite electron velocity, which is a non-physical condition. There is always a finite amount of charge injected into the collector, given by $n_c = J_c/qv_s$, where v_s is the saturation velocity in the collector. As J_c increases, injected charge in the collector increases, compensating the collector doping and decreasing the slope of electric field in the collector. Beyond a critical limit, J_{crit}, the electric field in the collector changes sign and the electric field at the base-edge decreases ultimately to *zero* at J_{KIRK}. Theoretically, if J_c is increased further, a situation arises whereby the direction of the electric field reverses in the region close to the base–collector edge, corresponding to the band diagram shown Fig. 2.5. However, for a homojunction BJT this is a non-physical situation as there is no blocking potential for the holes and, consequently, holes from the base move into the collector, resulting in a wider base. This effective base width extension results in a rapid increase in base recombination current, causing β degradation and f_τ reduction. In an HBT, a potential barrier exists that prevents holes from moving into the collector. Consequently, the holes pile up at the base–collector edge, resulting in an increased electron pile-up at the junction. The base carrier recombination accordingly increases degrading β, while the electron pile-up increases the base transit delay, degrading f_τ.

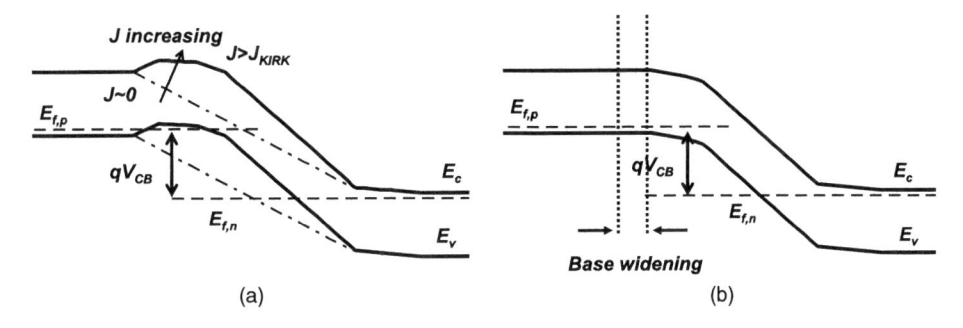

Fig. 2.5 (a) Band diagram for a BJT with no current flow and $J_C > J_{KIRK}$. The $J_C > J_{KIRK}$ band diagram is a non-physical situation as holes from the base will flow into the collector, resulting in (b) base widening.

2.2.2 Vertical scaling of HBTs

Enhanced device performance can be achieved by a reduction in each of the carrier delay components, which is typically accomplished by the reduction in the thickness of the vertical graded device profile, shown in Fig. 2.3. Additionally, the delay components can be reduced by several other methods, which will be discussed in the ensuing paragraphs.

The base–collector space charge region delay, τ_{CSCL}, is decreased by reducing the width of the SCR region, W_{CSCL}. In devices with a selectively implanted collector (SIC), this is realized by one or a combination of increasing the SIC dopant concentration and reducing the thickness of the grown un-doped collector layer between the graded base and the SIC. However, such methods inherently increase the parasitic collector–base capacitance, C_{CB}, with a resultant increase in the RC charging delay of the device, degrading device performance ($\tau_{C_{CB}} = C_{CB} \cdot (R_C + R_E + 1/g_m)$). The collector scaling therefore requires a precise balance of τ_{CSCL} and C_{CB} for the optimal speed improvement.

The base transit delay, τ_B, is reduced by decreasing the neutral base width, W_B. In SiGe epitaxial base transistors the base is grown by epitaxy methods, which allows for the controlled growth of a narrow base. Inevitably, subsequent thermal processing during the device fabrication widens the base layer by dopant thermal diffusion, resulting in a device which features a much wider base than was initially grown. A variety of processing techniques are therefore used to reduce this inherent base widening, including the incorporation of carbon in the graded-base layer, reductions in thermal budget subsequent to base deposition, and reductions in the total in-situ doped boron content.

It is well known that boron diffuses by primarily an interstitial-mediated mechanism, in which it pairs with silicon interstitials to diffuse in the lattice [5–7]. Incorporated carbon in the graded-base SiGe layer competes with boron for the silicon interstitials, reducing the interstitial concentration that is available to pair with boron. In this way, significant reduction in the base boron diffusion is accomplished [8]. Another process measure by which the boron diffusion is limited is by minimizing the subsequent thermal processing to which the base is subjected after it has been epitaxially grown. Rapid thermal annealing (RTA) techniques have become commonplace in semiconductor processing and are typically utilized to attain very high processing temperatures in the fastest possible times. Fick's diffusion equations are shown below (Eqns (2.12) and (2.13)), which demonstrate the relation between diffusion and the dopant concentration, processing temperature, and processing time:

$$J = -D(T)\frac{\partial \phi}{\partial x} \qquad (2.12)$$

$$\frac{\partial \phi}{\partial t} = D\frac{\partial^2 \phi}{\partial x^2}, \qquad (2.13)$$

where D is the diffusion constant and ϕ is the concentration (flux) of the diffusing species. From the equations it is evident that reduced processing times result in less dopant diffusion. Hence the adoption of RTA techniques is beneficial for reduced base diffusion and base width control. Similarly, lower dopant concentrations result in reduced base diffusion. Since the adoption of such techniques effects increases in the base resistance, which negatively impacts the device's RF performance levels, process optimization is necessary to ensure the device performance metrics are achieved.

In addition to the processing techniques mentioned above, τ_B is also reduced by increasing the field factor, γ, which is achieved by increasing the germanium ramp through the base region. There are limits to the total germanium content that can be incorporated in the lattice that are associated with the total strain in the epitaxially grown film [9, 10]. When the germanium content exceeds a given concentration for a defined thickness, the material releases the strain and relaxes, resulting in dislocation formation and epitaxial growth challenges. As a result, the profile must be carefully designed to ensure a high-quality film.

Vertical scaling of the device via the reduction in τ_{CSCL} and τ_B decreases the collector resistance, R_C, but increases the parasitic capacitances, C_{BE} and C_{CB}, thus increasing the RC charging delays. Total device C_{CB} and C_{BE} can be reduced by laterally scaling the device, which reduces device area. However, reduction in device area increases total emitter resistance, R_E, which can be engineered back to lower values using novel device architectures and processes.

2.2.3 Lateral scaling of HBTs

In several cases, when used as an amplifier, devices can amplify power beyond the f_τ since the voltage gain can be achieved at frequencies higher than f_τ. The maximum frequency of operation beyond which the power gain is less than unity, or the maximum oscillation frequency, is called f_{MAX}. Once f_{MAX} is exceeded, the device becomes passive and the power dissipation exceeds the device output. In a well-designed HBT, f_{MAX} is generally higher than f_τ. f_{MAX} can be approximated by the formula

$$f_{MAX} = \sqrt{\frac{f_\tau}{8\pi (R_B C_{CB})_{eff}}}, \quad (2.14)$$

where f_τ is the current gain cut-off frequency, R_B is the base resistance, and C_{CB} is the collector–base capacitance. The product $(R_B C_{CB})_{eff}$ is the effective time constant of the base–collector junction [11]. The strong dependence of f_{MAX} on R_B and C_{CB} necessitates processes and device architectures that are designed specifically to reduce R_B and C_{CB}. Opposing physical requirements of the base and the collector regions are necessary to achieve both high f_τ and high f_{MAX}. Vertical scaling for high f_τ, in which W_{CSCL} is reduced, increases C_{CB}, while reduction in W_B for lower τ_B increases base pinch sheet resistance, increasing R_B. Lateral scaling of the device is therefore necessary to reduce these components in addition to novel integration schemes.

The device base resistance is dependent on several parameters related directly to the process and the device layout. In HBTs, which feature a raised extrinsic base process, the extrinsic base is typically a combination of polysilicon and silicon. In such cases the base resistance has an additional component associated with the polysilicon grain itself, which depends on carrier mobility degradation and grain-boundary diffusion [12]. The components of the total base resistance in a typical HBT include the silicided extrinsic base resistance, the silicon/polysilicon region resistance, which includes the extrinsic base to intrinsic base link resistance, the sheet resistance of the single-crystal SiGe base beneath the emitter spacer, and the pinched base sheet resistance. The process challenge lies in the ability to reduce the base resistance while maintaining a relatively thin intrinsic base. Similarly, the total collector–base capacitance, C_{CB}, is the sum of different individual components, namely, capacitance due to the selectively implanted collector, $C_{CB,i}$, capacitance between the base and sub-collector outside the SIC region, $C_{CB,ex}$, capacitance across the shallow trench isolation region, $C_{CB,ox}$, and capacitance at the end of the device, $C_{CB,end}$ [12, 13]. Several device structures and process innovations are therefore used in SiGe technologies to reduce the components of the total base resistance and collector–base capacitance.

2.2.4 Performance trends

Irrespective of the structural implementation, the emphasis in all device designs is on reduction of the parasitic capacitances, C_{BE} and C_{CB}, and resistances, R_E, R_C, and R_B, to achieve the highest power gain attainable at a given frequency. Since vertical scaling of the device, to increase the device f_τ, increases the device parasitics, it is particularly difficult to achieve higher power gain, f_{MAX}. In Fig. 2.6 the SiGe HBT power gain, f_{MAX}, versus the current gain cut-off frequency, f_τ, is shown for several SiGe HBT generations. Examination of the plot clearly reveals the difficulty associated with concurrently

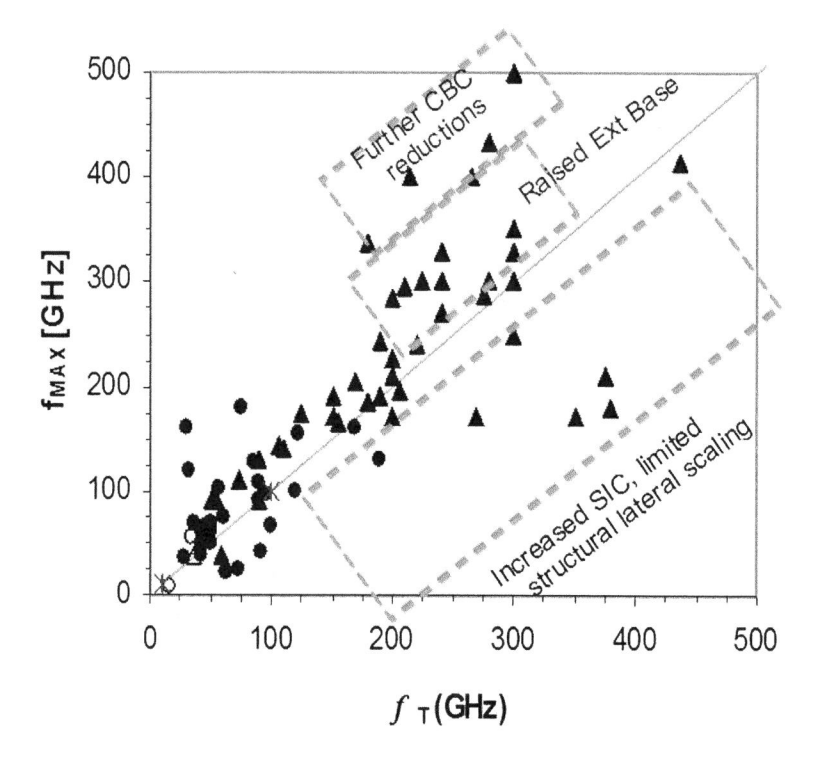

Fig. 2.6 Plot of f_τ and f_{MAX} for SiGe HBTs showing performance trends.

achieving f_τ and f_{MAX} above the 200 GHz level. Below 200 GHz f_τ and f_{MAX} can be seen to be approximately equal in value.

The next section will describe the processing technologies and various SiGe HBT transistor structures used in fabricating SiGe HBTs. SiGe HBT technologies employ a myriad of device architectures depending on performance requirements, cost, process capabilities, and manufacturer. Examination of the approaches used in the industry to address the fundamental device optimization challenges is an interesting aspect of this field.

2.3 SiGe HBT processing and structures

2.3.1 SiGe epitaxy

SiGe BiCMOS technology was the first silicon production technology that incorporated germanium in the silicon lattice to obtain a variable bandgap and improve device performance. In 1963, Kroemer proposed the graded-base heterojunction bipolar transistor at the Solid State DRC conference; however, it was not mass-manufactured until the 1990s because of difficulties associated with growing device quality epitaxial silicon–germanium layers. The epitaxy process remains the key process in achieving performance and a robust manufacturing capability for the SiGe HBT.

The primary process application for silicon epitaxy in bipolar technology up to 1980 was to grow a lightly doped layer subsequent to the n+ sub-collector formation, referred to as *collector-epi*. The epitaxial deposition tools entailed placing the wafers on an inductively heated susceptor to enable process uniformity and batch throughput, at a process temperature of approximately 1100 °C. The growth of collector-epitaxy is fundamentally a high-temperature process, largely due to an extensive high-temperature prebake, which is required to reduce the sub-collector arsenic dopant concentration at the surface and to remove all interfacial oxide to facilitate high-quality silicon growth for device-level quality silicon. As a result, the epitaxy systems were high-temperature batch processing systems. The associated large thermal mass restricted the ability to change the process temperature at a fast rate. Such high temperatures resulted in substantial dopant diffusion of the intrinsic base and other device dopants, which rendered the process unusable for growing the intrinsic SiGe graded-base BiCMOS. A low-temperature epitaxial technique with an ability to grow arbitrary profiles was therefore required.

The fundamental problem associated with achieving low-temperature silicon epitaxy is the removal of and maintenance of an oxide-free surface for growth [14]. A strict correlation between defect density and the oxygen content at the initial growth interface exists, as indicated in Fig. 2.7. Therefore, irrespective of the epitaxial growth technique employed, it is important to minimize interfacial oxygen levels and to manipulate the composition to an *atomic* level simultaneously. Ultrahigh-vacuum chemical vapor deposition systems (UHV/CVD) provided a solution to low-temperature epitaxy by utilizing a radiant heated furnace tube, by modifying the wafer preparation techniques and modifying the configuration of the furnace tube.

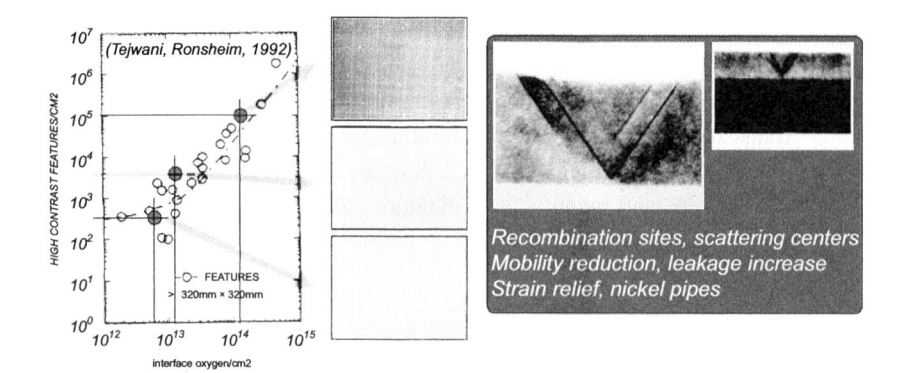

Fig. 2.7 Defect density in the epitaxially grown film as a function of oxygen concentration at the initial interface [14].

The wafer preparation technique involved an *HF last* step to remove the surface oxide and to form a hydrogen passivation layer that protects the surface from any further oxidation. This hydrogen layer is very stable at temperatures below 600 °C, hence by maintaining the epitaxy growth temperature below 600 °C the hydrogen passivation layer is maintained during the low-temperature growth. This is accomplished by the use of a disilane, SiH_4, source gas, which dissociates and grows silicon at the low temperature of 500 °C. Dramatic modifications to the furnace tube configuration were made to reduce oxygen contamination prior to and during epitaxy. These included the implementation of a load vacuum lock to prevent ambient air from entering the chamber. The furnace tube was also coated with silicon for growth and maintained at a high temperature to enable the hot clean silicon surface to react with any water or oxygen in the chamber, which provided the ability to *getter* or absorb any contaminating species. A dry turbo pump was also added to the system to facilitate low pressure in the mtorr range, in the chamber. The low-pressure system ensured a long mean free path, on the order of a meter, and prevented stagnant layers which limit growth. A double-walled tube was used to prevent contaminants from diffusing into the tube. When a silicon wafer is loaded into the system, SiH_4 decomposes on the silicon surface and replenishes the hydrogen passivation layer while growing silicon. This system configuration referred to as UHV/CVD was operated at a very low-temperature growth, with no prebake, and hence provided a solution to the low-temperature epitaxy problem. The use of the UHV/CVD technique provided a path for high-quality, device-level SiGe films.

Several other techniques were also available in the mid 1980s, including molecular beam epitaxy (MBE), limited reaction processing, and lamp-heated single-wafer tools. The lamp-heated single-wafer tools, in which an array of lamps heats the wafers, eventually became the technique of choice throughout the industry. This was partly driven by a trend in the industry to develop single-wafer tools to reduce the tooling footprint with increasing wafer sizes and flexibility of the tools. Improved heating lamps and reaction chamber uniformity, wafer temperature measurement, gas purification, and load locks to reduce contamination greatly improved the quality of epitaxial growth in these systems.

Additionally, the incorporation of improved pumps facilitated lower-pressure tool operation. The use of chlorine-containing silicon sources (for example, Si_2Cl_2) together with HCl enabled selective epitaxy growth. Selective epitaxy is an important capability extensively employed to reduce device parasitics. The flexibility of the single-wafer tools also made it much easier to tailor each wafer processing condition, which also resulted in significant reductions in process development time. New silicon sources that decompose at lower temperatures, like trisilane, have enabled further reduction in the growth temperature and extended the composition range and processing options.

2.3.2 Emitter–base alignment methodologies for SiGe HBT device structures

For commercial applications, the SiGe HBT must be integrated into a silicon CMOS process to produce a BiCMOS technology. Although many process integration flows have been demonstrated, the vast majority of them are a *base-after-gate* flow, which is illustrated in Fig. 2.8. This integration flow largely decouples the SiGe HBT manufacture from the CMOS production and therefore supports a variety of different SiGe HBT structures. The base-after-gate integration flow as described by Onge et al. [15] in 1999 is still applicable and will be described herein. The process begins with a high-temperature sub-collector implant, anneal, and drive-in followed by a lightly doped collector epitaxy step. The sub-collector formation is followed by an oxide-lined, polysilicon-filled deep trench (DT) to isolate the bipolar transistor from other devices. A shallow trench isolation (STI) fully compatible with the CMOS technology at the node is then formed and used to isolate FET devices from one another. A deep implant to contact the buried sub-collector from the surface often referred to as a *reach-thru* provides a low-resistance diffusion region to contact the sub-collector. Next the CMOS device

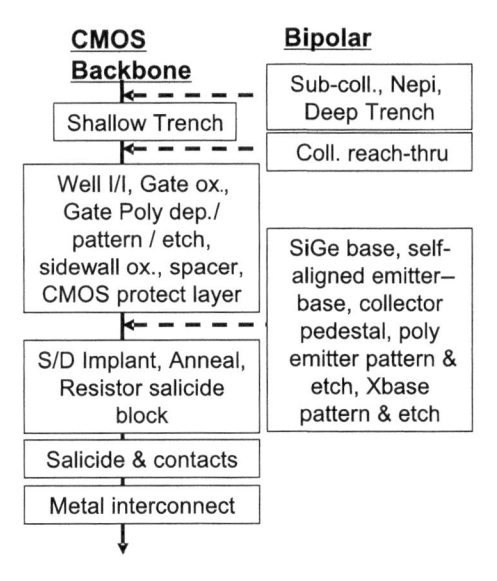

Fig. 2.8 Base-after-gate integration flow for SiGe HBTs.

structure is fully formed and subsequently covered with *protect* layers to protect it during the bipolar processing. The bipolar regions are then opened up and the entire bipolar structure is formed, including the growth of the intrinsic SiGe base. The protect layers are then removed to uncover the CMOS devices, so that the remaining processing steps required to complete the CMOS processing are conducted. An example may include the source–drain ion implantation step. A common *shared* high-temperature anneal concludes the *front-end* of the process. Silicide, contacts, and interconnects are then formed to complete the process.

The key design feature to attaining the SiGe HBT device performance in a BiCMOS process is a structure designed specifically to reduce the device parasitics. The formation of the emitter opening and the alignment of the extrinsic base contact of the intrinsic base to the emitter opening are key structural aspects that determine the SiGe HBT performance. The performance of the transistor structure also relies heavily on three other features: (1) the sub-collector and local isolation, (2) the extrinsic base growth and formation, and (3) the metal contacts to the emitter, base, and collector. These three facets will be covered in the following sections, emphasizing the importance of the integration.

While there are many variations of SiGe HBT structures in the published literature, it is possible to group all bipolar structures into four types of transistor structures: (1) simple non-self-aligned (SNSA), (2) cavity-fill self-aligned, (3) sacrificial mandrel, and (4) inner sidewall.

The simple non-self-aligned structure is commonly implemented when a very-low-cost SiGe BiCMOS process is required. In the low-cost processes the deep-trench isolation is replaced by junction isolation and the emitter–base alignment is a simple lithographic alignment. The key process steps for a non-self-aligned structure with a sub-collector and collector-epi are described by Joseph *et al.* [16] and shown schematically in Fig. 2.9. The process begins with the sub-collector and collector-epi, shallow trench isolation, and the *reach-thru* diffusion region to the sub-collector. An additional implant provides junction isolation between the bipolar transistors and the CMOS devices. In a base-after-gate process, the CMOS protect layers are removed from the bipolar area and the intrinsic SiGe epitaxial base layer is grown. This is followed by the deposition of dielectric layers, which are patterned and etched to define the emitter opening. This opening is then filled with highly doped polysilicon and finally patterned. The extrinsic base is then doped, patterned, and etched. Finally the CMOS protect layers are removed and the CMOS source/drain (S/D) implant steps performed. A final high-temperature anneal, silicide, contacts, and interconnects complete the process. An SEM cross section of such a transistor structure is shown in Fig. 2.10. The SNSA process is used throughout the industry for lower-cost derivatives for many commercial applications, but is not typically used in mm-wave processes where higher SiGe HBT performance is required.

The cavity-fill, self-aligned structure has many variations in production throughout the world and this is probably the most common self-aligned structure [17]. The essence of the structure is a selective SiGe epitaxial base growth into a cavity defined by the emitter opening which self-aligns the extrinsic base to the emitter. The schematic illustration

2.3 SiGe HBT processing and structures

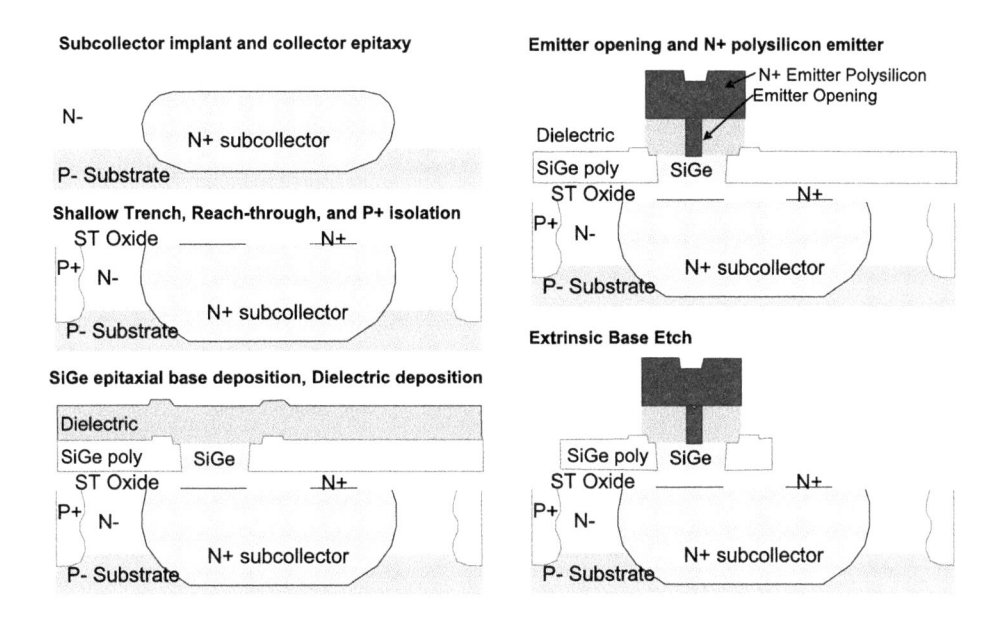

Fig. 2.9 Non-self-aligned epitaxial base transistor structure formation.

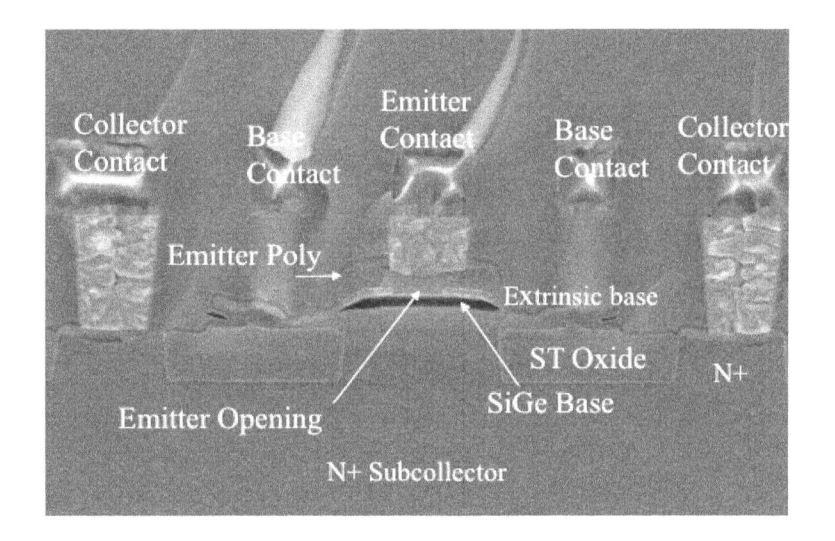

Fig. 2.10 SEM cross section of a non-self-aligned SiGe HBT.

of the key process steps are shown in Fig. 2.11. This structure is very similar in most of the processing steps to the ion-implanted double polysilicon bipolar transistor structure except for the modification to replace the ion-implanted base with a selective SiGe epitaxial layer. The overall process is similar to the previous SNSA process: the subcollector, shallow trench, and reach-thru, except that the p+ junction isolation is replaced with a polysilicon-filled deep trench. After the shallow trench and well implants the CMOS transistor structures are formed and then covered up with CMOS protect layer(s).

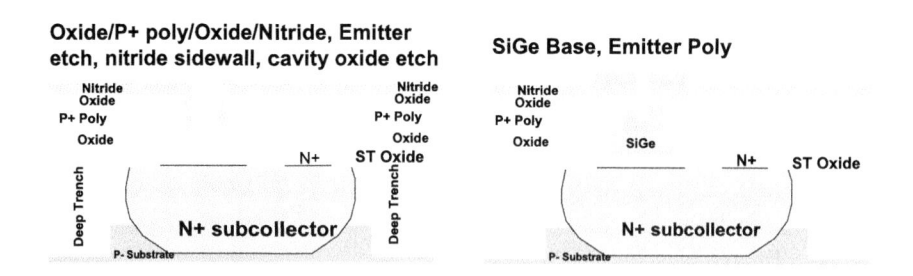

Fig. 2.11 Schematic illustration of the process to form a cavity-fill SiGe epitaxial base transistor structure.

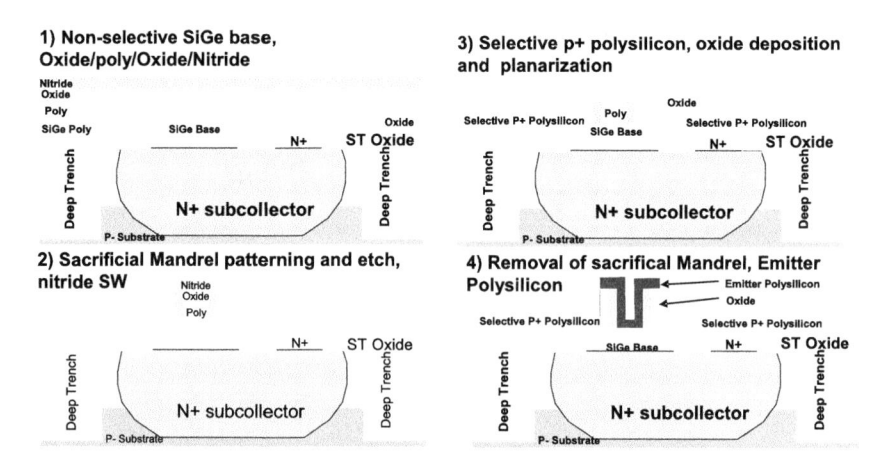

Fig. 2.12 Sacrifical emitter mandrel epitaxial base structure formation.

In a base-after-gate process the CMOS protect layers are removed from the bipolar area and the bipolar epitaxial base transistor structure is now formed with the following process steps. First a series of layers is deposited consisting of oxide, p+ polysilicon, oxide, and nitride. The emitter opening is then patterned and etched down to the first oxide layer followed by a silicon nitride sidewall formation. The first oxide is then wet etched to form the cavity undercutting the p+-polysilicon extrinsic base layer. This is followed by a selective epitaxial deposition of the intrinsic SiGe base which links up the extrinsic base during the growth. This may be followed by a second sidewall formation step. Subsequently the emitter polysilicon is deposited, doped, and patterned. The final anneal, silicide, contacts, and interconnects complete the process.

The sacrificial mandrel structure follows a non-selective deposition of the intrinsic SiGe base layer and selective deposition of extrinsic base [18]. The process consists of deposition of a series of sacrificial layers that are then patterned to form a free-standing structure perpendicular to the wafer which will ultimately define the emitter opening. For high-performance HBTs, the process may be identical through the sub-collector, collector epitaxy, deep trench, shallow trench, and reach-thru formation steps. An example of such a process is schematically shown in Fig. 2.12. The process begins by growing an oxide, depositing polysilicon, oxide, and nitride as in the case of polysilicon gate CMOS. The sacrificial mandrel stack is etched and sidewalls are formed. These steps are followed by selective epitaxy growth to form a raised extrinsic base in a similar fashion to raised source/drain for CMOS. After oxide deposition and planarization the sacrificial mandrel material is removed. This is followed by polysilicon emitter deposition and patterning.

The final structure is the *inner sidewall* structure in which the emitter–base alignment is achieved by etching an opening in the extrinsic base followed by an inner sidewall to self-align the emitter opening area to the extrinsic base. This technique has not been widely practiced and will therefore not be covered here.

2.3.3 Sub-collector and local isolation

Two important device parasitics in achieving the high-performance SiGe HBTs are the collector resistance (R_C) for f_τ (Eqn (2.9)) and low base–collector capacitance (C_{BC}) for both f_τ and f_{MAX} (see Eqn (2.14)). Optimizing R_C and C_{BC} depends on the sub-collector and local collector isolation structures.

The sub-collector formation is of two types. The first is a heavily doped sub-collector layer (implanted at $\sim 10^{16}$ atom/cm^2) in which the dopant, usually arsenic, is driven deep into the silicon ($>1.0\,\mu$m) by a high-temperature ($\sim 1100\,°$C) drive-in anneal step. This is followed by an un-doped silicon epitaxy step to separate the doped sub-collector layer from the active transistor, referred to as *collector-epi*. The second is an ion-implanted dopant layer in the native substrate, which is not followed by an epitaxy process and is referred to as a lower-concentration implanted sub-collector (LCIC). Since the LCIC is formed subsequent to key process steps such as the DT and STI formation, there are limitations to the total dose and depth of the implant. As a result these LCIC layers feature much higher sheet resistances than the epitaxial sub-collector counterpart. The LCIC is commonly found in lower-cost derivative technologies but may also be present in high-performance technologies if the geometry and final structures are formed such that the sub-collector contact is very close to the emitter opening, as in a *compact* transistor layer. Since this approach is less compatible with existing CMOS, it is generally not employed.

The local isolation can make a profound difference in the C_{CB} of the transistor. The more commonly used approach is to rely on the STI of the base CMOS. However, the active silicon region between isolation tends to be deeper and much larger than the emitter opening, which restricts the quantity of reduction in C_{CB} possible. In addition, facets are formed during the non-selective SiGe base epitaxy which increase the perimeter C_{CB} components [19]. In combination with local isolation, selective epitaxy processes may be effectively used to reduce C_{CB}. For example, in the cavity-fill epitaxial-base structure discussed earlier, the oxide underlying the extrinsic base polysilicon determines the selective SiGe base deposition area and the layer thickness and acts as a local isolation to reduce the C_{CB}. This is one of the main reasons why the cavity-fill device structure is commonly used. Another approach employed for local isolation to reduce C_{CB} is to deposit an oxide in regions where growth is undesirable. This approach enables silicon growth only in preferentially selected regions. The area of the silicon in such an approach may be tuned relative to the size of the emitter opening to gain substantial reductions in the capacitance. In this case the local isolation is not self-aligned to the emitter opening.

There have also been integration approaches in which the active area under the intrinsic SiGe base is removed with an etch in what has been referred to as *second shallow trench isolation* or *second STI* [12].

2.3.4 Extrinsic base formation

The nature of the extrinsic base contact has a significant impact on the power gain of the transistor because of its influence on both R_B and C_{CB}, which are inversely related to f_{MAX}, shown in Eqn (2.14). The objective of the device design for f_{MAX} improvements is to achieve as low an extrinsic-base resistance as possible. A commonly used process to form the extrinsic base is to implant boron into/through the intrinsic SiGe base away from the emitter contact sufficiently to allow for lateral diffusion. Often the emitter polysilicon masks the implant and positions the implant from the emitter opening as described previously in the SNSA emitter–base structure (Fig. 2.9). The implanted extrinsic base technique is fairly simple and reduces cost but the consequence is a deeper extrinsic base junction that increases C_{CB} and lowers the device breakdown voltage. An alternative method used is to diffuse boron from intrinsically doped polysilicon layers, which provided an improved method for the formation of shallow base contacts. This methodology is highly leveraged in the cavity-filled structures, seen in Fig. 2.11. However, the polysilicon layers are inherently of higher resistance than single-crystal material due to a lower carrier mobility. Hence this results in an increased resistance with the shallower junctions. To reduce the overall base resistance, the polysilicon to intrinsic base link-up region is typically self-aligned to the emitter opening and scaled. This inherently increases the contribution of the extrinsic base polysilicon resistance to the overall base resistance and it therefore becomes a limiting factor to the highest attainable f_{MAX} performance.

Several structural and process solutions have been presented to address this problem. Fox *et al.* [20] proposed a novel structural approach to reducing the extrinsic-base resistance in which the polysilicon layer was replaced with a single-crystal, heavily boron-doped selective silicon deposition. An alternative method is the use of high-temperature millisecond anneals preceded by low-temperature silicide and interconnect processing steps, to decrease the sheet resistance of the polysilicon layers. The high-temperature millisecond anneals enable higher dopant activation levels in the extrinsic base and subsequent dopant deactivation is prevented by the lower thermal budget of the *back-end* processing steps [12, 21]. At performance levels exceeding 300 GHz f_{MAX}, increasingly new innovations will be required. It is likely that selective epitaxy processes will become more commonplace in this important area.

2.3.5 Polysilicon emitters and contacts

In the 1970s, the emitter and base layers were formed in silicon BJTs by implantation of dopants and subsequent anneals into silicon. The combination of the implant

2.3 SiGe HBT processing and structures

and anneal cycles led to very wide bases and deep emitter junctions. There were high base currents in these transistors both from holes back injected into the emitter and because of recombination in the wide base layers. It was very difficult to achieve high-current-gain transistors with shallow junctions and narrow base widths. The polysilicon emitter dramatically transformed silicon bipolar technology because it made possible the formation of shallow emitter junctions. The polysilicon layer was doped with arsenic, usually by implantation, and annealed, allowing the arsenic to rapidly redistribute in the polysilicon layer and then diffuse into the single crystal. Arsenic tended to have a lower diffusion coefficient, compared with phosphorus, and high solubility, forming a highly doped and very abrupt diffusion profile. The presence of an interfacial *oxide* layer at the polysilicon/single-crystal interface was an unintended consequence of the processing recipes used, but it proved to be a valuable aspect of the process as the *interface* created a barrier to the base hole current and reduced the base current density. Therefore the current gain, β, of the transistor could be increased by increasing the thickness of the interfacial oxide of the polysilicon emitter process and increasing the barrier to holes, which reduced the base current. The location of the sub-monolayer-thick interfacial oxide is schematically represented in Fig. 2.13. Since the usual thickness of this interface oxide is less than a monolayer, it is not easy to get an image of the oxide. The thicker interfacial oxide impacted many other aspects of the transistor by also providing a barrier to arsenic diffusion and making the emitter–base junction shallower. Changing the emitter–base junction depth changed many other characteristics of the SiGe HBT. In particular, the SiGe HBT collector current density is decreased by a shallower emitter–base

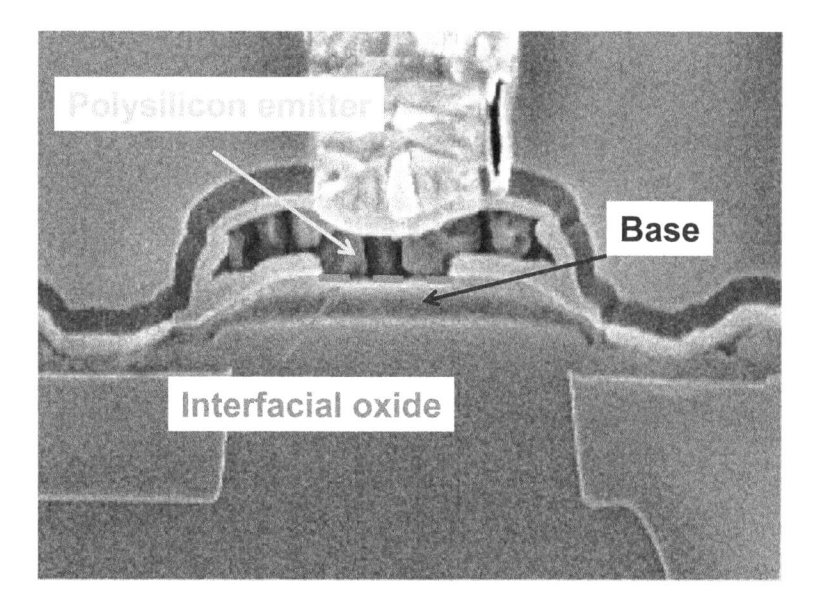

Fig. 2.13 Cross-sectional TEM of SiGe HBT showing the polysilicon emitter and representative location of interface oxide. Since the usual thickness of this interface oxide is less than a monolayer, very-high-resolution TEM is needed to image it.

junction depth by reducing the intercept of the junction with the Ge grade and lowering the intercept Ge% ($\Delta E_g(0)$), which reduces the collector current (see Eqn (2.3)).

Control of the emitter junction depth is also important because it controls the neutral base width and total base charge. The collector current is dependent on the base width and inversely related to the integrated base dopant through the *Gummel* relationship [1]. Therefore changes in the emitter junction depth change the collector current as well. A shallow emitter diffusion will increase the base layer width and total base charge, decreasing the collector current. The increased oxide interface also introduces a resistance to electron as well as hole current, increasing the emitter resistance, which lowers f_τ ($R_E \cdot C_{BC}$).

In all SiGe HBT technologies the control of the interfacial oxide is a key parameter for reliable manufacturing. Variations in this parameter lead to variations in electrical parameters as discussed above. In lower-performance SiGe HBT technologies (under 60 GHz peak f_τ), where the Ge profile is wider and the peak Ge concentration is low, the impact of variations in interfacial oxide is less severe but still an important aspect in manufacturing. However, in high-performance mm-wave SiGe HBT technologies (peak $f_\tau > 100$ GHz) where the total emitter–collector transit time is smaller, the performance is very sensitive to the interfacial oxide. In high-performance (HP) SiGe HBTs the peak Ge% is higher (15%–30%), and the Ge grade is steeper, increasing the sensitivity to emitter–base junction depth variations caused by the interfacial oxide. In addition, the $R_E \cdot C_{BC}$ delay time constant can become a large part of the delay if R_E becomes excessively high. The push for higher-performance SiGe HBT technologies will only make controlling these interfaces more critical. There is very little published material on the impact of emitter oxide interfacial layers on SiGe HBT device characteristics in general.

The total emitter resistance is also dependent on the metal contacts and bulk emitter polysilicon resistance. A schematic representation of a SiGe HBT emitter is shown in Fig. 2.14a taken from Cheng *et al.* [22]. Various parameters shown in this figure – emitter width and depth, polysilicon thickness, and contact regions – are very critical in determining the total emitter resistance as well as the f_τ and f_{MAX}. In an experimental lot, emitter width was varied for the same emitter-formation process. Measured f_τ and

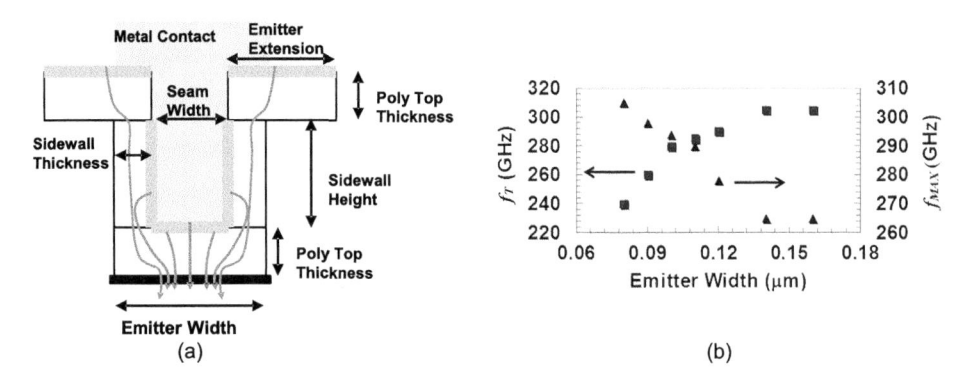

Fig. 2.14 (a) Analytical model of R_E for the unplugged U-shaped emitter. (b) The effect of lateral scaling on f_τ and f_{MAX}.

f_{MAX} as a function of emitter width are shown in Fig. 2.14b. In very-narrow-emitter SiGe HBTs, if the emitter becomes plugged and the metal contact is further away from the emitter polysilicon/single-crystal interface, the resistance components from the bulk polysilicon layer become significant. This results in lower f_τ for narrower emitter width, due to higher emitter resistance (Eqn (2.9)), posing a challenge for lateral device scaling. Reduction in emitter width reduces parasitic C_{CB} and R_B, which improves f_{MAX} (Eqn. (2.14)).

2.4 Key circuit design metrics

The key HBT RF parameters, f_τ and f_{MAX}, were discussed earlier in the chapter. In this section, additional important parameters that are critical for mm-wave designs will be discussed.

2.4.1 Noise figure

Inherent noise in the HBTs degrades the signal-to-noise ratio of a circuit's output signal and thus limits the sensitivities and data rates of ranging, communications, and imaging systems. Noise performance, in addition to f_τ and f_{MAX}, is therefore a key RF metric by which a technology is evaluated. To date, each generation of SiGe BiCMOS technology has demonstrated outstanding high-frequency noise performance when compared with competing silicon technologies.

There are multiple physical noise sources in a bipolar transistor. Shot noise is one example and is due to fluctuations associated with the DC current flow across a potential barrier. Electrons moving from the emitter to the base approach the p-n junction with a statistical distribution of direction and energies. Additionally, a probability of crossing the potential barrier exists. The result is shot noise in collector current. Holes moving in the opposite direction from the base to the emitter produce a similar base current shot noise. These two noise currents are correlated by a factor that is proportional to base and space charge region transit delay. Parasitic base resistance contributes to thermal noise. At high collector–base reverse bias, avalanche multiplication contributes an additional noise source. Further details of these different sources of noise can be found in [23, 24].

The noise figure for an HBT can be roughly estimated from the expression given in Eqn (2.15) below, in which each term has been defined previously [25, 26]:

$$NF_{min} \simeq 1 + \frac{1}{\beta} + \sqrt{\frac{2I_c}{V_T}(R_B + R_E)\left(\frac{f^2}{f_\tau^2} + \frac{1}{\beta}\right) + \frac{1}{\beta}}. \quad (2.15)$$

Figure 2.15 shows a typical plot of the measured NF_{min} as a function of frequency at a given applied bias for an HBT with an operating frequency f_τ of 300 GHz. At low frequencies, there is very little change in NF_{min}. This can be attributed to a much smaller f than f_τ, whereby NF_{min} is approximated to $\sqrt{2R_B I_B/V_T}$. At higher frequencies, f approaches f_τ and NF_{min} becomes proportional to frequency [27].

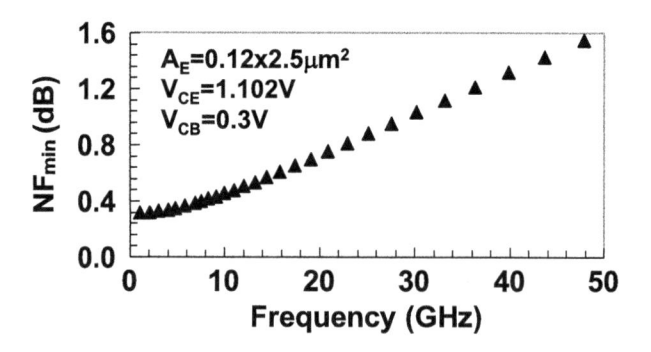

Fig. 2.15 Noise figure as a function of frequency for an HBT with peak f_τ of 300 GHz.

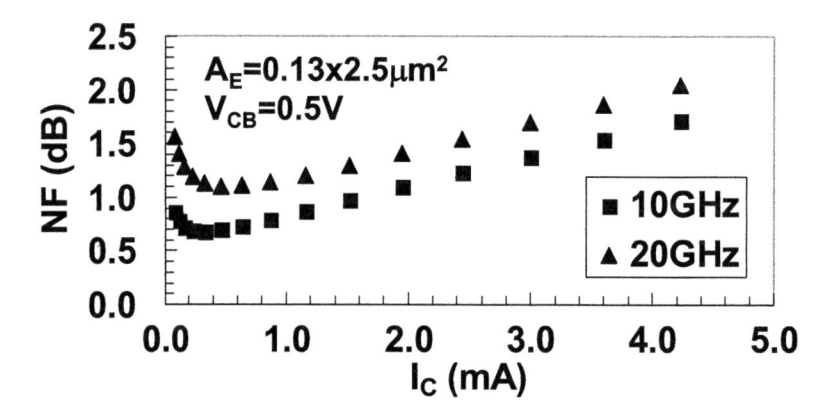

Fig. 2.16 Noise figure as a function of collector current (I_C) at two different frequencies of measurement.

Figure 2.16 shows the variation in NF_{min} with collector current at a constant $V_{CE} = 1.35$ V for an HBT with an emitter width of 130 nm. From the relation in Eqn (2.15), it can be seen that the minimum noise figure depends on R_B, C_{BE}, C_{CB}, and transit delay. For frequencies f much greater than $f_\tau/\sqrt{(\beta)}$, I_C for NF_{min} is independent of frequency or in other words NF_{min} occurs at the same I_C independent of frequency.

Figure 2.17 shows NF_{min} as a function of frequency for different generations of BiCMOS technologies ranging from an emitter width of 0.44 µm and peak f_τ of 43 GHz to an emitter width of 90 nm and peak f_τ of 300 GHz. NF has improved with each new generation and is expected to continue to decrease with future generations.

2.4.2 Low-frequency noise

Low-frequency noise predominantly affects frequency-generation and -conversion circuits, where circuit nonlinearities result in noise signals being up-converted into the band of interest. For example, in voltage-controlled oscillators, the phase-noise performance of the circuit within a few kHz of the carrier frequency is typically determined by the $1/f$ noise performance of the devices in the circuit [28, 29].

2.4 Key circuit design metrics

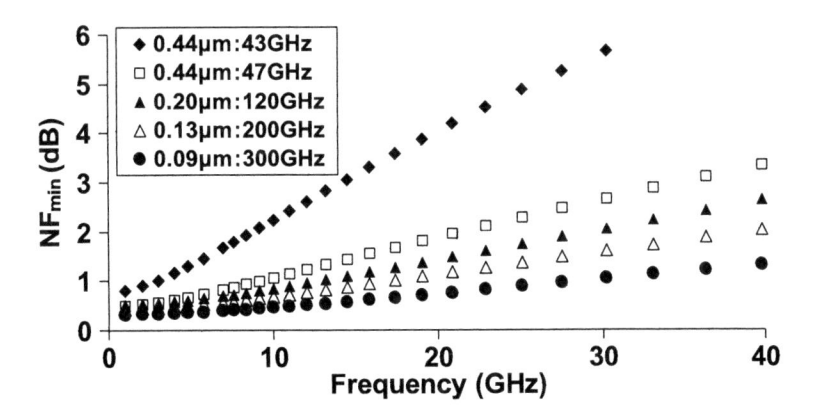

Fig. 2.17 Improvement in HBT NF_{min} with each BiCMOS generation.

The physical mechanisms responsible for $1/f$ noise arise from carrier capture and emission from traps located at the material interfaces in the device. These interfaces include the interface between the base and emitter regions of the bipolar, and also the passivation interfaces along the perimeter of the device regions. In an ideal device, the base–emitter junction would be formed within a uniform solid region of the device. In practical devices, the base and emitter are formed in different process steps, so a non-ideal interface results as discussed in earlier sections. Residual silicon oxide or other imperfections at the interface can give rise to carrier traps. The capture and emission of carriers from these traps modulates the currents flowing through the terminals of the device, producing a noise current which has a power spectral density that is inversely proportional to frequency.

The $1/f$ noise is typically modeled as a base-current component which has a $1/f$ noise spectrum (added in parallel with the base shot-noise source). The noise equation is given as

$$\overline{\langle i_b^2 \rangle} = \frac{k_f \cdot I_b{}^a}{f^b} \cdot \Delta f, \qquad (2.16)$$

where $\overline{\langle i_b^2 \rangle}$ is the base-current noise power-spectral-density, I_B is the base bias-current, and a, b, and k_f are fitting parameters.

Typical $1/f$ noise performance for a modern HBT device is shown in Fig. 2.18 for three different base currents. The collector-current noise power-spectral-density is shown to be nearly exactly proportional to $1/f$. The magnitude of the $1/f$ noise is also shown to increase with increasing bias current.

Typical $1/f$ noise variation with bias current for the devices above is shown in Fig. 2.19. The noise current increases linearly with bias, yielding a quadratic dependence of noise power-spectral-density on bias current. For these particular devices, this quadratic variation is an indication that modulation of recombination centers in the base–emitter junction, and along the passivated perimeter of the device, is primarily responsible for the $1/f$ noise [30]. Devices fabricated with other process may have a different variation of noise with bias.

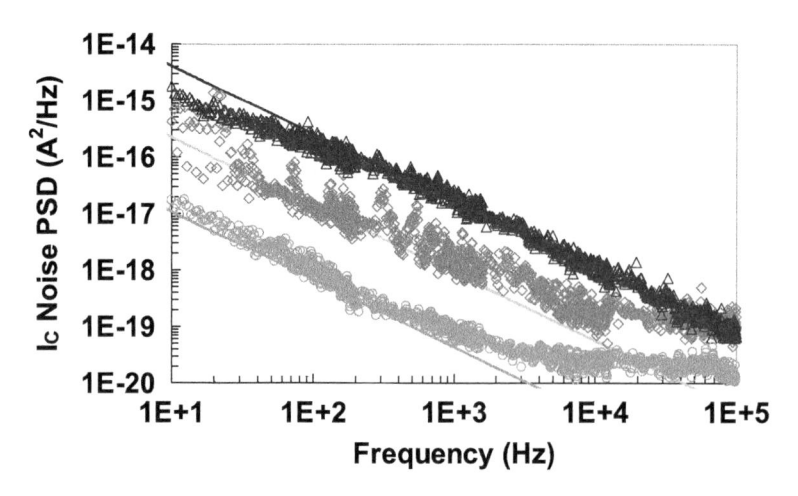

Fig. 2.18 Collector-current noise power-spectral-density (A^2/Hz) vs frequency for HBTs with $A_E = 1.1\,\mu m^2$ for $I_B = 2.5\,\mu A,\ 12\,\mu A,\ 75\,\mu A,\ V_{CE} = 3.0\,V$. The fitlines follow the expression $(freq)^{-1.2}$. The system noise floor influences the data shown here below 10^{-19}.

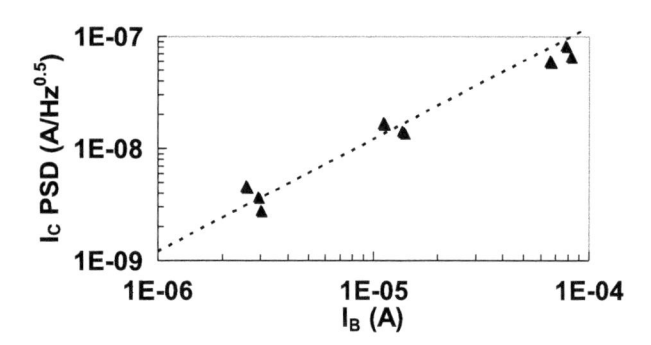

Fig. 2.19 Collector current-noise power-spectral-density as a function of I_B. Note the units here are current noise, which is the square root of noise PSD. Different points represent data taken from separate wafers. The dotted line represents the $n = 1$ fit depicting linear change in noise current with bias.

2.4.3 Avalanche multiplication

The collector–base junction in HBTs during the forward active mode of operation is usually reverse biased, which results in a large electric field in the collector space-charge region. For sufficiently high electric fields, electrons injected into the collector from the base gain sufficient energy from the electric field to create an electron–hole pair upon collision with the lattice. The result is extra carrier generation, and the process is called *impact ionization*. Electrons and holes formed in this manner can subsequently acquire enough energy to generate additional electron–hole pairs by impact ionization. This process results in a multiplicative impact ionization sequence and is known as *avalanche multiplication*. As a result of avalanche multiplication, the electron current leaving the collector space-charge region, I_{out}, is higher than the current injected into the collector, I_{in}, and the ratio of the two currents is known as the avalanche multiplication factor M

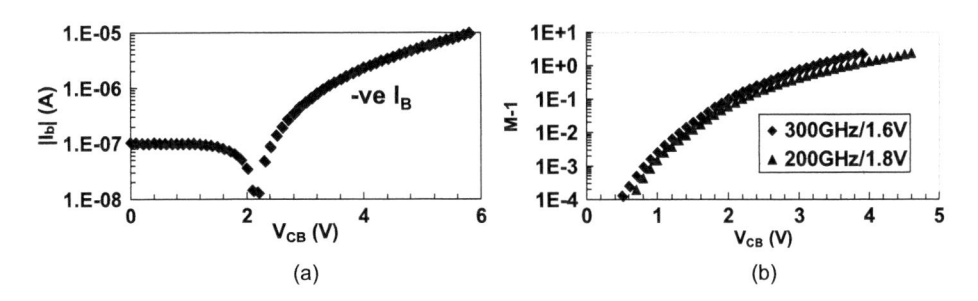

Fig. 2.20 (a) Base-current reversal for a SiGe high-performance HBT at high V_{CB} due to avalanche multiplication. (b) Measured value of $(M - 1)$ for two different HBTs having peak f_τ of 200 GHz and 300 GHz.

($M = I_{out}/I_{in}$). An equivalent number of holes are generated in the process, and these flow into the base of the device. This *extra* base current compensates for the base-hole current injected from the terminal into the base and, consequently, the base current at the base terminal decreases. For sufficiently high V_{CB}, when the avalanche multiplication rate is high, base-current reversal occurs. In Fig. 2.20a, it can be seen that the base current starts to decrease at $V_{CB} = 1.5$ V and base-current reversal occurs at 2.1 V (for a fixed $I_E = 10\,\mu\text{A}$).

The term $(M - 1)$ is generally reported for avalanche multiplication, instead of M. This is largely due to the fact that it more clearly shows the increase in collector current with bias. We can measure $(M - 1)$ by changing the V_{CB} at either a fixed V_{BE} or fixed I_E. The fixed V_{BE} method is sufficient for low V_{BE} or low J_C values, but can result in permanent damage to the device should thermal runaway occur at high-bias conditions due to the associated self-heating. This will be discussed in more detail in the next section. In the fixed I_E method, a current I_E is forced into the emitter, V_{CB} is swept and V_{BE} is measured [31, 32]. We can calculate $(M - 1)$ from the expression

$$M - 1 = \frac{I_c}{I_E - I_B(V_{BE})|_{V_{CB}=0}}. \tag{2.17}$$

Figure 2.20b shows the measured value $(M - 1)$ for two different transistors using the forced I_E method at $I_E = 10\,\mu\text{A}$. Note that $(M - 1)$ increases at a given V_{CB} as the collector–base space-charge region is scaled with each new technology for improved f_τ.

2.4.4 Breakdown voltage

HBT vertical scaling for improved f_τ adversely affects breakdown voltages BV_{CBO}, defined as the collector–base breakdown voltage for an open emitter, and BV_{CEO}, defined as the collector–emitter breakdown voltage for an open base. Each new generation of SiGe HBTs that feature increased f_τ generally has a decreased BV_{CEO} which reduces the flexibility available to circuit designers. As a result, HBTs which feature different trade-offs between breakdown voltages and f_τ are offered for varied applications. This is typically accomplished in the process by tailoring the SIC implant dose and energy and the sub-collector implant to meet the f_τ–BV_{CEO} needs.

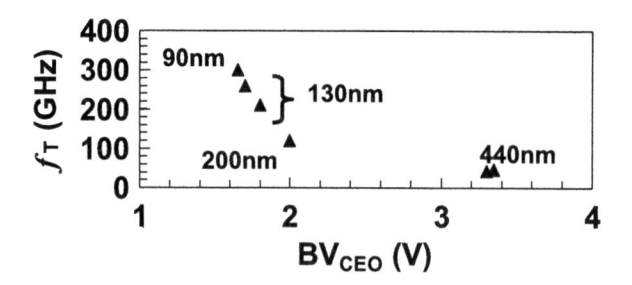

Fig. 2.21 Change in f_τ and BV_{CEO} for different generations of high-performance SiGe HBTs.

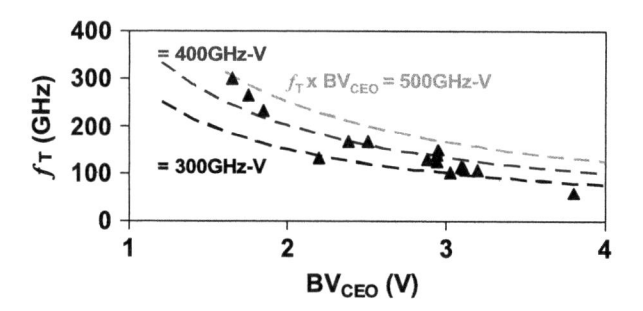

Fig. 2.22 Change in BV_{CEO} and f_τ for different collector and sub-collector implants for experimental 90 nm HBT technology.

Figure 2.21 shows the change in f_τ and BV_{CEO} for different generations of high-performance SiGe HBTs. Within a given technology, higher BV_{CEO} values are typically traded for lower f_τ performance devices by changing the collector and/or sub-collector implants. Figure 2.22 shows the change in BV_{CEO} and f_τ for different collector and sub-collector implants for experimental 90 nm HBT technology [33].

As collector–base reverse-bias voltage is increased for HBTs, avalanche multiplication occurs in the collector–base space-charge region, resulting in excess holes being swept into the base. For a common emitter configuration, driven by either a constant base current (I_B) or a constant emitter–base voltage (V_{BE}), this excess hole current is injected into the emitter and becomes amplified by the forward current gain, β, of the device. This is a positive-feedback mechanism, which increases collector current and avalanche current, leading to premature breakdown of the device. Thus the open-base breakdown voltage (BV_{CEO}) is much lower than the collector–base reverse-junction breakdown voltage (BV_{CBO}). For common base configuration designs, driven by a constant emitter current, the avalanche current (holes injected into the base from the collector–base junction) exits the base terminal, thereby eliminating the current amplification effect and positive-feedback mechanism due to β. As V_{CB} is increased, due to avalanche multiplication, the base current decreases and eventually changes sign, ultimately leading to a central current crowding or *pinch-in* point [34]. However, stable operation is still possible using the forced I_E method in regions of high avalanche at voltages much larger than BV_{CEO} and limited by BV_{CBO}. BV_{CBO} is the absolute voltage limit which can be

2.4 Key circuit design metrics

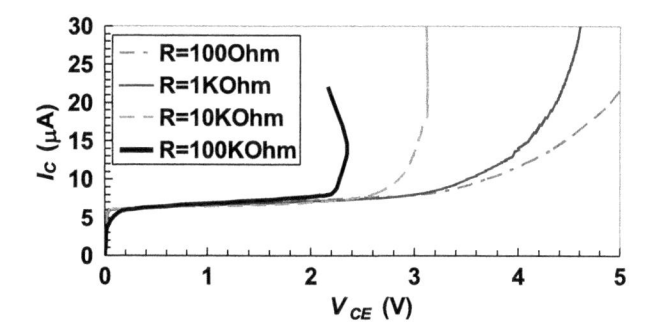

Fig. 2.23 Measured BV_{CER} (I_C vs V_{CE}) for a 0.13 μm HBT as a function of base resistance for applied V_{BE} of 0.72 V.

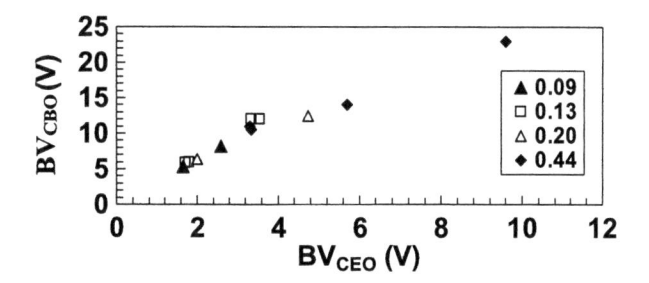

Fig. 2.24 BV_{CBO} and BV_{CEO} values for several SiGe HBT generations.

applied before breakdown occurs. However, the real voltage limit (BV_{CER}) lies somewhere between BV_{CEO} and BV_{CBO}, determined by the base resistance. Figure 2.23 shows measured BV_{CER} data for a 0.13 μm HBT as a function of base resistance for applied V_{BE} of 0.72 V. As expected, BV_{CER} reduces with increased R_B.

Several circuit designs have been demonstrated where devices are operating successfully above the BV_{CEO} limit [35, 36]. Figure 2.24 shows the BV_{CBO} for several SiGe HBT generations along with the BV_{CEO} values.

2.4.5 Device self-heating

Junction temperature is a critical parameter in determining the reliability and performance of HBTs and their operation bias limits in circuits. As devices are vertically scaled for improved f_τ, J_{KIRK} increases due to a thinner space-charge region, which increases the power dissipation in the device at peak performance given by

$$P = V_{CE} \cdot I_C + V_{BE} \cdot I_B \sim V_{CE} \cdot I_C. \tag{2.18}$$

This results in an increase in the device junction temperature $\Delta T = P \cdot R_{th}$, where R_{th} is the thermal resistance of the HBT. High junction temperature affects the DC as well as RF performance of the device adversely. In Fig. 2.25a the degradation in device f_τ and f_{MAX} with increase in ambient temperature for a 130 nm HBT is shown. Thus, device self-heating and thermal management is a key issue in practical circuit designs. As the

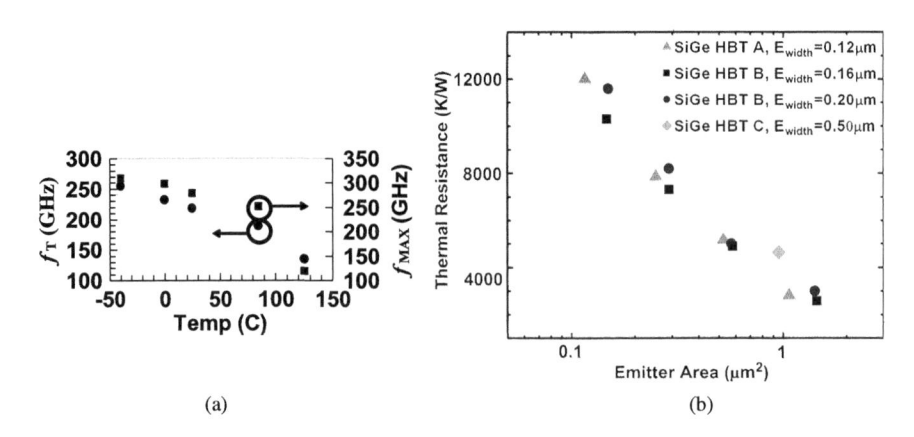

44 SiGe HBT technologies for mm-wave power amplifiers

Fig. 2.25 (a) Degradation in HBT f_T (circle symbol) and f_{MAX} (square symbol) with increase in ambient temperature for a 130 nm HBT. (b) R_{th} of SiGe HBTs as a function of emitter area for different emitter widths.

device area is scaled for improved performance, it is imperative that thermal effects are considered in device designs for improved reliability.

The thermal resistance of an HBT can be extracted from the relationship between the change in V_{BE}, at constant I_C, with power dissipation and ambient temperature [37]:

$$R_{th} = \frac{dV_{BE}}{dP}\bigg|_T \bigg/ \frac{dV_{BE}}{dT}\bigg|_P. \qquad (2.19)$$

Figure 2.25b shows the thermal resistance of SiGe HBTs as a function of emitter area for different emitter widths [38]. Lateral and vertical scaling of HBTs for improvements in the device performance levels increases both the thermal resistance, due to lateral scaling, and the power dissipation, due to vertical scaling. These factors significantly impact the reliability of devices due to the sharp temperature rise. Novel device layout and circuit design schemes are utilized as a means to reduce the impact of the rise in junction temperature.

In the literature several models for the thermal resistance of isolated devices exist [39, 40], which have been incorporated into PDKs and device models for BiCMOS technologies. Thermal modeling is necessary to enable reasonable predictions of R_{th} and assist in the device design optimization and layout required for R_{th} improvements. Rieh et al. [39, 40] proposed an analytical and predictive thermal model for arbitrarily shaped device structures, utilizing a simpler geometry-based approach. The model assumes that R_{th} can be expressed as a series combination of the partial thermal resistances from individual thin slabs of area A and thickness d along the heat flow path. The model has been applied to modern SiGe HBTs and assumes that the oxide-based trenches behave as perfect heat insulators due to a low thermal conductivity. The model further assumes that the heat flux is confined to a cone with a 45° heat spreading angle. The partial thermal resistances were expressed in analytic form based on the geometry of the trenches and the thermal conductivity of the substrate material, and thus an analytic expression for the total thermal resistance ($R_{th,int}$) was obtained (Fig. 2.26). However, the model falls

2.4 Key circuit design metrics

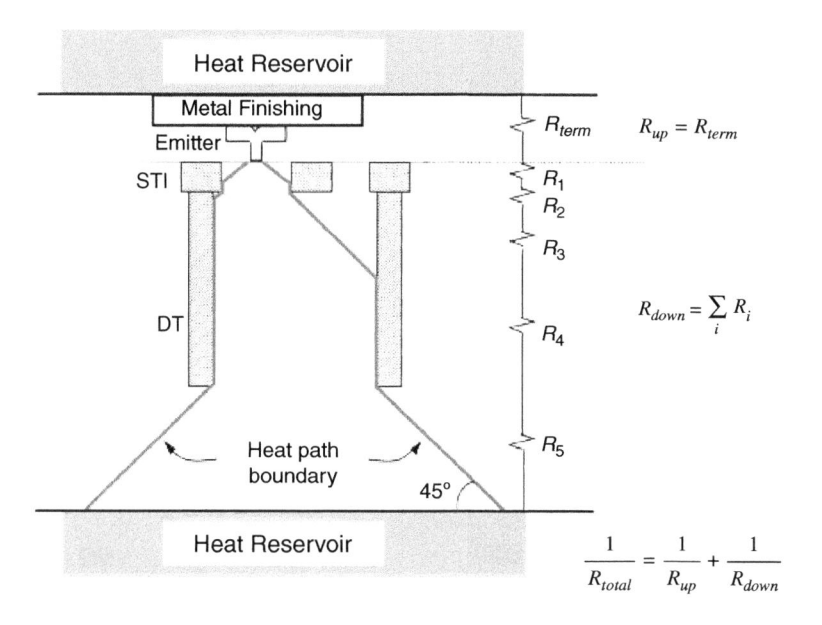

Fig. 2.26 Cross-sectional schematic of a typical HBT along with the heat dissipation path and partial thermal-resistance components.

short in that oxide is not a perfect heat insulator. In reality, a finite heat leakage occurs, particularly associated with polysilicon-filled deep trenches.

In addition to the heat flow into the substrate, there is a finite amount of heat dissipation through the metal wiring stack connected to the device. The thermal resistance of the metal stack ($R_{th,metal}$) can be roughly estimated using

$$R_{th,metal} = \frac{H_m}{A_m \cdot \sigma_m}, \qquad (2.20)$$

where H_m and A_m are the height and cross-sectional area respectively of the metal/via layers and σ_m is the bulk metal thermal conductivity. The top metal layer connected to the probes is considered the heat reservoir. The total thermal resistance of the device can then be computed from

$$\frac{1}{R_{th}} = \frac{1}{R_{th,int}} + \frac{1}{R_{th,metal}}. \qquad (2.21)$$

Figure 2.27 shows the change in device R_{th} with emitter length and deep-trench depth for a 90-nm-wide emitter and compares the measured data with the predicted R_{th} value from the analytical model. A good fit between the analytical model and measured data is obtained. The variation observed between the measured and predicted values for the DT depths investigated may be attributed to the model assumption that DT is a perfect insulator, as previously discussed.

Since heat is dissipated both into the substrate and out of the metal stack connected to the device, the thermal resistance of an HBT can be reduced through improved heat dissipation out of the metal stack. This has been proven experimentally and is shown

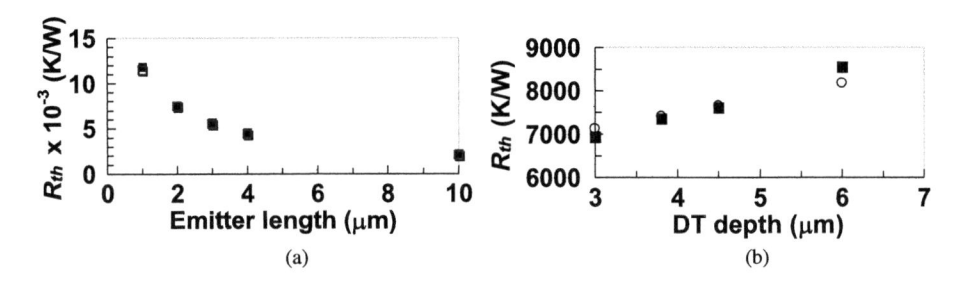

Fig. 2.27 (a) Measured (solid symbols) and calculated (hollow symbols) R_{th} for 0.1-μm-wide SiGe HBTs as a function of emitter length. (b) Measured (solid square) and calculated (hollow circles) R_{th} for an HBT with emitter area of $0.1 \times 2\,\mu m^2$ as a function of DT depth.

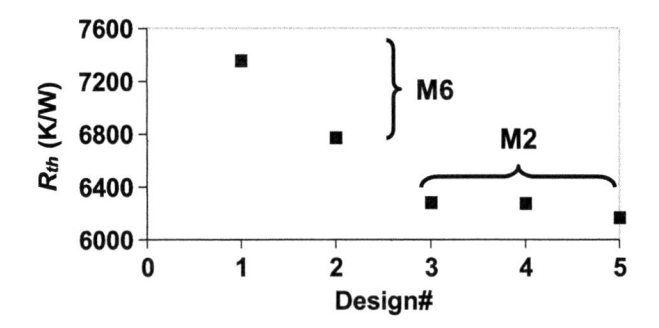

Fig. 2.28 Change in measured R_{th} for an HBT with emitter area of $0.1 \times 2\,\mu m^2$ as the heat reservoir metal level is brought closer to the device from M6 to M2. The device with lower R_{th} at M6 (#2) has a mesh of dummy M1 around the device for better heat dissipation. Different devices wired out at M2 (#3–5) have different widths of dummy metal stack surrounding the device.

in Fig. 2.28. Measured R_{th} for a $0.1 \times 2\,\mu m^2$ SiGe HBT decreases as the heat reservoir metal level is brought closer to the device from M6 to M2. R_{th} can be lowered further by using dummy metal stacks around the HBT for improved heat flow. The device with lower R_{th} at M6 (#2) has a mesh of dummy M1 around the device for better heat dissipation. Different devices wired out at M2 (#3–5) have different widths of dummy metal stack surrounding the device. However, when the HBT is wired at M2, the impact of the dummy metal stack around the device on R_{th} is not significant [41].

2.4.6 Mutual thermal resistance

With increase in packaging density, thermal coupling in circuits between active HBT and other power-dissipating elements is becoming more critical. Figure 2.29a shows the layout of an isolated HBT in a collector–base–emitter or *CBE* configuration. The HBT is surrounded by a ring of 24 HBTs such that the adjacent devices share the same DT isolation. This is a dense configuration and allows for the minimum possible distance between two adjacent HBT devices. The Gummel characteristics of an isolated HBT are directly impacted by the power dissipated by the HBTs surrounding it, due to thermal coupling and the associated rise in junction temperature. The corresponding Gummel curves are shown Fig. 2.29b. The mutual thermal resistance ($R_{th,mutual}$) is computed using

2.5 BiCMOS passive devices and features

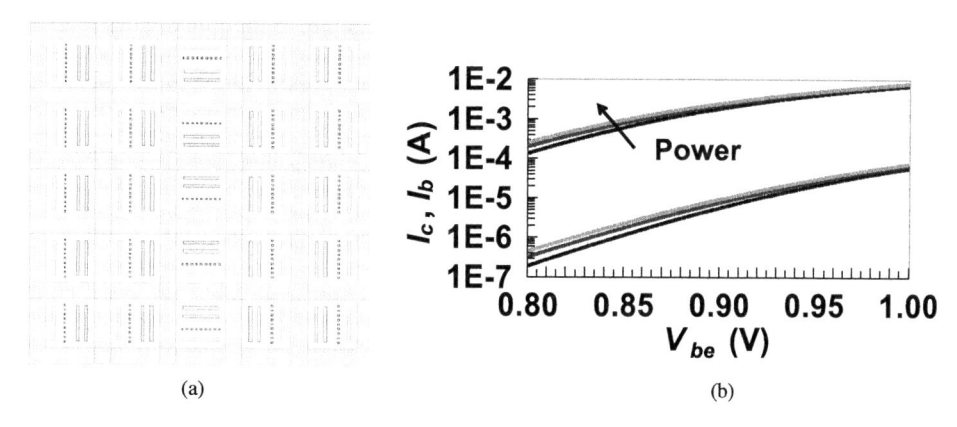

Fig. 2.29 (a) Layout of a ring of 24 HBTs in CBE configuration surrounding an isolated CBE HBT in the center for calculation of mutual thermal coupling (dense configuration). The image shows a mesh of DT and contacts to emitter, base, and collector. (b) Change in measured I_C and I_B for the solitary HBT as the power in the outer ring of HBTs is increased.

$$R_{th} = \left. \frac{dV_{BE,single}}{dP_{array}} \right|_T \bigg/ \left. \frac{dV_{BE,single}}{dT} \right|_P. \qquad (2.22)$$

For this configuration of HBTs, $R_{th,mutual}$ was determined to be approximately 190 K/W. Thus, if all transistors in the ring are operating at $V_{CE} = 1.2$ V and $I_C = 2$ mA, the central HBT will experience an increase in ambient temperature of approximately 11 K in addition to the self-heating effects. This is a significant value for practical circuit designs, like power amplifier blocks where multiple PAs are stacked together, and needs to be incorporated in the relevant models for improved hardware to simulation correlation.

Multiple emitter finger layouts are preferred for PA designs since they are more suitable for high-frequency, high-power operation aided by low base resistance. The central finger is generally hotter than the adjacent fingers due to self- and mutual heating. In addition, for stacked PA designs, the edge finger might be hotter due to the presence of DT and/or low-thermal-conductivity packaging material. Several references exist in the literature related to modeling of these effects [42–44].

2.5 BiCMOS passive devices and features

2.5.1 Schottky barrier diodes

As discussed earlier in the chapter, improvement in the HBT f_τ performance level degrades the breakdown voltage, making it challenging to handle power at high frequencies. Integrating passive diodes like the Schottky barrier diode (SBD) and p-i-n diode (PIN) with SiGe BiCMOS, therefore, provides a valuable addition to the circuit designer's toolbox. These diodes can be added to the BiCMOS process flow without additional mask requirement or processing steps, which is cost-effective and easy to implement.

SBDs are used in various circuit applications including mm-wave frequency-doublers and sub-harmonic mixers [45, 46], power detectors [47, 48], low-noise voltage-controlled oscillators [49, 50], and high-speed data-converter front-ends [51]. With cut-off frequencies beyond 1.0 THz, Schottky diodes integrated in a BiCMOS technology provide a means to extend the range of power handling, without significant distortion or additional noise.

SBDs with cut-off frequency (f_c) approaching 2.0 THz have been reported in the literature [52–54]. Two critical parameters for the SBD are series resistance, R_{on}, and off-state capacitance, C_{off}. Minimization of these parasitic resistances and capacitances is critical for improved f_c. C_{off} and R_{on} can be extracted utilizing either DC or AC measurements. From S-parameter data, C_{off} is generally extracted at 0 V or in reverse bias using

$$C_{off} = -\frac{1}{\Im(1/\omega Y_{11})}, \qquad (2.23)$$

and R_{on} is extracted at sufficiently large forward bias using

$$R_{on} = \Re(-1/Y_{12}). \qquad (2.24)$$

The cut-off frequency (f_c) is calculated from these values using the equation

$$f_c = \frac{1}{2\pi R_{on} C_{off}}. \qquad (2.25)$$

The quality factor (Q) for the diodes is computed from

$$Q = \frac{\Im(-Y_{11})}{\Re(Y_{11})}. \qquad (2.26)$$

Figure 2.30a shows a schematic cross section of an SBD integrated into a 90 nm BiCMOS process flow. R_{on} includes the resistance of the *reach-thru* region, the buried cathode layer sheet resistance, the intrinsic device resistance, and the contact and wiring

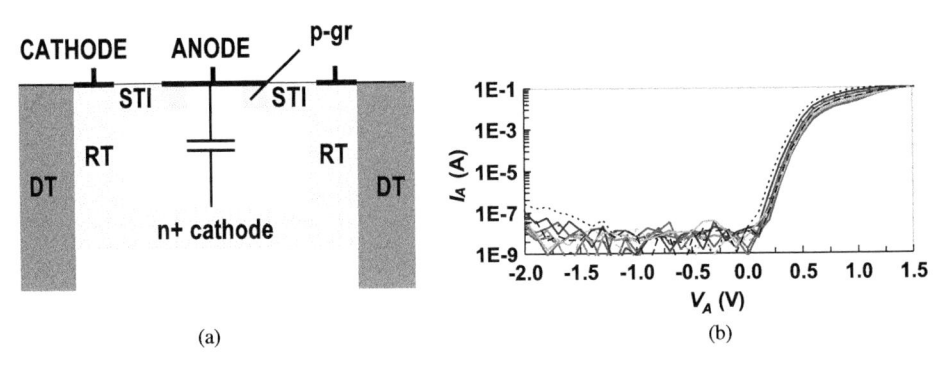

Fig. 2.30 (a) Cross-sectional schematic of a Schottky barrier diode (SBD) depicting the deep-trench (DT) isolation, shallow-trench isolation (STI), n+ cathode, cathode reach-thru region (RT), p-guardrings (p-gr), and cathode and anode contacts. (b) DC current–voltage characteristics (*I–V*) of SBDs with different anode areas.

2.5 BiCMOS passive devices and features

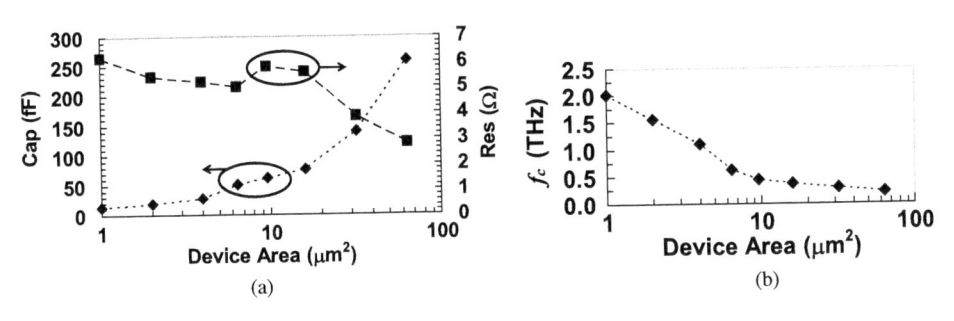

Fig. 2.31 (a) SBD resistance (R_{on}) and capacitance (C_{off}) extracted from S-parameters as a function of anode area. (b) SBD cut-off frequency (f_c) as a function of anode area.

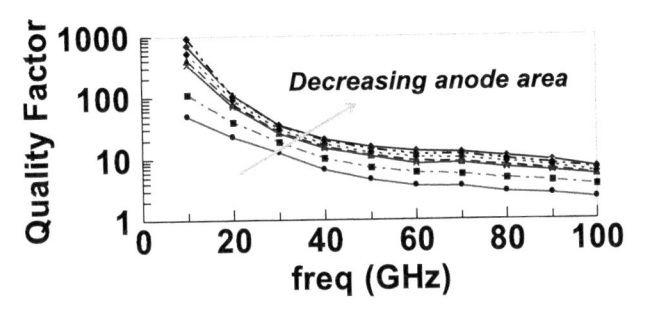

Fig. 2.32 SBD quality factor (Q) as a function of frequency for different anode area devices.

resistances. C_{off} is the accumulative capacitance between the anode and the cathode, the device and the substrate, and the p-guardrings and buried cathode layer. The anode-to-cathode capacitance includes the capacitance associated with the anode silicide and buried cathode layer, anode silicide and reach-thru region, and the anode and cathode wiring. The p-type guardring is used to reduce reverse bias leakage in the SBD and for improved device reliability. DC current voltage curves ($I-V$) are shown in Fig. 2.30b. Figures 2.31 and 2.32 show the measured R_{on}, C_{off}, f_c, and quality factor for different SBDs implemented in 90 nm BiCMOS process as a function of device area. For an SBD, in general, C_{off} increases with anode area as $C_{off} \propto A_{Anode}$ while R_{on} decreases as $R_{on} \propto A_{Anode}^{-0.5}$. Hence device f_c is reduced with increase in anode area. It is possible to achieve a higher f_c by utilizing a smaller diode area and to achieve high-current operation by stacking multiple devices in parallel.

2.5.2 p-i-n (PIN) diodes

PIN diodes are similar in cross section to SBDs except for the resistive anode contact on the p+ layer rather than a Schottky contact. Similar parasitic resistances and capacitances are present in PINs as in SBDs. Two important figures of merit for a PIN diode are isolation loss, which is a measure of leakage in the *off* state, analogous to C_{off}, and insertion loss, which is a measure of signal loss in the *on* state, analogous to R_{on}.

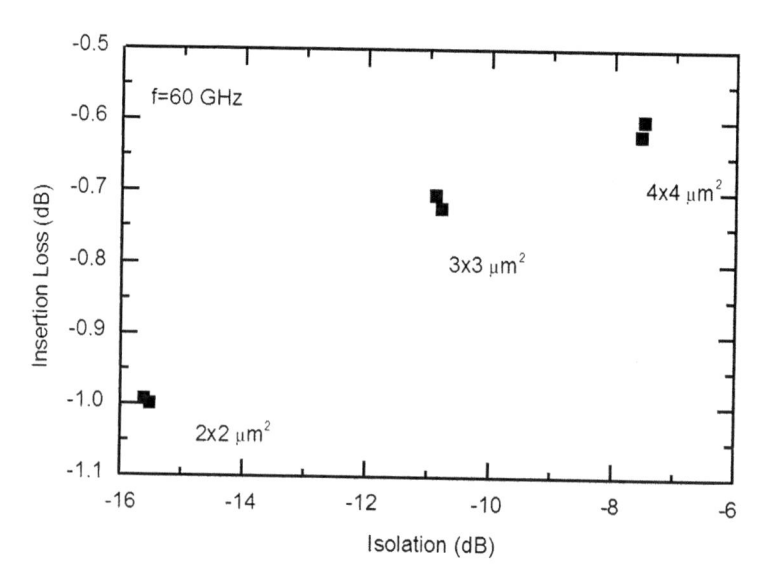

Fig. 2.33 Insertion vs isolation loss for p-i-n diodes in 130 nm BiCMOS technology as a function of anode area at 60 GHz.

Figure 2.33 shows the trade-off between insertion loss and isolation loss for PINs integrated into a 130 nm BiCMOS technology [55]. A circuit designer has the flexibility to balance insertion loss against isolation requirements by changing the device layout.

Both PINs and SBDs serve similar purposes in a circuit; however, their selection criteria are dependent on several circuit application factors. PINs are preferred over SBDs in applications where the reverse leakage is a concern, since SBDs generally have a relatively higher reverse leakage or isolation loss than PINs. However, switching speed for SBDs is generally faster than that for PINs. This is attributed to electron injection across the Schottky barrier in SBDs compared with the recombination of minority injected carriers, electrons, for PINs. SBDs also have a better ideality factor than PINs as there is essentially no recombination in the depletion region for SBDs, while PINs have high recombination in the wide depletion region arising due to the presence of a low-doped intrinsic region.

2.5.3 Other passive elements

In addition to the PIN and Schottky diodes, BiCMOS technologies offer a rich suite of passive elements including resistors, capacitors, inductors, and transmission lines to facilitate mm-wave designs. Comprehensive models of transmission lines, single and coupled microstrip and co-planar transmission lines, and distributed passive devices are available to reduce the need for electromagnetic simulations [56–58]. Several transmission line discontinuities are modeled accurately as discrete models and are available in the design kits. Such elements have been used to design complex mm-wave transceivers in SiGe technologies [59].

Multiple resistor designs are also supported, with each design kit ranging from sheet resistance of $9\,\Omega/\square$ to $1700\,\Omega/\square$ to provide flexibility to circuit designers. These resistors can be formed in the FEOL process as polysilicon resistors or in the BEOL process as TaN. Thin and thick MOS varactors with different capacitance density and HA junction varactors are also offered. Metal–insulator–metal (MIM) capacitors are also supported, with different capacitance density offerings for BEOL wiring capacitors [60, 61].

2.5.4 Through-silicon-via (TSV)

As alluded to previously in the chapter, power gain at a given frequency is one of the most important design criteria for any power-amplifier application. Power gain of a bipolar transistor in a common emitter configuration with very low intrinsic resistance is dominated by the ground inductance and is approximately given by [62]

$$Gain \propto \frac{1}{\omega^2 C_{CB} L_{Emit}}, \tag{2.27}$$

where ω is the operating frequency, C_{CB} is the collector–base capacitance, and L_{emit} is the emitter inductance to ground. In advanced SiGe HBT designs, C_{CB} has been significantly reduced for improved gain, but ground inductance is still a limiting factor to improved performance. One method employed to reduce the emitter inductance is the use of a large number of emitter wirebonds in parallel. However, this is expensive and has little benefit since mutual inductance results in a relatively high inductance. Another option is the use of solder bumps in a flip-chip package, which is very effective for reducing emitter inductance. However, for power amplifiers, a flip-chip package is undesirable due to device self-heating problems [63].

Through-silicon-vias (TSVs) are now being used in SiGe BiCMOS technologies to provide a low-inductance path to emitter ground for large power cells at the output (Fig. 2.34). The low loss and low inductance of the TSV helps to improve the gain and power-added efficiency (PAE) at high frequencies. It also helps improve the gain flatness

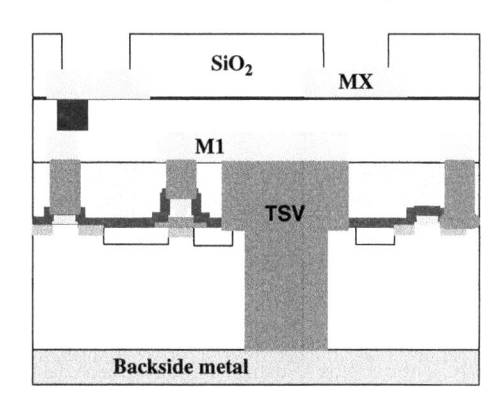

Fig. 2.34 Schematic cross section of a through-silicon-via (TSV) depicting the backside metallization.

for both active and passive circuits such as filters. TSVs used in these processes provide a good ground connection and are not used for power or signal lines. Hence this is a cost-effective method of improving the ground inductance and providing effective heat dissipation by utilizing a material with excellent thermal properties. The TSV formation is realized by using a BOSCH deep silicon etch process [64]. This process provides excellent selectivity to photoresist and enables the creation of very-high-aspect-ratio features, approximately 50:1. The timed silicon etch controls the TSV depth to approximately 100 μm within ±10%. The TSVs are filled with tungsten because its coefficient of thermal expansion closely matches that of silicon. Another advantage of tungsten is its filling properties. It has the ability to fill features with high aspect ratios as in the TSV. Subsequent to filling, chemical mechanical polishing (CMP) techniques are employed to planarize the silicon surface. Finally, the wafers are thinned to expose the TSV and a backside metal is deposited on the silicon, which serves to connect all TSVs forming the ground plane [65–68].

TSV models consisting of self-inductance, mutual inductance and resistance, and a layout parametrized cell are available in design kits to enable low-inductance Si/SiGe PA designs [69, 70].

2.6 SiGe HBT modeling and characterization

Accurate compact models for advanced HBT devices are critical to the successful utilization of modern SiGe technologies. Compact models are the basis for SPICE simulation, allowing circuit designers to optimize and validate their designs across several dimensions prior to committing to expensive hardware fabrication. The primary goal of a compact model is to provide a statistical physics-based scalable model that provides a combination of accuracy and flexibility, thus enabling aggressive RF/mixed-signal design while minimizing design cycle time.

In PA applications the HBT can swing into extreme operation regimes under large-signal conditions. This requires a compact model to accurately describe high-current and high-frequency effects such as self-heating, avalanche breakdown, quasi-saturation, non-quasi-static effects, emitter current crowding, and base width narrowing. It is especially important that the model correctly describes I–V and C–V nonlinearities. Dominant nonlinearities are I_C and I_B V_{BE} dependence, the avalanche multiplication current, depletion capacitances of the BE and BC junction, and bias-dependent diffusion capacitances. It is desirable for a compact model to maintain a strong physics-based background to ensure a correct model fit across a wide range of bias, frequency, and temperature while at the same time keeping the model equations simple enough to reduce simulation time and increase robustness.

The most common models available in all mainstream commercial circuit simulators are the SPICE Gummel–Poon Model (SGP), the Vertical Bipolar Inter Company Model (VBIC), the HIgh CUrrent Model (HICUM), and the Most EXquisite TRAnsistor Model (MEXTRAM). Until the mid 1990s, the semiconductor industry relied almost exclusively on the SGP model for BJT circuit design. SGP consisted of equations based on the more complete Integral Charge-Control Relation (ICCR) [71]. But with the increasing applications of BJT and HBT technology in high-speed communications and RF, the SGP model was found to be inadequate and needed to be revised to include more accurate modeling of the physical effects found in high-speed devices operating at high current densities. These effects include better Early effect modeling (output conductance), quasi-saturation, avalanche multiplication, thermal self-heating, and accurate transit-time modeling. This led to the creation of the VBIC model formally presented in 1995 [72]. In addition to SGP, VBIC modeled effects that include a parasitic p-n-p, self-heating, bias-dependent Early voltages, temperature scaling, a Kull-based model for quasi-saturation [73], and additional parasitic capacitances found in aggressively scaled modern devices. The VBIC model works well, but has known deficiencies in several areas, including poor modeling of strong quasi-saturation, output conductance, and avalanche breakdown. Additionally, Si-based compact models do not contain the basis for a physically accurate model of the SiGe HBT base charge. Inclusion of the Ge grading layer modifies the base bandgap and therefore the intrinsic base carrier concentration, altering the critical charge-storage components in the charge control relations.

For these reasons, two new HBT models, HICUM and MEXTRAM, are more common for high-performance HBT modeling. MEXTRAM extends the modified Kull model by adding the effects of velocity saturation in the collector at high current densities, correctly predicting quasi-saturation and the onset of the Kirk effect [74]. It also takes into account the bandgap grading in SiGe devices and has an extra parameter to model the changes in I_B due to neutral-base recombination.

The current state-of-the-art in HBT modeling centers on use of the industry-standard HICUM core model equations. HICUM is available in all mainstream commercial circuit simulators and is used by leading foundries worldwide to model high-speed HBTs for performance RF/mixed-signal design, including mm-wave, radar, and imaging applications. The Compact Modeling Council (CMC) has adopted the use of HICUM Level 2 as the standard for modeling of high-performance SiGe HBTs. In addition to HICUM, MEXTRAM is also recognized by CMC as a standard HBT model.

A comparison of important features for the SGP, VBIC, HICUM, and MEXTRAM models is given in Table 2.1 [75].

HICUM was developed by the University of Technology in Dresden, Germany, to improve the modeling of bipolar transistors at high frequencies and high current densities using Si, SiGe, or III–V processes. The model equations are based on the generalized charge control relation (GICCR) providing a very physical description of the device behavior [76]. The HICUM equations enable accurate modeling of physical effects such as self-heating, impact ionization, noise correlation, non-quasi-static behavior, base–emitter tunneling, bias-dependent transit time, distributed parasitics, and more. In addition, recent HICUM versions have been augmented to correctly describe the strong barrier effect related to Ge in the base as well as to model the impact of the strong Ge gradient in the BE junction due to high Ge concentration and thin base in advanced SiGe technologies ($f_\tau > 200$ GHz). HICUM is the first and currently the only compact model

Table 2.1 Comparison of important features for the SGP, VBIC, HICUM, and MEXTRAM models.

Feature	SGP	VBIC	HICUM	MEXTRAM
HBT/SiGe modeling	—	✓	✓✓	✓✓
Quasi-saturation	—	✓	✓✓	✓✓
f_τ modeling	—	✓	✓✓✓	✓✓
Self-heating	—	✓	✓	✓
Substrate modeling	—	✓	✓	—
Emitter–base breakdown	—	✓	✓	—
Parasitic PNP	—	✓✓	✓	✓✓
# Internal nodes	3	7	5	5
# Parameters	35	80+	90+	65+

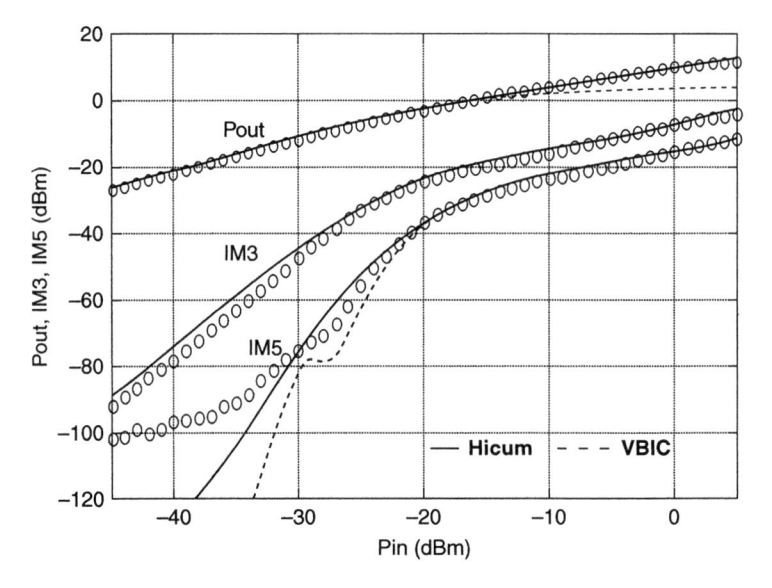

Fig. 2.35 P_{out}, IM3, and IM5 at $V_{BE} = 0.91$ V, $V_{CE} = 2$ V, $f = 900$ MHz, $\Delta = 0.1$ MHz of $0.32\,\mu\text{m} \times 16.8\,\mu\text{m}$ transistor.

for bipolar transistors that was specifically developed to be geometry scalable. The physical nature of this core model, when coupled with the appropriate parameter extraction, allows for the creation of a single SPICE model deck that can accurately predict HBT device behavior over a wide range of bias, frequency, and temperature. HICUM has an improved model for the bias-dependent diffusion capacitances, thus providing a better modeling of nonlinearities and improving the fit of f_τ/f_{MAX} dependence on bias. It is also more accurate in predicting the high-frequency and large-signal behavior of HBTs. Owing to all this complexity, in circuit-level simulations VBIC still exceeds HICUM in terms of simulation time and stability.

Model-to-hardware correlation plots of VBIC and HICUM models of a $0.32\,\mu\text{m} \times 16.8\,\mu\text{m}$ SiGe HBT are shown in Figures 2.35 and 2.36. It can be seen that HICUM

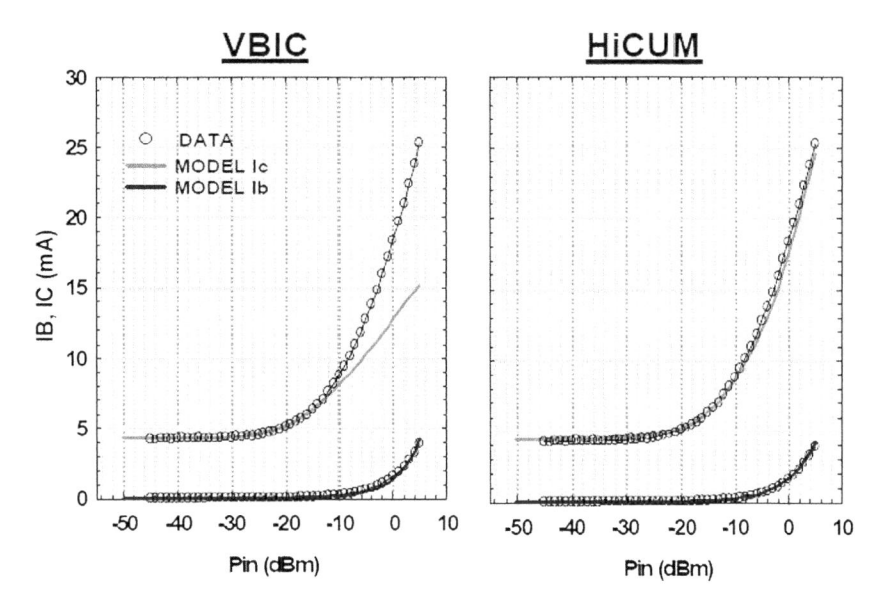

Fig. 2.36 I_C and I_B at $f = 900$ MHz, $\Delta = 0.1$ MHz of a $0.32\,\mu$m $\times$ $16.8\,\mu$m transistor.

provides a better fit of large-signal characteristics at high input power. The improved model fit makes HICUM the preferred model for SiGe HBTs used for PA applications.

The HICUM model describes charge storage in all five transistor regions: base–emitter and base–collector junction capacitance as well as in the neutral emitter, base, and collector. Model parameters related to charge storage are extracted from S-parameter data across bias at fixed frequency or across different frequency ranges. Measured HBT S-parameter is generally followed by de-embedding to remove the impact of wiring capacitances and inductances. The models take into account the parasitic metal-to-metal capacitances and M1 resistances that are part of the p-cell. These parasitic elements as well as parasitic capacitances of the nitride spacers around the emitter and oxide capacitances of the shallow trenches are estimated from the device layout. The device resistances are extracted from test structures specially designed to return accurate parameter values. Once the device's resistances and parasitic elements are known their impact on the S-parameter can be mathematically removed, enabling an accurate extraction of the charges stored in the different transistor regions. Junction capacitances are extracted from S-parameter measurements at low frequency and at a DC bias where the device current is either zero or close to zero (device is not active). This data is usually referred to as cold S-parameter data [77]. Stored charges in the neutral device regions are a function of collector current and become important when the transistor is biased near the f_τ/f_{MAX} peak. Consequently related model parameters are extracted at high DC bias. A more detailed explanation about model parameter extraction can be found in [78–80]. Figure 2.37 shows a schematic representation of the parasitics in the HICUM model and their physical location in the HBT structure. A more complete schematic for the HICUM model is shown in Fig. 2.38

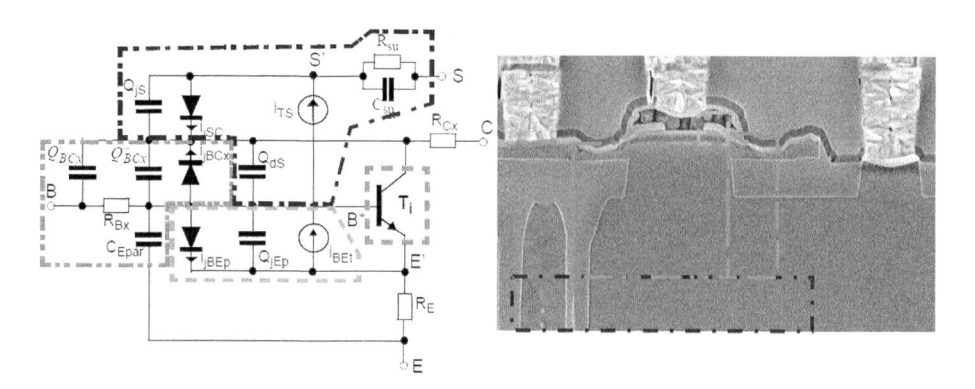

Fig. 2.37 Model of the parasitics associated with SiGe HBT and their physical location in a representative HBT cross-sectional TEM.

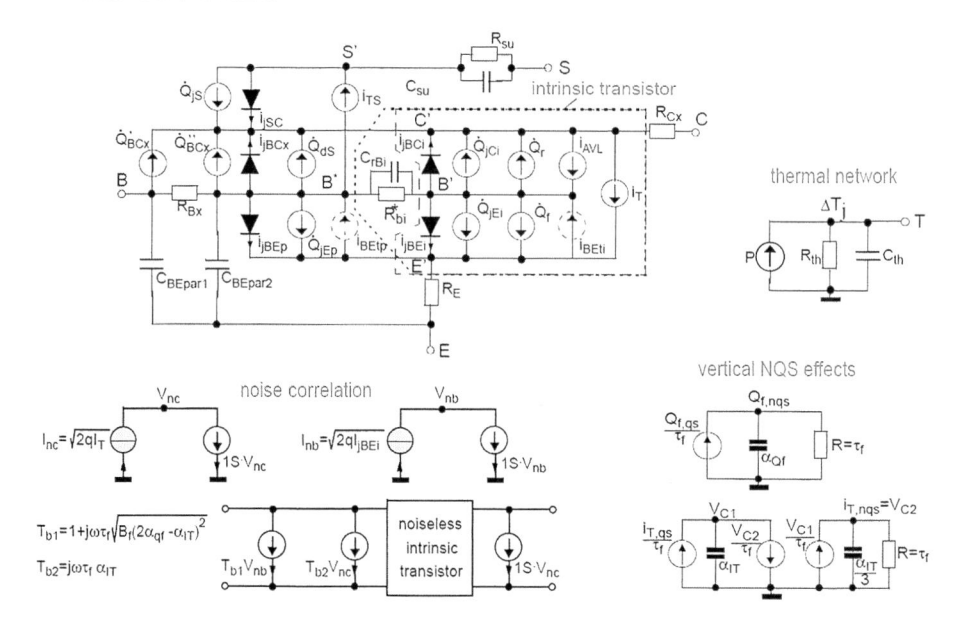

Fig. 2.38 A complete model schematic for the HICUM model; this figure also depicts the model parameters for thermal network, noise correlation, and vertical non-quasi-static effects.

depicting the intrinsic transistor and parasitic components associated with the structure including the substrate effects [81]. Model parameters associated with thermal network, noise correlation, and vertical non-quasi-static effects are also shown in the figure.

Figures 2.39 and 2.40 show representative model (HICUM)-to-hardware correlation plots for a high-performance HBT with emitter area of $0.12\,\mu m \times 2.5\,\mu m$. The extracted models are a representation of the device up to the first metal (M1) level. This metal level is included in the nominal HBT footprint (p-cell). Model parameters are extracted from a set of standard devices with varying emitter dimensions and special test struc-

2.6 SiGe HBT modeling and characterization

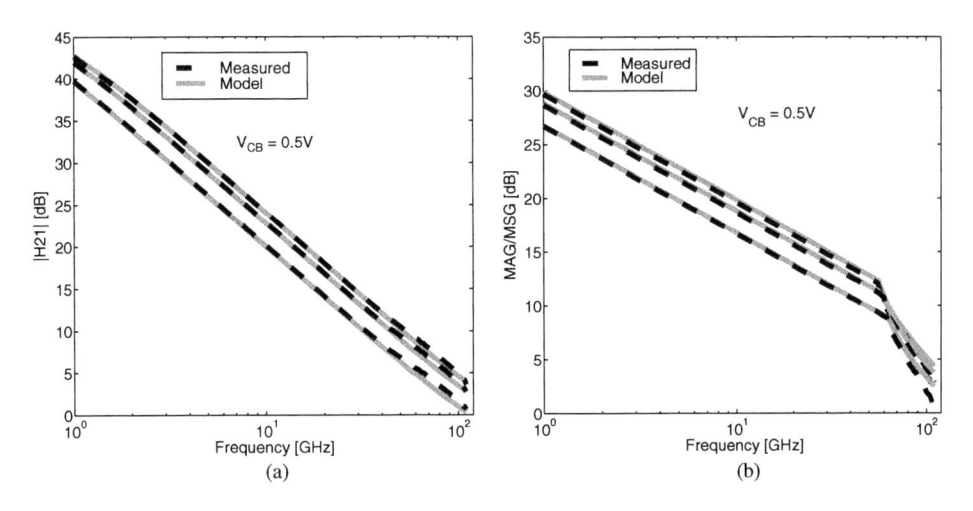

Fig. 2.39 HICUM model-to-hardware correlation plots of (a) H_{21} and (b) MAG for a 0.12 μm × 2.5 μm SiGe HBT at $V_{CB} = 0.5$ V for $J_C = 1.5$, 2.7, and 3.8 mA/μm^2.

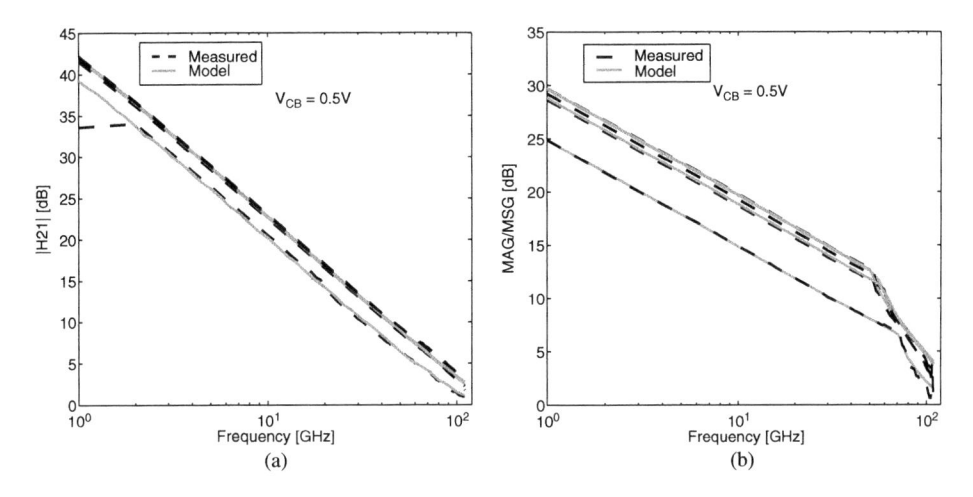

Fig. 2.40 HICUM model-to-hardware correlation plots of (a) H_{21} and (b) MAG for a 0.12 μm-wide HBT at $V_{CB} = 0.5$ V for $J_C = 2.7$ mA/μm^2 for emitter lengths of 0.52, 2.5, 12, and 18 μm.

tures for the extraction of parasitic device resistances and capacitances. The frequency response of the transistor depends on the DC operating point, stored charges in the different transistor regions, and parasitic elements.

The final compact model represents a unified description of the device behavior, bringing together comprehensive characterization results, process statistics, physical design dimensions, and device physics into a single statistical physics-based scalable model.

2.7 Reliability

The reliable operation of integrated circuits is a key requirement toward reducing system downtime in the field. Therefore, test and qualification protocols should address all possible failure mechanisms. However, the methods for evaluating silicon bipolar device reliability have not significantly changed over the past 30 years. In spite of being in production since 1996, there have been no publications of a failure mechanism related to the SiGe layer itself, so all of the failure mechanisms are due to fundamental silicon bipolar junction transistor technology issues and are not heterojunction specific. All the stresses are evaluated in DC mode and tests are typically conducted at elevated temperature and increased voltage bias to accelerate the reliability tests [82]. The assumption is that these accelerated testing conditions do not introduce new mechanisms but in fact actually represent the same mechanisms found in normal operation [83]. Integrated circuits and individual transistors may be stressed either at the wafer level or within a package for extended periods of time using dedicated test equipment and facilities. A much-needed area is the evaluation of robustness for AC operation of the SiGe HBT. In this section, we will discuss the reliability aspects of the SiGe HBT under reverse-bias emitter–base junction stress and forward-bias current stress. This is followed by a discussion on electro-migration. The section will end with a discussion of the safe operating area (SOA) of the transistor, which defines the bias current and voltage range for safe operation of the transistor.

2.7.1 Scaling impact on reliability

Scaling impacts the robustness of high-performance SiGe HBTs because of the more heavily doped junctions and higher electric fields. In fact it was scaling-driven reliability concerns in the 1980s with ion-implanted base Si BJTs that led to the invention of epitaxial base transistors. Doping the intrinsic base via boron implantation results in the formation of a channeling tail that determines the bipolar transistor's base width. This occurs as ion implantation is not a perfect process and the implanted species can channel along the crystalline axis. To obtain a narrower base, the implant energy needs to be decreased, leading to higher dopant concentrations at the silicon surface. In polysilicon emitter transistors, the emitter region is formed via arsenic diffusion from the arsenic-doped emitter polysilicon region. The arsenic diffusion forms a shallow, heavily doped profile. Doping concentrations at the emitter–base junction of $\sim 1 \times 10^{19}$ atoms/cm^2 create a tunneling junction, leading to emitter–base leakage. This results in poor reliability when the junction is reverse biased. The invention of in-situ doped epitaxial-base Si BJTs and SiGe HBTs allowed control of the location of the boron dopant profile from the silicon surface. Careful engineering of the in-situ base dopant profile can be used to decrease the doping concentration at the EB junction, thereby increasing the reliability associated with the intrinsic EB junction fields [84]. However, the same controlled base dopant profile creates a retarding electric field for the electrons transport injected into the base, substantially decreasing the switching speed of the Si BJT. In SiGe HBTs the quasi-electric field from the graded germanium profile in the base is larger than the dopant-profile-induced retarding field, thereby reducing the base transit delay. This is an important aspect of the design of reliable high-performance SiGe HBTs.

There are other aspects of silicon bipolar technology that may also increase the EB junction electric field in the device perimeter as opposed to the intrinsic center of the device. To reduce the extrinsic components of the base resistance, the extrinsic base is diffused laterally to form a low-resistance link to the intrinsic base, forming a link-up region. The spacing between the emitter opening and heavily doped extrinsic-base region can be reduced for low base link-up resistance. Consequently, the lateral *perimeter* emitter–base electric fields may be larger than the vertical intrinsic profile emitter–base electric fields, resulting in a lower emitter–base breakdown voltage (BV_{EBO}) and higher reverse-bias EB leakage.

2.7.2 Reverse-bias emitter–base stress

Reverse-bias emitter–base (EB) stress evaluates the robustness of both the passivation oxide over the emitter–base junction and the emitter–base junction electric field strength [85]. This stress test is conducted by applying a reverse bias to the emitter–base junction which exceeds the rated conditions of the transistor. The stress takes place at elevated temperature with the collector terminal either open or shorted to the base. The Gummel characteristic of a SiGe HBT after emitter–base stressing is shown in Fig. 2.41. The figure shows an increase in base current leakage (dashed lines) with increase in reverse-bias emitter–base stress at low V_{BE}. The high electric fields create traps in the passivation oxide at the perimeter of the emitter–base junction. These traps increase the generation–recombination component of the base current, thereby leading to a non-ideal base current at low V_{BE}. The degradation mechanisms are well behaved and a fit to an analytical expression is used to predict the low-bias leakage behavior as a function of voltage, temperature, and time. It should be noted that the low-bias leakage currents do not impact AC performance at higher bias. Therefore this failure mechanism has implications for the long-term stability of circuits with low-bias current density but is relatively transparent in regimes of high current density. Interestingly, these traps may also be produced via impact-ionized carriers in the base–collector junction traveling to

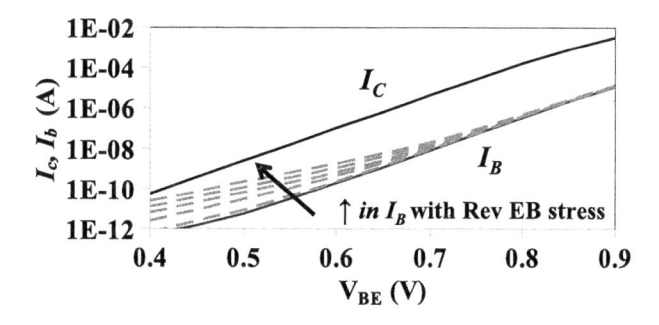

Fig. 2.41 Emitter–base reverse-bias stress results for a 130 nm SiGe HBT, showing degradation in base current at low V_{BE} (dashed lines).

the emitter–base oxide [86] and also in the base–collector junction oxides such as the shallow trench.

There are some reports on silicon BJTs that discuss recovery mechanisms after reverse EB stress, including activation and rate [87–90]. Although similar investigations have not been performed on SiGe HBTs, the physical recovery mechanisms are potentially the same for BJTs and HBTs.

2.7.3 Forward current stress

Forward-biased stressing is a high-current-density stress utilizing currents at or above peak f_τ current at elevated temperature (85 °C–240 °C) for extended periods of time. During these stresses the collector–base junction is biased at a low value, much below BV_{CBO}.

In SiGe HBTs there is a close correlation between higher f_τ transistors and higher peak f_τ current densities. For example, a typical SiGe HBT with f_τ of 300 GHz may have peak f_τ current densities approaching 20 mA/μm² [17]. These relatively high peak f_τ current densities of SiGe HBTs are one of the major disadvantages for the manufacturing of high-performance SiGe technologies. Failures from forward-biased stressing can come from a variety of mechanisms.

A typical failure mechanism with forward-biased long-term stress degradation of the SiGe HBT can be characterized as a change in base current across all V_{BE}. This represents a parallel shift in the $\log(I_B)$ vs V_{BE} plot. This shift has been attributed to the modification of the very thin oxide layer between the single-crystal and polycrystalline emitter regions of the device with high current stress. Both an increase in I_B and a decrease in I_B have been observed [34, 91–94]. In the case of increasing I_B, it has been hypothesized that the generation or de-passivation of traps with high current density enhances the trap-assisted tunneling through the oxide material, causing the increase in I_B. In the case of a reduction in I_B, again the interfacial oxide is implicated, with a hypothesis that the hydrogen passivation of the interface states causes a reduced hole-recombination velocity at the interface. Figure 2.42 shows the beta stability for IBM's 0.13 μm SiGe technology (f_τ = 200 GHz) over several thousands of hours of stress at a 125 °C ambient. Freeman *et al.* [95, 96] have also reported robust reliability performance of these HBTs under forward stress at $I_C = 2 \times I_{C,peakfT}$ and ambient temperature of 180 °C (junction temperature ∼239 °C).

Low-activation-energy failure mechanisms are failures that occur with extended time at normal operating conditions and are revealed with high current stressing at elevated temperatures for extended periods of time. These results differentiate Si-based HBTs from III–V HBTs, which do have low-activation-energy failures.

2.7.4 Mixed-mode stress

The last stressing technique is mixed-mode testing. Mixed-mode testing is when stressing is done with both high current density and high voltage. In mixed-mode stressing, impact-ionized carriers are generated in the base–collector junction and travel to oxide interfaces, creating traps. When operating the transistor in inverse mode the effect of the

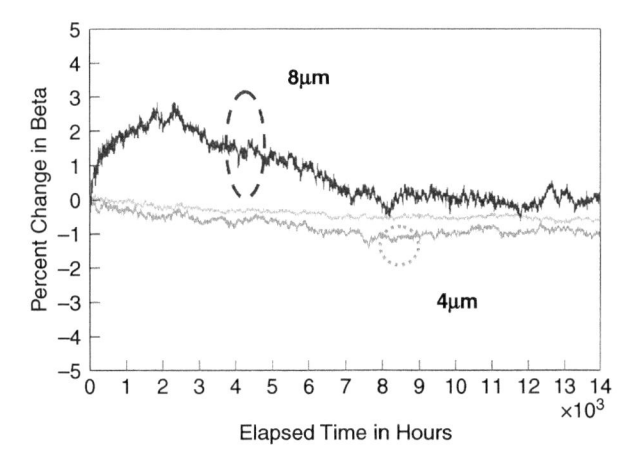

Fig. 2.42 Current gain stability with forward-bias stressing at elevated temperature (125 °C) for 0.13 μm SiGe HBTs having length = 2, 4, and 8 μm with f_τ = 200 GHz at J_c = 12 mA/μm².

traps can be seen on base-current leakage due to the collector–base junction similar to the leakage created by emitter–base reverse-bias stressing. The results of the stress are dependent on the spacing of the shallow trench to the emitter–base junction. Degradation takes place by ionized carriers creating traps at the oxide interfaces [97, 98].

2.7.5 Electromigration

A major failure mechanism from high-current-density stresses is associated with electromigration. In SiGe BiCMOS, the lower-level metals are all taken from base CMOS technology to efficiently leverage existing libraries and functionality. With high peak f_τ current densities it may be difficult to wire the transistor in a manner that avoids electromigration failures at the interconnect level. These electromigration wiring concerns are not independent of the operating point of the SiGe HBT since, to achieve high performance, the HBTs operate at simultaneously high DC current and voltage bias, resulting in high junction temperature due to self-heating. Based on the Black equation [99], the electromigration limited current density degrades with temperature by the relationship expressed in Eqn (2.28):

$$J_{EM} = A \exp\left(\frac{-\Delta H}{kT}\right). \quad (2.28)$$

Therefore, the transistor electromigration current density limits will be dependent on the ambient temperature and the power consumption of the transistor (Fig. 2.43). It is therefore critical that high-performance SiGe HBTS are wired such that they pass electromigration guidelines across all operating conditions. There are several wiring techniques that can be used to relax the electromigration constraints, including use of parallel metal layers, thicker metal layers, and wider straps for the emitter and collector contacts. This mechanism determines the safe-operating-area current limit to be discussed in the next section.

Another improvement is gained by the selection of interconnect metals. For example, copper has around a ∼1000× higher electromigration limit failure time than aluminum

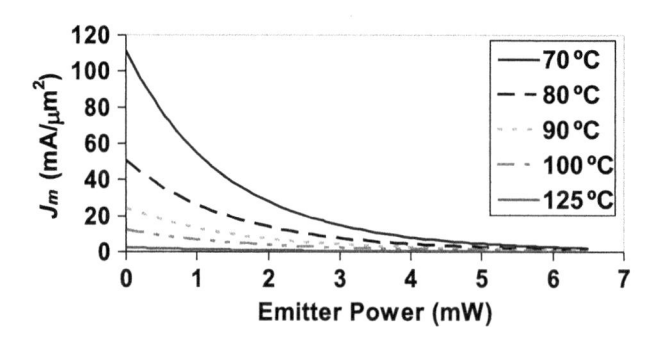

Fig. 2.43 Electromigration current limits are a function of ambient temperature and self-heating (emitter power).

interconnect under the same stress conditions [100, 101]. Within a given metal system there are doping and other techniques designed to further limit metal ion movement and improve electromigration resistance [102–104]. These techniques often come at the expense of higher metal resistance and a more complex process integration.

The electromigration current limits can be calculated based on the interconnect temperature, which is a function of the ambient temperature, thermal resistance of HBT, power consumption of HBT, and the rated current limit at a given temperature. An example of such a calculation result for a 90 nm experimental SiGe HBT is shown in Fig. 2.43. Clearly there is a strong dependence on ambient temperature and SiGe HBT power consumption. Experimentally, the electromigration current limit for a given SiGe HBT is determined from accelerated tests using high DC current at elevated temperatures. Based on the test results, the DC current limit allowed for reliability operation under a particular use condition can then be projected, with the acceleration factor described by Eqn (2.29):

$$Acceleration\ factor = \left(\frac{J_{stress}}{J_{use}}\right)^n \cdot \exp\left(\left(\frac{\Delta H}{k}\right)\left(\frac{1}{T_{use}} - \frac{1}{T_{stress}}\right)\right), \qquad (2.29)$$

where J_{stress} and J_{use} are current densities under stress and use conditions, respectively; n is the current exponent; ΔH is the activation energy associated with metal diffusion; and T_{stress} and T_{use} are the interconnect temperatures under stress and use conditions, respectively [105]. The reported values of ΔH and n are 0.9 eV and 1.1 for Cu and 0.85 eV and 1.7 for Al(Cu), respectively [105].

Another factor in determining the current operation limit is the failure rate tolerance, which is the total number of malfunctioning chips due to electromigration wearout within the expected product lifetime. For example, for a chip containing a million interconnects, a failure rate tolerance of 10^{-12} allows only one electromigration failure out of a million chips during the expected lifetime. The current limit can be adjusted according to the total number of interconnects in a particular product chip, as well as the product reliability tolerance for reduced electromigration failure rate [106, 107]. Furthermore, the current operation limit can also be adjusted for individual interconnects by considering their duty cycle during circuit operation.

There is also another electromigration-influenced failure mechanism that is a result of pile-up of metal atoms at the emitter contact. Compressive stress induced by metal pile-up at the emitter changes the silicon band structure, resulting in an enhancement of hole transport which increases the base current. However, it is possible to completely recover β after an anneal, indicating that the electromigration-induced compressive stress is relieved by creep in an aluminum-based metallization [108].

2.7.6 Safe operating area (SOA)

The safe operating area (SOA) of the transistor defines the bias current and voltage range for safe operation of the transistor. The bias condition of a transistor is often chosen in such a way as to optimize the device performance for a given application. However, there exists a boundary for current voltage operation beyond which the device may experience electrical or thermal issues which may permanently damage the transistor or limit its useful lifetime. This is not just an HBT issue but exists in virtually all transistors and particularly so in power amplifier applications. Since the thermal and breakdown voltage operation limits of HBTs have already been discussed in previous sections, they will be briefly reviewed here. SOA is particularly important for power amplifiers due to large signal voltage swing operation.

The voltage operation limit is generally dictated by the avalanche multiplication process within the device and thus can be represented by the avalanche breakdown voltages. As discussed previously, the voltage limit of a device cannot be represented by one single parameter because its avalanche behavior strongly depends on the external bias configurations and bias levels. The actual breakdown for this configuration is determined by the effective external base resistance seen from the base as it modulates the strength of the positive feedback. At current levels beyond peak f_τ a lateral current instability stemming from the *pinch-in* effect may also result in lowered voltage operation limits [96, 109]. Some examples of SiGe HBT SOA can be found in [110–112]

The current operation limit of the SiGe HBTs is mostly determined by the electromigration effect of metal lines and/or vias connected to the devices, since the device reliability is closely related to this effect. As described in the electromigration section above, the electromigration limits depend on the ambient temperature, specific wiring and transistor layout, and interconnect properties which determine the self-heating effect and interconnect temperature. Hence, accurate temperature information at the metal interconnects becomes critical for the current limit estimation, causing the self-heating effect crucial for this purpose. The limit decreases with increasing voltage applied (larger power dissipation) and with decreasing emitter length (larger thermal resistance).

A relatively unexplored area is the AC operating limits of the transistor. Designers operate the transistor above BV_{CEO} and closer to BV_{CBO}, based on bias resistors [35, 36]. This is an area for further exploration.

The failure mechanisms of the SiGe HBT have been discussed in this section. It has been observed that there are no specific failure mechanisms related to the presence of the graded SiGe base profile. All of the failure mechanisms are related to silicon bipolar junction transistor failure mechanisms. Scaling exacerbates the current and voltage limits of the transistor operation through increased peak f_τ current density and lower breakdown voltage operation. Data prove that SiGe BiCMOS technology is a very robust technology. It is possible to address scaling limitations on reliability by incorporating the learning from other silicon technology like deeply scaled CMOS at small lithographic nodes.

2.8 Performance limits of SiGe HBTs

Several investigations have been conducted that aim to predict the ultimate performance limits of SiGe HBTs, largely because of the myriad of potential applications that exist. Traditionally, the Johnson limit [113] was applied for silicon-based BJTs in which one-dimensional, drift-diffusion models were utilized and suggested a f_τ performance limit of 200 GHz, which has since been exceeded by silicon HBTs.

In recent years, with the rise in the number of potential applications for SiGe HBTs in the terahertz (THz) frequency regime, there have been several attempts at determining the device performance limits. Such predictions have been made utilizing technology computer-aided design techniques, commonly known as TCAD, whereby two-dimensional device simulations that exploit drift-diffusion (DD), hydrodynamic (HD), and Boltzmann transport (BTE) models were used. The drift-diffusion models are known to result in an underestimation of the maximum predicted operation frequency, since the electron velocity is limited to a maximum value of the carrier velocity saturation; while with the hydrodynamic transport models an electron velocity overshoot is known to occur, which over-predicts the maximum oscillation frequency. Boltzmann transport equations, which enable simulation of the full SiGe band structure, are best suited for modeling the carrier transport in SiGe HBTs. However, such simulations are very computationally demanding and extremely slow. Hence an alternative approach has been developed, in which the HD model parameters are calibrated to the BTE models for relevant conditions and utilized for a much more efficient simulation.

Recently, within the framework of the European DOTFIVE project, SiGe HBTs demonstrating 300/500 GHz f_τ/f_{MAX} have been fabricated, which is the highest reported performance to date at room temperature [17, 114]. In another instance, experimental results at room and cryogenic temperatures were used to derive scaling laws, which formed the basis of a projected performance of 782/910 GHz f_τ/f_{MAX} [115]. Utilizing TCAD techniques, Shi and Nui proposed f_τ/f_{MAX} of 760/1090 GHz via a two-dimensional HD simulation [116]. However, in these studies ideal dopant profiles were assumed, a single-value energy relaxation energy was used, and the device self-heating was not taken into account, while the f_{MAX} was determined from the simple standard equation. In another report in which process simulations were utilized to generate the baseline for the final device dopant profiles, 630 GHz was proposed as the maximum value of the device f_τ [117]. The reported performance was determined using a conservative two-dimensional DD approach. In such an instance, the simulations were conducted by using a process and device calibration for an existing technology. Such an approach provided a means by which to verify the process models and hence account for the effects of processing on dopant diffusion, resulting in a more realistic device dopant profile. An attempt was also made in this work, at predicting the simultaneous limits of f_τ and f_{MAX}, for an aggressively scaled device in both the vertical and lateral dimension. In this instance feasibility studies were utilized to optimize both figures of merit and determine reasonable values for a balanced device design. The reported f_τ/f_{MAX} optimum values were reported at 500/480 GHz.

More recently, a more rigorous approach for the device simulation that utilized the above-mentioned combined BTE-calibrated HD two-dimensional simulations was used, in which the relevant physical effects were taken into account. However, as in many of the previous reports, an idealized profile was arrived at utilizing a one-dimensional device simulation. For an aggressively scaled vertical structure the effects of base punch-through and forward-tunneling effects such as band-to-band and trap-assisted tunneling become relevant and were included along with self-heating effects. The estimated maximum predicted oscillation frequency, f_τ, for this idealized, isothermal, one-dimensional case was reported to be 1.5 THz, with an open base breakdown voltage, BV_{CEO}, of 1 V [118, 119]. However, in a three-dimensional device, it is expected that the associated capacitances, resistances, perimeter current injection, collector current spreading, and self-heating would degrade the device performance. Utilizing compact modeling techniques for scaling the device laterally for the optimized vertical profile, the trade-off between the figures of merit f_τ and f_{MAX} was balanced. The proposed ultimate performance limit was found to be f_τ/f_{MAX} of 1.0/2.0 THz. It should be noted that there is no significant reduction in HBT BV_{CEO} with improved f_τ, thereby increasing the $f_\tau \times BV_{CEO}$ product limit predicted for SiGe HBTs [120].

Practically, there have been no reports in the literature of a manufactured device demonstrating f_τ performance levels exceeding 300 GHz levels to date. This hints at the level of difficulty associated with achieving a scaled vertical profile which reflects those that have been theoretically proposed. The achievement of highly doped, abrupt base and collector profiles requires the ability to epitaxially grow a highly doped boron profile and limit the diffusion of these regions by minimizing the subsequent thermal diffusion. Hence once the base has been deposited it should not undergo significant diffusion during the remainder of the process, which limits the subsequent thermal processing.

As outlined in Section 2.3.2 the HBT base and emitter integration into the BiCMOS process is conducted after the FET gate formation, in what is commonly called the base-after-gate process. This integration approach allows for a significant portion of the thermal cycle associated with the FET formation process to be eliminated prior to the HBT base and emitter formation. Once the base has been deposited, the emitter is formed. The current temperature range of the processes associated with the emitter formation process coincides with the temperature range at which transient enhanced dopant diffusion is observed. This is a key driver for base and collector diffusion. As a result, once a highly doped base layer has been deposited, the emitter formation thermal cycle, prior to the final source/drain activation anneal, should be limited to a maximum temperature of 650 °C, to avoid TED and limit the base diffusion. The thermal budget of the source/drain anneal should also be minimized. Currently, rapid thermal anneal processes are employed in the BiCMOS processes in which the time frames of source/drain anneals are on the order of seconds. However, if a minimum base width is to be attained to enable faster device operation and performance, additional reductions in processing times are necessary.

Grown base profiles exhibiting peak concentrations of $3.5 \times 10^{20} \mathrm{cm}^{-2}$ and widths of 1.2 nm have been demonstrated [121]. The ideal dopant profile proposed in [118] required for >1 THz performance level was 8.5 nm wide. Hence once the base has been deposited it should not undergo significant diffusion during the remainder of the process, which limits the subsequent thermal processing steps. Several studies [12, 122] have explored the use of millisecond anneal techniques to limit the dopant diffusion and simultaneously obtain close to theoretical solid solubility dopant levels in the device. The maximum reported performance levels of 300/420 GHz f_T/f_{MAX} were obtained for such studies and demonstrate that millisecond anneal techniques are in fact viable approaches to minimizing the base diffusion. However, in those experiments the millisecond anneals were performed subsequent to the source–drain anneals that defined the base width. But, for terahertz-level device performance, the thermal cycle would need to be considerably less and would require alternative HBT integration approaches which do not feature the current high-thermal-cycle emitter and base formation processes.

Another aspect of the process that would need to be addressed is the maximum germanium concentration needed in the base. High Ge concentrations on the order of 30% germanium were quoted as a requirement for THz performance. Reliable growth of the SiGe layers without dislocation formation in the active region of the device is a necessity for an operational device.

Clearly there is much innovation that may yet be applied to SiGe HBTs in large lithographic nodes as far as thermal cycles and structures are concerned. There is much that can be borrowed from the innovations in advanced CMOS processing. Significant innovation in process integration is still required for improved HBT performance.

2.9 Summary and conclusions

The intention of this chapter was to introduce the reader to SiGe HBT (BiCMOS) technology for mm-wave applications and discuss some of the more important aspects of the technology. The chapter began with a general analysis of SiGe HBTs with specific discussion on differences between HBTs and silicon BJTs. Advantages of introducing SiGe epitaxial growth layer in the process were also discussed. The chapter then reviewed some of the key aspects of SiGe HBT process technology, including SiGe epitaxy and epitaxial-base transistor structures, and how these processes influence the mm-wave transistor performance. This was followed by an extensive discussion on the mm-wave characteristics of the SiGe HBT including speed, power gain, low-frequency, and broadband noise characteristics. Passives are an important aspect of circuit design in SiGe BiCMOS technology, and that was also covered, including FEOL mm-wave passives like SBDs and PIN diodes. An important differentiator for the SiGe HBT is the robustness of these HBTs, which was presented in the section on reliability. The ultimate limits of the technology were then discussed as a concluding section.

The SiGe HBT is a reliable highly versatile technology, which still has plenty of performance yet to be had through scaling. Key HBT device parameters across different IBM SiGe BiCMOS technologies are compared in Table 2.2 [105, 123]. More detailed description can be found in [124].

2.9 Summary and conclusions

Table 2.2 Comparison of key HBT parametrics across different IBM SiGe BiCMOS technology offerings; HP = high-performance and HB = high-breakdown HBT.

Features	5HPE	6HP	7WL	8HP	8WL	8XP	9HP
Lithography (μm)	0.35	0.25	0.18	0.13	0.13	0.13	0.09
HP f_T/f_{MAX} (GHz)	43/79	47/88	60/120	210/265	100/200	250/330	300/360
HP BV_{CEO}/BV_{CBO} (V)	3.3/11	3.35/10.5	3.3/11	1.8/6.0	2.5/8.75	1.65/5.65	1.7/5.30
HB f_T/f_{MAX} (GHz)	17/38	27/70	29/90	60/225	54/170	67/225	135/350
HB BV_{CEO}/BV_{CBO} (V)	9.6/23.0	5.7/14.0	6.0/16.0	3.55/12	4.7/16.0	3.30/11.7	2.5/8.0
CMOS V_{DD} (V)	3.3/5.0	2.5/3.3	1.8/3.3	1.2/2.5	1.2/2.5	1.2/2.5	1.2/2.5/3.3
Gates density (Kgates/mm^2)		40	100	200	200	200	400
BEOL Mx Metal	Al	Al	Cu/Al	Cu	Cu	Cu	Cu

With improvement in device performance of SiGe HBTs, circuit applications continue to expand [125, 126]. SiGe BiCMOS enables circuit designers to develop highly integrated designs at mm-wave frequencies while preserving the overall NF, phase noise, output power, power dissipation, and linearity of the system. SiGe BiCMOS has been used in highly integrated 60 GHz [59] and 94 GHz [127] SOC phased array radars for communications and imaging [128–130]. The technology is widely used for 10G/40G/100G optical transceivers, microwave-backhaul, eband communications, imaging, and ADCs and DACs [131–135].

References

[1] H. Kroemer, "Two integral relations pertaining to the electron transport through a bipolar transistor with a nonuniform energy gap in the base region," *Solid-State Electronics*, vol. **28**, no. 11, pp. 1101–1103, 1985.

[2] H. Kroemer, "Zur Theorie des Diffusions- und des Drifttransistors III," *Archiv der elektrischen Ubertragung*, vol. **8**, pp. 499–504, 1954.

[3] H. Kroemer, "Quasi-electric and quasi-magnetic fields in nonuniform semiconductors," *RCA Review*, vol. **18**, pp. 332–342, 1957.

[4] J. D. Cressler and G. Niu, *Silicon–Germanium Heterojunction Bipolar Transistors*. Artech House, 2003.

[5] A. E. Michel, W. Rausch, P. A. Ronsheim, and R. H. Kastl, "Rapid annealing and the anomalous diffusion of ion implanted boron into silicon," *Applied Physics Letters*, vol. **50**, no. 7, pp. 416–418, 1987.

[6] T. O. Sedgwick, in *Proceedings of the Symposium of Reduced Temperature Processing for VLSI*, 1985.

[7] G. S. Oehrlein, R. Gbez, J. D. Fehribach, *et al.*, "Diffusion of ion-implanted boron and phosphorus during rapid thermal annealing of silicon," in *13th International Conference on Defects in Semiconductors*, pp. 539–546, 1985.

[8] P. A. Stolk, D. J. Eaglesham, H. J. Gossman, and J. M. Poate, "Carbon incorporation in silicon for suppressing interstitial-enhanced boron diffusion," *Applied Physics Letters*, vol. **66**, no. 11, pp. 1370–1372, 1995.

[9] L. Freund and W. Nix, "A critical thickness condition for a strained compliant substrate/epitaxial film system," *Applied Physics Letters*, vol. **69**, no. 2, pp. 173–175, 1996.

[10] S. Stiffler, J. Comfort, C. Stanis, *et al.*, "The thermal stability of SiGe films deposited by ultrahigh-vacuum chemical vapor deposition," *Journal of Applied Physics*, vol. **70**, no. 3, pp. 1416–1420, 1991.

[11] M. Vaidyanathan and D. L. Pulfrey, "Extrapolated f_{max} of heterojunction bipolar transistors," *Electron Devices, IEEE Transactions on*, vol. **46**, no. 2, pp. 301–309, 1999.

[12] R. Camillo-Castillo, Q. Liu, J. Adkisson, *et al.*, "SiGe HBTs in 90 nm BiCMOS technology demonstrating 300 GHz/420 GHz f_t/f_{max} through reduced R_B and C_{CB} parasitics," in *Bipolar/BiCMOS Circuits and Technology Meeting (BCTM), 2013 IEEE*, pp. 227–230, IEEE, 2013.

[13] P. Cheng, Q. Liu, R. Camillo-Castillo, *et al.*, "A novel C_{CB} and R_B reduction technique for high-speed SiGe HBTs," in *Bipolar/BiCMOS Circuits and Technology Meeting (BCTM), 2012 IEEE*, pp. 1–4, IEEE, 2012.

[14] M. J. Tejwani and P. A. Ronsheim, "The dependence of etch pit density on the interfacial oxygen levels in thin silicon layers grown by ultra high vacuum chemical vapor deposition," *Soviet Physics Crystallography C/C of Kristallografiya*, vol. **259**, pp. 467–467, 1993.

[15] S. S. Onge, D. Harame, J. Dunn, et al., "A 0.24 μm SiGe BiCMOS mixed-signal RF production technology featuring a 47 GHz ft HBT and 0.18 μm l_{eff} CMOS," in *Proc. of the 1999 Bipolar/BiCMOS Circuits and Technology Meeting*, pp. 117–120, 1999.

[16] A. Joseph, Q. Liu, W. Hodge, et al., "A 0.35 μm SiGe BiCMOS technology for power amplifier applications," in *Bipolar/BiCMOS Circuits and Technology Meeting, 2007. BCTM '07. IEEE*, pp. 198–201, IEEE, 2007.

[17] P. Chevalier, T. Meister, B. Heinemann, et al., "Towards THz SiGe HBTs," in *Bipolar/BiCMOS Circuits and Technology Meeting (BCTM), 2011 IEEE*, pp. 57–65, IEEE, 2011.

[18] D. L. Harame, J. Comfort, J. Cressler, et al., "Si/SiGe epitaxial-base transistors. ii. Process integration and analog applications," *Electron Devices, IEEE Transactions on*, vol. **42**, no. 3, pp. 469–482, 1995.

[19] R. Camillo-Castillo, J. Johnson, Q. Liu, et al., "Understanding the effects of epitaxy artifacts on SiGe HBT performance through detailed process/device simulation," *ECS Transactions*, vol. **50**, no. 9, pp. 73–81, 2013.

[20] A. Fox, B. Heinemann, R. Barth, et al., "SiGe:C HBT architecture with epitaxial external base," in *Bipolar/BiCMOS Circuits and Technology Meeting (BCTM), 2011 IEEE*, pp. 70–73, IEEE, 2011.

[21] J. Adkisson, M. Khater, J. Gambino, et al., "Improved frequency response in a SiGe npn device through improved dopant activation," *ECS Transactions*, vol. **50**, no. 9, pp. 83–93, 2013.

[22] P. Cheng, M. Dahlstrom, Q. Liu et al., "Modeling of U-shaped and plugged emitter resistance of high speed SiGe HBTs," in *Bipolar/BiCMOS Circuits and Technology Meeting (BCTM), 2011 IEEE*, pp. 154–157, IEEE, 2011.

[23] A. Van der Ziel, *Noise in Solid State Devices and Circuits*. Wiley, 1986.

[24] C. Kittel and H. Kroemer, *Thermal Physics*. Macmillan, 1980.

[25] G. Niu, "Noise in SiGe HBT RF technology: physics, modeling, and circuit implications," *Proceedings of the IEEE*, vol. **93**, no. 9, pp. 1583–1597, 2005.

[26] G. Niu, J. D. Cressler, S. Zhang, A. Joseph, and D. Harame, "Noise-gain tradeoff in RF SiGe HBTs," *Solid-State Electronics*, vol. **46**, no. 9, pp. 1445–1451, 2002.

[27] J.-S. Rieh, D. Greenberg, A. Stricker, and G. Freeman, "Scaling of SiGe heterojunction bipolar transistors," *Proceedings of the IEEE*, vol. **93**, no. 9, pp. 1522–1538, 2005.

[28] D. B. Leeson, "A simple model of feedback oscillator noise spectrum," *Proceedings of the IEEE*, vol. **54**, no. 2, pp. 329–330, 1966.

[29] G. Niu, J. Tang, Z. Feng, A. J. Joseph, and D. L. Harame, "Scaling and technological limitations of $1/f$ noise and oscillator phase noise in SiGe HBTs," *Microwave Theory and Techniques, IEEE Transactions on*, vol. 53, no. 2, pp. 506–514, 2005.

[30] A. van der Ziel, X. Zhang, and A. H. Pawlikiewicz, "Location of $1/f$ noise sources in BJTs and HBJTSi theory," *Electron Devices, IEEE Transactions on*, vol. **33**, no. 9, pp. 1371–1376, 1986.

[31] G. Niu, J. D. Cressler, U. Gogineni, and D. L. Harame, "Collector–base junction avalanche multiplication effects in advanced UHV/CVD SiGe HBTs," *Electron Device Letters, IEEE*, vol. **19**, no. 8, pp. 288–290, 1998.

[32] G. Niu, J. D. Cressler, S. Zhang, U. Gogineni, and D. C. Ahlgren, "Measurement of collector–base junction avalanche multiplication effects in advanced UHV/CVD SiGe HBTs," *Electron Devices, IEEE Transactions on*, vol. **46**, no. 5, pp. 1007–1015, 1999.

[33] J. J. Pekarik, J. W. Adkisson, R. Camillo-Castillo, et al., "Co-integration of high-performance and high-breakdown SiGe HBTs in a BiCMOS technology," in *Bipolar/BiCMOS Circuits and Technology Meeting (BCTM), 2012 IEEE*, pp. 1–4, IEEE, 2012.

[34] J.-S. Rieh, K. M. Watson, F. Guarin, et al., "Reliability of high-speed SiGe heterojunction bipolar transistors under very high forward current density," *Device and Materials Reliability, IEEE Transactions on*, vol. **3**, no. 2, pp. 31–38, 2003.

[35] K. Datta, J. Roderick, and H. Hashemi, "A 20 dBm Q-band SiGe class-E power amplifier with 31% peak PAE," in *Custom Integrated Circuits Conference (CICC), 2012 IEEE*, pp. 1–4, IEEE, 2012.

[36] K. Datta, J. Roderick, and H. Hashemi, "Analysis, design and implementation of mm-wave SiGe stacked class-E power amplifiers," in *Radio Frequency Integrated Circuits Symposium (RFIC), 2013 IEEE*, pp. 275–278, IEEE, 2013.

[37] T. Vanhoucke, H. Boots, and W. Van Noort, "Revised method for extraction of the thermal resistance applied to bulk and SOI SiGe HBTs," *Electron Device Letters, IEEE*, vol. **25**, no. 3, pp. 150–152, 2004.

[38] P. Cheng, C. Zhu, A. Appaswamy, and J. D. Cressler, "A new current-sweep method for assessing the mixed-mode damage spectrum of SiGe HBTs," *Device and Materials Reliability, IEEE Transactions on*, vol. **7**, no. 3, pp. 479–487, 2007.

[39] J.-S. Rieh, J. Johnson, S. Furkay, et al., "Structural dependence of the thermal resistance of trench-isolated bipolar transistors," in *Bipolar/BiCMOS Circuits and Technology Meeting, 2002. Proceedings of the 2002*, pp. 100–103, IEEE, 2002.

[40] J.-S. Rieh, D. Greenberg, Q. Liu, et al., "Structure optimization of trench-isolated SiGe HBTs for simultaneous improvements in thermal and electrical performances," *Electron Devices, IEEE Transactions on*, vol. **52**, no. 12, pp. 2744–2752, 2005.

[41] V. Jain, B. Zetterlund, P. Cheng, et al., "Study of mutual and self-thermal resistance in 90 nm SiGe HBTs," in *Bipolar/BiCMOS Circuits and Technology Meeting (BCTM), 2013 IEEE*, pp. 17–20, IEEE, 2013.

[42] M. Weiß, A. K. Sahoo, C. Raya, et al., "Characterization of intra device mutual thermal coupling in multi finger SiGe:C HBTs," in *Electron Devices and Solid-State Circuits (EDSSC), 2013 IEEE International Conference on*, pp. 1–2, IEEE, 2013.

[43] M. Weiss, A. K. Sahoo, C. Maneux, S. Fregonese, and T. Zimmer, "Mutual thermal coupling in SiGe:C HBTs," in *Microelectronics Technology and Devices (SBMicro), 2013 Symposium on*, pp. 1–4, IEEE, 2013.

[44] J. M. Andrews, C. M. Grens, and J. D. Cressler, "Compact modeling of mutual thermal coupling for the optimal design of SiGe HBT power amplifiers," *Electron Devices, IEEE Transactions on*, vol. **56**, no. 7, pp. 1529–1532, 2009.

[45] U. R. Pfeiffer, C. Mishra, R. M. Rassel, S. Pinkett, and S. K. Reynolds, "Schottky barrier diode circuits in silicon for future millimeter-wave and terahertz applications," *Microwave Theory and Techniques, IEEE Transactions on*, vol. **56**, no. 2, pp. 364–371, 2008.

[46] B. Gaucher, S. Reynolds, B. Floyd, et al., "Progress in SiGe technology toward fully integrated mmwave ICs," in *SiGe Technology and Device Meeting, 2006. ISTDM 2006. Third International*, pp. 1–2, IEEE.

[47] J. W. May and G. M. Rebeiz, "High-performance w-band SiGe RFICs for passive millimeter-wave imaging," in *Radio Frequency Integrated Circuits Symposium, 2009. RFIC 2009. IEEE*, pp. 437–440, IEEE, 2009.

[48] S. Wane, R. van Heijster, and S. Bardy, "Integration of antenna-on-chip and signal detectors for applications from RF to THz frequency range in SiGe technology," in *Radio Frequency Integrated Circuits Symposium (RFIC), 2011 IEEE*, pp. 1–4, IEEE, 2011.

[49] J. Zohios, B. Kramer, and M. Ismail, "A fully integrated 1 GHz BiCMOS VCO," in *Electronics, Circuits and Systems, 1999. Proceedings of ICECS '99. The 6th IEEE International Conference on*, vol. **1**, pp. 193–196, IEEE, 1999.

[50] B. A. Floyd, S. K. Reynolds, U. R. Pfeiffer, *et al.*, "SiGe bipolar transceiver circuits operating at 60 GHz," *Solid-State Circuits, IEEE Journal of*, vol. **40**, no. 1, pp. 156–167, 2005.

[51] J. C. Jensen and L. E. Larson, "A broadband 10-GHz track-and-hold in Si/SiGe HBT technology," *Solid-State Circuits, IEEE Journal of*, vol. **36**, no. 3, pp. 325–330, 2001.

[52] S. Sankaran, C. Mao, E. Seok, *et al.*, "Towards terahertz operation of CMOS," in *Solid-State Circuits Conference – Digest of Technical Papers, 2009. ISSCC 2009. IEEE International*, pp. 202–203, IEEE, 2009.

[53] R. Rassel, J. Johnson, B. Orner, *et al.*, "Schottky barrier diodes for millimeter wave SiGe BiCMOS applications," in *Bipolar/BiCMOS Circuits and Technology Meeting, 2006*, pp. 1–4, IEEE, 2006.

[54] V. Jain, P. Cheng, B. Gross, *et al.*, "Schottky barrier diodes in 90 nm SiGe BiCMOS process operating near 2.0 THz cut-off frequency," in *Bipolar/BiCMOS Circuits and Technology Meeting (BCTM), 2013 IEEE*, pp. 73–76, IEEE, 2013.

[55] B. A. Orner, Q. Liu, J. Johnson, *et al.*, "p–i–n diodes for monolithic millimetre wave BiCMOS applications," *Semiconductor Science and Technology*, vol. **22**, no. 1, p. S208, 2007.

[56] D. L. Harame, D. C. Ahlgren, D. D. Coolbaugh, *et al.*, "Current status and future trends of SiGe BiCMOS technology," *Electron Devices, IEEE Transactions on*, vol. **48**, no. 11, pp. 2575–2594, 2001.

[57] D. Coolbaugh, E. Eshun, R. Groves, *et al.*, "Advanced passive devices for enhanced integrated RF circuit performance," in *Radio Frequency Integrated Circuits (RFIC) Symposium, 2002 IEEE*, pp. 341–344, IEEE, 2002.

[58] J. N. Burghartz, M. Soyuer, K. A. Jenkins, *et al.*, "Integrated RF components in a SiGe bipolar technology," *Solid-State Circuits, IEEE Journal of*, vol. **32**, no. 9, pp. 1440–1445, 1997.

[59] A. Valdes-Garcia, S. T. Nicolson, J.-W. Lai, *et al.*, "A fully integrated 16-element phased-array transmitter in SiGe BiCMOS for 60-GHz communications," *Solid-State Circuits, IEEE Journal of*, vol. **45**, no. 12, pp. 2757–2773, 2010.

[60] P. Candra, V. Jain, P. Cheng, *et al.*, "A 130 nm SiGe BiCMOS technology for mm-wave applications featuring HBT with f_T/f_{MAX} of 260/320 GHz," in *Radio Frequency Integrated Circuits Symposium (RFIC), 2013 IEEE*, pp. 381–384, IEEE, 2013.

[61] B. Orner, Q. Liu, B. Rainey, *et al.*, "A 0.13 μm BiCMOS technology featuring a 200/280 GHz f_T/f_{MAX} SiGe HBT," in *Bipolar/BiCMOS Circuits and Technology Meeting, 2003. Proceedings of the*, pp. 203–206, IEEE, 2003.

[62] P. Magnee, F. Van Rijs, R. Dekker, *et al.*, "Enhanced RF power gain by eliminating the emitter bondwire inductance in emitter plug grounded mounted bipolar transistors,"

in *Bipolar/BiCMOS Circuits and Technology Meeting, 2000. Proceedings of the 2000*, pp. 199–202, IEEE, 2000.

[63] J. Gambino, T. Doan, J. Trapasso, et al., "Through-silicon-via process control in manufacturing for SiGe power amplifiers," in *Electronic Components and Technology Conference (ECTC), 2013 IEEE 63rd*, pp. 221–226, IEEE, 2013.

[64] F. Laermer and A. Urban, "Milestones in deep reactive ion etching," in *Solid-State Sensors, Actuators and Microsystems, 2005. Digest of Technical Papers. TRANSDUCERS '05. The 13th International Conference on*, vol. **2**, pp. 1118–1121, IEEE, 2005.

[65] A. Joseph, J. Gillis, M. Doherty, et al., "Through-silicon vias enable next-generation SiGe power amplifiers for wireless communications," *IBM Journal of Research and Development*, vol. **52**, no. 6, pp. 635–648, 2008.

[66] A. Joseph, Q. Liu, W. Hodge, et al., "A 0.35 μm SiGe BiCMOS technology for power amplifier applications," in *Bipolar/BiCMOS Circuits and Technology Meeting, 2007. BCTM '07. IEEE*, pp. 198–201, IEEE, 2007.

[67] R. M. Malladi, A. Joseph, P. Lindgren, et al., "3D integration techniques applied to SiGe power amplifiers," *ECS Transactions*, vol. **16**, no. 10, pp. 1053–1067, 2008.

[68] A. Joseph, C.-W. Huang, A. Stamper, et al., "Through-silicon via technology for RF and millimetre-wave applications," in *Ultra-thin Chip Technology and Applications*, pp. 445–453, Springer, 2011.

[69] J. Zhang, D. Wang, H. Ding, et al., "SiGe HBT power amplifier design using 0.35 μm BiCMOS technology with through-silicon-via," in *ASIC (ASICON), 2011 IEEE 9th International Conference on*, pp. 1082–1085, IEEE, 2011.

[70] C.-W. Huang, M. Doherty, P. Antognetti, L. Lam, and W. Vaillancourt, "A highly integrated dual band SiGe BiCMOS power amplifier that simplifies dual-band WLAN and MIMO front-end circuit designs," in *Microwave Symposium Digest (MTT), 2010 IEEE MTT-S International*, pp. 256–259, IEEE, 2010.

[71] H. Gummel, "A charge control relation for bipolar transistors," *Bell Systems Technical Journal*, pp. 115–120, 1970.

[72] C. C. McAndrew, J. A. Seitchik, D. F. Bowers, et al., "VBIC95, the vertical bipolar intercompany model," *Solid-State Circuits, IEEE Journal of*, vol. **31**, no. 10, pp. 1476–1483, 1996.

[73] G. M. Kull, L. W. Nagel, S.-W. Lee, et al., "A unified circuit model for bipolar transistors including quasi-saturation effects," *Electron Devices, IEEE Transactions on*, vol. **32**, no. 6, pp. 1103–1113, 1985.

[74] H. De Graaff, W. Kloosterman, J. Geelen, and M. Koolen, "Experience with the new compact MEXTRAM model for bipolar transistors," in *Bipolar Circuits and Technology Meeting, 1989, Proceedings of the 1989*, pp. 246–249, IEEE, 1989.

[75] J. Berkner, "Compact models for bipolar transistors," in *European IC-CAP Device Modeling Workshop*, pp. 7–8, 2002.

[76] M. Schroter, M. Friedrich, and H.-M. Rein, "A generalized integral charge-control relation and its application to compact models for silicon-based HBTs," *Electron Devices, IEEE Transactions on*, vol. **40**, no. 11, pp. 2036–2046, 1993.

[77] B. Ardouin, T. Zimmer, H. Mnif, and P. Fouillat, "Direct method for bipolar base–emitter and base–collector capacitance splitting using high frequency measurements," in *Bipolar/BiCMOS Circuits and Technology Meeting, Proceedings of the 2001*, pp. 114–117, IEEE, 2001.

[78] M. Schroter and A. Chakravorty, *Compact Hierarchical Bipolar Transistor Modeling with HICUM*. World Scientific, 2010.

[79] M. Schroter, A. Pawlak, and A. Mukherjee, "Hicum/l2 a geometry scalable physics-based compact bipolar transistor model," 2015. https://www.iee.et.tu-dresden.de/iee/eb/forsch/Hicum_PD/Hicum23/hicum_L2V2p33_manual.pdf

[80] A. Pawlak, M. Schroter, J. Krause, D. Céli, and N. Derrier, "Hicum/2 v2.3 parameter extraction for advanced SiGe-heterojunction bipolar transistors," in *Bipolar/BiCMOS Circuits and Technology Meeting (BCTM), 2011 IEEE*, pp. 195–198, IEEE, 2011.

[81] http://www.iee.et.tu-dresden.de/iee/eb/forsch/hicum_pd/hicum23/hicum_l2_manual.pdf

[82] Z. Ma, J.-S. Rieh, P. Bhattacharya, *et al.*, "Long-term reliability of Si–Si$_{0.7}$Ge$_{0.3}$–Si HBTs from accelerated lifetime testing," in *Silicon Monolithic Integrated Circuits in RF Systems, 2001. Digest of Papers. 2001 Topical Meeting on*, pp. 122–130, IEEE, 2001.

[83] J. D. Cressler and H. A. Mantooth, *Extreme Environment Electronics*, vol. **10**. CRC Press, 2012.

[84] D. L. Harame and B. S. Meyerson, "The early history of IBM's SiGe mixed signal technology," *Electron Devices, IEEE Transactions on*, vol. **48**, no. 11, pp. 2555–2567, 2001.

[85] D.-L. Tang and E. Hackbarth, "Junction degradation in bipolar transistors and the reliability imposed constraints to scaling and design," *Electron Devices, IEEE Transactions on*, vol. **35**, no. 12, pp. 2101–2107, 1988.

[86] A. Neugroschel, C.-T. Sah, and M. Carroll, "Degradation of bipolar transistor current gain by hot holes during reverse emitter–base bias stress," *Electron Devices, IEEE Transactions on*, vol. **43**, no. 8, pp. 1286–1290, 1996.

[87] A. Neugroschel, R.-Y. Sah, M. S. Carroll, and K. G. Pfaff, "Base current relaxation transient in reverse emitter–base bias stressed silicon bipolar junction transistors," *Electron Devices, IEEE Transactions on*, vol. **44**, no. 5, pp. 792–800, 1997.

[88] H. Wurzer, R. Mahnkopf, and H. Klose, "Annealing of degraded npn-transistors – mechanisms and modeling," *Electron Devices, IEEE Transactions on*, vol. **41**, no. 4, pp. 533–538, 1994.

[89] L. Vendrame, P. Pavan, G. Corva, *et al.*, "Degradation mechanisms in polysilicon emitter bipolar junction transistors for digital applications," *Microelectronics Reliability*, vol. **40**, no. 2, pp. 207–230, 2000.

[90] C. J. Sun, D. Reinhard, T. Grotjohn, C.-J. Huang, and C.-C. Yu, "Hot-electron-induced degradation and post-stress recovery of bipolar transistor gain and noise characteristics," *Electron Devices, IEEE Transactions on*, vol. **39**, no. 9, pp. 2178–2180, 1992.

[91] M. S. Carroll, A. Neugroschel, and C.-T. Sah, "Degradation of silicon bipolar junction transistors at high forward current densities," *Electron Devices, IEEE Transactions on*, vol. **44**, no. 1, pp. 110–117, 1997.

[92] T. Chen, C. Kaya, M. Ketchen, and T. Ning, "Reliability analysis of self-aligned bipolar transistor under forward active current stress," in *Electron Devices Meeting, 1986 International*, vol. **32**, pp. 650–653, IEEE, 1986.

[93] K. Hofmann, G. Bruegmann, and M. Seck, "Impact of inter metal dielectric on the reliability of SiGe npn HBTs after high temperature electrical operation," in *Silicon Monolithic Integrated Circuits in RF Systems, 2003. Digest of Papers. 2003 Topical Meeting on*, pp. 126–129, IEEE, 2003.

[94] J. Zhao, G. Li, K. Liao, *et al.*, "Resolving the mechanisms of current gain increase under forward current stress in poly emitter npn transistors," *Electron Device Letters, IEEE*, vol. **14**, no. 5, pp. 252–255, 1993.

[95] G. Freeman, J.-S. Rieh, B. Jagannathan, *et al.*, "(Invited) SiGe HBT performance and reliability trends through f_t of 350 GHz," in *IEEE International Reliability Physics Symposium Proceedings*, pp. 332–338, IEEE; 1999, 2003.

[96] G. Freeman, J.-S. Rieh, Z. Yang, and F. Guarin, "Reliability and performance scaling of very high speed SiGe HBTs," *Microelectronics Reliability*, vol. **44**, no. 3, pp. 397–410, 2004.

[97] C. Zhu, Q. Liang, R. Al-Huq, *et al.*, "An investigation of the damage mechanisms in impact ionization-induced," in *Electron Devices Meeting, 2003. IEDM '03 Technical Digest. IEEE International*, pp. 7–8, IEEE, 2003.

[98] P. Cheng, C. Zhu, J. D. Cressler, and A. Joseph, "The mixed-mode damage spectrum of SiGe HBTs," in *Reliability Physics Symposium, 2007. Proceedings. 45th Annual. IEEE International*, pp. 566–567, IEEE, 2007.

[99] J. R. Black, "Electromigration failure modes in aluminum metallization for semiconductor devices," *Proceedings of the IEEE*, vol. **57**, no. 9, pp. 1587–1594, 1969.

[100] D. Edelstein, J. Heidenreich, R. Goldblatt, *et al.*, "Full copper wiring in a sub-0.25 µm CMOS ULSI technology," in *Electron Devices Meeting, 1997. IEDM '97. Technical Digest, International*, pp. 773–776, IEEE, 1997.

[101] C.-K. Hu, R. Rosenberg, H. Rathore, D. Nguyen, and B. Agarwala, "Scaling effect on electromigration in on-chip Cu wiring," in *Interconnect Technology, 1999. IEEE International Conference*, pp. 267–269, IEEE, 1999.

[102] O. Aubel, C. Hennesthal, M. Hauschildt, *et al.*, "Backend-of-line reliability improvement options for 28 nm node technologies and beyond," in *Interconnect Technology Conference and 2011 Materials for Advanced Metallization (IITC/MAM), 2011 IEEE International*, pp. 1–3, IEEE, 2011.

[103] O. Aubel, C. Hennesthal, M. Hauschildt, *et al.*, "Comparison of process options for improving backend-of-line reliability in 28 nm node technologies and beyond," *Japanese Journal of Applied Physics*, vol. **50**, no. 5, 2011.

[104] C. Christiansen, B. Li, and J. Gill, "Blech effect and lifetime projection for Cu/low-K interconnects," in *Interconnect Technology Conference, 2008. IITC 2008. International*, pp. 114–116, IEEE, 2008.

[105] J. S. Dunn, D. C. Ahlgren, D. D. Coolbaugh, *et al.*, "Foundation of RF CMOS and SiGe BiCMOS technologies," *IBM Journal of Research and Development*, vol. **47**, no. 2.3, pp. 101–138, 2003.

[106] B. Li, P. S. McLaughlin, J. P. Bickford, *et al.*, "Statistical evaluation of electromigration reliability at chip level," *Device and Materials Reliability, IEEE Transactions on*, vol. **11**, no. 1, pp. 86–91, 2011.

[107] B. Li, T. D. Sullivan, T. C. Lee, and D. Badami, "Reliability challenges for copper interconnects," *Microelectronics Reliability*, vol. **44**, no. 3, pp. 365–380, 2004.

[108] R. Hemmert, G. Prokop, J. Lloyd, P. Smith, and G. Calabrese, "The relationship among electromigration, passivation thickness, and common-emitter current gain degradation within shallow junction npn bipolar transistors," *Journal of Applied Physics*, vol. **53**, no. 6, pp. 4456–4462, 1982.

[109] C. M. Grens, J. D. Cressler, and A. J. Joseph, "On common-base avalanche instabilities in SiGe HBTs," *Electron Devices, IEEE Transactions on*, vol. **55**, no. 6, pp. 1276–1285, 2008.

[110] C. M. Grens, P. Cheng, and J. D. Cressler, "Reliability of SiGe HBTs for power amplifiers. Part I: Large-signal RF performance and operating limits," *Device and Materials Reliability, IEEE Transactions on*, vol. **9**, no. 3, pp. 431–439, 2009.

[111] C. M. Grens, P. Cheng, and J. D. Cressler, "An investigation of the large-signal RF safe-operating-area on aggressively-biased cascode SiGe HBTs for power amplifier applications," in *Silicon Monolithic Integrated Circuits in RF Systems, 2009. SiRF '09. IEEE Topical Meeting on*, pp. 1–4, IEEE, 2009.

[112] P. Cheng, C. M. Grens, and J. D. Cressler, "Reliability of SiGe HBTs for power amplifiers. Part II: Underlying physics and damage modeling," *Device and Materials Reliability, IEEE Transactions on*, vol. **9**, no. 3, pp. 440–448, 2009.

[113] E. Johnson, "Physical limitations on frequency and power parameters of transistors," in *IRE International Convention Record*, vol. **13**, pp. 27–34, IEEE, 1965.

[114] H. Rucker, B. Heinemann, and A. Fox, "Half-terahertz SiGe BiCMOS technology," in *Silicon Monolithic Integrated Circuits in RF Systems (SiRF), 2012 IEEE 12th Topical Meeting on*, pp. 133–136, IEEE, 2012.

[115] J. Yuan, J. D. Cressler, R. Krithivasan, et al., "On the performance limit of cryogenically operated SiGe HBTs and its relation to scaling for terahertz speeds," *Electron Devices, IEEE Transactions on*, vol. **56**, no. 5, pp. 1007–1019, 2009.

[116] Y. Shi and G. Niu, "2-d analysis of device parasitics for 800/1000 GHz f_τ/f_{MAX} SiGe HBT," in *Bipolar/BiCMOS Circuits and Technology Meeting, 2005. Proceedings of the*, pp. 252–255, IEEE, 2005.

[117] R. Camillo-Castillo, A. Stricker, J. B. Johnson, et al., "Technology computer-aided design (TCAD) feasibility study of scaling SiGe HBTs," *ECS Transactions*, vol. **33**, no. 6, pp. 319–329, 2010.

[118] M. Schroter, G. Wedel, B. Heinemann, et al., "Physical and electrical performance limits of high-speed SiGeC HBTs part i: Vertical scaling," *Electron Devices, IEEE Transactions on*, vol. **58**, no. 11, pp. 3687–3696, 2011.

[119] M. Schroter, J. Krause, N. Rinaldi, et al., "Physical and electrical performance limits of high-speed SiGeC HBTs part ii: Lateral scaling," *Electron Devices, IEEE Transactions on*, vol. **58**, no. 11, pp. 3697–3706, 2011.

[120] K. K. Ng, M. R. Frei, and C. A. King, "Reevaluation of the f_τ BV_{CEO} limit on Si bipolar transistors," *Electron Devices, IEEE Transactions on*, vol. **45**, no. 8, pp. 1854–1855, 1998.

[121] T. Tominari, S. Wada, K. Tokunaga, et al., "Study on extremely thin base SiGe:C HBTs featuring sub 5-ps ECL gate delay," in *Bipolar/BiCMOS Circuits and Technology Meeting, 2003. Proceedings of the*, pp. 107–110, IEEE, 2003.

[122] D. Bolze, B. Heinemann, J. Gelpey, S. McCoy, and W. Lerch, "Millisecond annealing of high-performance SiGe HBTs," in *Advanced Thermal Processing of Semiconductors, 2009. RTP '09. 17th International Conference on*, pp. 1–11, IEEE, 2009.

[123] A. J. Joseph, D. L. Harame, B. Jagannathan, et al., "Status and direction of communication technologies – SiGe BiCMOS and RFCMOS," *Proceedings of the IEEE*, vol. **93**, no. 9, pp. 1539–1558, 2005.

[124] http://www-03.ibm.com/technology/foundry/.

[125] A. J. Joseph, J. Dunn, G. Freeman, et al., "Product applications and technology directions with SiGe BiCMOS," *Solid-State Circuits, IEEE Journal of*, vol. **38**, no. 9, pp. 1471–1478, 2003.

[126] D. Harame, S. Koester, G. Freeman, et al., "The revolution in SiGe: Impact on device electronics," *Applied Surface Science*, vol. **224**, no. 1, pp. 9–17, 2004.

[127] A. Valdes-Garcia, A. Natarajan, D. Liu, et al., "A fully-integrated dual-polarization 16-element w-band phased-array transceiver in SiGe BiCMOS," in *Radio Frequency Integrated Circuits Symposium (RFIC), 2013 IEEE*, pp. 375–378, IEEE, 2013.

[128] S. Y. Kim, O. Inac, C.-Y. Kim, and G. M. Rebeiz, "A 76–84 GHz 16-element phased array receiver with a chip-level built-in-self-test system," in *Radio Frequency Integrated Circuits Symposium (RFIC), 2012 IEEE*, pp. 127–130, IEEE, 2012.

[129] F. Golcuk, T. Kanar, and G. M. Rebeiz, "A 90–100-GHz 4, 4 SiGe BiCMOS polarimetric transmit/receive phased array with simultaneous receive-beams capabilities," *Microwave Theory and Techniques, IEEE Transactions on*, vol. **61**, no. 8, pp. 3099–3114, 2013.

[130] D. Shin, C.-Y. Kim, D.-W. Kang, and G. M. Rebeiz, "A high-power packaged four-element-band phased-array transmitter in CMOS for radar and communication systems," *Microwave Theory and Techniques, IEEE Transactions on*, vol. **61**, no. 8, pp. 3060–3071, 2013.

[131] W. Cheng, W. Ali, M.-J. Choi, *et al.*, "A 3b 40 gs/s ADC–DAC in 0.12 μm SiGe," in *Solid-State Circuits Conference, 2004. Digest of Technical Papers. ISSCC. 2004 IEEE International*, pp. 262–263, IEEE, 2004.

[132] M. Chu, P. Jacob, J.-W. Kim, *et al.*, "A 40 gs/s time interleaved ADC using SiGe BiCMOS technology," *Solid-State Circuits, IEEE Journal of*, vol. **45**, no. 2, pp. 380–390, 2010.

[133] Y. Yao, X. Yu, D. Yang, *et al.*, "A 3-bit 20 gs/s interleaved flash analog-to-digital converter in SiGe technology," in *Solid-State Circuits Conference, 2007. ASSCC '07. IEEE Asian*, pp. 420–423, IEEE, 2007.

[134] Y. Yang, S. Cacina, and G. M. Rebeiz, "A SiGe BiCMOS w-band LNA with 5.1 dB NF at 90 GHz," in *Compound Semiconductor Integrated Circuit Symposium (CSICS), 2013 IEEE*, pp. 1–4, IEEE, 2013.

[135] H.-C. Lin and G. M. Rebeiz, "A 200–245 GHz balanced frequency doubler with peak output power of +2 dBm," in *Compound Semiconductor Integrated Circuit Symposium (CSICS), 2013 IEEE*, pp. 1–4, IEEE, 2013.

3 Characteristics, performance, modeling, and reliability of CMOS technologies for mm-wave power amplifiers

Antonino Scuderi and Egidio Ragonese

3.1 Introduction

CMOS technology is a new player in the scenario of mm-wave technologies. Only at the beginning of the 1990s did the scientific community start to consider CMOS for RF applications and it was only at the beginning of the new millennium that aggressive dimensional scaling allowed us to figure out its exploitation at mm-waves. The dimensional scaling produced higher unity current-gain cut-off frequency (f_T) and maximum oscillation frequency (f_{max}), at the expense of a lower breakdown voltage. Classical small-signal high-frequency building blocks (i.e. low-noise amplifiers) can manage the reduced supply voltage as the main impact is related to the available voltage dynamic. Different considerations have to be made for power amplifiers (PAs) as the inherent power performance is directly connected with the allowed voltage swing. Indeed, the generation of a 1-W power level on a 50-ohm load would require a sinusoidal voltage of 10 V (i.e. 20-v_{pp} swing) that is not compliant with the typical breakdown voltage of submicron CMOS technologies. In order to reduce the swing on active devices, impedance transformation is needed. The problem with classical matching networks is that, when increasing the transformation ratio, very low impedance at the transistor level has to be managed. This would require a very-high-quality matching network to avoid losses [1, 2]. Passive components, mainly spiral inductors and transformers, cannot achieve very high quality factors (Q-factor), especially on CMOS conductive bulk. Silicon-on-insulator (SOI) processes can help in giving a solution for compact and high-performance integration of matching networks at the expense of higher costs.

A second basic constraint for the design of a CMOS amplifier is the gain limitation due to high input/output capacitances (i.e. intrinsic f_{max} limitation) and source inductive parasitics. Finally, the dimensional scaling also causes a relative increase of resistive parasitics with a direct impact on the amplifier efficiency. Indeed, since impedance transformation is related to the conversion of voltage swing into current swing, any resistive contribution would generate higher losses on high-current paths.

Table 3.1 High-frequency material characteristics.

Material	Bandgap [eV]	Mobility [cm^2(V · s)]	Saturated drift velocity [cm/s]	Critical E_{cr} [V/cm]	Thermal conductivity [W/(cm · K)]
Si	1.12	1300	$0.7 \cdot 10^{-7}$	$2.5 \cdot 10^5$	1.5
GaAs	1.42	5000	$1 \cdot 10^{-7}$	$3 \cdot 10^5$	0.49
GaN	3.39	1500	$2.5 \cdot 10^{-7}$	$30 \cdot 10^5$	2.2
InP	1.35	4500	$1 \cdot 10^{-7}$	$3 \cdot 10^5$	0.68
Ge	0.66	3900	$0.6 \cdot 10^{-7}$	$2 \cdot 10^5$	0.58

The nonlinear parasitics as well as the knee voltage (i.e. the voltage at which there occurs the transition from "linear" to "saturation" in the $I_{DS}-V_{DS}$ characteristic) and oxide traps also impact the overall device linearity performance that is of utmost importance in modern communication systems. On the other hand, the actual big advantage of CMOS platforms is the large availability of devices in quantity and kind, allowing the implementation of highly integrated complex systems at very low cost. The complexity is also increasingly required in power amplifiers, as the need for linearity and efficiency calls for pre-distortion techniques and customized control circuitries that can be easily managed only using CMOS technologies. This is the real strength of such a powerful technology that is in direct competition with silicon–germanium (SiGe) and compound semiconductor technologies, inherently characterized by higher intrinsic performance. In order to overcome some intrinsic device limits, specific techniques allowing high-voltage-swing management and power combining will be discussed in the next few chapters. Here, we will go through the physics of the device in order to get a basic understanding of the critical mechanisms that act in high-power, high-frequency CMOS transistors.

3.2 Materials for high frequency: CMOS and its evolution

High-frequency transistors are typically fabricated by exploiting compound semiconductor materials. Gallium arsenide (GaAs), indium phosphide (InP), and gallium nitride (GaN) are the most common ones traditionally adopted for mm-wave applications. They are used as the basis for the fabrication of field-effect transistor devices (FETs), heterojunction transistors, such as high-electron-mobility transistors (HEMTs), and heterojunction bipolar transistors (HBTs). The advantages of such materials in comparison with silicon and germanium (adopted for high-performance SiGe devices) are clearly represented in Table 3.1.

In order to compare devices manufactured with different materials, several figures of merit (FoM) have been proposed, among which are the power-frequency-squared limit [3] and the Johnson FoM [3, 4] that are reported in Eqns (3.1) and (3.2), respectively:

$$Pf^2 \sim \left(\frac{E_{cr}V_{sat}}{2\pi}\right)\frac{1}{X_c} \tag{3.1}$$

3.3 CMOS active devices

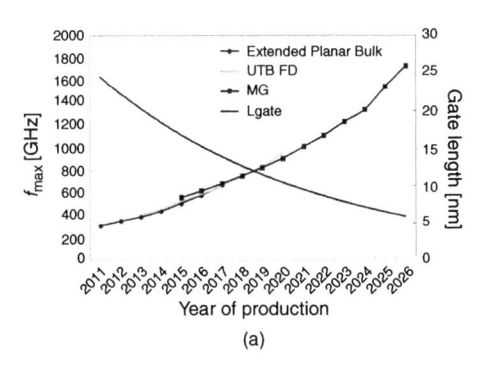

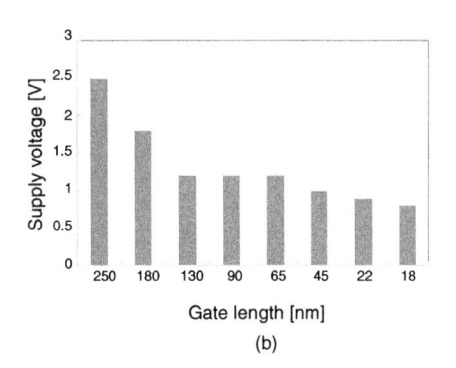

Fig. 3.1 (a) Prediction of f_{max} increase and gate length scaling for extended planar bulk, ultra-thin body fully depleted SOI (UTB FD) and multi-gate (MG). Source: ITRS 2012 (markers represent known production techniques for the corresponding performance). (b) Supply voltage versus gate length.

$$JFoM = \left(\frac{E_{cr} v_{sat}}{2\pi}\right)^2, \qquad (3.2)$$

where E_{cr} is the electric field before breakdown, v_{sat} is the saturated electron drift velocity, and X_c is the device impedance level. As E_{cr} and v_{sat} are higher in III–V semiconductors, at the same frequency, the power capability is higher in such semiconductors than in silicon. Moreover, higher mobility means better gain, while the insulating property of the substrate improves device isolation and the Q-factor of passive components. Although CMOS platforms exhibit limited FoM values, they have the strength of exploiting the most mature, cheap, characterized, modeled, and understood material, i.e. silicon. Silicon is very stable, with an excellent native oxide, and has been aggressively scaled in the past 10 years from 0.13 μm down to 22 nm with device reengineering from 2D to 3D structures. This enlarged the application coverage for such a process, as performance has been improved, attaining an f_{max} increase of around 5×, as reported in Fig. 3.1. On the other hand, the device scaling produces a significant supply voltage reduction (Fig. 3.1) and an increase of parasitic resistance, thus making more challenging the achievement of high output power and efficiency, respectively. Growing performance, extraordinary integration level, low cost, and size reduction have already allowed CMOS to enter the mm-wave application area with great success. Moreover, increased performance provided by the SOI option and now the opportunity of mixing compound GaN-based HEMTs within the CMOS process flow make the CMOS roadmap more and more attractive, envisioning it as the technology race winner even at mm-waves and even for the most demanding blocks, such as power amplifiers.

3.3 CMOS active devices

Complementary MOS technology allows for both n-type and p-type device availability on the same silicon platform. This feature gives high flexibility in designing complex circuitry. As carrier mobility is a critical parameter for high-speed devices, it is evident

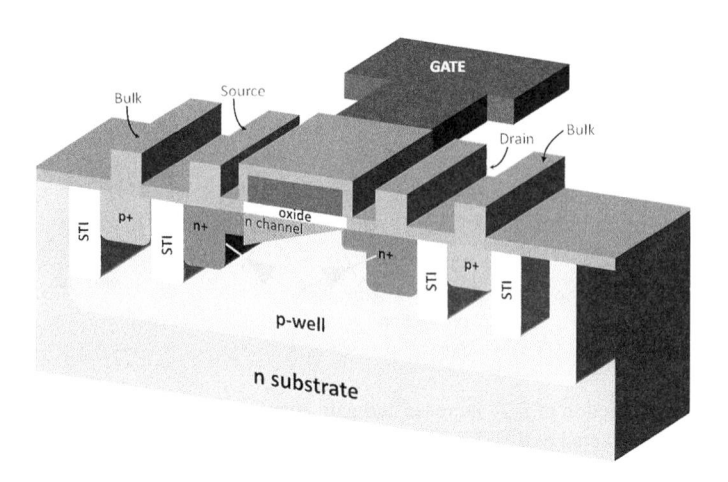

Fig. 3.2 Typical MOS cross section.

that n-type devices will be the core components for signal processing at mm-wave, while p-type ones will be mainly exploited for support functions. CMOS devices are field-effect transistors. A simplified cross section of an MOS transistor is shown in Fig. 3.2. An MOS transistor consists of four terminals (i.e. source, gate, drain, and substrate), a charge-controlled intrinsic device, source/drain-bulk diodes, and parasitic resistances. Focusing on the intrinsic n-type FET, at zero gate bias no current flow is allowed between drain and source since two series back-to-back diodes hinder the conduction. When a positive voltage is applied to the gate terminal, a negative charge is induced in the channel, namely the inversion layer, thus providing a conductive path between the drain and source. The current can flow between the drain and the source terminals, controlled by the gate voltage. Thus, in an MOS device the input voltage (i.e. the gate voltage) controls the output current (i.e. the drain current). The detailed device physics is widely described in several publications [5–8]. Here we will simply review some behavior of specific interest for high-frequency power amplifiers, with the main focus on enhancement-mode n-MOS devices.

3.3.1 Physics of nanoscaled MOS transistors

The basic structure of an MOS device has been maintained through the technology scaling down to 22 nm. Of course, short-channel effects have become more and more evident, thus requiring proper compensation techniques. Referring to Fig. 3.2, two heavily doped n+ regions, fabricated on a p-type substrate, are the source and drain of the n-MOS device. Shallow-trench isolation (STI) is exploited for device isolation instead of local oxidation of silicon (LOCOS), assuring a better planarity. The core of the MOS transistor is the metal–oxide–semiconductor capacitor structure. A thin SiO_2 layer is fabricated on top of the silicon surface and a metal layer or a polysilicon one covers the oxide. This layer is the first plate of the capacitor, while the second one is the substrate or the induced inversion layer, depending on the applied gate voltage. In ultra-scaled devices, the MOS capacitance interlayer distance is no longer the oxide thickness. Indeed, two nanoscale effects have to be taken into account, i.e. the quantum confinement effect (QCE) [9, 10] and the polysilicon depletion effect (PDE) [9]. The QCE is related to the high vertical electric field present in the ultra-scaled MOS. Indeed, a potential well is established in the inversion layer and electrons experience a quantum confinement effect away from the oxide interface, increasing the real spacing between the metal gate and the channel charge. The PDE is again related to the high vertical electric field, which induces a depletion region in the polysilicon if it is exploited as a metal gate. Both the QCE and the PDE have the effect of reducing the gate capacitance since the effective dielectric thickness, i.e. the distance between the charge in the channel and the charge in the polysilicon, is increased. The capacitive equivalent thickness CET can be calculated as follows [8]:

$$CET \cong EOT + \frac{\epsilon_{Si}}{\epsilon_{ox}}(X_{gd} + \Delta d), \tag{3.3}$$

where X_{gd} is the depth of the polysilicon depletion region, Δd is the spatial offset of the charge in the inversion channel, and EOT is the effective oxide thickness used in the case of high-k dielectrics ($EOT = t_{ox} \cdot \epsilon_{SiO_2}/\epsilon_{ox}$). In advanced CMOS technology nodes, EOT of 1.2 nm are generally exploited while CET can be up to 50% higher. This produces relevant deviations of the classical calculation of C_{ox}, explaining the relevance of the above considerations at mm-wave. Moving back to the MOS capacitor structure, when a positive gate voltage is applied the positive charge on the gate will induce a depletion region under the oxide, resulting in the elimination of holes from this region. The associated operation region when a drain-to-source voltage is applied is reported as the sub-threshold region and here the I–V relation shows an exponential behavior. When a silicon potential of twice the Fermi level Φ_F ($kT/q \cdot \ln(N_a/n_i)$) is reached, the inversion occurs and the related gate voltage is called the threshold voltage, V_T. It can be expressed as

$$V_T = \phi_{MS} + \frac{Q_b}{C_{ox}} + 2\Phi_F, \tag{3.4}$$

where ϕ_{MS} is the potential due to the different work functions of gate, metal, and silicon, and Q_b is the charge in the depletion layer less the charge in the oxide (at the silicon interface). Indeed, the channel-length reduction has the effect of reducing the effective V_T with respect to the value given by (3.4), which means a loss of channel control. This can be understood by considering that the charge Q_b is associated with acceptors and donors in the depletion region. As the gate length is reduced, a larger part of this charge is incorporated in the source and in the drain, giving a Q_b reduction and then a V_T reduction. A more accurate expression of V_T is given by

$$V_T = V_{FB} + \frac{\sqrt{4\epsilon_s q N_B \Phi_F}}{C_{ox}} \left(1 - \frac{x_j}{L}\sqrt{1 + \frac{2x_{dmax}}{x_j}}\right) - 2\phi_F, \tag{3.5}$$

where V_{FB} is the flat-band voltage, N_B is the dopant concentration, x_j is the depth of the drain and source region, and x_{dmax} is that of the channel depletion region. Indeed, at short gate length, the source–bulk and drain–bulk depletion regions become very close and the channel doping concentration should be increased to avoid punch-through. Since an increment of the channel doping concentration would have a negative effect on mobility, the doping level is increased only at the edge sections of the channel toward the source and drain terminals. However, the resulting effect is an increase of N_B and then an increase of V_T according to (3.5). Therefore, by reducing the gate length, the resulting transistor will show an increase of V_T at first and then the classical short-channel V_T reduction effect. This V_T variation impacts IC design as the gate length (L) must be kept constant in order to get V_T-matched devices. Finally, at high V_{DS} voltages, electrons in the source see a lower barrier potential, thus suffering from the so-called drain-induced barrier-lowering effect (DIBL). This results in V_T reduction as V_{DS} increases. These effects are particularly interesting for PA designers as the wide V_{DS} swing has to be managed. More generally, they need to be taken into account in the design of the biasing circuitry.

Coming back again to the MOS operation, a further increase of the gate voltage will induce the generation of an inversion layer whose charge is controlled by the gate voltage. It is interesting to analyze the MOS capacitance–voltage relationship that is shown in Fig. 3.3. At low frequency and very negative gate voltage, the hole concentration under the gate is increased, defining a perfect C_{ox} capacitor with the metal gate covering the oxide. When the gate voltage approaches "zero," the depletion region begins to appear and the total capacitance is the series connection of C_{ox} and C_{cb} (i.e. the channel to bulk capacitance associated with the depletion layer). Therefore, the total capacitance decreases to a minimum that happens at the maximum depletion extension, when the gate voltage V_G is equal to V_T. A further increase to above the threshold voltage will generate the inversion layer that shields the bulk and then the total capacitance moves back to C_{ox}. Indeed, it is expected that, at high frequency, the variation of gate capacitance does not happen because the time constant of the generation–combination center is very long. Unfortunately, when an MOS device is fabricated, highly doped n+ regions at the drain and source terminals are required in order to obtain good contacts and these

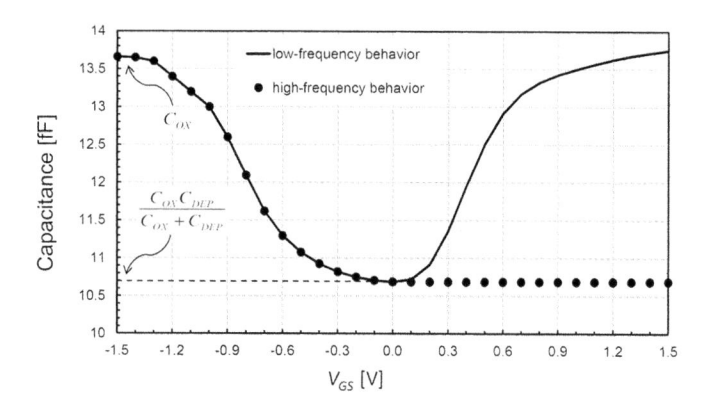

Fig. 3.3 Capacitance–voltage relationship for a 65-nm MOS capacitor. Adapted from Fig. 1.6 of P. Aaen, J. A. Plá, and J. Wood, *Modeling and Characterization of RF and Microwave Power FETs*, Cambridge University Press, 2011.

regions act as very good minority carrier generators, thus keeping the non-constant gate capacitance–voltage behavior also at high frequency. This effect is very relevant for power-amplifier designer since it directly impacts the linearity behavior, as discussed in the next section.

When the inversion layer is established, by applying a positive voltage between drain and source, a current will flow through the channel. The current expression can be calculated by expressing the channel charge as a function of the local potential and integrating along the channel, thus obtaining the following:

$$I_D = \mu_S C_{ox} \frac{W}{L}\left[\left(V_G - V_T\right)V_D - \frac{1}{2}V_D^2\right], \quad (3.6)$$

where μ_S is the surface or channel electron mobility, V_D is the drain voltage, and W and L are the width and length of the gate, respectively. The mobility μ_S is affected by scattering phenomena at the oxide interface as well as quantum confinement effects, being a function of the effective vertical field E_{EFF}, and then it is largely reduced with respect to the silicon mobility reported in Table 3.1, as shown in Fig. 3.4. Indeed, while in long-channel devices the device mobility is related to the acoustic-phonon-limited mobility that decays as $E_{EFF}^{-0.3}$, in a short-channel MOS it is dominated by the surface mobility that decays with E_{EFF}^{-1}. The vertical electric field has a primary role in such devices and it prevails on the lateral field effects limiting the velocity saturation. An empirical expression showing the mobility dependence on V_{GS} is reported in [11] and described as

$$\mu(V_{GS}) = \frac{\mu_0}{1 + \theta(V_{GS} - V_T)}, \quad (3.7)$$

where μ_0 is the mobility at $V_{GS} - V_T = 0$, and θ is a fitting parameter. In power amplifier design, the reduced impact of the lateral field explains the relatively higher supply voltage that nanoscaled devices can tolerate. Indeed, the critical lateral field ($E_C = v_{sat}/\mu$) increases as the carrier mobility degrades [12], delaying the onset of velocity saturation. To the same end, it is possible to consider the electron velocity as a function of the lateral field $v = \mu E_{lat} = \mu V_{DS}/L$. Reduced mobility allows a higher V_{DS} voltage with the same electron velocity. Indeed, when the drain–source voltage is increased above the triode region where (3.6) is valid, the device reaches the saturation region. The related drain voltage is called $V_{sat} = V_{GS} - V_T$ and, above this voltage, the drain current is ideally independent of V_D and expressed as

$$I_D = \frac{1}{2}\frac{W}{L}C_{ox}\mu_s(V_G - V_T)^2. \quad (3.8)$$

Unfortunately, in short-channel devices the combined effect of vertical and lateral fields limits the region where the device shows real quadratic behavior and the current in the saturation region is better approximated by linear I–V dependence [8]:

$$I_D = W v_{xo} C_{ox}^{inv}(V_{GS} - V_T)(1 + \lambda V_{DS}), \quad (3.9)$$

where v_{xo} is the effective carrier velocity at the intrinsic source, C_{ox}^{inv} is the effective oxide capacitance at the inversion, and λ is the channel length modulation parameter. This last

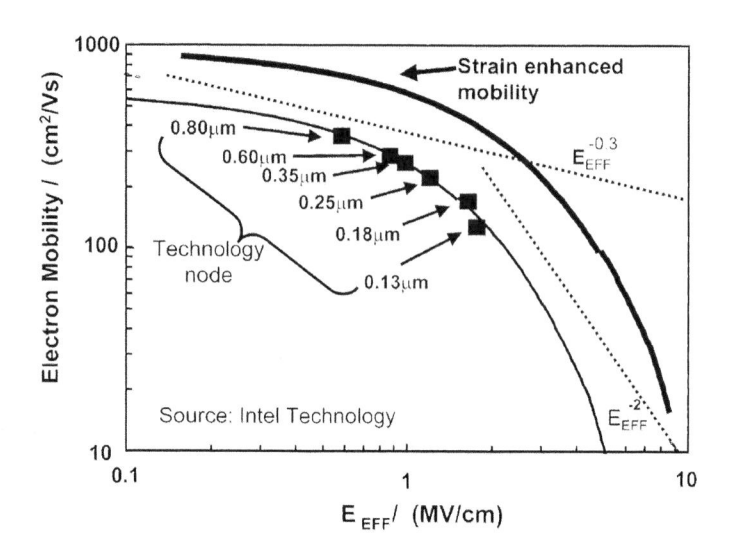

Fig. 3.4 Mobility dependence as a function of node scaling ©[2004] IEEE. Reprinted, with permission, from S. E. Thompson *et al.*, "A 90 nm logic technology featuring strained-silicon," *IEEE Trans. Electron Devices*, vol. **51**, pp. 1790–1797, Nov. 2004.

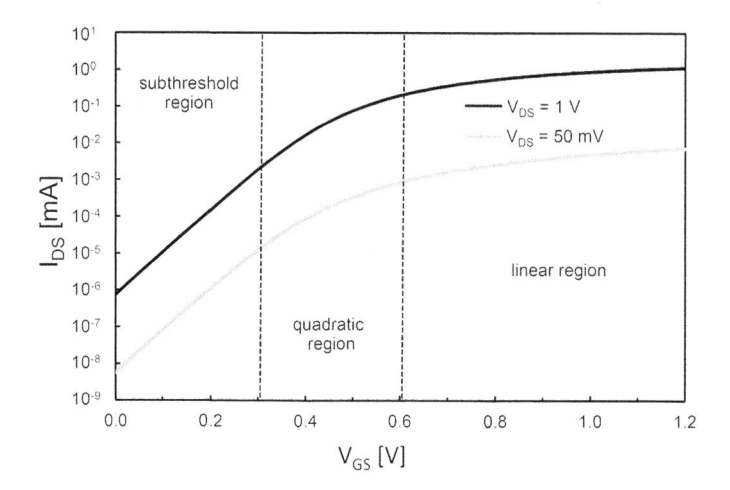

Fig. 3.5 I_{DS} vs V_{GS} for a typical 65-nm MOS transistor with 10 gate fingers ($W_f = 1$ μm).

parameter takes into account the finite DC output resistance; it is a linear function of the effective channel length and an increasing function of the doping level in the channel [12]. In Figs. 3.5 and 3.6, the I_{DS} vs V_{GS} relation and the classical output behavior are reported, respectively, for a typical 65-nm MOS. The output conductance should be as low as possible, rising to flat I–V behavior, while the knee voltage should be as low as possible because it limits the available voltage swing of the output voltage in a power amplifier [13], impacting efficiency and linearity performance.

Finally, the DC high-field effect has to considered, especially for power-amplifier applications:

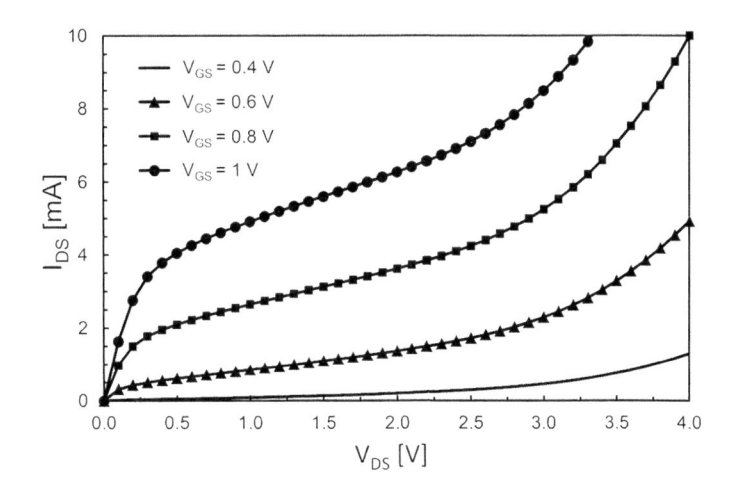

Fig. 3.6 Drain-current output characteristics for a 65-nm MOS with 10 gate fingers ($W_f = 1\ \mu\text{m}$).

- at high drain voltages, a breakdown caused by the high voltage across the oxide at the drain end of the gate generates device failure;
- attention also has to be paid to the p-n junction at the drain with the body as well as gate-voltage-induced breakdown in the gate, especially for very-short-channel devices;
- at high current with high electric field, hot-carrier injection, due to very high electron acceleration, can generate trapped charge in the oxide as well as in the body, modifying the device behavior and eventually moving the device toward its failure.

3.3.2 Small-signal parameters, parasitic effects, and large-signal modeling

Figure 3.7 depicts the cross section of the n-MOS device with the small-signal model summarizing intrinsic and parasitic resistances, capacitances, and transconductances g_m and g_{mb}. In particular, g_{mb} is included in the model to account for the so-called body effect, i.e. the body acts as a second gate for the electrons in the channel. Table 3.2 reports the common expressions for the above small-signal parameters together with the transition frequency, f_T, and the maximum oscillation frequency, f_{max}. The reported transconductance is the small-signal approximation extracted in saturation. A more detailed analysis is reported in the large-signal section. In Table 3.2, C_{ov} refers to the parasitic capacitance caused by the overlap of gate with source and drain; C_{fring} is the gate-to-drain/source fringing capacitance and it is a function of the poly (or metal) gate thickness with respect to the oxide thickness. Finally, $R_{G,i}$ (Fig. 3.7) is the intrinsic gate-channel-to-source resistance.

As far as large-signal modeling is concerned, BSIM models (i.e. BSIM3v3 or BSIM4) are widely used. By adding inductive parasitics (i.e. L_G, L_D, L_S) and a voltage-dependant substrate network, such as the one depicted in Fig. 3.7, these models are able to predict both DC nonlinearities and high-frequency effects. Moreover, better accuracy could

Table 3.2 Small-signal parameters.

Parameter	Expression
Saturated transconductance (g_m)	$\mu_n C_{ox} \frac{W}{L}(V_{GS} - V_T) = \sqrt{2\mu_n C_{ox} \frac{W}{L} I_D}$
g_m with gate and source resistance ($g_{m,R}$)	$g_m / [1 + g_m(R_S + R_G)]$
Gate–source capacitance in triode (C_{gs})	$C_{ov} + \frac{1}{2} \cdot (W \cdot L \cdot C_{ox})$
Gate–source capacitance in saturation (C_{gs})	$\frac{2}{3} \cdot (W \cdot L \cdot C_{ox})$
Gate–drain capacitance in triode (C_{gd})	$C_{ov} + \frac{1}{2} \cdot (W \cdot L \cdot C_{ox})$
Gate–drain capacitance in saturation (C_{gd})	$C_{ov} + C_{fring}$
Output resistance (r_o)	$\frac{1}{g_o} = \frac{1}{\lambda I_D} = \frac{V_A}{I_D}$
Junction capacitance (C_{xb})	$C_{xb0} / \sqrt{1 + \frac{V_{XB}}{\Phi_0}}$ (x = source or drain)
f_T (simple)	$\frac{g_m}{2\pi(C_{gs}+C_{gd}+C_{gb})}$
f_T (full) [14]	$\frac{1}{2\pi}\left[\frac{(C_{gs}+C_{gd})}{g_m} + (R_S+R_D)C_{gd} + \left[(R_S+R_D)(C_{gs}+C_{gd})\frac{g_o}{g_m}\right]\right]^{-1}$
f_{max} (simple)	$\frac{f_T}{2}\sqrt{\frac{1+g_m R_S}{R_S+R_G g_o}}$
f_{max} (full) [15]	$\frac{0.5 \cdot f_T}{\sqrt{(R_{gs}+R_S+R_D)g_o + 2\pi f_T R_G C_{gd}}}$

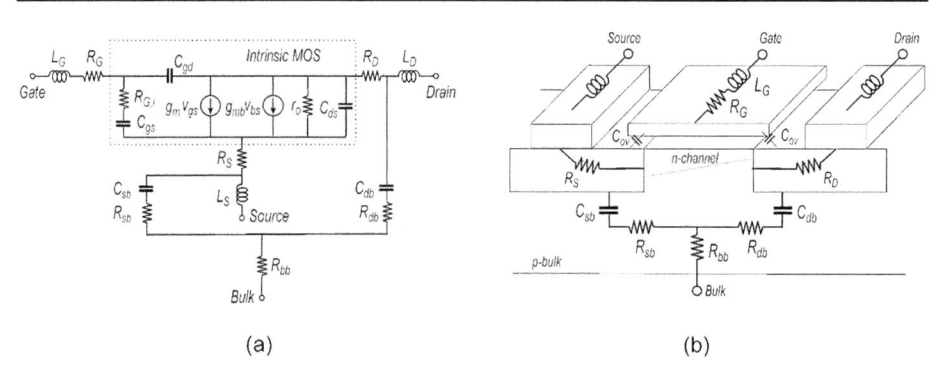

Fig. 3.7 (a) Small-signal equivalent model and (b) MOS cross section.

be achieved on a specific device by properly fitting BSIM parameters with device measurement data.

3.3.3 Resistive and capacitive parasitics

Any intrinsic or extrinsic resistance will generate losses. Both series drain and source resistances (R_D and R_S) are very important in power-amplifier design as they are placed in high-current paths, impacting the device R_{on} and then efficiency, as well as the g_m and f_{max} (they also impact f_T, but this parameter is not very relevant in PA design). The intrinsic drain/source resistances are inversely proportional to the gate width and dependant on the source, drain, and channel engineering. For a given technology, the PA designer can modify only the external resistive contribution due to the contact vias and the back-end metals. These interconnections could largely impact circuit design and generally the PA designer spends more time in optimizing the device layout than in doing circuit simulations. Commercial 3D EM CAD tools help during the extraction of effective parasitic contributions, but basic hand-made calculations are always a good way to screen the layout architecture to be adopted. Similar concepts can be applied to the gate resistance that directly impacts f_{max} and the output power of power amplifiers. A basic layout technique exploited to reduce the gate resistance is to contact the gate fingers at the two ends. Assuming only one gate finger of width W_f with resistance R_{sheet} and W_{ext} as the external path, it is possible to reduce the gate resistance from

$$R_{G(single\text{-}extraction)} = \frac{R_{sheet}}{L}\left(\frac{W_f}{3} + W_{ext}\right) \quad (3.10)$$

to

$$R_{G(double\text{-}extraction)} = 0.5\frac{R_{sheet}}{L}\left(\frac{W_f}{6} + W_{ext}\right) \quad (3.11)$$

with a ratio of around 4 times. Indeed, it has to be noted that the double gate connection will increase the parasitic gate capacitance, giving a reduction of f_T, as reported in Fig. 3.8a. However, a better f_{max} translates to a higher output power capability and higher efficiency at high frequency (above 14 GHz for the analyzed 45-nm process), as shown in Fig. 3.8b [16].

Parasitic drain–bulk and source–bulk capacitances, reported in Table 3.2, are indeed layout dependent. These capacitances are equal if an odd number of fingers are exploited. However, mainly for power amplifiers, it is suggested to keep the drain–bulk capacitance as low as possible, even if the source–bulk capacitance is increased. To this end, layouts with source-first access are suggested for PA design, as sketched in Fig. 3.9. Moreover, even if small width will improve the gain (due to better intrinsic device f_T) it is generally preferred to exploit a long-width device with double-gate connection because these arrangements reduce the number of parallel interconnected devices and related parasites. This choice has two drawbacks, i.e. the reduction of f_{max} and the potential latch-up problems, especially for SOI CMOS. This last aspect will be discussed later. The choice of the device width will be always a complex trade-off to be optimized case by case.

Moreover, the high current level experienced in power amplifiers makes it not possible to have a planar single-metal-layer drain/source interconnection. Generally, several vias are directly placed on top of the source and drain fingers, thus superposing several metal layers and distributing the current to higher and more conductive metal layers. On one hand, this technique allows the extraction of higher current values with consequent reduction of the series resistances; on the other hand, additional capacitive effects will be

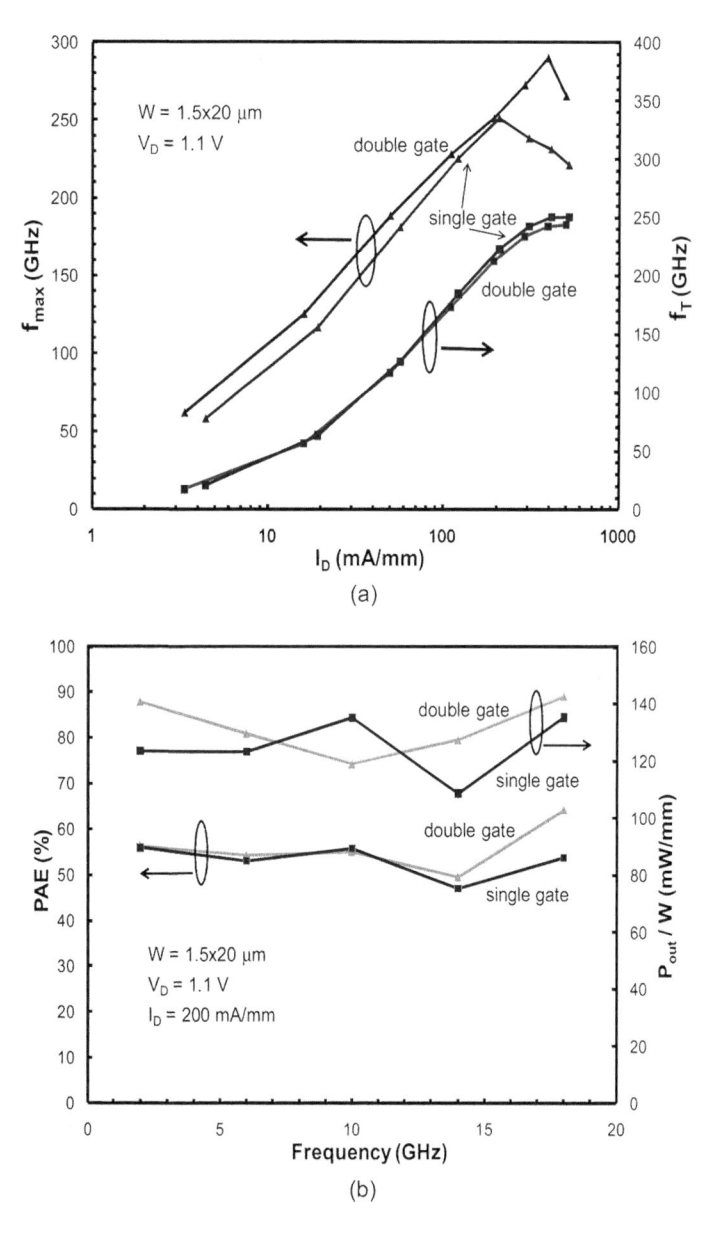

Fig. 3.8 Power potential at 45 nm. (a) f_T and f_{max} as a function of drain current density for single- and double-gate contact structures; (b) PAE and normalized output power as a function of measurement frequency for single-gate and double-gate contact structures ©[2010] IEEE. Reprinted, with permission, from U. Gogineni, J. A. del Alamo, and C. Putnam, "RF power potential of 45 nm CMOS technology," in *Proc. of Topical Meeting on Silicon Monolithic Integrated Circuits in RF Systems (SiRF)*, pp. 204–207, Jan. 2010.

experienced due to the fringing capacitive parasitics. Drain–source and drain–gate parasitic capacitances should be minimized in order to avoid impact on the output impedance and on the gain, respectively. A typical and widespread structure for source/drain metal extraction is reported in Fig. 3.10. It is based on ladder metal-stack configurations that

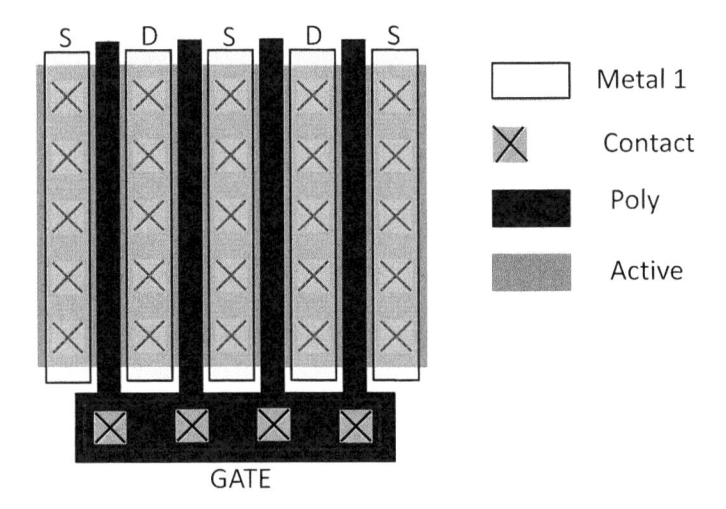

Fig. 3.9 Simplified 2D planar layout of n-MOS device.

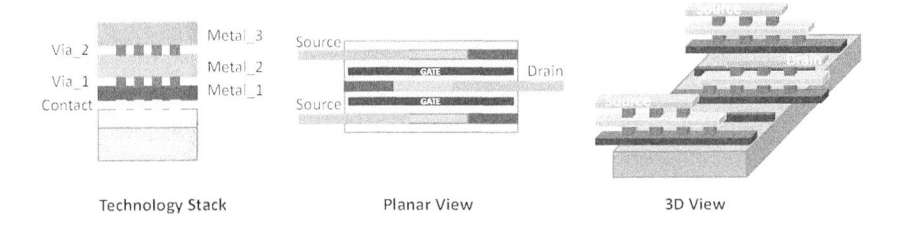

Fig. 3.10 Shifted metal-stacking architecture.

are properly placed on both source/drain fingers to avoid having face-to-face metals and thus minimize the fringing effects, especially for the higher metal layers (typically the thicker ones). In any case, a customized design will require its own specific optimization.

3.3.4 High-frequency performance parameters: f_T and f_{max}

The transition frequency is defined as the frequency for which the current gain value in common-source (CS) configuration becomes equal to one. Ignoring the feedback capacitance and the series resistances, the current gain can be written as

$$A_I = \frac{g_m V_{gs}}{C_{gs}\frac{dV_{gs}}{dt}} = \frac{g_m}{j\omega C_{gs}} \quad (3.12)$$

for a sinusoidal excitation and, applying the definition, i.e. $A_I = 1$,

$$f_T = \frac{g_m}{2\pi C_{gs}}. \quad (3.13)$$

By adding the gate-capacitive parasitics and the resistive ones, it is possible to obtain the simplified and complete expressions reported in Table 3.2, respectively. It is worth noting that, assuming classical quadratic saturation behavior, it is possible to write the f_T expression reporting the L^{-2} dependence as

$$f_T = 1.5 \frac{\mu}{2\pi L^2}(V_{GS} - V_T). \qquad (3.14)$$

Indeed, at very short channel dimension, the velocity saturation of electrons in the channel is largely affected by the electric field as described in [7]. For large electric fields $v_{sat} \simeq \mu E_{crit} = \mu V_{ds}/L$ (where V_{ds} is the intrinsic drain-to-source voltage under the gate), and thus

$$V_{DS} = v_{sat}\frac{L}{\mu}. \qquad (3.15)$$

Manipulating the above expressions, a secondary f_T scaling effect is found:

$$f_T = \frac{g_m}{2\pi C_{gs}} \approx \frac{v_{sat}}{L}. \qquad (3.16)$$

As already discussed, in terms of departure from the quadratic law for nanoscaled CMOS devices, f_T increases as L^{-1} rather than L^{-2}.

A more useful parameter for PA designers is the maximum oscillation frequency f_{max}. This is because f_{max} is defined as the frequency for which the maximum available power gain is equal to 1. Both f_T and f_{max} are generally very high and therefore cannot be directly measured. Indeed, f_T is extrapolated from the 20 dB-dec slope of the current gain. On the other hand, f_{max} does not have a constant slope vs frequency, but it can be evaluated from the unilateral power gain U_g slope because the 0-dB crossing of both curves happens at the same frequency and U_g shows a constant slope of 10 dB-dec. Focusing on f_{max}, the analysis of its dependence on the bias current density is of utmost importance. Indeed, this allows the optimum bias current density to be identified for maximum power gain and best linearity, as well. A typical relation for a 65-nm device is reported in Fig. 3.11. Increasing the current density, f_{max} will increase up to a flat region that corresponds to a high-linearity region. The further this region extends versus current the more linear the device. However, PAs are generally large-signal devices and the above consideration starts to fail when large signals are applied to a nanoscaled

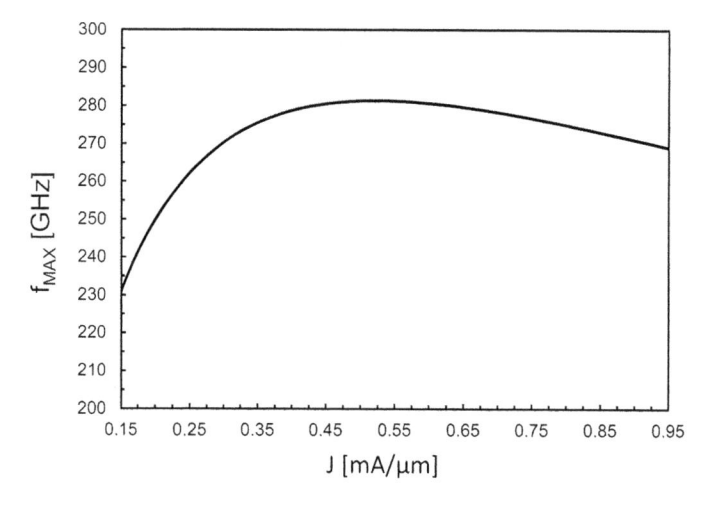

Fig. 3.11 Plot of f_{max} vs current density J_D for a 65-nm nMOS transitor.

device. In this case, a reduced bias level could generate better linear behavior. Finally, thermal effects must be taken into account as they can greatly reduce f_{max} (by about 20%–30%).

3.4 Nonlinearities

To obtain the best performance in terms of output power and efficiency, a PA designer has to manage wide signals that typically overcome the device dynamic capability. In such a regime, the PA designer is interested in large-signal parameters, such as power gain, saturated power and maximum linear power, drain efficiency, and linearity. Typical parameters to evaluate the linearity of a PA are the intermodulation distortion (IMD) and the adjacent channel power, used when two-tone and multi-tone analysis is performed, respectively, while the error vector magnitude is exploited when modulated signals are applied. When large signals are applied to the device, the small-signal approximation fails since such devices start to show wide deviation from classical behavior predictions.

(1) Transconductance g_m is not constant and its variation greatly affects the device linearity.
(2) Output conductance is not constant and it can contribute to AM–AM distortion.
(3) Gate capacitance is considerably nonlinear with respect to the gate voltage, thus affecting the device linearity (since it controls the voltage-to-current behavior).
(4) Junction capacitances are functions of the voltage swing applied to the related junction.
(5) Feedback components (mainly C_{gd} and source degeneration L_s) contribute to second-harmonic remix with the fundamental at the gate, generating third-order nonlinearities.
(6) Large signals can also induce device failure due to oxide breakdown and/or hot electrons or at least a degradation of performance over time.

In this section the above phenomena will be briefly discussed.

3.4.1 g_m nonlinearity

The transconductance models the dependence of drain current with respect to the gate voltage at a drain current bias. As discussed above, the I_{DS}–V_{GS} relation shows an exponential behavior for low V_{GS} values, then it turns into a small quadratic region and becomes linear at very high V_{GS} values. In Fig. 3.12 the g_m vs V_{GS} relation is reported together with its higher-order derivative components, as discussed in [17].

A classical technique to evaluate device linearity is based on the exploitation of a two-tone signal at frequencies ω_1 and ω_2, closely spaced above and below the carrier frequency. Nonlinearities will generate mixing products of the two-tone signal, referred to as intermodulation products, IMD. The third-order intermodulation tones, at $2\omega_1 - \omega_2$ and $2\omega_2 - \omega_1$, produce the third-order intermodulation product, IM3. This parameter is exploited to evaluate the device nonlinearity by means of current-to-voltage dependence

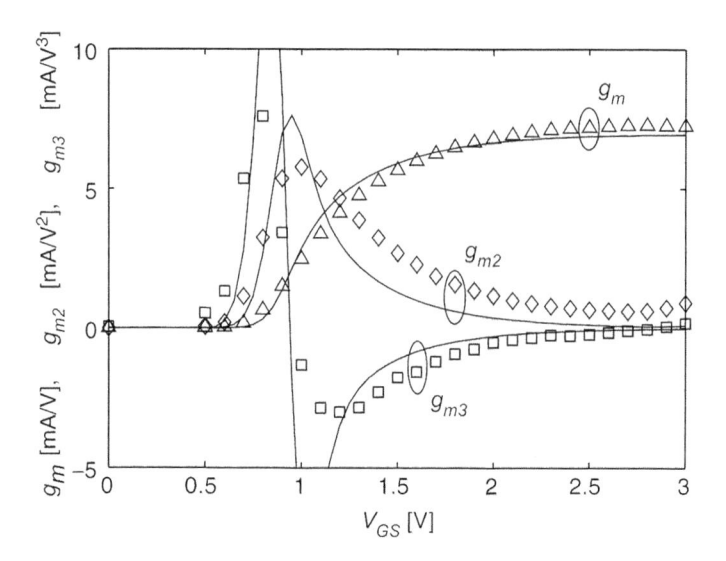

Fig. 3.12 Small-signal g_m and derivative components ©[2004] IEEE. Reprinted, with permission, from C. Fager et al., "A comprehensive analysis of IMD behavior in RF CMOS power amplifiers," *IEEE J. Solid-State Circuits*, vol. **39**, pp. 24–34, Jan. 2004.

expansion using the Taylor series (in the small-signal regime) or Volterra series (in the large-signal regime, below compression) or with "describing functions" [18] (around compression and above).

Using Taylor series [19] with only the dominant terms, it is possible to express the drain current as:

$$i_{DS}(v_{GS}, v_{DS}) = i_{DS}(V_{GS}, V_{DS}) + \frac{\delta i_{DS}}{\delta v_{GS}} v_{gs} + \frac{\delta i_{DS}}{\delta v_{DS}} v_{ds} + \frac{1}{2} \frac{\delta^2 i_{ds}}{\delta v_{GS}^2} v_{gs}^2 + \frac{1}{6} \frac{\delta^3 i_{DS}}{\delta v_{GS}^3} v_{GS}^3$$

$$= i_{DS}(V_{GS}, V_{DS}) + g_m v_{gs} + g_o v_{ds} + g_{m2} v_{gs}^2 + g_{m3} v_{gs}^3. \quad (3.17)$$

Even operating with small signals, the second and third derivatives of g_m have an important role in the device linearity. Moreover, although g_{m2} is always positive, as reported in Fig. 3.12, g_{m3} has a null in the exponential-to-saturated transition region. When large-signal behavior is taken into account, more realistic curves are obtained, as reported in Fig. 3.13. For large-signal operation, the main difference can be noted at high-input voltages where the g_m collapses and a second g_{m3} null emerges. The latter is usually exploited for the so-called IM3 sweet spot. This is a notch in the IM3 behavior close to the compression point. It is related to the bias-current level and harmonic termination impedances, as well. Approximating g_m with a piecewise linear function, g_{m3} can be discretized by a series of pulses with amplitude K_i at voltages V_i, as sketched in Fig. 3.14. Exploiting the transfer function representation [19], the third-order I_{ds} contribution due to a sinusoidal signal with amplitude A can be evaluated as follows [17]:

$$I_{ds,3} = \frac{2}{15\pi} \text{Re} \sum_{i=1}^{N} \frac{k_i (A^2 - (V_i - V_{GS})^2)^{5/2}}{A^3}. \quad (3.18)$$

3.4 Nonlinearities

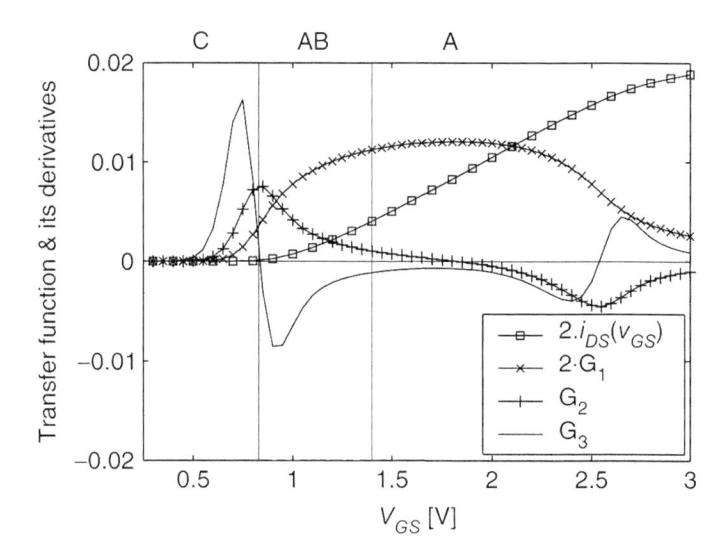

Fig. 3.13 Large-signal transconductance components ©[2004] IEEE. Reprinted, with permission, from C. Fager et al., "A comprehensive analysis of IMD behavior in RF CMOS power amplifiers," *IEEE J. Solid-State Circuits*, vol. **39**, pp. 24–34, Jan. 2004.

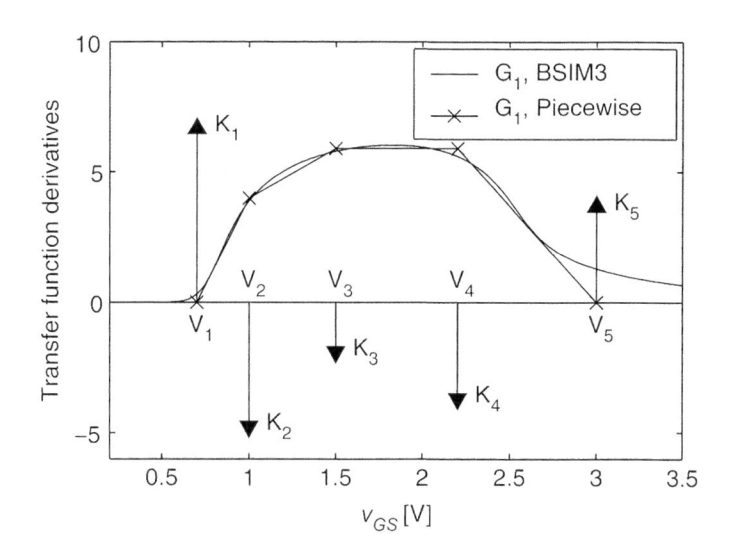

Fig. 3.14 Piecewise g_m approximation ©[2004] IEEE. Reprinted, with permission, from C. Fager et al., "A comprehensive analysis of IMD behavior in RF CMOS power amplifiers," *IEEE J. Solid-State Circuits*, vol. **39**, pp. 24–34, Jan. 2004.

This expression gives an accurate estimation of the device nonlinearity. Indeed, for small-input signals (i.e. for low v_{gs} values) no g_{m3} pulses will be considered and the IMD is zero. On increasing the signal amplitude, the contributions to $I_{ds,3}$ rise and the shape of IMD is a function of the PA class of operation [17] and baseband/second-harmonic impedance terminations, as well [20]. Analytical formulations of IM3 in dBc (assuming zero baseband drain termination $Z_d(\Delta\omega) = 0$) and of the third-order output voltage, V_{d3}, at the intermodulation frequencies as functions of termination impedances are provided for FET circuits in [20] and [21], respectively:

$$\text{IM3} = 20 \log \left| V_s^2 \frac{[c_1 + c_2 Z_0(2\omega_c)]}{G_m} \right| \quad (3.19)$$

$$V_{d3} = V_s^3 Z_0(\omega_c) r^2 r^* [c_0 Z_0(\Delta\omega) + c_1 + c_2 Z_0(2\omega_c)]. \quad (3.20)$$

Here r (a function of the gate impedance), c_0, c_1, and c_2 are defined in [21], V_s is the tone amplitude, ω_c is the fundamental tone frequency (under the assumption $\omega_1 \cong \omega_2 = \omega_c$), Z_0 is the drain conductance in parallel with the drain termination

$$Z_0(\omega) = \frac{Z_d(\omega)}{1 + G_d Z_d(\omega)}, \quad (3.21)$$

and G_m and G_d are the first-order transconductance and drain conductance, respectively.

The above relations not only show the baseband and second-harmonic dependence of the intermodulation, but also demonstrate that in power amplifiers the usual technique of setting the baseband impedance to zero is not always optimal: indeed, a resistor in series with the drain inductor can improve IM3. However, even if the gain of the device does not change, this technique has two drawbacks: the baseband impedance will vary with both tone spacing and dynamic swing; and efficiency loss can greatly impact the device performance, especially in a power amplifier. If second-harmonic tuning is feasible, it could be more effective at reducing the baseband impedance and optimizing the second-harmonic termination in order to minimize IM3. This will not affect efficiency since, with c_1 being generally negative and c_2 positive, the second-harmonic termination is required to be inductive, fitting the typical high-efficiency termination needs. The expression (3.18) also allows the introduction of an easy technique for g_m linearization, namely derivative superposition [22]. Referring to Fig. 3.14, the $I_{ds,3}$ value is a function of the g_{m3} Dirac pulses. For very large signals (i.e. for high v_{gs} values), as usual in PAs, all five Dirac pulses have effect. Pulses at V_1 and V_4 are always contrary in sign. This property can be properly exploited by means of the simple scheme shown in Fig. 3.15a, with the aim of reducing g_{m3} in a wide V_{gs} region, as depicted in Fig. 3.15b. Indeed, two currents are generated by similar devices (M_A and M_B), driven by two equal input signals properly offset-biased with optimum V_{Off}. This technique and the effect on g_m shape were introduced by Webster [22, 23] and analytically discussed in [20, 24]. A heuristic explanation can be provided by means of the diagram in Fig. 3.16. Here, the original g_m shape of the device B is shown in light gray while the shifted g_m of the auxiliary device A is depicted in dark gray. It is worth noting that the dark gray g_m curve is also compressed, due to the reduced gate bias and to the adoption of similar device area. The pulse V_{1_a} is shifted in order to correspond to V_{4_b}. The current combination will be equivalent to an extension of the flat g_m region and it is depicted in black. This technique is able to improve the device linearity, pushing the compression point very close to the power saturation level, especially if the imaginary component of fundamental drain impedance is negligible [20]. Moreover, the intermodulation distortions

3.4 Nonlinearities

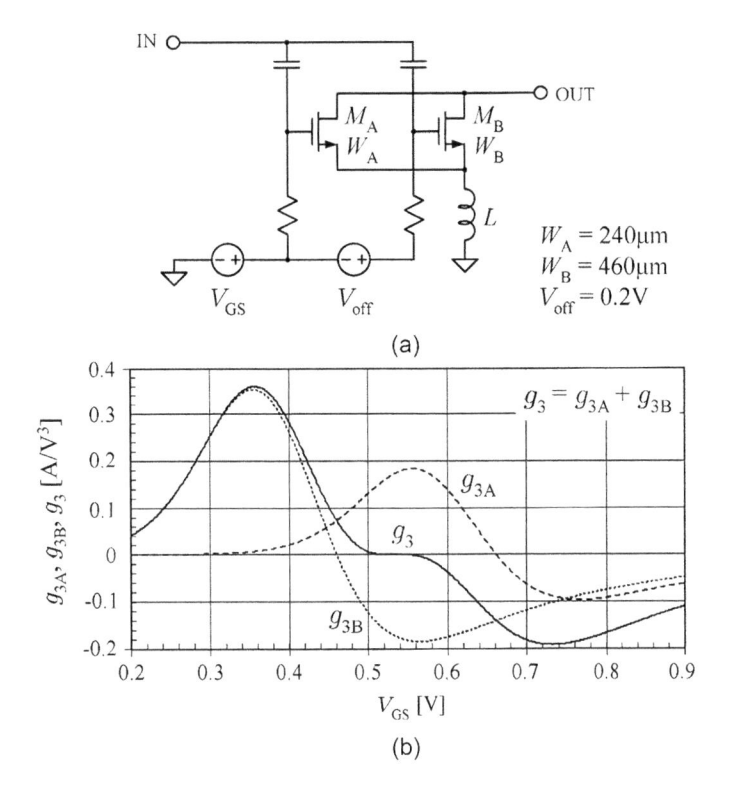

Fig. 3.15 Offset-based linearization techniques: (a) schematic and (b) effect on transconductance components ©[2005] IEEE. Reprinted, with permission, from V. Aparin and L. E. Larson, "Modified derivative superposition method for linearizing FET low-noise amplifier," *IEEE Trans. Microwave Theory Tech.*, vol. 53, pp. 571–581, Feb. 2005.

can be entirely associated with the $g_{m,3}$ component if the source degeneration is zero or very low. When the effect of source degeneration, L_S, is included, the contribution of second-order nonlinearity has to be considered in the intermodulation evaluation. For the weakly nonlinear case, an easy expression for the third-order intercept point (IIP3) is reported in [24]:

$$\text{IIP}_3 = \frac{4g_m^2 \omega^2 L_S C_{gs}}{3 |\epsilon|} \tag{3.22a}$$

$$\epsilon = g_{m3} - \frac{2g_{m2}^2}{3\left(g_m + \frac{1}{j2\omega L_S} + j2\omega C_{gs} + Z_1(2\omega)\frac{C_{gs}}{L_S}\right)}, \tag{3.22b}$$

where Z_1 is the gate termination. In order to increase the IIP3, ϵ has to be minimized with respect to the second-order contribution and then $Z_1(2\omega)$ must be increased. As a consequence, the drain-to-gate feedback due to C_{gd} (not included in (3.22)) starts to contribute, as well. Therefore, an optimum design of both gate and drain terminations is needed to obtain the minimum distortion corresponding to the bias level. Indeed, if

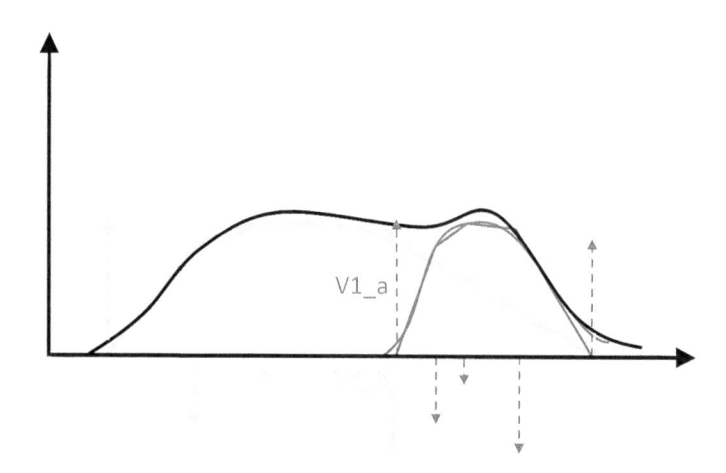

Fig. 3.16 Plot of the g_m superposition effect of offset-biased linearized CMOS PA.

the source degeneration inductance can be a free design parameter, it will be possible to modify the derivative superposition method by dividing the inductance into two series inductors and connecting the auxiliary device source only to the second section of the degeneration inductance. The idea is to use the second-order nonlinearity to reduce the third-order one by managing the phase of the above contributions. By exploiting this technique, a 20-dB IIP3 improvement has been already demonstrated in [24].

3.4.2 IMD asymmetry

The intermodulation distortion is usually analyzed by means of third-order upper, $(2\omega_2 - \omega_1)$, or lower, $(2\omega_1 - \omega_2)$, IMD tones, assuming $\omega_2 > \omega_1$. In the above discussions, these tones have been considered equal, but in actual amplifiers IMD-tone asymmetry is often observed. A first analytic analysis of this effect has been reported in [25, 26], demonstrating that the IMD asymmetry is mainly caused by the reactive part of the baseband termination impedances and its interaction with the second-harmonic termination impedances. More comprehensive studies can be found in [20, 27, 28].

To better understand this phenomenon, it has to be considered that each IMD tone is the result of the vector combination of several distortion contributions corresponding to different distortion orders, whose mixing produces effects at the specific IMD tone. Even considering only the distortion terms up to the third order, the second-order tones can easily find a feedback path (i.e. by means of C_{dg} and L_s) and then generate IMD-tone components due to the mixing with the fundamental tones. A similar consideration also applies to the baseband ($\Delta\omega = \omega_2 - \omega_1$) contribution. Owing to the typical pass-band behavior of matching networks in high-frequency amplifiers, these mixing effects result in a small asymmetry, due to different filtering suffered by the above components. A simplified Volterra series analysis allows one to express the upper and lower IMD tone for weak nonlinearities as follows [25, 26]:

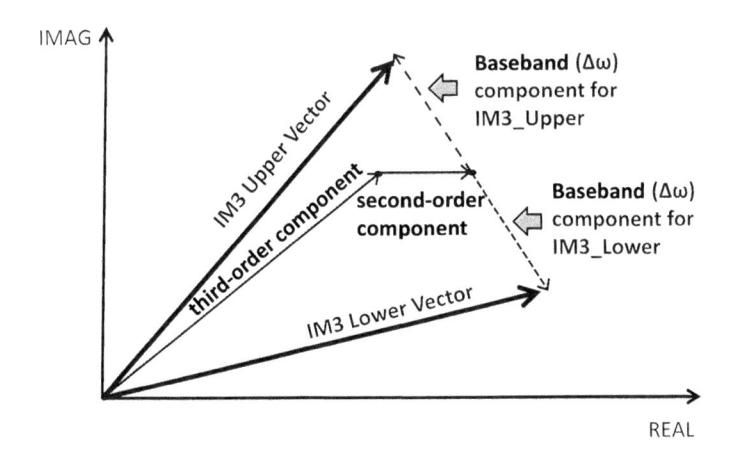

Fig. 3.17 IMD asymmetry: representation with graphical vector addiction.

$$V_{IM3}^{Upper} = \frac{1}{2}\Re\left\{\begin{array}{l} H_2(\omega_2 - \omega_1)H_1(\omega_2)\exp[j(\omega_2 - \omega_1)]\exp(j\omega_2) + \\ H_2(2\omega_2)H_1(-\omega_1)\exp(2j\omega_2)\exp(-j\omega_1) + \\ H_3(2\omega_2 - \omega_1)\exp[j(2\omega_2 - \omega_1)] \end{array}\right\} \quad (3.23a)$$

$$V_{IM3}^{Lower} = \frac{1}{2}\Re\left\{\begin{array}{l} H_2^*(\omega_2 - \omega_1)H_1(\omega_1)\exp[j(\omega_1 - \omega_2)]\exp(j\omega_1) + \\ H_2(2\omega_1)H_1(-\omega_2)\exp(2j\omega_1)\exp(-j\omega_2) + \\ H_3(2\omega_1 - \omega_2)\exp[j(2\omega_1 - \omega_2)] \end{array}\right\}, \quad (3.23b)$$

where H_i refer to the first-order, second-order, and third-order nonlinear transfer functions ($i = 1, 2, 3$). It is worth noting that the relation $H(-f) = H^*(f)$ has been exploited in the lower tone. From the above relations it is evident that the baseband contribution is 180° out of phase in the lower IMD tone with respect to the upper IMD tone. A very intuitive graphical representation of such vector magnitude additions is shown in Fig. 3.17.

Although the analytical study in [28] refers only to weak nonlinearities, it confirms the relevance of the termination impedances at $\Delta\omega$ and at the second harmonics, as well as the effect of feedback paths (i.e. the source inductance or the gate–drain capacitance) that must be minimized. In [27], the IMD asymmetry in large-signal conditions has been analyzed, drawing the IMD3 asymptotic saturation at 9.4 dB below the fundamental signals. Moreover, it has been demonstrated that large-signal IMD asymmetries may take place only where the dominant real part of IMD components is canceled out. This corresponds to the IMD sweet spots region, provided that baseband and second-order harmonic terminations are reactive.

3.4.3 Other nonlinearities

(a) C_{gs} **nonlinearity** Gate–source capacitance nonlinearity has already been introduced. An easy way to balance its variation is to include a complementary gate–source capacitance able to compensate the nonlinearity. To this end, a p-MOS device can be exploited. This technique has the disadvantage of requiring large-area devices [29].

(b) g_o **nonlinearity** The output conductance is subject to variations associated with voltage, current, and power level in the channel. Generally, the effect associated with these variations is AM/AM distortion.

(c) **Large-signal parasitic capacitance dependence** In the bulk CMOS, all the parasitic capacitances are junction capacitances and hence they are dependent on the applied voltage. The variations of the output capacitance, especially, produce phase distortion. This issue is one of the reasons why CMOS SOI is greatly preferred at very high frequency. Indeed, it allows lower parasitic capacitances that are almost constant with the applied voltages, especially if trap-rich SOI substrates are exploited, as will be discussed later.

(d) **Thermal memory effect description** The thermal time constant is in the range of the baseband modulation and it can induce nonlinearities and dispersion effects, as discussed later. The result is a non-constant linear behavior versus the two-tone spacing frequency. This dependence is hampered by bias-induced memory that provides a baseband-dependent impedance.

3.5 Noise

In electronic devices, noise determines the signal lower limit that can be processed without losses in signal quality [30]. Since power amplifiers are usually operated in non-linear mode, the active devices spread both noise and low-frequency contributions due to bias and conditioning circuitry and up-convert them around higher signal harmonics. At the system level, the effect is an increased difficulty in transmitting and then receiving without interfering with nearby channels and/or other communication systems.

Electronic devices suffer mainly from four types of noise (the related models are summarized in Table 3.3).

(1) Shot noise [18, 31–33]: this is associated to the conduction in a p-n junction (as in a diode); the current is not a continuous flow of electrons but the combination of successive casual electron jumps between the potential barrier of the junction. This phenomenon produces a small fluctuation of the current around its average value. A white noise for frequency lower than $1/\tau$ is generally assumed, τ being the carrier transit time in the depletion region (generally very short).

(2) Thermal noise [18, 31, 32]: this is associated with the thermal random movements of the electrons. It is a white noise and is proportional to the absolute temperature. It is present in any resistor.

(3) Flicker noise [34–36]: this is associated with traps and crystal defects that act as centers of carrier generation and recombination. The spectral density is proportional to $1/f$ and it is classically assumed to be a low-frequency noise.

(4) Burst noise [35]: this is associated with metal-ion contamination and it is proportional to $1/f^2$.

3.5.1 Noise in MOS transistor

The main noise source in an MOS device comes from the resistive channel and therefore it has thermal-noise behavior. The flicker noise is also associated with the current in the

3.5 Noise

Table 3.3 Noise contributions.

Noise type	Noise model	Notes
Shot	$\overline{i^2} = 2qI_D\Delta f$	$q = 1.6 \cdot e^{-19}$
Thermal	$\overline{v^2} = 4KTR\Delta f$	
	$\overline{i^2} = \frac{4KT}{R}\Delta f$	$4KT = 1.66 \cdot e^{-20}$
Flicker	$\overline{i^2} = k_1 \frac{I^a}{f} \Delta f$	K_1 is a device constant $a \in [0.5, 2]$
Burst	$\overline{i^2} = k_2 \frac{I^c}{1+\left(\frac{f}{f_c}\right)^2} \Delta f$	f_c is the noise characteristic frequency K_2 is a device constant $c \in [0.5, 2]$

channel, and globally these contributions are generally represented by a current-noise generator placed between drain and source terminals:

$$\overline{i_d^2} = 4KT\left(\frac{2}{3}g_m\right)\Delta f + k_1 \frac{I_D^a}{f}\Delta f \simeq 4KTg_m P\Delta f, \tag{3.24}$$

where P is a proportionality constant.

Although no DC current flows through the gate, it is characterized by channel induced thermal noise due to the capacitive coupling with the channel. Using R as a proportionality constant, it is represented by the following noise-current generator:

$$\overline{i_g^2} = 4KT\omega^2 \frac{(C_{gs} + C_{gd})^2}{g_m} R\Delta f. \tag{3.25}$$

Moreover, a shot noise is injected in the gate due to random electrons crossing the oxide that are associated with the gate leakage current, I_G. It is represented by a noise generator between drain and gate terminals:

$$\overline{i_g^2} = 2qI_G\Delta f. \tag{3.26}$$

Finally, any physical resistive parasitic component will increase the overall device noise due to thermal-noise contributions. Among others, the gate resistance r_g (due to its not-negligible value, especially in advanced technology nodes) and the source resistance r_s (due to the out-of-phase feedback effect) have to be considered by taking into account the following contributions:

$$\overline{v_g^2} = 4KTr_g\Delta f, \quad \overline{v_s^2} = 4KTr_s\Delta f. \tag{3.27}$$

Minimization of r_g can be accomplished by using multi-gate finger devices, while r_s can be greatly reduced thanks to proper arrangements of contacts and metal stack. Both actions will reduce the external-noise contributions, thus also reducing the total device noise.

The above noise modeling is known as the PRC noise model and it was introduced for MESFETs [37, 38] and HEMTs [39] by modifying the basic concepts introduced by Van der Ziel [40, 41]. For the sake of clarity, C is a third parameter and represents the correlation coefficient linking i_g and i_d [36]. The PRC noise model also works for

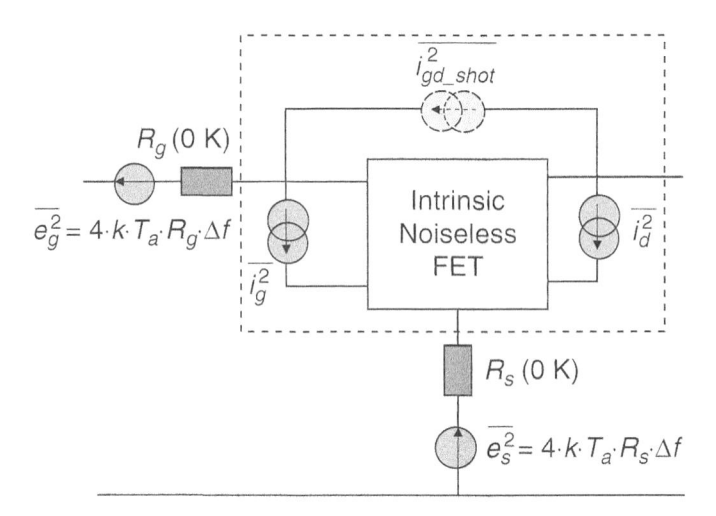

Fig. 3.18 Complete-noise FET model ©[2010] IEEE. Reprinted, with permission, from F. Danneville, "Microwave noise and FET devices," *IEEE Microwave Magazine*, vol. 11, pp. 53–60, Oct. 2010.

MOS devices on taking into consideration that the intrinsic FET behavior is greatly reduced. The induced gate-noise power, being correlated with the drain current through the parameter C, is partially subtracted from the drain-noise power [39]. Indeed, the weak-coupling C factor in an ultra-scaled MOS reduces this benefit.

A complete small-signal model with the above-discussed noise sources is depicted in Fig. 3.18. With easy manipulations and some simplification [30], it is possible to represent the MOS device as an ideal noiseless equivalent circuit with the addition of voltage and current equivalent input-noise generators, both having an inverse dependence on g_m:

$$\overline{v_i^2} = 4KT \frac{2}{3g_m}\Delta f + k_1 \frac{I_D^a}{f g_m^2}\Delta f \tag{3.28}$$

$$\overline{i_i^2} = 2qI_G\Delta f + \frac{\omega^2 C_{gs}^2}{g_m^2}\left(4KT\frac{2}{3}g_m + k_2\frac{I_D^a}{f}\right)\Delta f. \tag{3.29}$$

From the above relations, limited to the small-signal behavior, it is evident that

- high current levels (high g_m) give the benefit of a low-voltage noise effect;
- gain compression (g_m lowering) causes a rise in the noise contributions.

The main reason why we are interested in noise for power amplifiers is the effect that is produced in the adjacent TX channels (i.e. close to the channel exploited for transmission) or in the received bands. Indeed, noise is strictly related to the nonlinear behavior of a PA that at high-power level works more like a mixer than a small-signal amplifier, thus generating noise frequency translation of low-frequency noise around the amplified carrier.

From this point of view the above observations on noise should be totally reversed. Indeed, the "small-signal noise" is quite low, while, on increasing the input power, the PA starts to mix its bias noise, thus starting to generate relevant noise contributions all around the carrier (and its harmonics). Even more interestingly, on approaching the compression the noise does not rise. On the contrary, noise is reduced since the nonlinear gain of the PA is lower. In this scenario, baseband bias noise starts to be more and more relevant.

3.5.2 Discussion on low-frequency noise

Understanding low-frequency and $1/f$ noise is also very important for mm-wave devices, and especially for power amplifiers, since the baseband noise is generally up-converted by the nonlinear operation of the active device, as shown in Fig. 3.19.

Low-frequency noise is produced by the flowing of current in the device. Although it is commonly accepted that one can relate noise sources to the DC current value, some noise sources show a dependence on instantaneous current values, rather than on the average current value, as described in [42]. In this case, such noise sources are referred to as cyclo-stationary noise sources [43]. One of these cyclo-stationary noise sources is related to the activity of generation/recombination (GR) centers that are an important cause of low-frequency noise. Here, electrons are continuously captured and released after a certain time, thus generating current fluctuations. The macroscopic effect produced by the GR centers is a low-frequency noise with a corner frequency that is a function of the decay time. When the number of free electrons and then the current level increase, the trapping–detrapping phenomena will increase as well, producing a higher level of low-frequency noise. At high frequency (i.e. above the corner frequency) dependence on the instantaneous current should not happen. However, the trapping–detrapping phenomenon itself is inherently an instantaneous event that is strictly related to the instantaneously available free electrons. The trapping action time will depend on the specific trap nature, thus giving a specific time constant to the event. As a consequence, this phenomenon, although it appears as a low-frequency noise, is strictly related to single events following the instantaneous current values. This means that the quantity of channel electrons is controlled by the instantaneous current values, rather than the average value. For this reason, GR-induced noise must be treated as a low-frequency instantaneous-driven source.

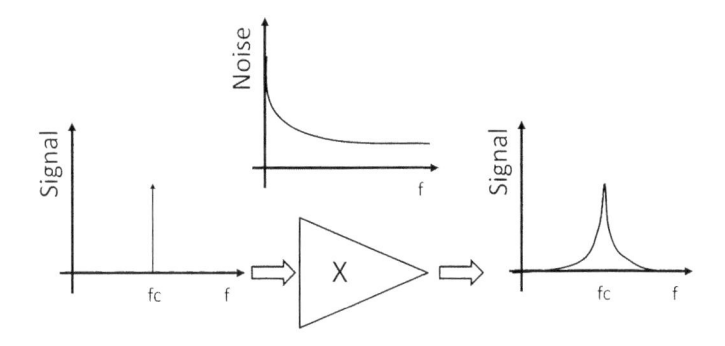

Fig. 3.19 Baseband noise up-conversion.

Such modeling is generally not included in classical device models and therefore designers should be aware that, in nonlinear operation, measurements will show noise not only at the baseband but also as sidebands at signal harmonics. Accurate noise modeling, which has to be developed on the basis of extensive nonlinear measurements, is required to allow noise-optimized design to be achieved.

Finally, all the above considerations should drive the designer to accurately analyze all the noise sources, putting special emphasis on low-frequency/baseband noise coming from the bias networks. The classical addition of capacitors or resonators in the bias networks could help in reducing noise at the cost of possible detrimental effects on the linearity performance. Indeed, the baseband impedance should be constant in a wide frequency range, but low-frequency filtering action could generate large-phase distortions, thus impacting the overall device linearity. These phenomena are generally known as memory effects.

3.6 Thermal effect

Any device for power amplification has to manage wide voltage/current swings and hence high power levels, which implies large thermal variations compared with room temperature. Almost all physical parameters are affected by temperature variations: threshold voltage, breakdown voltage, mobility saturation velocity, oxide thickness (due to thermal expansion), bandgap, etc. The dependence on temperature is a second-order effect for several parameters (e.g. oxide thickness and bandgap), but many other parameters are very sensitive to thermal issues. For instance, the threshold voltage reduces with increasing temperature and modifies the device bias point, thus compelling the adoption of customized bias circuitry to keep the bias point over the whole operative temperature range. Again, the electron velocity in the channel is reduced when the temperature increases, due to a combined effect on mobility and saturated velocity reduction, thus producing a compression of drain current at large gate voltages and high temperature. The latter effect is demonstrated by the wide differences between DC and pulsed measurements [44], as shown in Fig. 3.20. Indeed, an easy way to experiment and model thermal effects is to adopt pulsed measurements. More specifically, drain and gate voltages and HF signal are synchronized to apply the drain voltage first, then the gate and at the end the HF signal as reported in Fig. 3.21. Some margin for non-overlapping operation is exploited in order to take into account the delay introduced by bias-T or bias network adopted for on-wafer or on-PCB measurements, respectively.

With reference again to Fig. 3.20, it is worth noting that the combined effect of threshold voltage and electron velocity variations due to temperature increments produces an overall useful thermal effect. Indeed, at low gate voltage the V_T-dependence dominates, thus producing higher overdrive voltage, while at high gate voltage the velocity limitation is dominant and the drain current will be reduced, avoiding thermal run-away. In any case, for the minimization of thermal effects on small-signal parameters, a current-mode bias network is generally preferred since it allows setting of the current in the power device. Of course, special care must be taken in the layout positioning of bias

3.6 Thermal effect

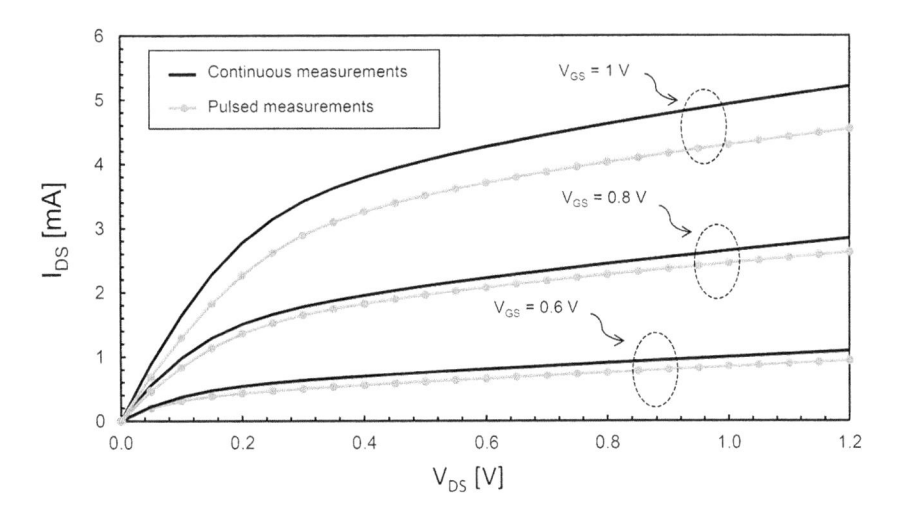

Fig. 3.20 Continuous and pulsed output I–V behavior.

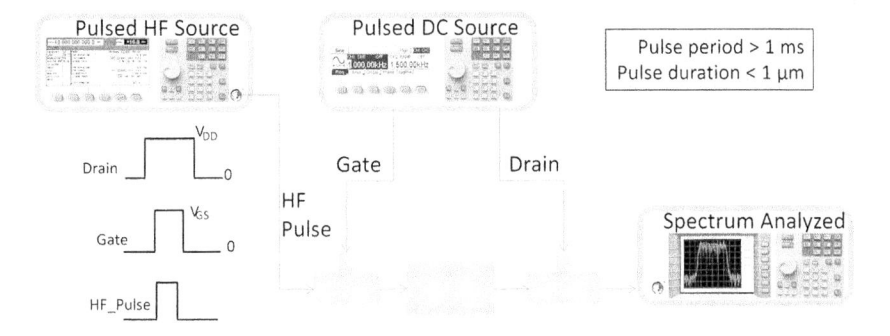

Fig. 3.21 Pulsed test bench and signal synchronization.

devices in order to assure almost the same temperature of the power device. The basic technique is illustrated in Fig. 3.22, where a current mirror is implemented by interposing the bias transistor Q_1 to Q_{2a} and Q_{2b}, where $Q_{2a} + Q_{2b} = Q_2$, with Q_2 the power transistor.

Another important thermal effect is related to the R_{on} increment with temperature, which considerably degrades efficiency in a similar way to an external parasitic resistance. Finally, the self-heating effect reduces the mean time before failure and increases the possibility of burning the device. On the other hand, high temperatures can assure higher breakdown voltage, which is achieved due to the decreasing of the impact ionization rate.

All thermal effects should be included in a model to take into account the PA thermal budget. In particular, some advanced models include a thermal terminal that can be exploited for the connection of a proper thermal network. Generally, a thermal network includes thermal resistances and capacitances that are functions of geometrical parameters and material characteristics, as shown in Fig. 3.23. Such networks allow the

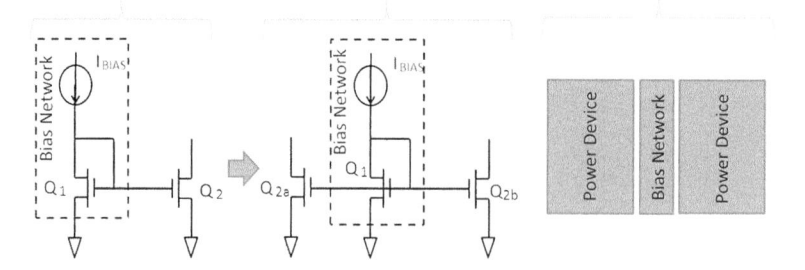

Fig. 3.22 Thermal enhanced layout architecture.

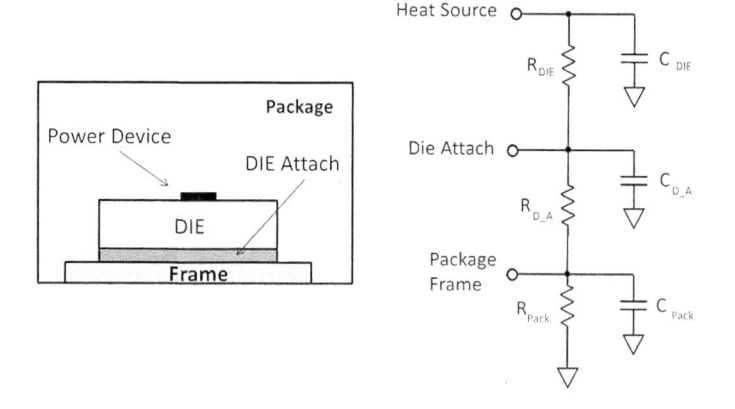

Fig. 3.23 Thermal model.

simulation of the thermal effects due to the silicon substrate, the die attach, and the package. In some cases (e.g. small PCB, high working temperature, etc.), it could be required to include also the thermal effects of PCB and/or eventual heat sinks.

It is well known that, at high frequency, it is a common practice to exploit flip-chip devices to reduce the parasitic effects of the bond wires. In this case, the thermal-flow mechanism could be very different from the classical one for bulk-soldered devices. Indeed, the electrical connections toward external components (i.e. the bumps) become the primary thermal-flow path. This is due to the relatively high thermal resistance of the molding compound surrounding the flip-chip device (if a package is used). In this case, the thermal resistance associated with electrical drain/source contacts, metallization, vias, and bumps will set the thermal behavior of the device. On the other hand, if no package is exploited, the substrate will contribute to the dissipation. In this case it is not trivial to build a thermal model. Typically, specific CAD simulations need to be set to evaluate the operative temperature of the device as a function of both room temperature and the operating-power level. Thermal effects, due to typical time constants, are long-term memory effects that fall within the bandwidth of common modulated signals. The result is a variable response of the transistor based on the signal history. In Figs. 3.24a, b, the variation of the output conductance and transconductance, g_o and g_m, due to temperature is highlighted by the different behaviors of continuous and pulsed

3.7 Large-signal performance degradation and reliability

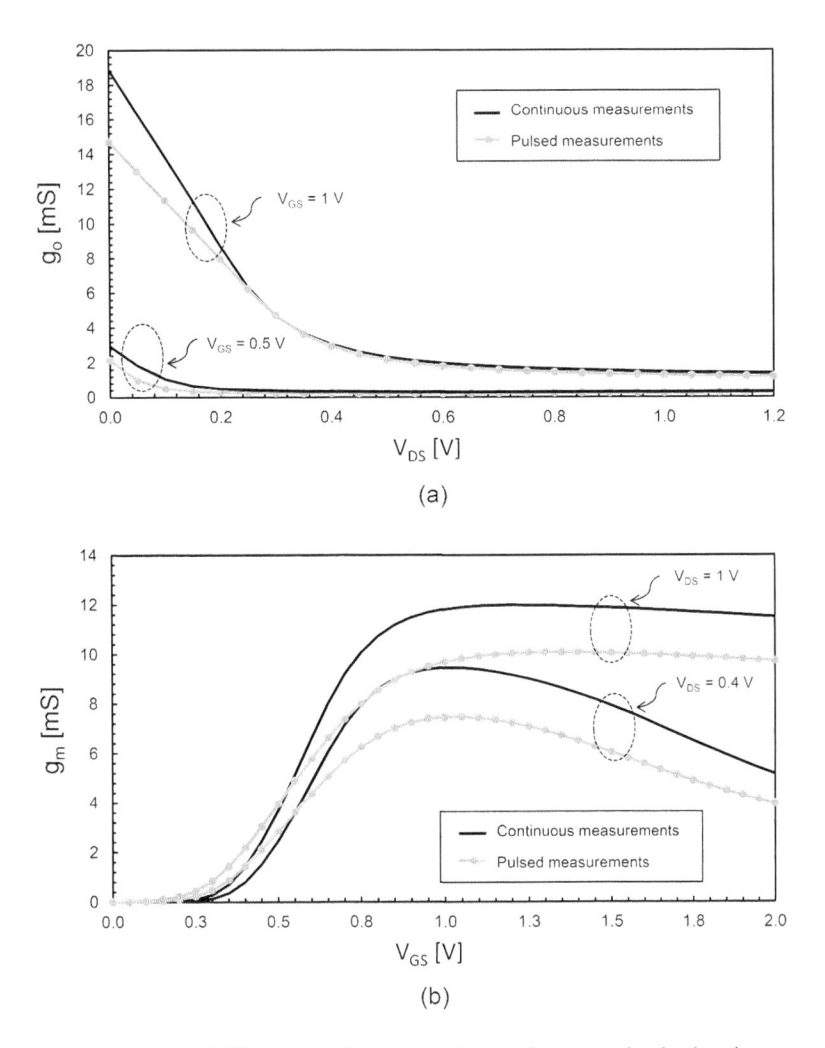

Fig. 3.24 (a) Output conductance and (b) transconductance under continuous and pulsed regimes.

curves. Figures. 3.25a, b show the AM–AM and AM–PM responses to a modulated signal due to thermal effects. In this case, and mainly for small-power signals, the gain and the phase of the output signal are greatly affected by the signal history (i.e. the previous instantaneous power level that is statistically different due to the modulation), giving rise to a wide dispersion of gain and phase at the same input power, while at high power the device saturation reduces this effect.

3.7 Large-signal performance degradation and reliability

The wide signal ranging of power amplifiers is related to the delivered output power and to the wave shaping needed for optimum efficiency [13]. Moreover, if the load impedance changes, mismatch conditions could determine a wider signal swing even

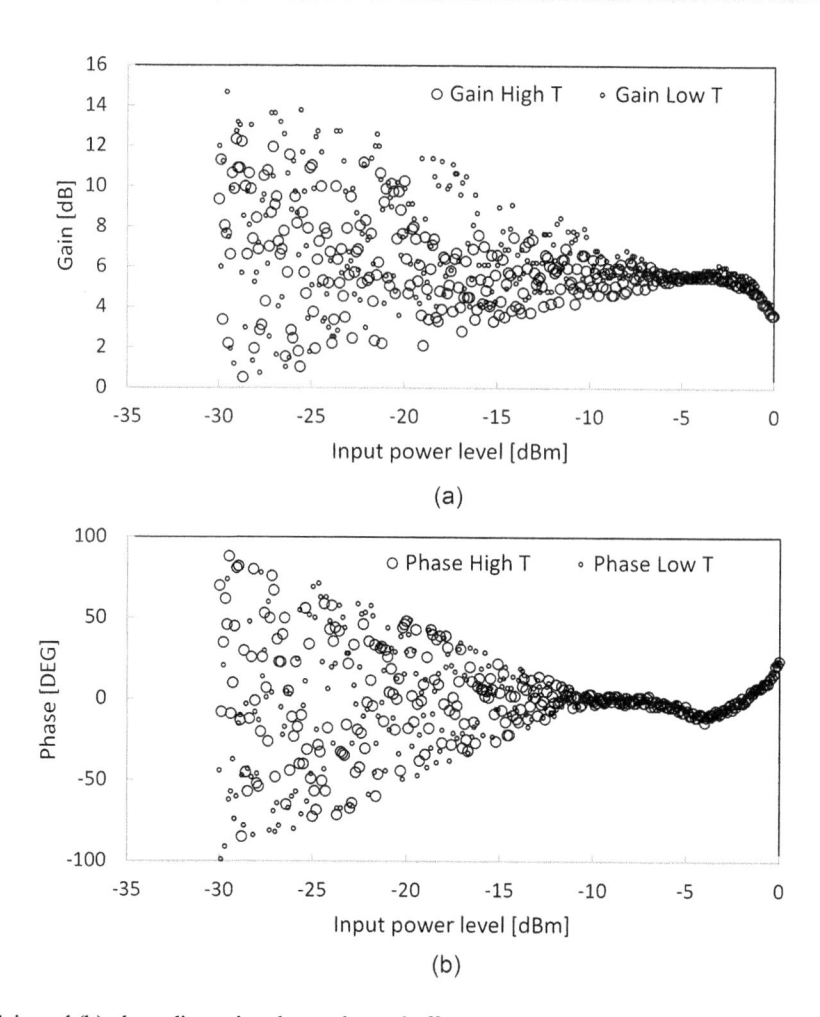

Fig. 3.25 (a) Gain and (b) phase dispersion due to thermal effects.

over the allowed device dynamic. In Fig. 3.26, output shaped V–I curves of a high-efficiency PA are reported in matched and mismatched conditions at 2.5-V supply voltage. Operation at such wide amplitudes impacts the device reliability, inducing a premature degradation of transistor parameters. Several studies have been done on such degradation effects [45–49]. The transistor is usually stressed with DC [46, 48, 49] or high-frequency inverter-like [47] voltage waveforms (i.e. square-like waveforms). However, the stress conditions, as shown in Fig. 3.26, are definitely not DC and are far from being inverter-like in shape. It has been reported that, in such severe stress conditions, the output power decreases with time [50]. This has been ascribed to channel hot carrier (HC) stress [50] or Fowler–Nordheim (F-N) gate oxide wearout [48].

In the following a study [51] comparing the classical F-N and HC stress effects based on a quasi-static model and experimental stress under optimal and suboptimal load conditions is discussed.

3.7 Large-signal performance degradation and reliability

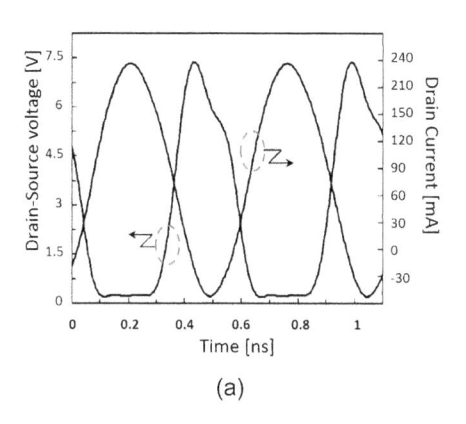

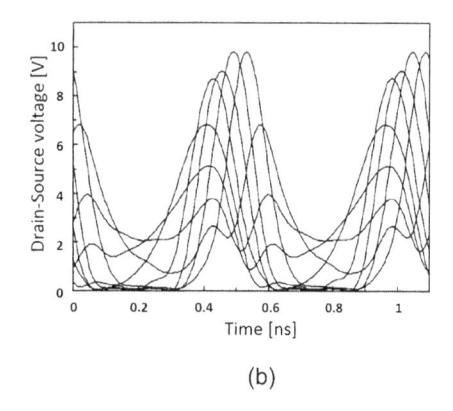

Fig. 3.26 (a) Matched and (b) 10:1 mismatched voltage swing of a PA at 2.5 V supply voltage.

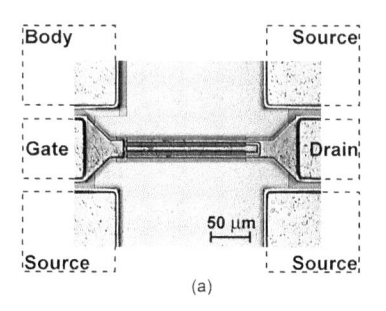

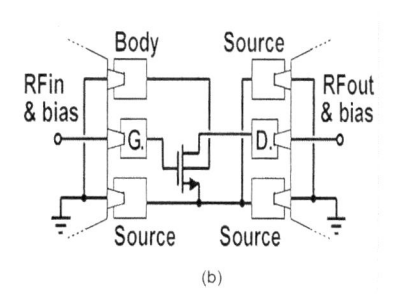

Fig. 3.27 Measurement setup of an n-MOS power transistor ©[2007] IEEE. Reprinted, with permission, from C. D. Presti, F. Carrara, A. Scuderi, S. Lombardo, and G. Palmisano, "Degradation mechanisms in CMOS power amplifiers subject to radio-frequency stress and comparison to the DC case," in *IEEE Proc. International Reliability Physics Symp. (IRPS)*, April 2007, pp. 86–92.

Generally, nanoscale CMOS devices show very low breakdown voltage. As a consequence, the high-voltage management, especially for power amplifiers, is obtained by cascoding a thin-oxide device with a thick-oxide one. Indeed, the first transistor guarantees a high enough f_T and consequently g_m at the operative frequency, while the second MOS improves the overall robustness of the PA. The classical cascode bias technique will set the drain of the thin-oxide device at an almost constant value (note that in stacked architecture a signal in phase with the output voltage will be managed also by the thin-oxide device and special care has to be taken in such cases in order to avoid stress effects or failure also in the thin-oxide device). In this case, only the thick-oxide device has to manage the high voltage swing.

For the following analysis a 5-nm-thick gate oxide device with gate finger of 5 μm has been chosen, as reported in Fig. 3.27a. It is worth noting that while the source terminal is firmly grounded, the body of the transistor is grounded only when an input ground–signal–ground (GSG) probe is lowered onto the wafer (see the schematic in Fig. 3.27b). All four transistor terminals are available when DC measurements are performed. The measurement technique is based on the application of several cyclic stress-and-sample

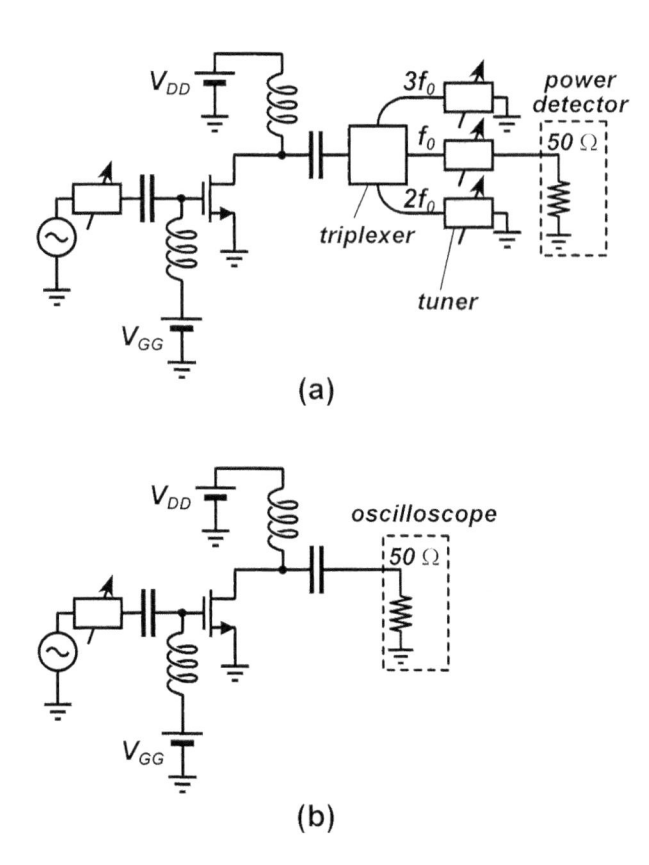

Fig. 3.28 (a) Load-pull bench for optimum load condition, (b) sub-optimal load condition ©[2007] IEEE. Reprinted, with permission, from C. D. Presti, F. Carrara, A. Scuderi, S. Lombardo, and G. Palmisano, "Degradation mechanisms in CMOS power amplifiers subject to radio-frequency stress and comparison to the DC case," in *IEEE Proc. International Reliability Physics Symp. (IRPS)*, April 2007, pp. 86–92.

measurements. The stress is applied by generating the desired waveform at the drain source terminal, while the sample measurement is based on the recording of the DC sub-threshold drain current.

The generation of a typical "high-efficiency shaped" waveform (optimum load) is accomplished by exploiting a load-pull bench, shown in Fig. 3.28a, consisting of a triplexer and a set of three harmonic tuners. The device is set to operate at the 3-dB compression point. The gate is kept constant and the drain voltage is increased while tuning the load in order to optimize the output power and the drain efficiency, η, of the PA, at each bias condition.

On the other hand, the same device is operated under sub-optimal load conditions, as shown in Fig. 3.28b, without any harmonic termination, by simply connecting the drain terminal to a 50-ohm load. In these conditions, the efficiency η is lowered.

As regards the DC sampling phase, by recording the sub-threshold drain current after each stress phase, the threshold voltage and sub-threshold slope can be extracted [7]

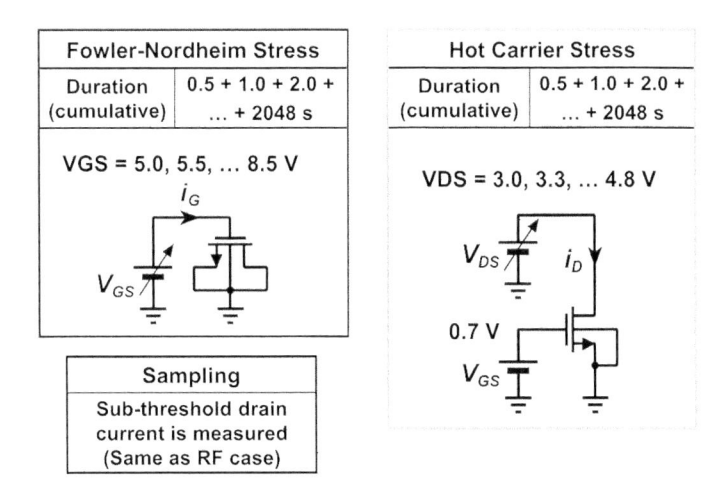

Fig. 3.29 N-F and HC measurement setup.

and used as indicators of transistor wearout. A best-fit algorithm is used to match the exponential region of I_D to the well-known sub-threshold MOS equation

$$I_D = I_{D0} \exp\left(\frac{V_{GS} - V_t}{\zeta V_{th}}\right), \qquad (3.30)$$

where ζ is the sub-threshold slope factor, V_t is the threshold voltage, V_{th} is the thermal voltage, and I_{D0} is a current level depending on geometric, physical, and electric quantities [7]. Indeed, the threshold voltage variation is correlated to the amount of trapped charge in the oxide [52], whereas the sub-threshold slope factor ζ is an indicator of the damage at the Si surface, since it increases as soon as traps are created at the Si–Si-oxide interface.

As a comparison, classic DC stress-and-sample measurements can be performed as well, by using DC probes connected to the four device terminals. The oxide characterization under F-N stress is performed by grounding drain, source, and body, and stressing the device by driving the gate. On the other hand, channel hot carrier stress requires that current is present in the channel; then a $V_{GS} = V_{GS\text{-}ON}$ must be exploited, while the drain is increased up to the device failure. Classical setups for both the stressing tests are depicted in Fig. 3.29.

As discussed in [51], the stress-induced damage is located close to the drain and can be accurately monitored either by inverting drain and source or by using a small V_{DS} (small compared with the thermal voltage V_{th}). The swapping of the terminals is a time-consuming task; therefore, the latter option is preferred for the implementation of an automatic test bench.

3.7.1 Dynamics of degradation and influence of the load

In order to investigate the mechanisms lying beneath the degradation due to signal stress, device performance can be compared under both optimal (maximum efficiency by controlling harmonics for E-class operation) and sub-optimal (50-ohm without controlling harmonics) load termination.

Under optimal load conditions, it is possible to study the supply dependence of the damage severity by increasing the supply voltage V_{DD} (in this study it was changed from 1.8 V through 2.2 V), while the load is optimized for each device and operating point. The maximum dissipated power on the device is kept constant to about 50 mW. Each device underwent a cumulative signal stress, whose duration started from 0.5 s and doubled up to 2048 s. DC measurements are performed on the fresh device and after each stress phase.

The results of these experiments under optimal load conditions are shown in Fig. 3.30. The threshold voltage variation $\Delta V_t = V_t - V_t(0)$ (where $V_t(0)$ is the device threshold voltage before stress) is plotted against total signal stress time in Fig. 3.30a. After the first 0.5-s stress, V_t often decreases by a few tens of mV (this negative variation corresponds to the missing points in the logarithmic graph), indicating the trapping of positive charge in the oxide [52]. Then V_t begins to increase, indicating the trapping of negative charge, with a log–log slope close to $\frac{1}{2}$ (dashed line in Fig. 3.30). This indicates that the charge increases with the square root of time. No indication of saturation was recorded, not even in a longer 10-h experiment, where the $\frac{1}{2}$ slope was maintained. At a fixed stress time, the accumulated V_t variation tends to increase with the supply voltage.

Analogous results can be found if the sub-threshold slope-factor variation $\Delta \xi = \xi - \xi(0)$ is analyzed, as shown in Fig. 3.30b. The same square-root dependence on time was found in the long term. However, in the first phase of the stress experiment, this quantity is more sensitive to the damage, since it is not affected by the charge compensation mechanisms which tend to hide the contemporaneous presence of trapped holes and electrons.

Now let us consider another set of devices that are loaded with the 50-ohm oscilloscope, i.e. the sub-optimal load. Similar stress conditions have been applied to the device and a dissipated power of around 80 mW has been obtained (the sub-optimal load provides a worse efficiency and hence a bit more dissipated power).

The results are shown in Fig. 3.31. By observing the measured variations, it is possible to notice that a much higher degradation was found in this load condition. This is really evident if one compares the first effects of degradation in the 2.9-V 50-ohm test with those in the 2.2-V optimal-load test. The former is characterized by a V_t shift of $\sim$80 mV (to be compared with $\sim$1 mV) after just 10 s. This much more severe damage happens in spite of the similar average dissipated power (i.e. 80 mW for the suboptimal-load versus 50 mW for the optimal load). The difference is even more pronounced if the variations of ξ are considered. The comparison of Fig. 3.30 and Fig. 3.31, even if still qualitative, is clear evidence of the dramatic influence of the voltage and current waveforms on CMOS PA reliability.

3.7.2 Degradation mechanisms

After the experimental demonstration of the impact of both dissipated power and harmonic termination in determining the wearout dynamics, it is interesting to study in

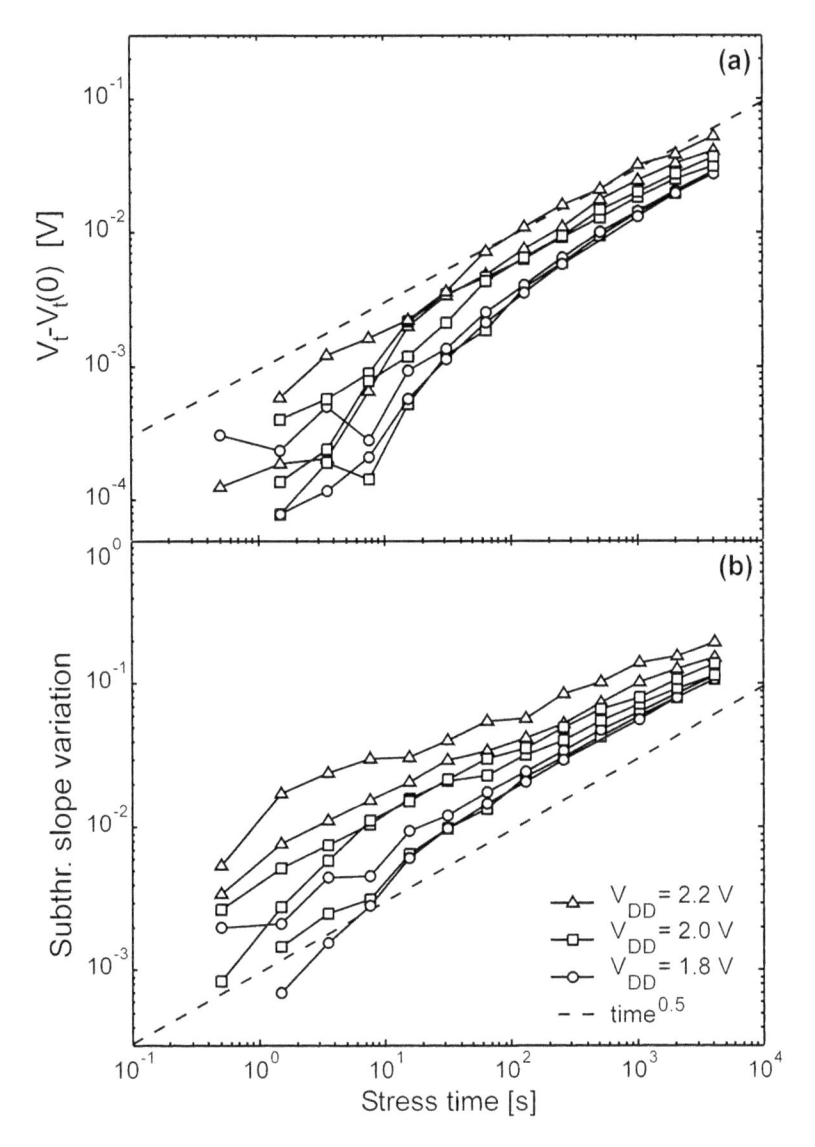

3.7 Large-signal performance degradation and reliability

Fig. 3.30 (a) Threshold voltage shift $\Delta V_t = V_t - V_t(0)$ and (b) sub-threshold slope variation $\Delta \xi = \xi - \xi(0)$ versus stress time, at various V_{DD}, under optimal load ©[2007] IEEE. Reprinted, with permission, from C. D. Presti, F. Carrara, A. Scuderi, S. Lombardo, and G. Palmisano, "Degradation mechanisms in CMOS power amplifiers subject to radio-frequency stress and comparison to the DC case," in *IEEE Proc. International Reliability Physics Symp. (IRPS)*, April 2007, pp. 86–92.

detail the most pronounced differences found in the two analyzed load situations. The aim is to get some insights into the physics of the degradation mechanisms. In particular, it would be very desirable to discern whether the transistor degradation is caused by channel hot carriers (HC) or Fowler–Nordheim (F-N) conduction through the oxide.

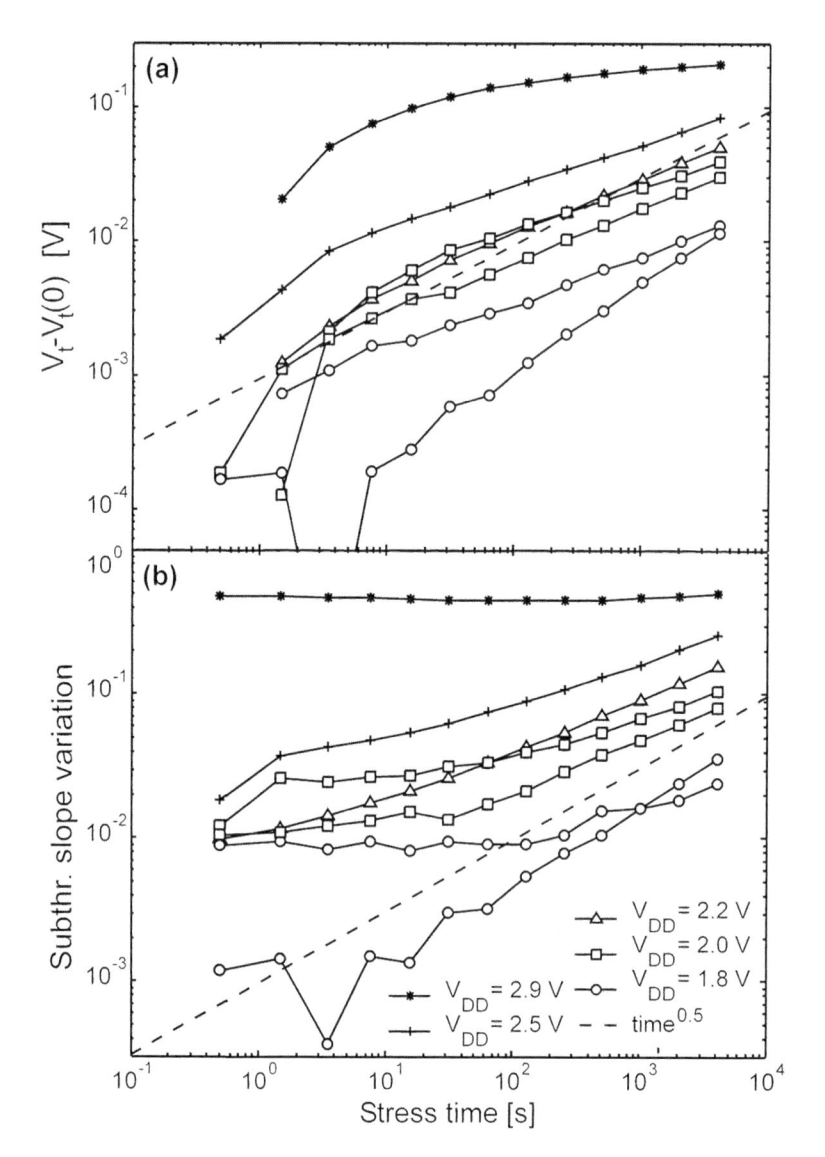

Fig. 3.31 (a) Threshold voltage shift $\Delta V_t = V_t - V_t(0)$ and (b) sub-threshold slope variation $\Delta \xi = \xi - \xi(0)$ versus stress time, at various V_{DD}, under sub-optimal load conditions ©[2007] IEEE. Reprinted, with permission, from C. D. Presti, F. Carrara, A. Scuderi, S. Lombardo, and G. Palmisano, "Degradation mechanisms in CMOS power amplifiers subject to radio-frequency stress and comparison to the DC case," in *IEEE Proc. International Reliability Physics Symp. (IRPS)*, April 2007, pp. 86–92.

It should be noticed that a large fraction of the drain current is actually capacitive, due to the high frequency of the carrier, and flows through the parasitic capacitances. However, transistor damage is caused by the carriers flowing through the channel (HC) and/or the gate oxide (F-N). Therefore, taking the total drain current rather than only its real part (i.e. that part associated with carriers flowing through channels and/or gate oxide)

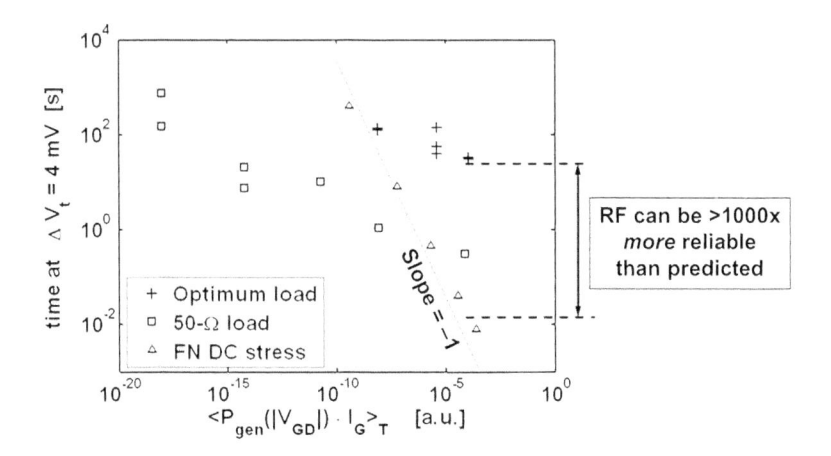

3.7 Large-signal performance degradation and reliability 113

Fig. 3.32 Time needed to reach $\Delta V_t = 4$ mV vs average theoretical F-N damage intensity, according to the quasi-static picture (dotted line has unitary slope) ©[2007] IEEE. Reprinted, with permission, from C. D. Presti, F. Carrara, A. Scuderi, S. Lombardo, and G. Palmisano, "Degradation mechanisms in CMOS power amplifiers subject to radio-frequency stress and comparison to the DC case," in *IEEE Proc. International Reliability Physics Symp. (IRPS)*, April 2007, pp. 86–92.

would lead to misleading results in this degradation mechanism analysis. By exploiting both measurements and simulations, paper [51] demonstrated that with an HF signal applied to the device, F-N stress is more severe under optimal load conditions, whereas HC injection at high voltage occurs if no harmonic termination is used. Nonetheless, it is not possible to exclude HC effects in the high-efficiency mode.

Moreover, by performing F-N and HC DC characterizations according to the schemes reported in Fig. 3.29, and comparing the degradation data by a quasi-static approach, it is possible to monitor the degradation effect simply by considering the time necessary to reach a small threshold voltage shift ΔV_t of 4 mV. Indeed, the threshold voltage shift is believed to be proportional to accumulated oxide damage, at least for small shift amounts. The time at $\Delta V_t = 4$ mV is plotted against the total average F-N stress intensity in Fig. 3.32, for both HF measurements setting (i.e. optimum load and 50-ohm load) and the F-N DC case. It is possible to notice that the DC data points lie along a straight line whose slope is approximately equal to -1. This is not true for the HF stressed devices. On the one hand, the sub-optimally loaded transistors reveal a premature reaching of the 4-mV threshold shift relative to the DC case and then they suffer from strong HC injection. On the other hand, most of the optimally loaded devices reveal an amount of damage lower than predicted by the quasi-static theory. This anomaly is quantitatively very high and indicates an interesting transient effect. These results prove the inadequacy of the quasi-static approach for a reliable analysis.

The results reported in Fig. 3.32 indicated that HC injection should be better investigated to explain the strong degradation of those devices stressed at HF under sub-optimal load conditions, extending the quasi-static approach to the HC phenomenon investigation.

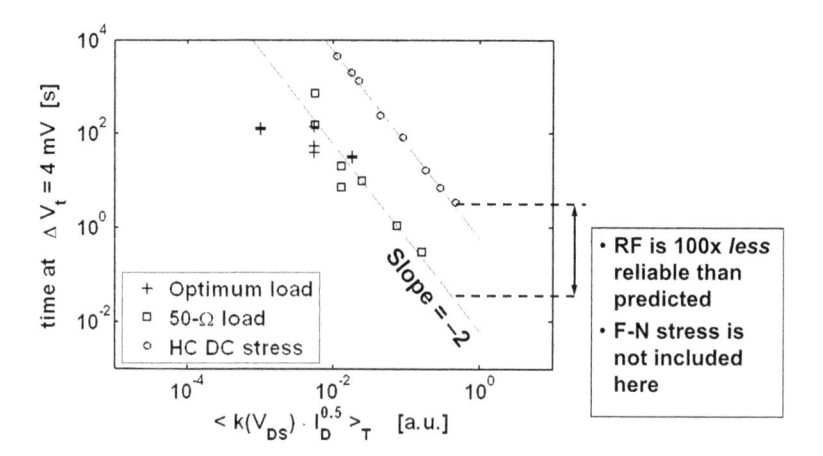

Fig. 3.33 Time needed to reach $\Delta V_t = 4$ mV vs average theoretical HC damage intensity ©[2007] IEEE. Reprinted, with permission, from C. D. Presti, F. Carrara, A. Scuderi, S. Lombardo, and G. Palmisano, "Degradation mechanisms in CMOS power amplifiers subject to radio-frequency stress and comparison to the DC case," in *IEEE Proc. International Reliability Physics Symp. (IRPS)*, April 2007, pp. 86–92.

Applying again the analysis approach also for the HC injection [51], the time at $\Delta V_t = 4$ mV is plotted in Fig. 3.33 against the average HC stress intensity for both HF measurements settings and for the HC DC case. By construction, the DC data are aligned on a straight line whose slope is equal to -2.

The HF stressed devices exhibit a higher damage growth rate than expected from the quasi-static approach. It must be noticed that the devices in the optimal load condition underwent a strong F-N stress, and this is not included in the abscissa of Fig. 3.33. On the other hand, it is known that the F-N stress is negligible for most 50-ohm loaded devices, as already shown. Nonetheless, their wearout is more than one order of magnitude faster than what is predicted from the quasi-static approach, even if the same sub-linear behavior ($a = 0.5$) is found.

This is another experimental indication that transistor stress cannot be successfully explained by a quasi-static theory of degradation.

In summary, a large quantitative discrepancy exists between measurement results and quasi-static-theory predictions as reported in Figs. 3.34 and 3.35. Owing to the direct impact of voltage swing in power-amplifier performance, custom reliability tests are recommended in order to comply with the safe voltage swing region, while at the same time maximizing the device performance.

3.8 CMOS passive devices

In mm-wave applications, integrated passive components are essential and can considerably affect circuit performance, especially in transmitters. Attention is particularly focused on both capacitive and inductive devices since they are widely used by designers

3.8 CMOS passive devices

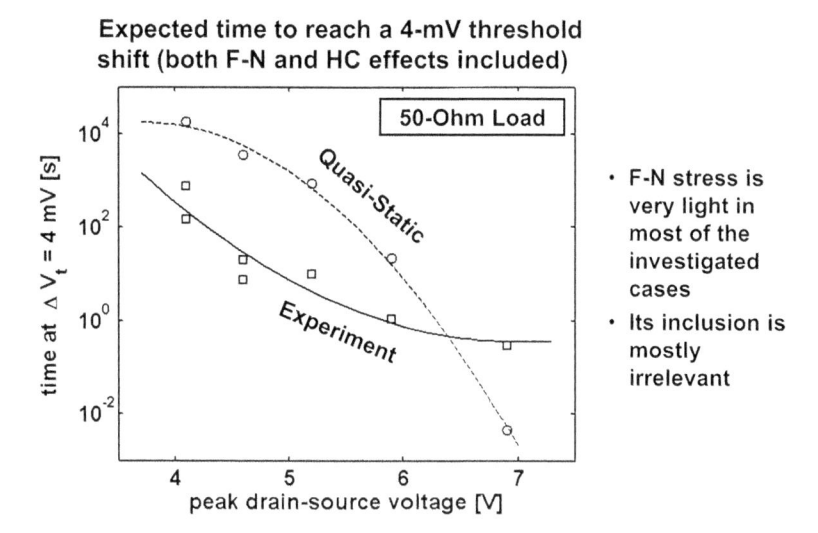

Fig. 3.34 F-N and HC experiment vs quasi-static prediction under sub-optimal load.

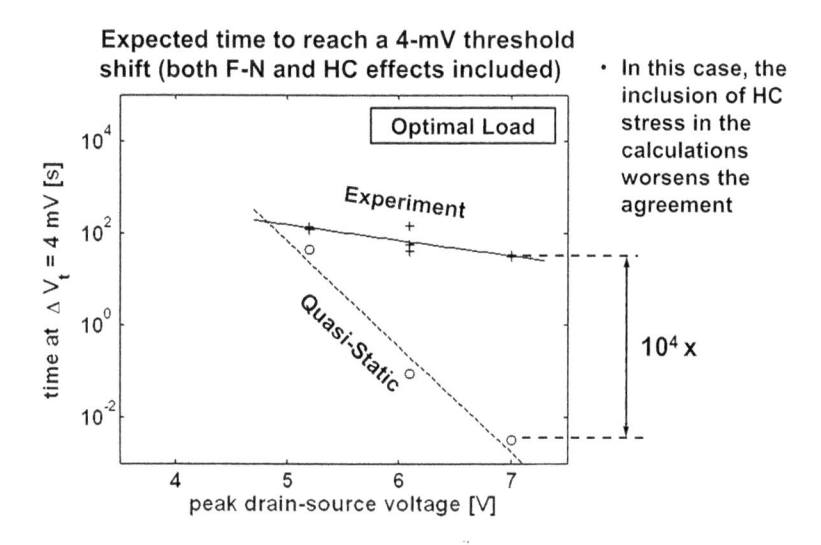

Fig. 3.35 F-N and HC experiment vs quasi-static prediction under optimal load.

to enable filtering capabilities, matching networks, resonant loads, single-ended-to-differential conversions, etc.

As far as capacitors are concerned, the scenario is quite clear. Indeed, in modern CMOS platforms for mm-wave applications two different kinds of devices are generally made available, i.e. metal–insulator–metal (MIM) and metal–oxide–metal (MOM) capacitors. The simplified structures of MIM and MOM capacitors are depicted in Fig. 3.36.

An MIM device is a parallel-plate capacitor that consists in two metal planes separated by a very thin (usually high-K to boost the capacitor value) dielectric. Of course,

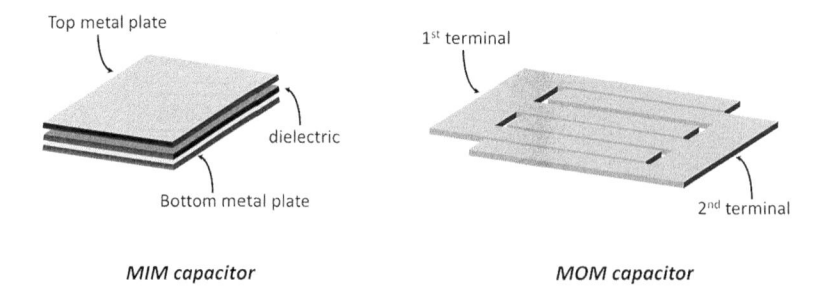

Fig. 3.36 Simplified 3D structures of MIM and MOM capacitors.

the fabrication of an MIM capacitor requires extra mask and photo-steps to implement a very thin and well-controlled insulator layer between the top and bottom metal planes. This is important not only to increase the density per unit of area (in general in the range of a few fF/μm^2), but also to obtain good accuracy. MIM capacitors are built in the upper metallization layers to take advantage of lower capacitive parasitics on the bottom plate and higher quality factors. Thanks to their structure, they are also very area effective and suitable to implement values of a few tens of fF, as are typically used at mm-wave frequencies.

On the other hand, an MOM device is an interdigitated multi-finger structure that exploits the intrinsic fringing capacitance between metals. It is typically used to implement very low capacitive values that are below the minimum guaranteed by MIM capacitors. To improve the density per unit of area, MOM capacitors often adopt multiple metal layers properly connected by vias, although the exploitation of lower metallizations can increase parasitics, thus affecting both the Q-factor and the self-resonance frequency (f_{SR}). It is worth noting that, for deep sub-micron (<100 nm) platforms with narrow minimum horizontal spacing between metals, the interdigitated MOM capacitors can be as area efficient as the MIM ones, with the advantage of not requiring extra process steps. Indeed, MOM devices are natural capacitors in any integrated platform. Generally, MOM capacitors have higher breakdown voltage than MIM devices. Finally, SOI technologies can considerably improve the performance of MOM capacitors since lower metal layers became available to boost the density per unit of area, without affecting the f_{SR} and the losses of the device.

The picture for integrated inductive components (i.e. inductors, transformers, and microstrips) is more complex. Indeed, both geometrical (e.g. spiral layout, metal slotting, spiral coupling configurations, etc.) and vertical parameters (e.g. multi-layer structures, ground shields, etc.) have to be carefully customized by the designer according to the figures of merit of main interest (e.g. inductance, Q-factor, ωQL product, series resistance, coupling factor, transformer characteristic resistance, insertion loss, maximum available gain) [53]. The job is even harder since DK availability and component modeling are often very scarce. For the mm-wave designer, a deep knowledge of the main loss mechanisms is required to address the most suitable structure, while an extensive use of 2D/3D EM simulations is suggested for structure optimization. It is well known that the most common way to implement inductors is using spiral-like lumped structures, whose low-frequency inductance increases on increasing the number

3.8 CMOS passive devices

of turns (n) and the inner diameter (d_{in}) and slightly on decreasing the metal width (w) [54]. At mm-wave frequencies, typical values for inductors are on the order of a few tens of pH. Therefore, single-turn spiral inductors and small d_{in} (<50 µm) are generally used. Narrow metal w (<5 µm) is also adopted to reduce the capacitance towards the substrate, provided that electromigration limitations are fulfilled. When inductors have to tolerate high currents (hundreds of mA) slotted metals are used in order to limit the w-reduction due to the skin/proximity effects. Circular or polygonal shapes are usually preferred to take advantage of slightly higher quality factors in comparison with the ones achieved by a square geometry. As far as the vertical structure of an integrated inductor is concerned, the availability of an optimized BEOL is crucial since top copper metals and thick inter-metal oxides are mandatory to lower ohmic losses in the metals and reduce the parasitic capacitance towards the substrate, respectively. Since CMOS technologies make available several metal layers but thick metals are not so common, shunting two or more metals to obtain an equivalent thick layer is very effective. Of course, the number of shunted metals has to be properly defined, since the use of lower metals generally degrades the f_{SR} of the component. The losses in the substrate are commonly reduced by means of polysilicon patterned ground shields (PGS) [55], although only SOI technologies are actually substrate-loss free.

As far as integrated transformers are concerned, the most important feature in circuit applications is the capability to transfer power from the input to the output port. This power transfer is related not only to the transformer itself but to the impedances connected to its input and output, including the corresponding matching networks [56]. Traditional transformer configurations are the interleaved and stacked ones [57]. At mm-wave frequencies, interleaved configurations can be profitably used to implement symmetric windings with coupling factor (k) as high as 0.5 by using the top metals available in the process, thus maximizing both the f_{SR} and the Q-factor at the expense of the area consumption. On the other hand, stacked transformers achieve higher k and a better area exploitation, but exhibit lower f_{SR} due to the increased parasitic capacitances. Recently, mixed stacked and interleaved transformer structures (namely interstacked) were proposed with the aim of improving k of traditional symmetric interleaved transformers by means of the stacked coupling [58–60].

As an example, a typical multi-layer inductor and a stacked transformer for 60-GHz applications are shown in Fig. 3.37, along with main geometrical and electrical data. The components were designed in a 65-nm CMOS technology [61]. The inductor adopts a three-layer structure consisting of two Cu metals and a top alucap layer for an equivalent thickness of 5 µm. The stacked configuration takes advantage of a thin (i.e. less than 1 µm) oxide between primary and secondary windings, which both exploit a two-layer structure. Coil widths were properly optimized to trade off series resistance and f_{SR}. The transformer insertion loss is about 0.8 dB. Both inductor and transformer exploit a polysilicon PGS under spirals to reduce substrate losses. Of course, inductive component performance would be greatly improved in CMOS SOI technologies, thanks to the almost complete elimination of substrate losses and capacitive parasitics [62].

Finally, special attention should be paid to the development of simple lumped scalable models for inductive devices [60, 63, 64] with the aim of helping the designer

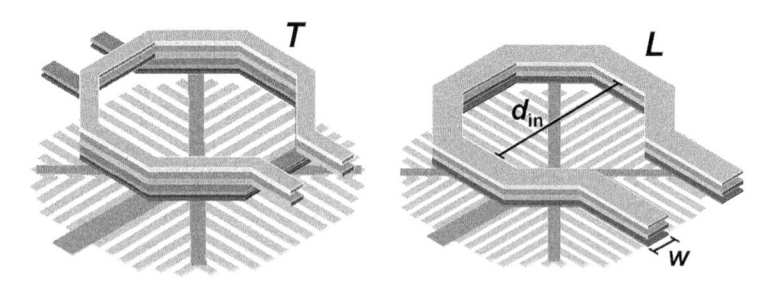

	L_{60GHz} [pH]	Q_{60GHz}	f_{SR} [GHz]	k	w [μm]	d_{in} [μm]	length [μm]
L_P/L_S (T)	90/90	14.5/14.5	145	0.7	5	41	165
L_1	80	14.5	155	–	5	35	141
L_2	115	15.2	125	–	5	48	185
L_r	85	14.5	150	–	5	38	150

Fig. 3.37 3D views, geometrical, and electrical data of a typical single-turn stacked transformer (T) and inductor (L) in a nanoscale CMOS platform. Source: V. Giammello, E. Ragonese, and G. Palmisano, "Transformer-coupled cascode stage for mm-wave power amplifiers in sub-μm CMOS technology," *Springer Analog Integrated Circuits and Signal Processing*, vol. **66**, pp. 449–453, Mar. 2011. With kind permission from Springer Science and Business Media.

in co-design circuit optimization, leaving EM simulations for refinement and accurate modeling.

3.9 Measurement and modeling issues

No simulation can be more accurate of the modeled device that in turn cannot be more precise than the measurement done for its extraction.

Measuring devices at RF/mm-wave, especially power devices, is not trivial, mainly for the two following reasons.

(a) Microwave: the functionality at such fast speeds implies that any small capacitance or small inductance will be responsible for current and voltage at their terminals, respectively.
(b) Power: power transistors convert DC energy to microwave frequency and also heat, and sometimes a microwave power amplifier is more a heater than an electronic amplifier.

The above aspects will require accurate instrument calibration, parasitic de-embedding and specific measurement techniques to allow the modeler to provide designers with a precise design kit.

Device characterization for modeling is generally performed on-wafer. This helps to keep the surrounding parasites as low as possible. Indeed, any interconnection to the intrinsic device has detrimental effects that must be removed. The device metallization

3.9 Measurement and modeling issues

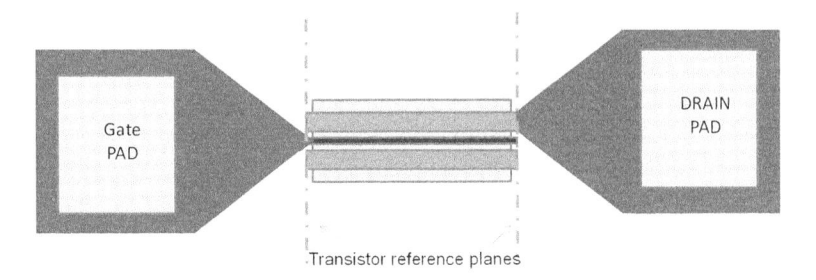

Fig. 3.38 Reference planes for on-wafer measurements.

of gate, source, drain, and bulk terminals in the active area has to be minimized to allow the characterization of the core device. This approach gives the designer a larger freedom in the device layout strategy that can be properly customized.

The basic characterization procedure moves from a DC characteristic under continuous and pulsed excitations to a frequency response based on S-parameters, again under continuous and pulsed excitation, measured as a function of the DC bias. The pulsed excitation (i.e. short pulses in the range of 1 μs) is exploited to avoid the rising of thermal dynamics that, generally, are slower (i.e. in the range of ms). In this way, a better understanding of pure electric and pure thermal effects can be achieved. Finally, nonlinear large-signal behavior should be investigated by means of large-signal vector analyzer and load-pull techniques. Such data are indeed fundamental to building an accurate device model. To get the complexity of the task, models like the BSIM level 4 are exploited [65]. This contains around 400 extractable parameters.

All the measures are needed at the device reference plane, but the instrumentation is largely far from the device. Typically GSG structures and GSG probes are exploited to allow on-wafer measurements at high frequency. These probes touch the wafer in custom pads that are still far from the device reference planes, as depicted in Fig. 3.38. In order to move the reference plane to the desired transistor reference planes procedures called calibration and de-embedding have to be implemented. Several calibration techniques are exploited in conjunction with vector network analyzers and DC parametric analyzers.

Among others, the most common calibration techniques are listed below:

1. SOL: Short, Open Load Calibration [66];
2. SOLT: Short Open Load Through Calibration [67];
3. QSOLT: Quick Short Open Load Thru [68, 69];
4. SOLR: Short Open Load Thru Reciprocal [70];
5. TRL: Through Reflect Line Calibration [71];
6. LRM: Line Reflect Match Calibration [72, 73];
7. LRRM: Line Reflect Reflect Match [74].

A discussion on calibration techniques and measurement issues is not within the scope of this chapter. However, any of these techniques has its own advantages and drawbacks. In the following are some considerations:

(a) TRL is frequency limited but exploitable on-wafer;
(b) LRM is broadband but useful only if very accurate 50-ohm loads are available, which are difficult to obtain on-wafer;
(c) SOLR does not require knowledge of the "through" [70];
(d) QSOLT simplifies the calibration at the cost of accuracy [68, 69];
(e) LRRM has been developed specifically for on-wafer measurement [75].

Moreover, the reader should consider that getting accurate measurements is not only a matter of the exploited calibration techniques. Indeed, probe quality, as well as contact repeatability, on-wafer pad metal vs probe metal (it can greatly impact the contact resistance), instrument uncertainty, quality of cables and their connections, surrounding shielding, wafer back grounding, and so on, are all aspects that must be taken into consideration if accurate measurements are desired.

Finally, after on-wafer probe calibration exploiting commercial or custom calibration kits, specific structures have to be placed on-wafer to de-embed the parasitic between the probe plane and the device plane: this will assure the placement of reference planes at the real device terminals [76, 77].

DC measurements

DC parametric analyzers allow us to get I_D–V_{DS} behavior versus V_G under continuous and pulsed excitation.

S-parameter

In order to get information about voltage and current at microwave for any device port, the plain choice is to collect the S-parameters. Of course, they have to be measured at the desired ports and for any combination of gate and drain voltages. Using the Root modeling approach [78], converting the S-parameters into Y-parameters, they will represent the partial derivative of the charge storage functions. Starting from these values it is possible to rebuild the storage components of the device model. The reader should note that the data collection requires the contemporaneous application of DC and microwave signals and then bias-tees are exploited. These bias components have low-frequency behavior that should be taken into account especially when linear performance is tested, since they can affect the baseband termination impedances and IMD. Moreover, to isolate the self-heating effects, very-short-pulse S-parameter measurements are needed, while bias-dependent S-parameters are collected [79]. This will impact even more on the choice or design of optimal bias-tees, but it will also enable the extraction of isothermal model parameters [80]. If the reduction of the pulse width will reduce the amount of energy the device has to suffer and then its thermal response will be less evident, a reduction in dynamic range is experimented, as discussed in [81].

Even if it is not common at mm-wave, when high-power devices have to be tested, it could be useful to perform non-50-ohm calibration. Indeed, the required power level implies very low device impedance with very high reflection coefficients. Then, the adoption of low-impedance calibration, based on the adoption of transforms or pre-match tuners, could be necessary to reduce measurement errors. TRL and LRM techniques can be exploited in this case.

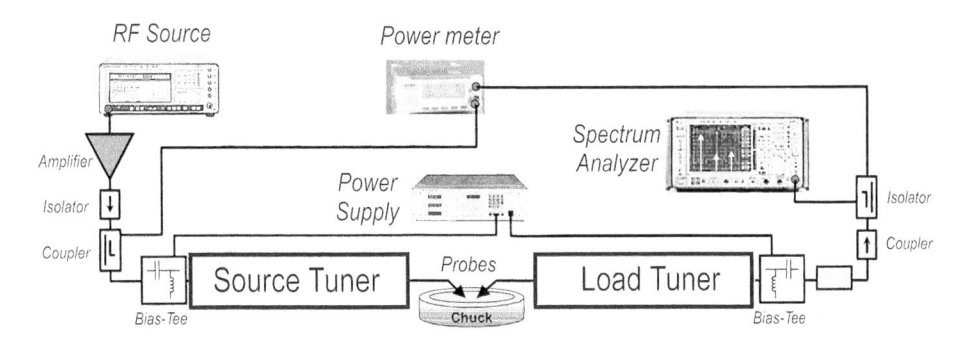

Fig. 3.39 Typical scheme of a source-load-pull bench.

Load-pull measurement

The set of measurements discussed above allows only the development of a linear model that is not sufficient for power amplifiers since they work with large signals and are largely nonlinear driven and termination-impedance dependent.

In order to get a valid model for large-signal operations, it is possible to characterize the device with source-load-pull techniques [82], as shown in Fig. 3.39.

The analysis of the device behavior is based on some key figures that are evaluated with several combinations of source/load impedances, while the input signal is increased up to saturation. Owing to the complexity of the testing equipment, it is possible to set the impedance at the fundamental frequency and harmonics. The evaluated key figures are then plotted in a Smith chart as contours of equal performance, as shown in Fig. 3.40. In some cases, it is helpful to use several figures of merit, like power level and efficiency, in order to better set the right termination impedances. Recently, more complex figures of merit, such as adjacent channel power ratio (ACPR), peak-to-average ratio (PAR), and error vector magnitude (EVM) as functions of the digital-modulated input signal have been taken into consideration to picture the device behavior.

Finally, large-signal network analyzers (LSNAs) are today being widely exploited [75, 83], which allow one to perform more complex measurements than the classical VNA ones. Indeed, operating in large-signal conditions, such measurements take into account harmonic generation by evaluating the correlated amplitude/phase values at each of them. This also enables the time-domain waveform calculation (while the measurement is done in frequency) and it is largely used to tune and/or verify the device model [84], as well as to optimize its performance.

3.10 CMOS trends: SOI

The most attractive evolution of CMOS technology for mm-wave applications is the silicon-on-insulator or SOI CMOS, an option that is gaining more and more interest for several reasons.

(1) SOI wafers are widely available with very good quality and acceptable cost.
(2) Development and production lines are the same as bulk CMOS.

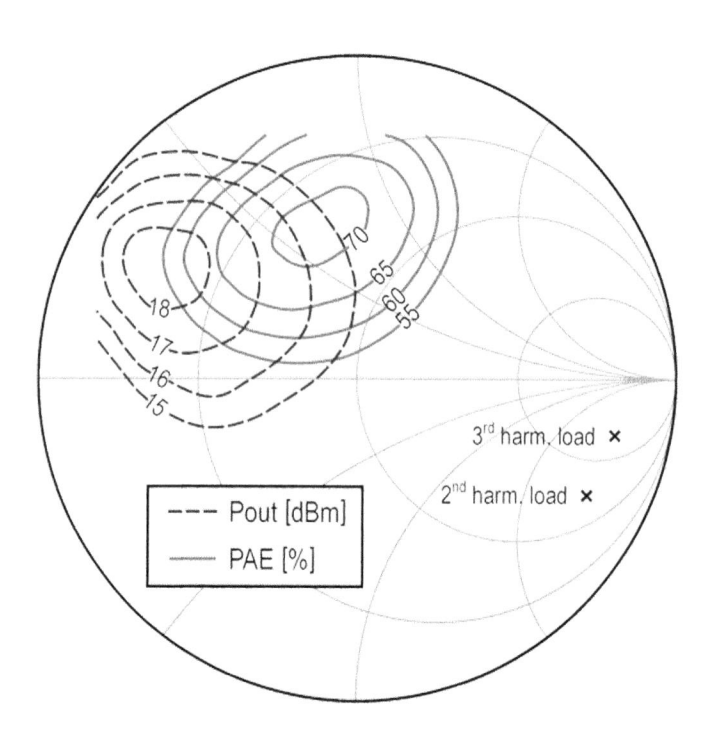

Fig. 3.40 Typical contour plot of output power and PAE of a power device on a Smith chart. Reprinted from F. Carrara, C. D. Presti, A. Scuderi, and G. Palmisano, "Single-transistor latch-up and large-signal reliability in SOI CMOS RF power transistors," *Elsevier Solid-State Electronics*, vol. **54**, pp. 957–964, Sep. 2010. Copyright (2010), with permission from Elsevier.

(3) SOI CMOS guarantees performance improvements compared with bulk CMOS (i.e. very low leakage currents, less capacitive parasites, lower threshold voltage, and reduced body effect).

The starting material for SOI device fabrication is generally a high-resistivity silicon wafer with buried oxide (BOX). Two possible transistor structures can be fabricated: fully depleted (FD) and partially depleted (PD) transistors, depending on the silicon thickness fabricated on top of the BOX. Both transistor structures are depicted in Fig. 3.41. The STI, along with the BOX, generates a fully isolated well where the FETs are fabricated. By controlling the silicon thickness t_{Si} and its doping level, it is possible to obtain the depletion region extending into the body and not totally depleting the body charge, thus obtaining the so-called PD SOI MOS [85]. In particular, this happens if $t_{Si} > 2y_{dmax}$ with

$$y_{dmax} = \sqrt{\frac{4\epsilon\phi_f}{qN_A}}, \quad (3.31)$$

where N_A is the doping level, ϕ_f is the Fermi potential, and ϵ is the dielectric permittivity of silicon.

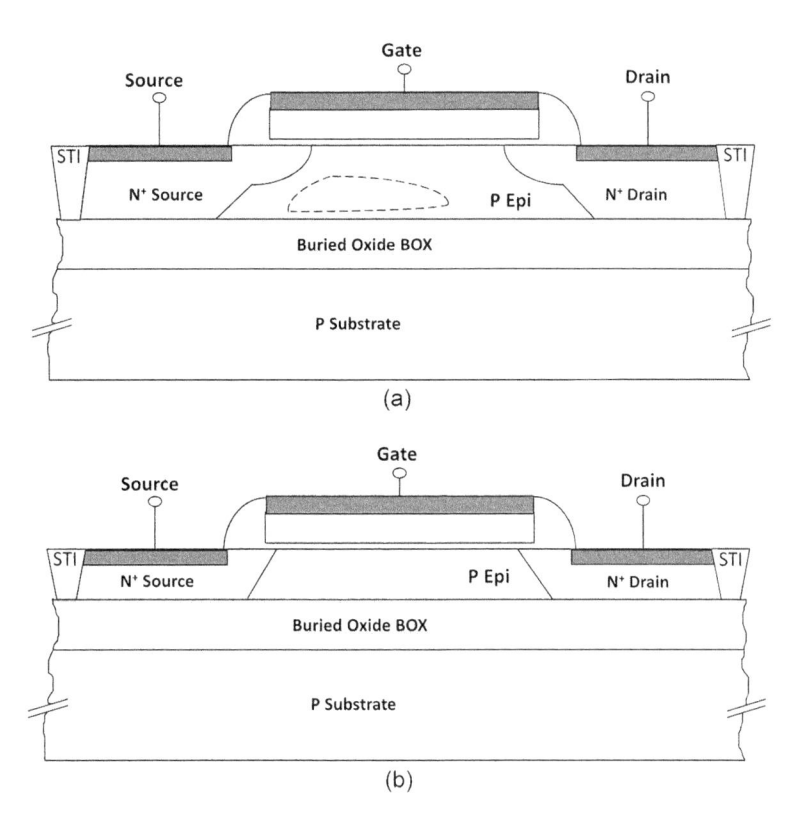

Fig. 3.41 (a) Partially depleted and (b) fully depleted SOI MOS cross sections.

On the other hand, when $t_{Si} < 2y_{dmax}$ no charge is located in the body and the resulting device is an FD SOI MOS.

In PD SOI technology, the availability of charge in the body allows the control of the body voltage. It is worth noting that the body resistance is largely affected by the drain–source voltage, as it affects the extension of the depletion region. Such effects can generate a relevant nonlinear device behavior at very high voltage swings. However, the PD devices have very low parasitic capacitance and the possibility to keep floating the body.

FD SOI devices are very similar to the PD SOI except for the silicon thickness that is lower. This means that all the mobile charges of the body are depleted, thus making more critical the control of the threshold voltage, which must be controlled by means of the thickness t_{Si} of the epitaxial silicon layer. Compared with PD devices, they exhibit higher g_m and lower short-channel effects with a quasi-ideal sub-threshold slope. The main disadvantage lies in the higher self-heating behavior and in the very accurate epi-thickness control.

For both PD and FD SOI MOS transistors, the structure, based on an isolated well, allows two relevant benefits with respect to traditional bulk CMOS.

(1) Reduced capacitive parasitic effects, as the body, drain, and source junction capacitances of bulk CMOS are now substituted by lower oxide-controlled parasitic capacitances.

(2) Independent bias levels for any device: in bulk CMOS, triple-well structures with well-controlled potential must be adopted; in contrast, SOI devices can be easily stacked without any additional well or complex bias techniques, allowing the management of large signals, as in the case of power amplifiers.

The reduced capacitive effects are essential for mm-wave operation. On the other hand, the stacked architecture is one of the most interesting solutions for power amplifier design since it allows the voltage swing to be divided among the stacked devices. This technique will be discussed in the next chapters.

In PD SOI platforms, devices can be also exploited either with a floating body (FB) or with the body contacted (BC) to the source. This last solution results in extra parasitic capacitance with respect to the floating-body device due to physical connection. The main application environment for floating-body devices is the low-loss switch. Indeed, a floating-body PD MOS can be stacked with minimum wiring, with excellent performance in terms of both R_{on} and off-state capacitance. However, as described in [86], for a high number of FB switches, considerable distortion with large-signal waveforms occurs in off-mode, due to the imbalance experienced by the stacked structure. This imbalance is due to the capacitive parasites toward the substrate along the stacked structure that modify the current distribution and hence the voltage partitioning among devices. Indeed, top-most devices are subjected to larger voltage swings, while a smaller swing develops in the grounded device. Therefore, the devices with larger voltage swings can suffer a potential forward biasing of the drain-to-body and source-to-body diodes with clipping effects and harmonic distortion generation. Moreover, the imbalance can produce gate-induced drain leakage [87], further enhanced by the parasitic BJT (see the next section). Finally, as will be discussed, the parasitic capacitance toward the substrate through the BOX is a nonlinear function of the applied voltage and thus contributes to the overall nonlinearity generation, especially if the body is floating. In contrast, if the body is maintained at the same potential as the gate, the above distortions are reduced.

Also in the on-state, the FB devices show worse linearity than BC due to the so-called kink effect. Let us consider the PD floating-body device sketched in Fig. 3.42.

The impact ionization in the channel generates holes that will accumulate in the lower-potential region, i.e. the body. This charge will increase the body potential, thus lowering the equivalent threshold voltage that becomes

$$V_T = V_{T0} + \gamma\left(\sqrt{2\Phi_P - V_{BS}} - \sqrt{2\Phi_P}\right), \tag{3.32}$$

and ultimately it increases the current. The effect on the DC behavior is depicted in Fig. 3.43. The depicted behavior shows potential AM/AM distortion during the on-mode. The lower off-capacitance associated with FB devices (due to the lack of body contact area) guarantees a better $R_{on} \cdot C_{off}$ figure of merit, which is commonly exploited for switch applications. On the other hand, in terms of noise, FB devices show higher low-frequency noise than the BC one.

All these considerations suggest the adoption of BC MOS for analog design, while FB devices can be favorably adopted for nonlinear functions, such as switches. However,

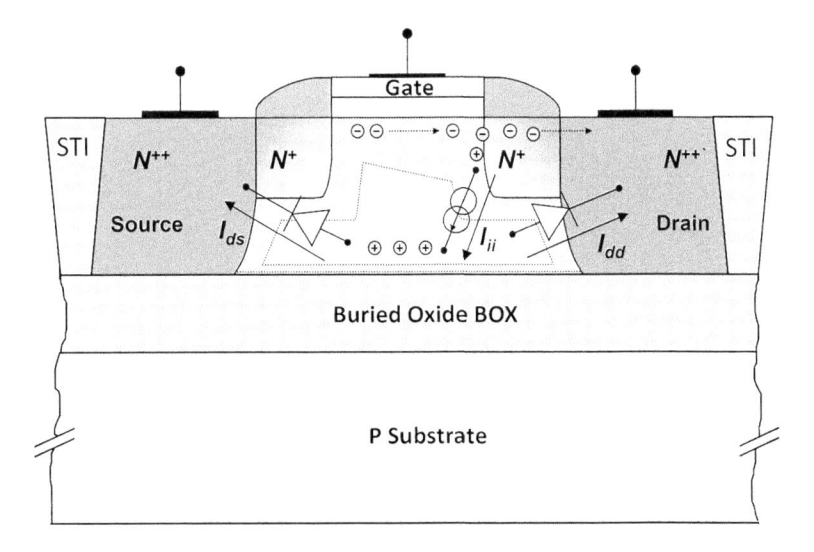

Fig. 3.42 PD floating-body SOI MOS.

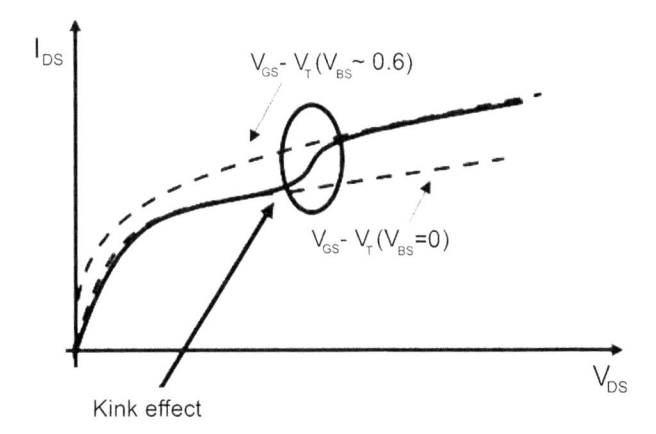

Fig. 3.43 Kink effect.

due to very demanding linearity, latest-generation switches also exploit BC devices, especially if large-throw-count switches are designed [86].

3.10.1 Large signals and parasitic BJT in SOI platforms

In large-signal operation, another relevant phenomenon must be analyzed for accurate mm-wave PA design, i.e. single-transistor latch-up [88]. This effect can be illustrated by referring to Fig. 3.44, which represents a simplified layout and cross section of an SOI BC n-MOS transistor. Owing to the narrow, thin shape of the body finger, a large parasitic body resistance R_{BB} is formed between the intrinsic body and the body contact (much larger than in bulk CMOS transistors). The body contact should short-out the base–emitter junction of the parasitic n-p-n BJT (see Fig. 3.45), thus keeping it off.

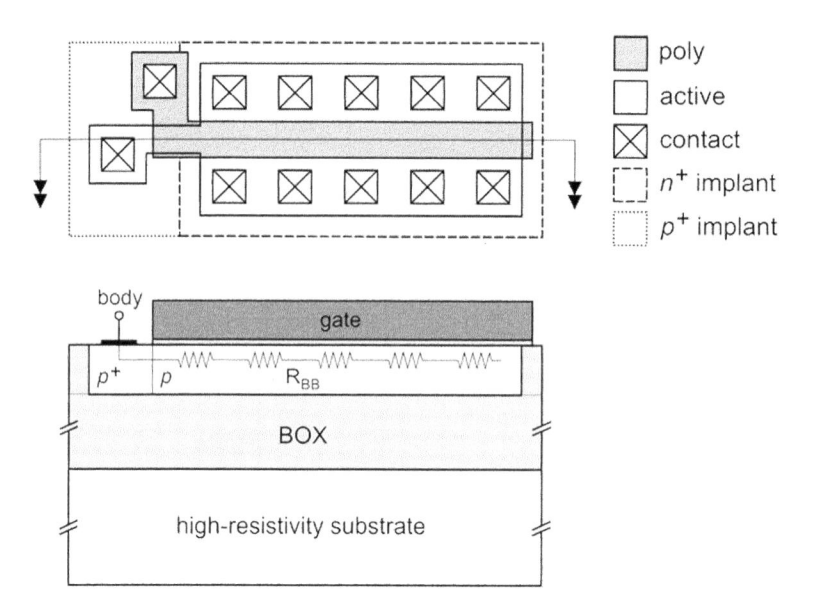

Fig. 3.44 Simplified layout and cross section of an SOI BC n-MOS transistor. Reprinted from F. Carrara, C. D. Presti, A. Scuderi, and G. Palmisano, "Single-transistor latch-up and large-signal reliability in SOI CMOS RF power transistors," *Elsevier Solid-State Electronics*, vol. **54**, pp. 957–964, Sep. 2010. Copyright (2010), with permission from Elsevier.

However, if a substantial channel current i_{CH} flows through the MOSFET and a large drain voltage is present at the same time, direct carrier generation (impact ionization) occurs close to the channel pinch-off region, generating the hole current

$$i_H = (M - 1) \cdot (i_{CH} + i_{BJT}), \tag{3.33}$$

where M is the drain current multiplication factor. If R_{BB} is large, a substantial fraction of i_H will flow through the body–source (base–emitter) junction, thus effectively turning the parasitic BJT on. The resulting collector current i_{BJT} will eventually increase the drain current, hence triggering a positive feedback which leads to latch-up. As is well known, drain-current run-away can be avoided by ensuring that the overall gain of the feedback loop is lower than unity. A more empirical (though conservative) methodology can be adopted by ensuring that the base–emitter voltage of the BJT is lower than the value needed to turn it on:

$$v_{BS} < V_{BS(ON)} = 0.7 \text{ V}. \tag{3.34}$$

Maximum v_{BS} is attained at the transistor region that is farthest from the body contact. Integral calculations give

$$v_{BS,max} = \frac{1}{2} j_H r_{BB} W_f^2, \tag{3.35}$$

where j_H is the hole current per unit length, R_{BB} is the specific body resistance (9 kΩ/μm for the technology adopted in [88]), and W_f is the length of the gate/body

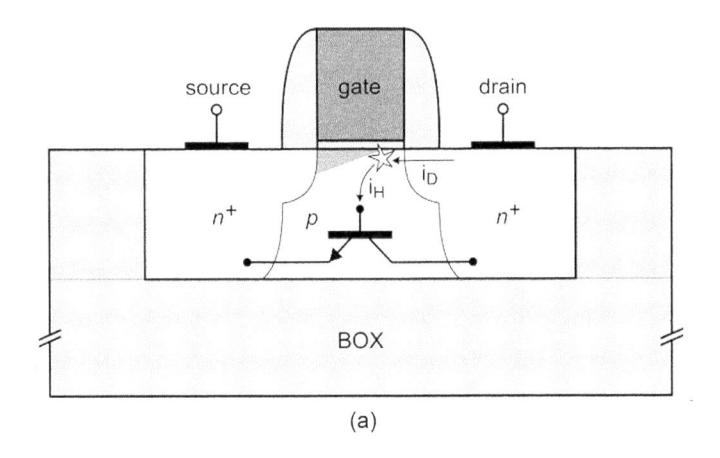

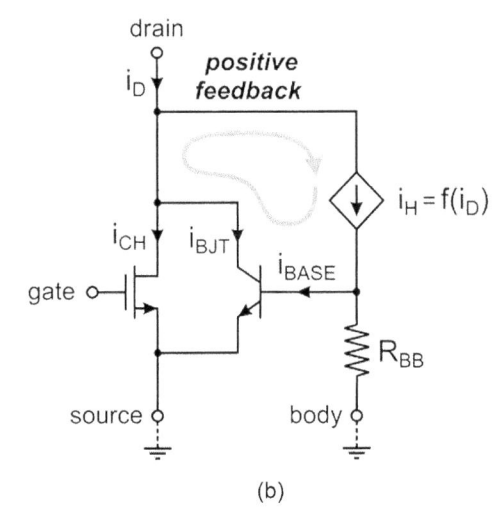

Fig. 3.45 Single-transistor latch-up in SOI BC n-MOS transistors. (a) Parasitic n-p-n bipolar transistor. (b) Equivalent electrical model. Reprinted from F. Carrara, C. D. Presti, A. Scuderi, and G. Palmisano, "Single-transistor latch-up and large-signal reliability in SOI CMOS RF power transistors," *Elsevier Solid-State Electronics*, vol. **54**, pp. 957–964, Sep. 2010. Copyright (2010), with permission from Elsevier.

finger. Hence, device latch-up is effectively avoided through the use of short gate fingers. To determine the maximum allowed value of W_f to comply with (3.34), j_H must be estimated. This can be done by recourse to high-frequency circuit simulations where the output capacitive current is removed from the evaluation. Then, by analyzing the quasi-static voltage and current waveforms as reported in Fig. 3.46a for a 1-mm saturated power device at 1.9 GHz and 2 V supply voltage, 2 V maximum impact ionization can be estimated by assuming that the worst case corresponds to the peak of the instantaneous power dissipation (Fig. 3.46b), where both the high-voltage and high-current conditions simultaneously occur. For this design example, a drain current of 60 mA at $V_{DS} = 2$ V must be safely sustained. Finally, from (3.33) we get

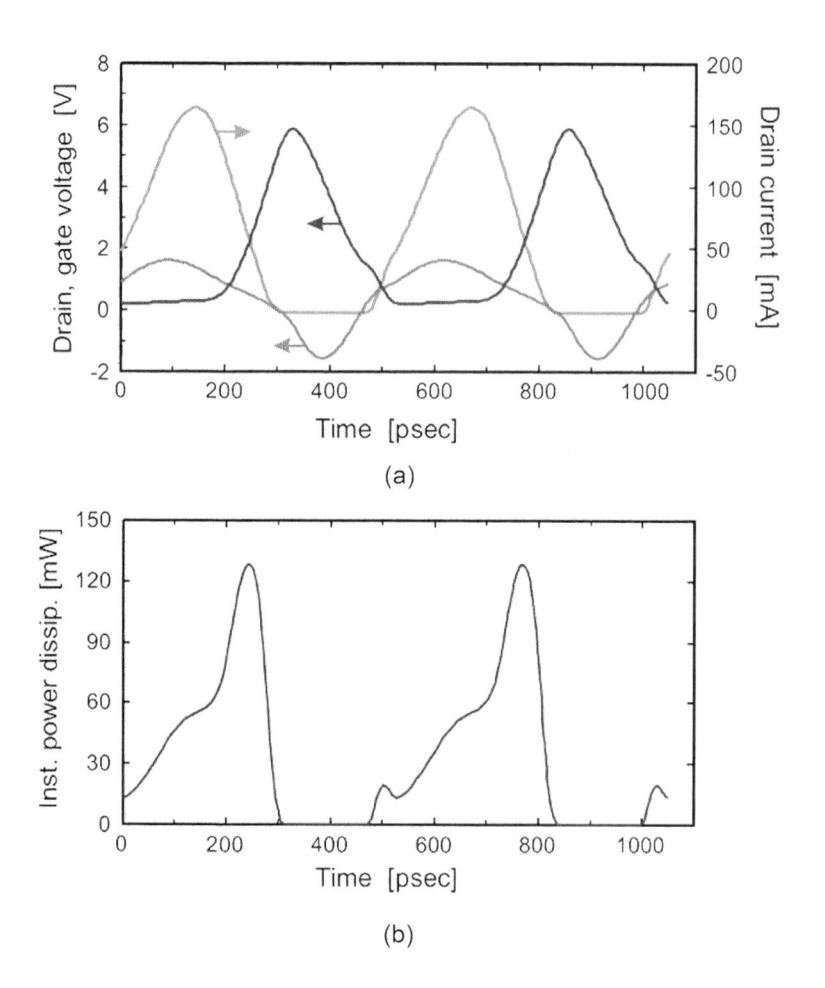

Fig. 3.46 Simulated voltage and current waveforms at saturated output power level ($W = 1\,\mu\text{m}$, $L = 0.28\,\mu\text{m}$, $V_{DD} = 2$ V, $V_{GG} = 0.45$ V, $f = 1.9$ GHz, $\Gamma_S = -0.23 + j0.57$, $\Gamma_{L1} = -0.38 + j0.42$, $\Gamma_{L2} = 0.72 - j0.36$, $\Gamma_{L3} = 0.81 - j0.17$). (a) Quasi-static extrapolation. (b) Instantaneous power dissipation, $v_{DS} \cdot i_D$. Reprinted from F. Carrara, C. D. Presti, A. Scuderi, and G. Palmisano, "Single-transistor latch-up and large-signal reliability in SOI CMOS RF power transistors," *Elsevier Solid-State Electronics*, vol. **54**, pp. 957–964, Sep. 2010. Copyright (2010), with permission from Elsevier.

$$J_H = (M-1)\frac{i_{CH} + i_{BJT}}{W}, \qquad (3.36)$$

where W is the overall device width. According to (3.36), a hole current density of 18 $\mu\text{A}/\mu\text{m}$ is estimated, having assumed a typical impact ionization factor $(M-1)$ of 0.3. By substituting (3.36) into (3.35), we obtain the following condition: $W_f < 2.9\,\mu\text{m}$.

As discussed in [88], the above concept has been demonstrated on real SOI devices with W_f of 2.5 μm and 5 μm, and the same overall width of 1 mm.

Load-pull contours at 1.7 V for the $W_f = 5\,\mu\text{m}$ device demonstrate ordinary device operation with well-behaved power characteristics. When the supply voltage is further

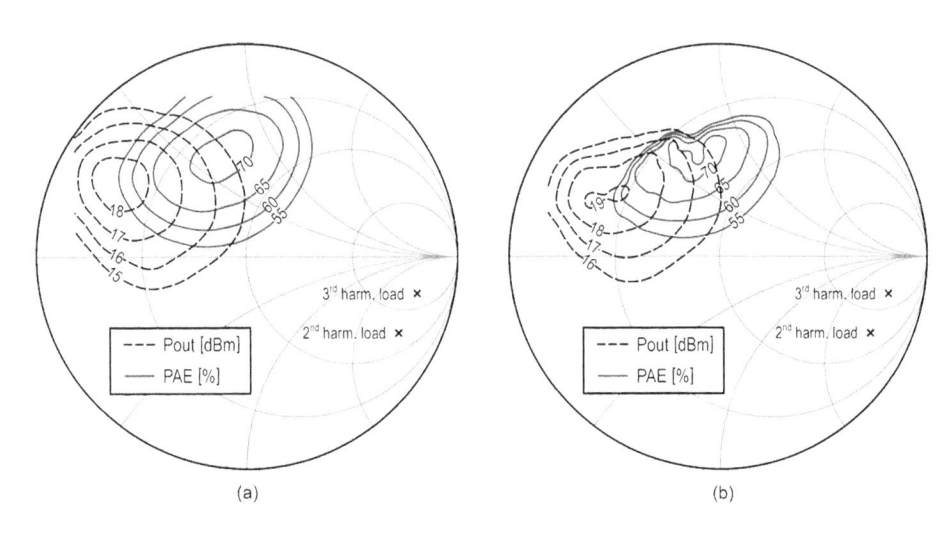

Fig. 3.47 (a) Output power and PAE load-pull contours at $V_{DD} = 1.7$ V ($W = 1\,\mu\text{m}$, $L = 0.28\,\mu\text{m}$, $V_{GG} = 0.45$ V, $f = 1.9$ GHz, $P_{in(av)} = 2$ dBm, single-tone CW input). (b) Output power and PAE load-pull contours at $V_{DD} = 2$ V ($W = 1\,\mu\text{m}$, $L = 0.28\,\mu\text{m}$, $GG = 0.45$ V, $f = 1.9$ GHz, $P_{in(av)} = 2$ dBm, single-tone CW input). Reprinted from F. Carrara, C. D. Presti, A. Scuderi, and G. Palmisano, "Single-transistor latch-up and large-signal reliability in SOI CMOS RF power transistors," *Elsevier Solid-State Electronics*, vol. **54**, pp. 957–964, Sep. 2010. Copyright (2010), with permission from Elsevier.

increased to 2 V, and even more so at 2.3 V, the PAE contours are considerably deteriorated and a non-destructive auto-sustaining current run-away can be reported, demonstrating the latch-up effect, as shown in Fig. 3.47.

When the 2.5-μm elementary finger device is exploited, no latch-up is experienced, confirming the discussed design procedure.

3.10.2 Substrate nonlinearity effect

Another important effect of devices fabricated on SOI wafers is related to the high-resistivity substrate, of resistivity typically around 1–4 kΩ·cm. At this low doping level, the small positive charge in the silicon–oxide interface is enough to invert the Si surface, generating a parasitic conduction layer [89]. On top of the BOX the CMOS device will experience variable electric fields that modulate the inversion layer on the bottom of the BOX. The result is a nonlinear capacitive effect (Cbox), as shown in Fig. 3.48, that generates a large amount of harmonics. As described in [90], a trap-rich layer, fabricated by deposition of an undoped polycrystalline silicon film on the high-resistivity silicon, stabilizes the interface and reduces the harmonic distortion, as reported in Fig. 3.49. Indeed, the high-density traps will fix the potential at the interface and DC or RF/mm-wave signals do not impact the carrier distribution.

For the sake of completeness, be aware of the following issues.

(1) This effect is very similar to what happens with GaAs, fabricated on semi-insulating material; the distortion is low as the GaAs interface has a trap-rich layer.

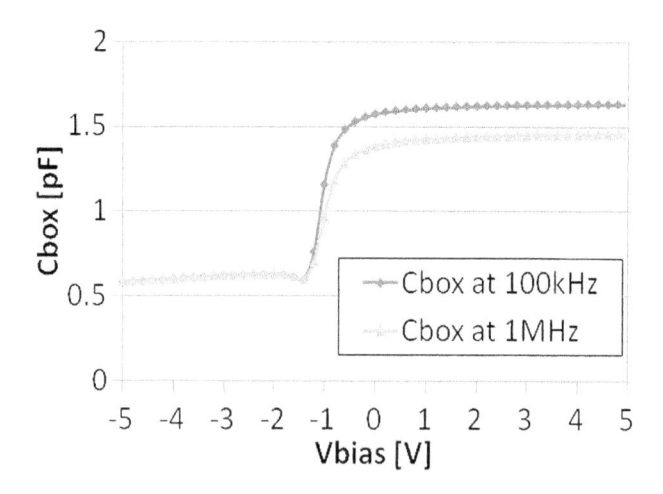

Fig. 3.48 Nonlinear Cbox behavior ©[2010] IEEE. Reprinted, with permission, from T.-Y. Lee and S. Lee, "Modeling of SOI FET for RF switch applications," in *IEEE Radio Frequency Integrated Circuits Symp. Dig. (RFIC)* May 2010, pp. 479–482.

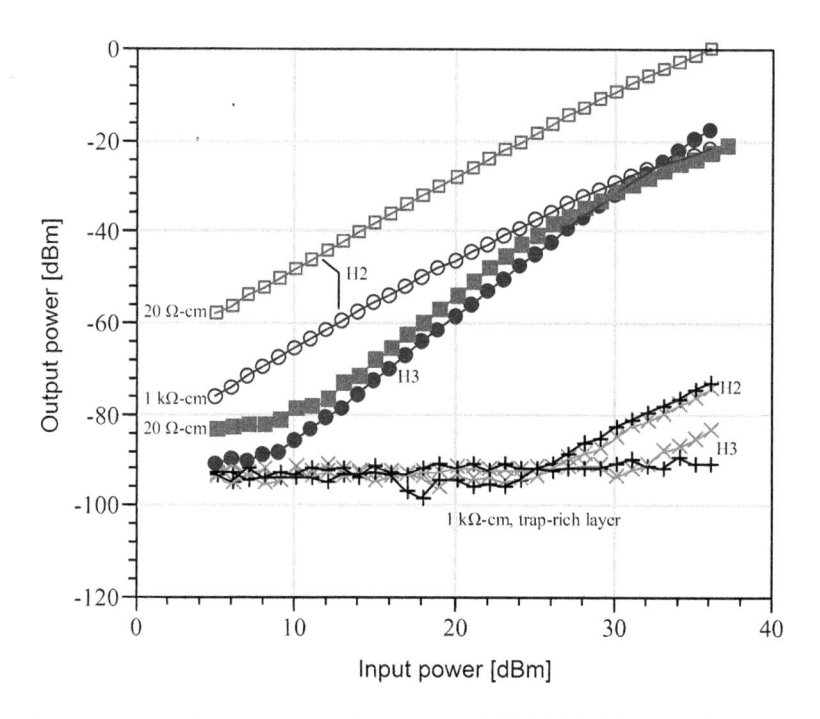

Fig. 3.49 Harmonic improvement due to trap-rich SOI substrate ©[2008] IEEE. Reprinted, with permission, from D. C. Kerr *et al.*, "Identification of RF harmonic distortion on Si substrate and its reduction using a trap-rich layer," in *Proc. IEEE SiRF*, Jan. 2008, pp. 151–154.

(2) Silicon-on-sapphire (SOS) CMOS devices do not suffer this effect since sapphire is a dielectric.
(3) The idea that, at high frequency, the majority carriers do not respond in the high-resistivity substrate, due to their long dielectric relaxation time, is wrong. Indeed, thanks to the entire interface inversion, the carriers can also move laterally following the high-frequency signals.
(4) The classical BSIM model does not take into account the above-discussed substrate effect [87]. The PSP SOI model [87] has been recognized to be one of the most suited models for SOI FETs since it is able to take into account this kind of substrate-dependent nonlinearity effect.

3.10.3 Fully depleted SOI

During the past few years, the technology improvement based on an accurate control of the epi-thickness and the possibility of back-gating provided by FD platforms has gained large interest on FD devices, down to the 28-nm technology nodes [91] (functionality has been already demonstrated down to 20 nm [92]). A very thin BOX, i.e. 25 nm deep reduced to only 7 nm after some process steps, is exploited in conjunction with custom implantation below the BOX. This helps in controlling the threshold voltage, while the channel is kept un-doped. The thin BOX allows very good short-channel effect control, thus allowing the decreasing of channel length at 28 nm while the leakage remains almost the same as its bulk equivalent. As described in [91], the DIBL and sub-threshold slope are lower than 100 mV and 85 mV/dec, respectively, while the device is able to achieve an I_{on} of 1070 $\mu A/\mu m$ at V_{DD} of 1 V with 32% and 84% speed boost at 1 V and 0.6 V, respectively, compared with the 28-nm bulk technology. Further improvements can be obtained by exploiting tensile strained SOI [93]. This technique allows the enhancement of carrier mobility that provides a gain of 20% in performance.

The FDSOI process is therefore very promising and reliability results are also aligned to, or even better than, bulk equivalent platforms. However, as usual, the limited voltage swing will impact the exploitation of a single common-source stage for power amplifier and advanced architectures (stacked, cascode, transformer-based).

3.10.4 ESD diodes in SOI platforms

Diodes are basic devices for ESD protection and for voltage/current references. In SOI a diode can be build like an FET with source and drain reverse implanted, as shown in Fig. 3.50. The main difference with respect to bulk diodes is that here the conduction is a function of the diode perimeter instead of being dependent on the area. So in order to assure the same current and thus protection level, bigger diodes are needed in SOI. Moreover, the gate can control the current capability of the diode and thus these devices are referred to as gated diodes. Generally, when gated diodes are available as ESD diodes, the gate is connected to the cathode to avoid effects on current capability.

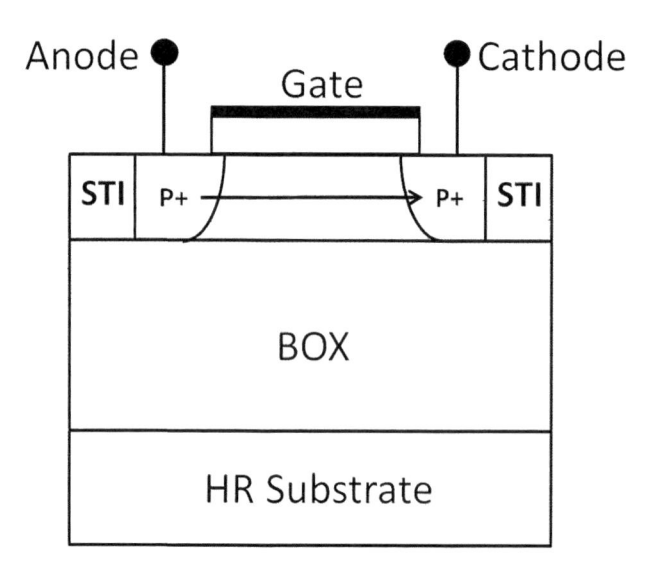

Fig. 3.50 ESD gated diode in SOI technology.

3.11 Conclusions

CMOS processes had, have, and will have a relevant impact in all our life actions, as they are pervasively adopted in almost any device we use. The success of this technology and the huge investment on its roadmaps are evidence of its future evolution and large exploitation. The device scaling, forecasted down to 7 nm, will allow the development of faster and more compact devices. However, as clearly discussed in this chapter, technologies for power amplifiers need to endure wide voltage swings. On top of architecture solution and SOI FD-enabling performance, further CMOS-based improvements are achievable by exploiting double-gate fully depleted thin-body devices [94]. This approach allows relaxation of the DIBL effect associated with large drain voltage swing and very low channel length. However, the allowable voltage swing can only be reduced with the device scaling as the transition frequency-breakdown product is material dependent. In this perspective, new materials have gained large interest, like GaN with its very high breakdown capability and a JFoM better than 5 THz · V [95]. The wide exploitation of CMOS and the performance of the GaN HEMT (high electron mobility transistor) focused the research of the integration of GaN devices on silicon substrate. First attempts [96] demonstrated the possibility of fabrication of GaN HEMTs on silicon with f_{max} of 40 GHz and an operating supply voltage of 40 V. Latest results [97] have shown f_{max} up to 200 GHz and an operating supply voltage around 15 V, while the GaN device is integrated on CMOS-compatible silicon substrate. These results and future optimization will gain more and more interest for the development of high-performance mm-wave power amplifiers [98–102].

We would like to close this chapter with the motivation that pushed designers like us to deeply investigate technology issues: optimum design at high frequency and large signal directly involves the physics and all the parasites associated with the symbol we ususally place in the simulator schematics. We have to recommend all designers to be more technology aware, since this is a sure key to success in optimum product development.

Acknowledgment

The authors thanks Francesco Carrara for helpful discussions.

References

[1] I. Aoki, S. Kee, D. Rutledge, and A. Hajimiri, "A 2.4-GHz, 2.2-W, 2-V fully integrated CMOS circular-geometry active-transformer power amplifier," in *Proc. IEEE Custom Integrated Circuits Conference (CICC)*, May 2001, pp. 57–60.

[2] I. Aoki, S. Kee, D. Rutledge, and A. Hajimiri, "Distributed active transformer: a new power combining and transformation technique," *IEEE Trans. Microwave Theory and Tech.*, vol. **50**, pp. 316–331, Jan. 2012.

[3] G. D. Vendelin, A. M. Pavio, and U. L. Rohde, *Microwave Circuit Design Using Linear and Nonlinear Techniques*, New York: John Wiley & Sons, 1990.

[4] A. Johnson, "Physical limitations on frequency and power parameters of transistors," *IRE International Convention Record*, vol. **13**, pp. 27–34, Mar. 1966.

[5] B. G. Streetman, *Solid-State Electronic Devices*, Englewood Cliffs, NJ: Prentice-Hall International, Inc., 1990.

[6] S. M. Sze, *Physics of Semiconductor Devices*, New York: John Wiley & Sons, 1969.

[7] R. S. Muller and T. I. Kamins, *Device Electronics for Integrated Circuits*, New York: Wiley, 1986.

[8] S. Voinigescu, *High Frequency Integrated Circuits*, Cambridge University Press, 2013.

[9] Y. Taur and T. H. Ning, *Fundamentals of Modern VLSI Devices*, 2nd edition, Cambridge University Press, 2009.

[10] F. Stern and W. E. Howard, "Properties of semiconductor surface inversion layers in the electric quantum limit," *Phys. Rev.* , vol. **163**, no. 3, pp. 816–835, Nov. 1967.

[11] T. O. Dickson *et al.*, "The invariance of the characteristic current densities in nanoscale MOSFETs and its impact on algorithmic design methodologies and design porting of Si(Ge) (Bi)CMOS high-speed building blocks," *IEEE J. Solid-State Circuits*, vol. **41**, pp. 1830–1845, Aug. 206.

[12] D. Frohman-Bentchkowsky and A. S. Grove, "Conductance of MOS transistors in saturation," *IEEE Trans. Electron Devices*, vol. **16**, pp. 108–113, Jan. 1969.

[13] S. C. Cripps, *RF Power Amplifiers for Wireless Communications*, 2nd edition, Norwood, MA: Artech House, 2006.

[14] P. J. Tasker and B. Hughes, "Importance of source and drain resistance to the maximum f_T of millimeter-wave MODFETs, *IEEE Electronic Device Lett.*, vol. **10**, pp. 291–293, July, 1989.

[15] P. H. Ladbrooke, *MMIC Design: GaAs, FETs and HEMTs*, Boston, MA: Artech House, 1989.

[16] U. Gogineni, J. A. del Alamo, and C. Putnam, "RF power potential of 45 nm CMOS technology," in *Proc. of Topical Meeting on Silicon Monolithic Integrated Circuits in RF Systems (SiRF)*, pp. 204–207, Jan. 2010.

[17] C. Fager *et al.*, "A comprehensive analysis of IMD behavior in RF CMOS power amplifiers, *IEEE J. Solid-State Circuits*, vol. **39**, pp. 24–34, Jan. 2004.

[18] M. Schwartz, *Information Transmission, Modulation, and Noise*, New York: McGraw-Hill, 1959.

[19] N. B. Carvalho and J. C. Pedro, "Large- and small-signal IMD behavior of microwave power amplifiers, *IEEE Trans. Microwave Theory Tech.*, vol. **47**, pp. 2364–2374, Dec. 1999.

[20] J. Brinkhoff, A. E. Parker, and M. Leung, "Baseband impedance and linearization of FET circuits," *IEEE Trans. Microwave Theory Tech.*, vol. **51**, pp. 2523–2530, Dec. 2003.

[21] J. Brinkhoff and A. E. Parker, "Effect of baseband impedance ion FET intermodulation," *IEEE Trans. Microwave Theory Tech.*, vol. **51**, pp. 1045–1051, Mar. 2003.

[22] D. Webster, J. Scott, and D. Haigh, "Control of circuit distortion by the derivative superposition method," *IEEE Microwave and Guided Wave Lett.*, vol. **6**, pp. 123–125, Mar. 1996.

[23] D. R. Webster, D. G. Haigh, and A. E. Parker, "Novel circuit synthesis technique using short channel GaAs FETs giving reduced intermodulation distortion," in *Proc. IEEE Int. Symp. Circuits and Systems (ISCAS)*, Apr.–May 1995, pp. 1348–1351.

[24] V. Aparin and L. E. Larson, "Modified derivative superposition method for linearizing FET low-noise amplifier," *IEEE Trans. Microwave Theory Tech.*, vol. **53**, pp. 571–581, Feb. 2005.

[25] N. B de Carvalho and J. C. Pedro, "Two-tone IMD asymmetry in microwaver power amplifiers," in *IEEE MTT-S Int. Microwave Symp. Dig. (IMS)*, vol. **1**, June 2000, pp. 445–448.

[26] J. F. Sevic, K. L. Burguer, and M. B. Steer, "A novel envelope-termination load-pull method for ACPR optimization of RF/microwave power amplifiers," in *IEEE MTT-S Int. Microwave Symp. Dig. (IMS)*, vol. **2**, June 1998, pp. 723–726.

[27] N. B. de Carvalho and J. C. Pedro, "A comprehensive explanation of distortion sideband asymmetries," *IEEE Trans. Microwave Theory Tech.*, vol. **50**, pp. 2090–2101, Sep. 2002.

[28] N. Kim, V. Aparin, and L. E. Larson, "Analysis of IM3 asymmetry in MOSFET small-signal amplifiers," *IEEE Trans. Circuits and Systems I*, vol. **58**, pp. 668–676, Apr. 2011.

[29] C. Wang, M. Vaidyanathan, and L. E. Larson, "A capacitance-compensation technique for improved linearity in CMOS class-AB power amplifiers, *IEEE J. Solid-State Circuits*, vol. **39**, pp. 1927–1937, Nov. 2004.

[30] P. R. Gray and R. G. Meyer, *Analog Integrated Circuits*, New York: McGraw-Hill, 1994.

[31] W. B. Davenport and W. L. Root, *An Introduction to the Theory of Random Signals and Noise*, New York: McGraw-Hill, 1958.

[32] J. L. Lawson and G. E. Uhlenbeck, *Threshold Signals*, New York: McGraw-Hill, 1950.

[33] A. Van der Ziel, *Noise*, New York: Prentice-Hall, 1954.

[34] J. L. Plumb and E. R. Chenette, "Flicker noise in transistors, *IEEE Trans. Electron Devices*, vol. **10**, pp. 304–308, Sep. 1963.

[35] R. C. Jaeger and A. J. Broderson, "Low frequency noise sources in bipolar junction transistors, *IEEE Trans. Electron Devices*, vol. **17**, pp. 128–134, Feb. 1970.

[36] M. Nishida, "Effects of diffusion-induced dislocations on the excess low-frequency noise, *IEEE Trans. Electron Devices*, vol. **20**, pp. 221–226, Mar. 1973.

[37] R. A. Pucel, H. A. Haus, and H. Statz, "Signal and noise properties of gallium arsenide microwave field effect transistors," Marton L. (ed), *Advances in Electronics and Electron Physics*, Academic Press, vol. **38**, pp. 195–265, 1975.

[38] H. Statz, H. A. Haus, and R. A. Pucel, "Noise characteristics of gallium arsenide field-effect transistor," *IEEE Trans. Electron Devices*, vol. **21**, pp. 549–562, Sep. 1974.

[39] A. Cappy, "Noise modeling and measurement techniques,' *IEEE Trans. Microwave Theory Tech.*, vol. **36**, pp. 1–10, Jan. 1988.

[40] A. Van der Ziel, "Thermal noise in field effect transistor," *Proceedings of the IRE*, vol. **50**, pp. 1808–1812, Aug. 1962.

[41] A. Van der Ziel,"Gate noise in field effect transistors at moderately high frequencies," *Proceedings of the IEEE*, vol. **51**, pp. 461–467, Mar. 1963

[42] M. Rudolph and F. Bonani, "Low frequency noise in nonlinear systems," *IEEE Microwave Magazine*, vol. **10**, pp. 84–92, Feb. 2009.

[43] T. H. Lee and A. Hajimiri, "Oscillator phase noise: a tutorial," *IEEE J. Solid-State Circuits*, vol. **35**, pp. 326–336, Mar. 2000.

[44] J. Plá,"Characterization and modeling of high power RF semiconductor devices under constant and pulsed excitations," in *Proc. Fifth Annual Wireless Symposium*, Feb. 1997, pp. 467–472.

[45] Q. Li, J. Zhang, W. Li, *et al.*, "RF circuit performance degradation due to soft breakdown and HC effect in deep submicrometer CMOS technology," *IEEE Trans. Microwave Theory Tech.*, vol. **49**, pp. 1546–1551, Sep. 2001.

[46] J. P. Walko and B. Abadeer, "RF S-parameter degradation under hot carrier stress," in *Proc. IEEE 42nd Annual Int. Reliability Phys. Symp.*, Apr. 2004, pp. 422–425.

[47] C. Yu and J. S. Yuan, "MOS RF reliability subject to dynamic voltage stress modeling and analysis," *IEEE Trans. Electron Devices*, vol. **52**, pp. 1751–1758, Aug. 2005.

[48] W.-C. Lin, T.-C. Wu, Y.-H. Tsai, L.-J. Du, and Y.-C. King, "Reliability evaluation of class-E and class-A power amplifiers with nanoscaled CMOS technology," *IEEE Trans. Electron Devices*, vol. **52**, pp. 1478–1483, July 2005.

[49] E. Xiao, "Hot carrier effects on CMOS RF amplifiers," in *Proc. 43rd Annual Int. Reliability Phys. Symp.*, Apr. 2005, pp. 680–681.

[50] T. Sowlati and D. M. W. Leenaerts, "A 2.4-GHz 0.18-μm CMOS self-biased cascode power amplifier," *IEEE J. Solid-State Circuits*, vol. **38**, pp. 1318–1324, Aug. 2003.

[51] C. D. Presti, F. Carrara, A. Scuderi, S. Lombardo, and G. Palmisano, "Degradation mechanisms in CMOS power amplifiers subject to radio-frequency stress and comparison to the DC case," in *IEEE Proc. International Reliability Physics Symp. (IRPS)*, Apr. 2007, pp. 86–92.

[52] S. Lombardo, J. H. Stathis, B. P. Linder, *et al.*, "Dielectric breakdown mechanisms in gate oxides," *J. Applied Physics*, vol. **98**, pp. 121301–121301-36, Dec. 2005.

[53] A. Scuderi, T. Biondi, E. Ragonese, and G. Palmisano, *Integrated Inductors and Transformers: Characterization, Design and Modeling for RF and mm-Wave Applications*, Boca Raton, FL: CRC/Taylor and Francis, 2010.

[54] S. S. Mohan, M. del Mar Hershenson, S. P. Boyd, and T. H. Lee, "Simple accurate expressions for planar spiral inductances," *IEEE J. Solid-State Circuits*, vol. **34**, pp. 1419–1424, Oct. 1999.

[55] C. P. Yue and S. S. Wong, "On-chip spiral inductors with patterned ground shields for Si-based RF ICs," *IEEE J. Solid-State Circuits*, vol. **33**, pp. 743–752, May 1998.

[56] F. Carrara, A. Italia, E. Ragonese, and G. Palmisano, "Design methodology for the optimization of transformer loaded RF circuits, *IEEE Trans. Microwave Theory Tech.*, vol. **53**, pp. 761–768, Apr. 2006.

[57] J. R. Long, "Monolithic transformers for silicon RF IC design," *IEEE J. Solid-State Circuits*, vol. **35**, pp. 1368–1382, Sep. 2000.

[58] E. Ragonese, G. Sapone, and G. Palmisano, "High-performance interstacked transformers for mm-wave ICs," *Wiley Microwave and Optical Technology Letters*, vol. **52**, pp. 2160–2163, Sep. 2010.

[59] V. Giammello, E. Ragonese, and G. Palmisano, "A transformer-coupling current-reuse SiGe HBT power amplifier for 77-GHz automotive radar," *IEEE Trans. Microwave Theory Tech.*, vol. **60**, pp. 1676–1683, June 2012.

[60] E. Ragonese, G. Sapone., V. Giammello, and G. Palmisano, "Analysis and modeling of interstacked transformers for mm-wave applications," *Springer Analog Integrated Circuits and Signal Processing*, vol. **72**, pp. 121–128, June 2012.

[61] V. Giammello, E. Ragonese and G. Palmisano, "Transformer-coupled cascode stage for mm-wave power amplifiers in sub-μm CMOS technology," *Springer Analog Integrated Circuits and Signal Processing*, vol. **66**, pp. 449–453, Mar. 2011.

[62] K. T. Ng, B. Rejaei, and J. N. Burghartz, "Substrate effects in monolithic RF transformers on silicon," *IEEE Trans. Microwave Theory Tech.*, vol. **50**, pp. 377–383, Jan. 2002.

[63] T. Biondi, A. Scuderi, E. Ragonese, and G. Palmisano, "Analysis and modeling of layout scaling in silicon integrated stacked transformers," *IEEE Trans. Microwave Theory Tech.*, vol. **54**, pp. 2203–2210, May 2006.

[64] E. Ragonese, A. Scuderi, T. Biondi, and G. Palmisano, "Scalable lumped modeling of single-ended and differential inductors for RF IC design," *Wiley International Journal of RF and Microwave Computer-Aided Engineering*, vol. **19**, pp. 110–119, Jan. 2009.

[65] W. Liu, *MOSFET Models for SPICE Simulation, Including BSIM3v3 and BSIM4*, New York: Wiley-IEEE Press, 2001.

[66] W. Kruppa and K. F. Sodomsky, "An explicit solution for the scattering parameters of a linear two-port measured with an imperfect test set," *IEEE Trans. Microwave Theory Tech.*, vol. **19**, pp. 122–123, Jan. 1971.

[67] S. Padmanabhan, P. Kirby, J. Daniel, and L. Dunleavy, "Accurate broadband on-wafer SOLT calibrations with complex load and thru models," in *Proc. 61st ARFTG Microwave Measurements Conf. – Spring*, June 2003, pp. 5–10.

[68] A. Ferrero and U. Pisani, "QSOLT: a new fast calibration algorithm for two port S parameter measurements," in *Proc. 38th ARFTG Microwave Measurements Conf.*, Dec. 1991, pp. 15–24.

[69] H. J. Eul and B. Schiek, "Reducing the number of calibration standards for network analyzer calibration," *IEEE Trans. Instrum. Meas.*, vol. **40**, pp. 732–735, Aug. 1991.

[70] A. Ferrero and U. Pisani, "Two-port network analyzer calibration using an unknown thru," *IEEE Microwave and Guided Wave Lett.*, vol. **2**, pp. 505–507, Dec. 1992.

[71] G. F. Engen and C. A. Hoer, "Thru-reflect-line: an improved technique for calibrating the dual six-port automatic network analyzer," *IEEE Trans. Microwave Theory Tech.*, vol. **27**, pp. 987–993, Dec. 1979.

[72] H. J. Eul and B. Schiek, "Thru-match-reflect: one result of a rigorous theory for de-embedding and network analyzer calibration," in *Proc. 18th European Microwave Conf.*, Sep. 1988, pp. 909–914.

[73] D. F. Williams and R .B. Marks, "LRM probe-tip calibrations using nonideal standards," *IEEE Trans. Microwave Theory Tech.*, vol. **43**, pp. 466–469, Feb. 1995.

[74] A. Davidson, K. Jones, and E. Strid, "LRM and LRRM calibrations with automatic determination of load inductance," in *Proc. 36th ARFTG Microwave Measurements Conf.*, Nov. 1990, pp. 57–63.

[75] B. Hughes, A. Ferrero, and A. Cognata, "Accurate on-wafer power and harmonic measurements of mm-wave amplifiers and devices," in *IEEE MTT-S Int. Microwave Symp. Dig.*, June 1992, pp. 1019–1022.

[76] T. Biondi, A. Scuderi, E. Ragonese, and G. Palmisano, "Characterization and modeling of silicon integrated spiral inductors for high-frequency applications," *Springer Analog Integrated Circuits and Signal Processing*, vol. **51**, pp. 89–100, May 2007.

[77] A. Rumiantsev and N. Ridler, "VNA Calibration," *IEEE Microwave Magazine*, vol. **9**, pp. 86–99, June 2008.

[78] D. R. Root and S. Fan, "Experimental evaluation of large-signal modeling assumptions based on vector analysis of bias-dependent S-parameter data from MESFET and HEMTs," *IEEE MTT-S Int. Microwave Symp. Dig. (IMS)*, June 1992, pp. 61–110.

[79] L. Betts, "Tracking advances in pulsed S-parameter measurements," in *Microwave and RF Journal*, Sep. 2007.

[80] J. P. Teyssier, Ph. Bouysse, Z. Ouarch, T. Peyretaillade, and R. Quere, "40 GHz/150 ns versatile pulsed measurement system for microwave transistor isothermal characterization," *IEEE Trans. Microwave Theory Tech.*, vol. **46**, pp. 2043–2052, Dec. 1998.

[81] Agilent Technologies, "Agilent PNA microwave network analyzers, application note: 1408-12."

[82] B. Noori, et al.,"Load-pull measurements using modulated signals," in *Proc. 36th European Microwave Conf.*, Sep. 2006, pp. 1594–1597.

[83] A. Ferrero, F. Sanpietro, U. Pisani, and C. Beccari, "Novel hardware and software solutions for a complete linear and nonlinear microwave device characterization," *IEEE Trans. Instrum. Meas.*, vol. **43**, pp. 299–305, Apr. 1994.

[84] D. Schreurs and J. Verspecht, "Large-signal modeling and measuring go hand-in-hand: accurate alternatives to indirect S-parameter methods," *Wiley International Journal of RF and Microwave Computer-Aided Engineering*, vol. **10**, pp. 6–18, Jan. 2000.

[85] K. Bernstein and N. J. Rohrer, *SOI Circuit Design Concepts*, Springer, 2000.

[86] D. Wang et al., "High performance SOI RF switches for wireless applications," in *10th IEEE Int. Conf. Solid-State and Integrated Circuit Technology (ICSICT)*, Nov. 2010, pp. 611–614.

[87] T.-Y. Lee and S. Lee, "Modeling of SOI FET for RF switch applications," *IEEE Radio Frequency Integrated Circuits Symp. Dig. (RFIC)*, May 2010, pp. 479–482.

[88] F. Carrara, C. D. Presti, A. Scuderi, and G. Palmisano, "Single-transistor latch-up and large-signal reliability in SOI CMOS RF power transistors," *Elsevier Solid-State Electronics*, vol. **54**, pp. 957–964, Sep. 2010.

[89] Y. Wu et al., "SiO_2 interface layer effects on microwave loss of high-resistivity CPW line," *Microwave and Guided Wave Lett.*, vol. **9**, pp. 10–12, Jan. 1999.

[90] D. C. Kerr et al., "Identification of RF harmonic distortion on Si substrate and its reduction using a trap-rich layer", in *Proc. IEEE SiRF*, Jan. 2008, pp. 151–154.

[91] N. Planes et al., "28 nm FDSOI technology platform for high-speed low-voltage digital applications," in *Proc. IEEE International Symposium on VLSI Technology*, June 2012, pp. 133–134.

[92] O. Weber et al., "Junction engineering for FDSOI technology speed/power enhancement," in *Proc. IEEE International Symposium on VLSI Technology, Systems, and Applications*, April 2013, pp. 1–2.

[93] C. Fenouillet-Beranger et al., "Enhancement of devices performance of hybrid FDSOI/bulk technology by using UTBOX sSOI substrates," in *Proc. IEEE International Symposium on VLSI Technology*, June 2012, pp. 115–116.

[94] W. G. Neudeck, "An overview of double-gate MOSFETs," in *Proc. of the 15th Biennial University/Government/Industry Microelectronics Symposium*, July 2003, pp. 214–217.

[95] M. J. Rosker, J. D. Albrecht, E. Cohen, J. Hodiak, and Tsu-Hsi Chang, "DARPA's GaN technology thrust," in *IEEE MTT-S Int. Microwave Symp. Dig. (IMS)*, May 2010, pp. 1214–1217.

[96] V. Hoel et al., "High-power AlGaN/GaN HEMTs on resistive silicon substrate," *IEEE Electronics Lett.*, vol. **38**, no. 14, pp. 750–752, July 2002.

[97] F. Medjdoub, M. Zegaoui, B. Grimbert, et al., "First demonstration of high-power GaN-on-silicon transistors at 40 GHz," *IEEE Electron Device Lett.*, vol. **33**, pp. 1168–1170, Aug. 2012.

[98] E. Ballesteros, F. Perez, and J. Perez, "Analysis and design of microwave linearized amplifiers using active feedback," *IEEE Trans. Microwave Theory Tech.*, vol. **36**, pp. 499–504, Mar. 1988.

[99] D. Suh and J. G. Fossum, "The effect of body resistance on the breakdown characteristics of SOI MOSFETs," *IEEE Trans. Electron Devices*, vol. **41**, pp. 1063–1066, June 1994.

[100] S. C. Kuehne, A. B. Y. Chan, C. T. Nguyen, and S. S. Wong, "SOI MOSFET with buried body strap by wafer bonding," *IEEE Trans. Electron Devices*, vol. **45**, pp. 1084–1091, May 1998.

[101] K. K. Young and J. A. Barnes, "Avalanche-induced drain-source breakdown in silicon-on-insulator n MOSFETs," *IEEE Trans. Electron Devices*, vol. **35**, pp. 426–431, Apr. 1988.

[102] S. E. Thompson et al., "A 90 nm logic technology featuring strained-silicon," *IEEE Trans. Electron Devices*, vol. **51**, pp. 1790–1797, Nov. 2004.

4 Linear-mode mm-wave silicon power amplifiers

James Buckwalter

4.1 Why linear?

Two fundamental questions arise in the design of a power amplifier circuit. First, what is the maximum output power given a device technology? Second, what is the peak power-added efficiency (PAE) given a device technology? In this chapter, these fundamental limits on output power and efficiency are analyzed for single-stage, linear millimeter-wave (mm-wave) PAs based on device and circuit factors. Linear amplifiers are favored under several conditions. Foremost, a linear response might be demanded for complex signaling. Quadrature amplitude modulation (QAM) requires a linear response to support the peak-to-average power ratio (PAPR) of the signal envelope. As shown in Fig. 4.1, the in-phase and quadrature symbols – shown as a 64QAM constellation – are up-converted to an RF carrier and amplified by the power amplifier. The envelope of the modulated RF signal is shown at the output of the PA and demonstrates substantial variation in the output power. The ratio of the peak-to-average power indicates the linearity requirements of the power amplifier. Additionally, concerns about the out-of-channel interference, e.g. spectral leakage, also dictate requirements for system linearity. However, specification of adjacent channel interference is not well-established at mm-wave bands since the spatial selectivity of the communication system at these bands is considered to be higher than for RF systems.

In RF systems, linear amplifiers are rarely sufficiently linear and additional linearization can be introduced in the form of feedforward, feedback, or digital pre-distortion. The evolution of communication standards tends to favor more spectrally efficient communication and, in turn, motivates interest in linearization techniques for power amplifiers. In the case of mm-wave communication, early commercial standards have avoided the need for linear amplifiers by using constant envelope waveforms such as GMSK or simple modulation schemes that feature low PAPR such as QPSK. As more interest develops in mm-wave QAM, circuit and system techniques that offer a linear variation in the output power will return to the forefront.

A second motivation for linear amplifier designs is that this class of operation may offer competitive output power and efficiency at mm-wave bands to other classes of amplifiers. For instance, switching amplifiers require harmonic control that complicates the design of these amplifiers over wide bandwidth at mm-wave bands. Another desirable feature of linear amplifiers is that two PAs can be configured to be maximally

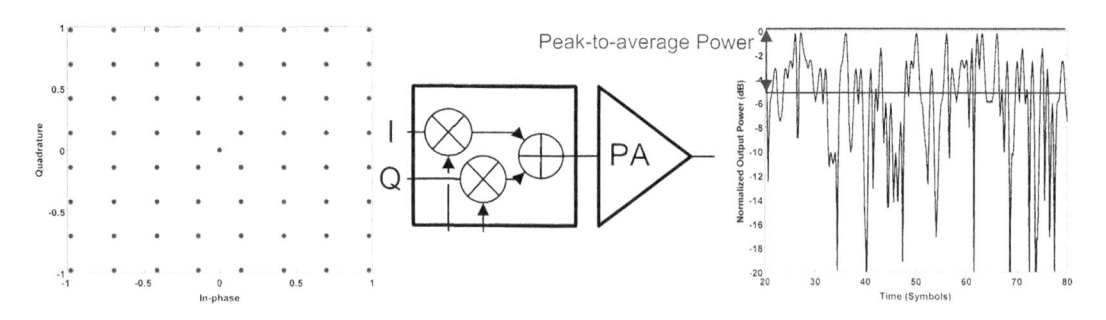

Fig. 4.1 Up-conversion of baseband 64QAM constellation to an RF envelope through a linear PA.

efficient at both the average and peak power levels. Recent research on mm-wave Doherty amplifiers has investigated the potential for high efficiency under high-PAPR conditions.

In this chapter, the power-added efficiency is decomposed into factors related to circuit and device technology. First, features of silicon device technologies are discussed to indicate the trade-offs between PAE and output power. Next, the layout of the mm-wave device is discussed in terms of the parasitic components that impact the device gain. From these considerations, the optimum class of operation for a linear amplifier can be demonstrated depending on the desired frequency and output power. Finally, the implementation of Doherty amplifiers for mm-wave bands is discussed.

4.1.1 Power-added efficiency

The power-added efficiency (PAE) of a power amplifier is defined by

$$PAE = \frac{P_{OUT} - P_{IN}}{P_{DC}} = \left(1 - \frac{P_{IN}}{P_{OUT}}\right)\frac{P_{OUT}}{P_{DC}} = \left(1 - \frac{1}{G}\right) \cdot \eta, \quad (4.1)$$

where the amplifier gain is G and η is the drain (or collector) efficiency of the amplifier. An upper bound on this gain is the maximum available gain of the device at the fundamental frequency. However, the small-signal gain overestimates the PAE since the gain of the amplifier compresses near the peak PAE. High gain suggests that the PAE converges to the drain efficiency of the transistor.

Unfortunately, transistors operating at mm-wave bands offer low gain, which tends to limit PAE. Figure 4.2 shows the degradation in the PAE relative to the drain efficiency as a function of gain. At 10 dB gain, the PAE will be 90% of the drain efficiency. Gain compression of 1 dB shows a marginal drop in the PAE. However, if the gain is only 6 dB, the efficiency drops more significantly as the amplifier enters compression. The PAE is only 75% of the available drain efficiency. As the gain compresses by 1 dB, the PAE drops by an additional 6%. If the gain compresses by 2 dB, the PAE drops by 15%. Low gain of mm-wave PAs poses a problem for efficient amplifier design regardless of the desired class of operation.

Drain efficiency introduces another dimension to the problem of improving PAE. For instance, drain efficiency can be improved through switching power-amplifier circuit techniques. However, if this comes at the expense of gain, then the solution may not

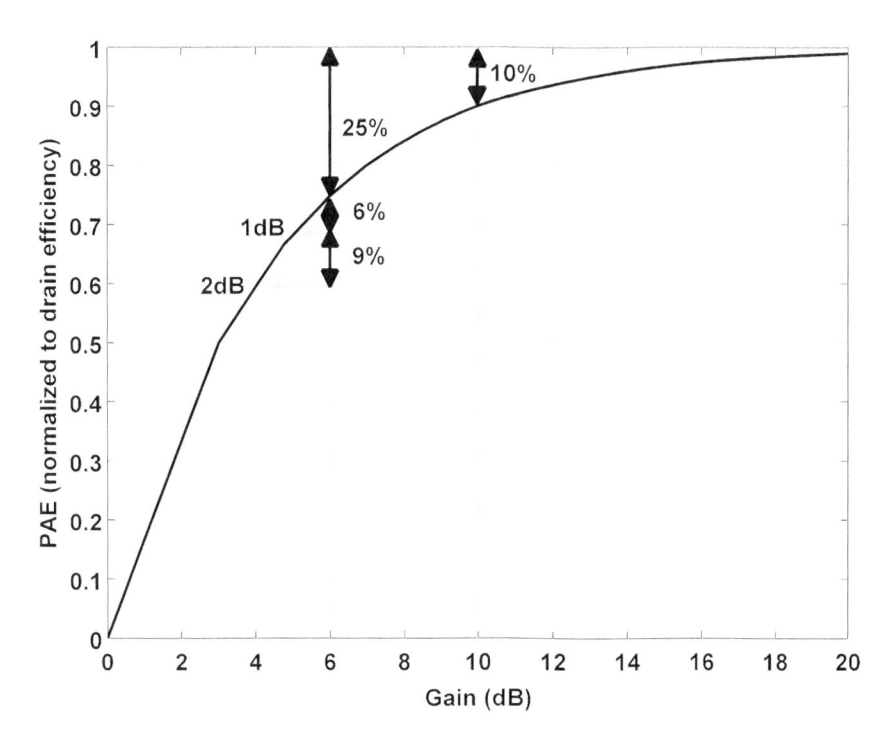

Fig. 4.2 PAE degradation due to low gain and gain compression for mm-wave power amplifiers.

actually improve PAE. The drain efficiency from (4.1) can be further decomposed into a combination of device and circuit design factors. Rudimentary understanding of the drain efficiency associates this with a class of operation, e.g. class-A, which will be discussed in detail later to show the relationship between the class of operation and the amplifier performance. Detailed analysis recognizes that a limit on the conduction of the power amplifier is imposed by the "knee voltage" V_K that represents a minimum limit on the swing across the transistor. In bipolar devices, this knee voltage is associated with the transistor entering saturation when the collector–emitter voltage is too low. In FETs, this knee voltage might be associated with the device being pushed into the deep triode region. In both cases, the device's speed – and hence gain – tend to be reduced when it is operated in these regions.

The knee voltage is found from the large-signal I_D–V_{DS} characterization of the transistor and is generally the minimum drain–source (or collector–emitter)[1] voltage that keeps the transistor in saturation. Keeping the transistor in saturation maintains the speed, e.g. f_T/f_{MAX}, as well as the loadline resistance. Thus, the drain efficiency is more aptly expressed as a product of $\eta = \left(1 - \frac{V_K}{V_{DD}}\right) \cdot S_F$, where S_F is the waveform shape factor associated with the class of operation and V_{DD} is the DC supply voltage on the transistor. Clearly, it is desirable to provide the largest supply voltage to the device without compromising the transistor reliability. Typically, fineline CMOS transistors feature thin oxide devices to provide the gain, but these devices also compel a

---

[1] The notation for the proceeding discussion will be associated with CMOS field-effect transistors.

lower supply voltage, which reduces the ratio of V_K/V_{DD} shown in the second term. To allow higher V_{DD}, circuits employ cascode or stacked FETs to increase the output power [1–3].

The shape factor ranges from one-half to one based on the operational class of the amplifier. For class-A operation, the S_F is one-half. Switch-mode amplifiers realize drain efficiencies approaching unity through harmonic tuning of the amplifier that allows zero-voltage switching. However, tuning the harmonics of a mm-wave signal requires gain at these harmonics. Since the transistor current gain is typically falling off at these harmonics and lower quality factor passives are available at the harmonic frequencies, the harmonic gain is low.

Finally, an output matching network is required between the loadline resistance of the transistor and the load impedance. In particular, mm-wave PAs most often implement this matching network on-chip. The loss factor of this output matching network LF_o reduces the power available from the drain. An idealized output matching network, i.e. one that is composed of passive elements with an infinite quality factor Q dissipates no power in the output network, and the loss factor is unity. In general, the loss factor is related to the transformation ratio between the output of the power amplifier and the load resistance.

Therefore, the PAE is more accurately expressed as a product of the device and circuit factors that will be explored in this chapter in the context of mm-wave power amplifiers operated in the linear mode [4]:

$$PAE \leq \left(1 - \frac{1}{G}\right) \cdot \left(1 - \frac{V_K}{V_{DD}}\right) \cdot S_F \cdot LF_o. \tag{4.2}$$

First, the large-signal device characterization is studied to quantify the biasing and knee-voltage relationship to the output power. Second, the small-signal gain of a mm-wave PA is characterized. Next, the waveform shape factor is quantified for linear classes of operation. Finally, the loss factor is discussed for realizable passives in silicon processes. Ultimately, approximations of each of these performance metrics show the fundamental PAE limits for linear amplifiers at mm-wave frequencies and the conditions under which linear amplifiers are favored.

4.2 Linear amplifier design: large-signal device characterization

Fineline CMOS devices present several challenges for mm-wave PA design. The characterization of I_D vs V_{DS} is shown in Fig. 4.3 for a 45-nm CMOS SOI process. The DC voltage supply for this process is intended to be 1 V. The knee voltage V_K is defined as the V_{DS} voltage at which the transistor is no longer operating in the saturation region. The knee voltage has the unfortunate consequence of restricting the voltage swing across the transistor. Once the transistor leaves the saturation region, the device speed is greatly diminished and the output impedance drops. Therefore, the knee voltage limits the peak RF voltage swing from V_{DD} to $V_{DD} - V_K$ and necessitates a lower output loadline impedance.

4.2 Linear amplifier design

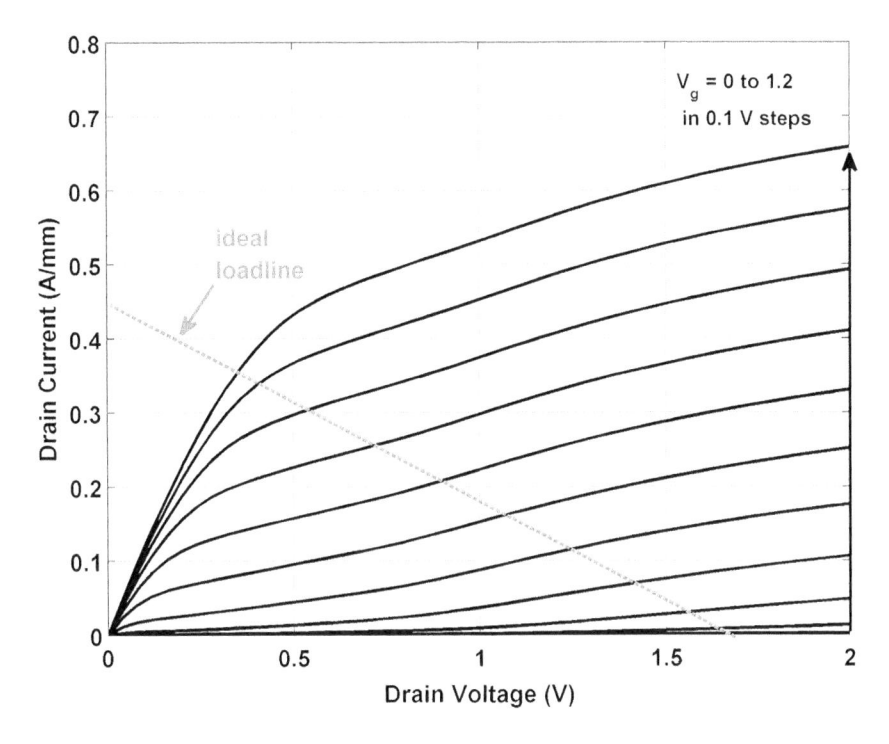

Fig. 4.3 Simulated I_D–V_{DS} characterization of an n-FET in a 45-nm CMOS SOI process with a notional loadline for matching. After [4]. Reprinted with permission from the IEEE.

The knee voltage is apparent from the I_D–V_{DS} curve in Fig. 4.3, where the knee voltage is roughly 0.4 V. Therefore, the peak voltage swing reduces to 1.6 V. The high knee voltage and low breakdown voltage of CMOS devices pose a significant challenge to the efficiency of silicon CMOS PAs. The factor in (4.2) suggests that the knee-voltage limit on the drain efficiency is $(1 - 0.4/1) = 0.6$. Therefore, the PA cannot achieve a PAE of more than 60% unless other methods are found to circumvent this limit. The loadline might be extended to 2 V to accommodate the peak RF swing. Typically, CMOS devices can handle large drain-to-source voltage swings reliably when the drain current at the peak drain swing is low [5, 6]. This minimizes degradation due to hot carrier injection. Based on the peak amplitude, the power is increased by either increasing the maximum current density of the transistor or increasing the device geometry.

SiGe HBTs are also candidates for mm-wave PA circuits. Two breakdown voltage limits exist for an HBT. The collector–emitter breakdown voltage is measured when the base is connected to an open circuit and is referred to as BV_{CEO}. However, BV_{CEO} is not the maximum breakdown voltage unless the HBT is biased with a fixed base current. Otherwise, the voltage breakdown for V_{CE} may be greater when the base current is provided from a low impedance bias. In this case, the second relevant breakdown voltage is the collector–base breakdown voltage, which is measured when the emitter is connected to an open circuit and is referred to as BV_{CBO}. The characterization of the collector current I_C vs collector–emitter voltage V_{CE} is shown in Figs. 4.4 and 4.5 under

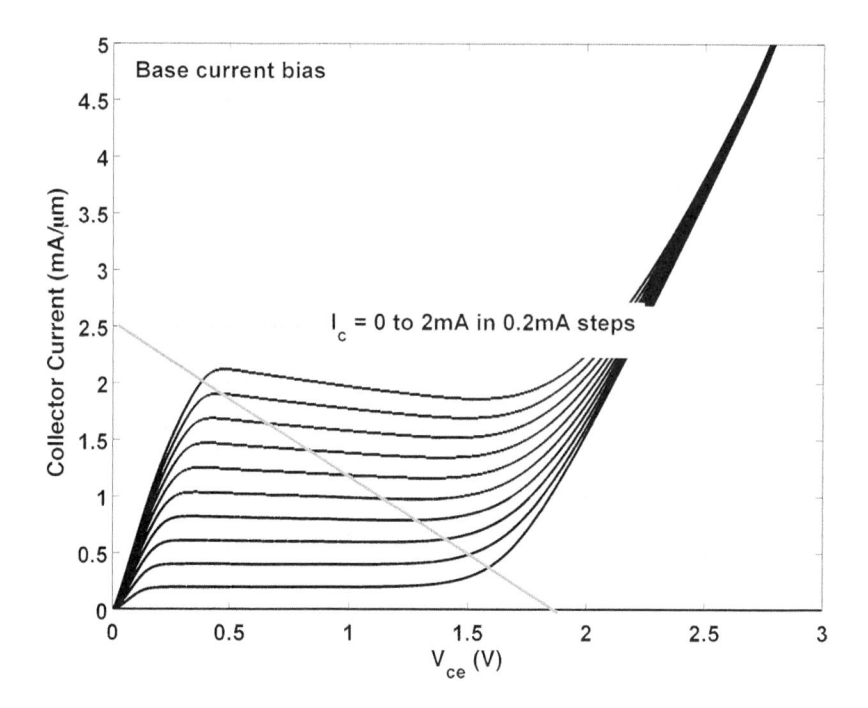

Fig. 4.4 Simulated I_C–V_{CE} characterization of an HBT in a 120-nm BiCMOS process under current biasing at the base.

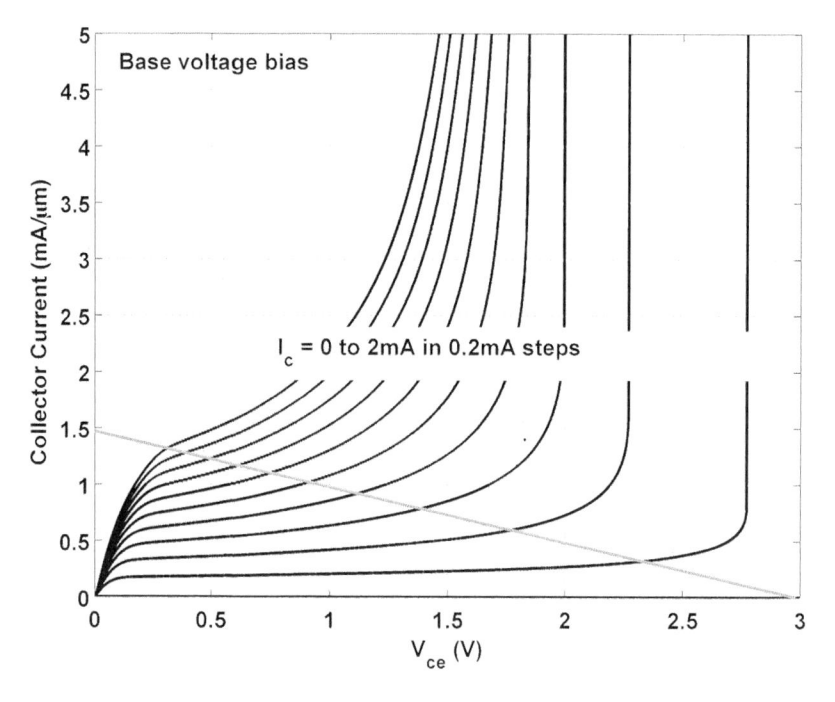

Fig. 4.5 Simulated I_C–V_{CE} characterization of an HBT in a 120-nm BiCMOS process under voltage biasing at the base.

each of these biasing conditions. In Fig. 4.4, the transistor is biased with a current source which supplies a constant current into the base of the transistor. The primary breakdown mechanism for the HBT is impact ionization, resulting from high-velocity electrons generating electron–hole pairs in the base region. These extra electrons contribute to current either into the emitter or into the base. In the case of a base current bias, the electrons are forced into the emitter and greatly degrade the device behavior. The lower base impedance allows these electrons to flow out of the base and leads to higher voltage limits. This breakdown is higher and allows for more RF swing at the collector node. For example, the 0.12-μm SiGe HBT offers $BV_{CEO} \approx 1.7$ V and $BV_{CBO} \approx 6.0$ V. From Fig. 4.5, the knee voltage of the HBT is also roughly 0.4 V, while this device can withstand higher voltage swings than the comparable CMOS transistor.

4.2.1 Determining device geometry

Figure 4.3 indicates that the maximum drain current density is 0.4 mA/μm for the FET at the knee voltage V_K of roughly 0.4 V. The optimal drain current density is somewhat process-invariant for CMOS and scaling of transistors does not significantly improve how much current density the transistor can handle. Higher current densities also tend to limit reliable operation and lower device performance.

The output power of the PA is now determined through the drain current. Under class-A operation, the power delivered to a matched load is

$$P_{DEL,A} = \frac{1}{4}(V_{DD} - V_K) I_{MAX} = \frac{1}{4}(V_{DD} - V_K) J_{MAX} W, \quad (4.3)$$

where I_{MAX} is the maximum drain current, J_{MAX} is the maximum current density per unit width, and W is the width of the FET. Based on the required output power and the limits on the voltage and current, the desired transistor width is determined as

$$W = \frac{4 P_{DEL,A}}{(V_{DD} - V_K) J_{MAX}}. \quad (4.4)$$

Therefore, the width of the FET required to deliver 20 dBm would be 1.6 mm! This basic trade-off lends itself to one of the most important features of mm-wave PA design. While this substantial transistor size might not pose significant issues for low-frequency (RF) design, the device layout introduces substantial practical issues at high frequency that will ultimately reduce the gain. The parasitic components associated with the layout of a large transistor will introduce parasitic resistances, capacitances, and inductances which degrade the available gain.

4.2.2 Loadline resistance

The peak output power (or the device size) also determines the first-order loadline matching of the FET. The conventional loadline – for class-A operation – is centered at the V_{DD} of the FET. Therefore,

$$R_{L,A} = \frac{2(V_{DD} - V_K)}{I_{MAX}} = \frac{2(V_{DD} - V_K)}{J_{MAX} W} = \frac{(V_{DD} - V_K)^2}{2 P_{DEL,A}}. \quad (4.5)$$

The knee voltage clearly reduces the loadline resistance for a given output power. For the transistor described in Fig. 4.3, the loadline matching impedance is roughly $3\Omega\cdot\text{mm}$. Therefore, the proposed 20 dBm PA from the previous section has a loadline resistance of $1.8\,\Omega$ based on a single transistor output stage. Low loadline resistance poses a design challenge to the circuit since a larger quality factor is required in order to match the loadline resistance to the load impedance. Higher quality factors introduce more loss in the output matching network.

4.2.3 Output matching

The output matching network is required to transform the loadline resistance to the load impedance, assumed here to be R_L. An ideal impedance transformation network would require a quality factor of

$$Q_o = \sqrt{\frac{R_L}{R_{L,A}} - 1} = \sqrt{\frac{2R_L P_{DEL,A}}{(V_{DD} - V_K)^2} - 1} \approx \frac{\sqrt{2R_L P_{DEL,A}}}{V_{DD} - V_K}. \qquad (4.6)$$

For an impedance transformation associated with the output network of Q_o, passives available in a monolithic CMOS or BiCMOS process typically have a Q of less than 20, which limits the effectiveness of using high-impedance transformations to increase the output power of the transistor. The loadline resistance of the transistor determines the required impedance transformation. Wider devices will reduce the loadline resistance and require more impedance transformation into a 50-Ω load, i.e. higher Q_o, and significant loss in the output matching network. As the loadline resistance becomes very small – as for Si CMOS – the output matching network Q_o increases. Unfortunately, the output matching network has losses associated with the passive components that are used in the network. For an LC matching network, the loss factor is

$$L_o = \frac{Q}{Q_o + Q}. \qquad (4.7)$$

For the proposed 20 dBm PA, the output impedance transformation is roughly five for the curve illustrated in Fig. 4.3. If Q of the passive elements is 10, the proposed 20 dBm PA will theoretically have a maximum PAE of 67% due to the output matching loss factor. If the Q of the passive elements is 20, the maximum PAE increases to 80%. Consequently, a significant factor in mm-wave design becomes the fundamental Q of the passive components and the complexity of the output matching network. For instance, if one considers the losses in a two-stage LC matching network the losses of the matching network are compounded and significantly limit the PAE. It is notable that lower output power requirements imply lower Q and the lower output power of each PA might be offset through power combining of several PAs. In this case, the losses of the output power combining network must be included to account for the loss of the entire combined PA.

4.2.4 Example: quality factor for $\lambda/4$ impedance transformers

Silicon CMOS and BiCMOS processes offer a variety of passive matching elements for mm-wave operation. Lumped passives such as inductors, transformers, and capacitors

4.2 Linear amplifier design

and distributed elements based on transmission lines can be implemented on-chip for matching and tuning networks. In mm-wave circuits, the Q of the matching elements must be compared carefully to minimize losses in the matching networks. To give realistic Q values, simulations and measurements of transmission structures are presented as an example.

The Q of a transmission line is highly dependent on the implementation of the transmission line and in particular the back-end-of-the-line metallization. Standard CMOS interconnect metallization is optimized for criteria determined by digital VLSI demands. In this case, a reverse scaling rule is observed, where the thickness of the metallization is increased for metal layers further from the substrate. Most notably, the distance from the top metal layer to the substrate is relatively small for standard CMOS processes. For RF and mm-wave inductors and transmission lines, this is not optimal. The highest Q passives are realized in a thick top metal layer far from the substrate, a feature sometimes found for silicon or silicon–germanium processes optimized for RF and mm-wave applications.

A cross section of a shielded coplanar transmission line is illustrated in Fig. 4.6. The transmission line consists of a signal line implemented in the topmost metal layer (2.2 μm thick) with coplanar ground lines on either side of the signal line. Additionally, a ground plane (255 nm thick) is placed below the signal line to shield the signal from the substrate. This ground plane is connected to the coplanar ground plane through a field of vias along the walls of the transmission line. A 50-Ω shielded coplanar transmission line

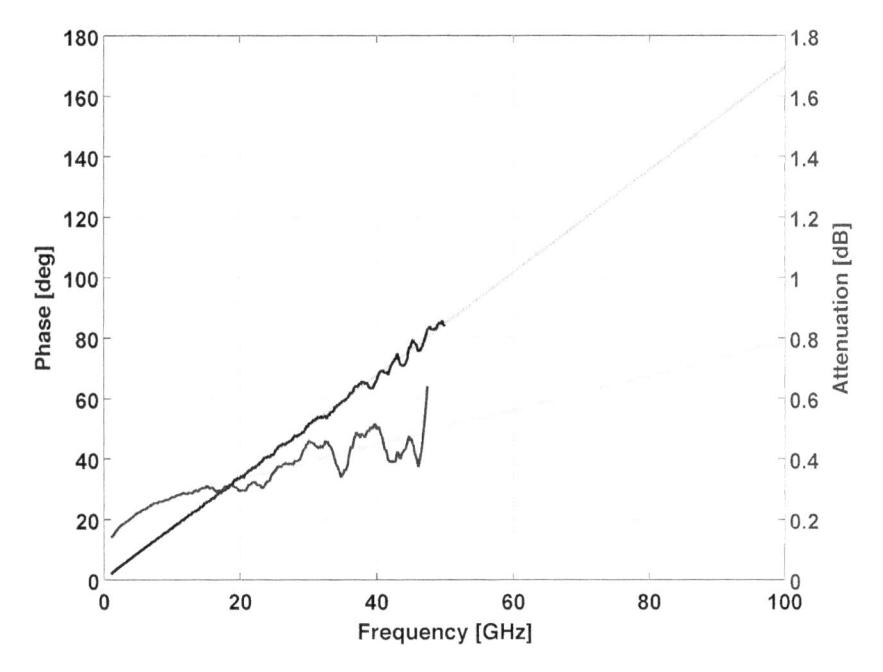

Fig. 4.6 Measured phase and attenuation vs frequency for a 670-μm (λ/4 at 45 GHz) transmission line in a 45-nm CMOS SOI process. After [4]. Reprinted with permission from the IEEE.

148 Linear-mode mm-wave silicon power amplifiers

design based on a CMOS-SOI process is demonstrated in Fig. 4.6 for a transmission line length of 670 μm. The distance between the signal and the side shield is 8 μm while the distance from the signal line to the bottom shield is 6.3 μm. The spacing of the signal line to these grounds determines the characteristic impedance of the transmission line.

The transmission line can be characterized by the attenuation constant α and phase constant β per unit length. Consequently, the Q for a $\lambda/4$ transmission line is given by

$$Q = \frac{\beta}{2\alpha} = \frac{\pi}{4\alpha l}. \tag{4.8}$$

The last equality holds when the length of the line is $\lambda/4$. When the phase shown in Fig. 4.6 is 90°, the attenuation is roughly 0.5 dB. Therefore, $Q = \frac{\pi}{4 \cdot 0.5} 8.6 = 13.5$. Work on slow-wave transmission lines has indicated similar Q between shielded CPW and slow-wave CPW structures [7]. Consequently, the implementation of impedance transformation networks illustrates the finite quality factor which ultimately limits the efficiency of the PA through the loss factor discussed in the previous section.

4.3 Gain of mm-wave amplifiers

The gain of a mm-wave amplifier is determined by the *intrinsic* characteristics of the device and the *extrinsic* parasitics of the interconnect that connects the device to a passive component such as a transmission line. For the purposes of mm-wave design, these extrinsic components are required to quantify the high-frequency gain. The intrinsic and extrinsic elements are shown in Fig. 4.7. The proposed model is conceptually simple to facilitate analytical conclusions.

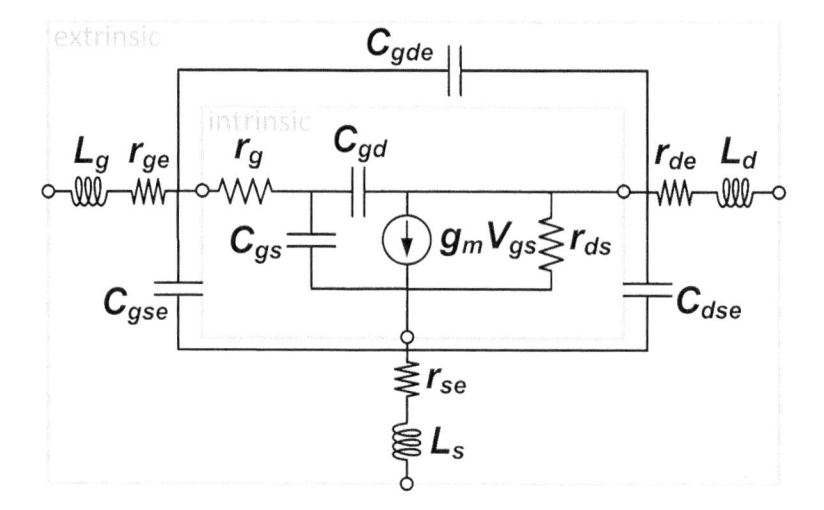

Fig. 4.7 A simplified small-signal model for a CMOS field-effect transistor for power-amplifier design decomposed into the intrinsic device found from the device model and the extrinsic parasitics resulting from the interconnection to the device.

4.3.1 Intrinsic device f_T/f_{MAX}

The gain of an amplifier is determined by the speed of the device. Power amplifiers are typically based on common-source transistor configurations. A well-known, first-order frequency limit on the current gain, h_{21}, of a common-source transistor is defined by

$$h_{21}(f_T) = \left|\frac{i_{out}}{i_{in}}\right|_{v_o=0} = 1, \quad (4.9)$$

where f_T is the unity current gain frequency of the transistor. The small-signal model of the intrinsic FET from Fig. 4.7 determines that the f_T is

$$f_T = \frac{1}{2\pi} \frac{g_m}{C_{gs} + C_{gd}}. \quad (4.10)$$

However, the f_T of the transistor is an oversimplified metric of circuit performance since the power amplifier must be matched to a load impedance.

The maximum frequency at which a transistor produces power gain is defined by f_{MAX}. This is commonly defined from the maximum available gain (MAG) of the transistor. The MAG of a mm-wave amplifier can be characterized as

$$MAG = \left(\frac{f_{MAX}}{f_o}\right)^2 = \frac{1}{4}\left(\frac{f_T}{f_o}\right)^2 \frac{R_O}{R_I}, \quad (4.11)$$

where R_I and R_O are respectively the small-signal input and output resistance. For mm-wave amplifiers, the low ratio f_{MAX}/f_o limits MAG. For operation at 60 GHz, the gain is on the order of 10 dB for a transistor f_{MAX} of 180 GHz.

Detailed calculations of MAG based on the intrinsic device model shown in Fig. 4.7 give the power available to the FET and power available to a load under matched conditions. The gate-to-drain capacitance C_{gd} defines the output matching of the transistor. Small-signal analysis of the intrinsic device model in Fig. 4.7 determines that the output resistance of the transistor is $R_O = \frac{1}{g_m}\frac{C_{gs}}{C_{gd}}$. Therefore, the existence of the reactive element C_{gd} implies the existence of a real output resistance that maximizes the small-signal gain. The input resistance of the intrinsic model of the transistor is only real if a resistive r_g exists. For the intrinsic device,

$$f_{MAX} = \sqrt{\frac{1}{16\pi^2} \frac{g_m}{C_{gs} C_{gd} r_g}} = \sqrt{\frac{1}{8\pi} \frac{f_T}{C_{gd} r_g}}. \quad (4.12)$$

This connects the transistor f_T to the f_{MAX} through an additional time constant related to the gate resistance and gate–drain capacitance. For CMOS processes, scaling does not tend to strongly impact either C_{gd} or r_g. Advancements in lithography allow scaling of features but the overlap of the drain diffusion makes a greater contribution to the gate–drain capacitance. In other words, the ratio C_{gd}/C_{gs} increases as gate lengths scale down. Furthermore, the gate resistance should increase as the pitch of the gate contact scales. The gate contact can be made from one side of the transistor (single contact) or, to compensate for high r_g, both sides of the transistor (double contact), with the latter offering the advantage of lower resistance but higher gate–drain capacitance.

Fineline CMOS transistors do not exhibit saturation behavior that is adequately described by a drain current that does not change with drain–source voltage. Referring to the I_D-V_{DS} curves in Fig. 4.3, the CMOS SOI FET exhibits relatively low drain–source resistance r_{ds} that results from channel modulation and also drain-induced barrier lowering (DIBL). The r_{ds} introduces an element that shunts the output resistance of the intrinsic device and therefore negatively impacts the achievable f_{MAX} of the transistor. An expression for the f_{MAX} that incorporates the impact of the channel conductance is

$$f_{MAX} = \sqrt{\frac{1}{8\pi} \frac{f_T}{r_g} \frac{g_m r_{ds}}{C_{gs} + (1 + g_m r_{ds}) C_{gd}}}. \tag{4.13}$$

The impact of r_{ds} appears through the intrinsic gain $g_m r_{ds}$ of the transistor. A plot of the degradation in f_{MAX} with intrinsic gain is shown in Fig. 4.8. As the r_{ds} approaches infinity, this result converges to the result in (4.12). As the intrinsic gain drops below around 10, a 10% drop in the f_{MAX} is observed. Since the MAG is generally simulated from the small-signal model, the degradation of the intrinsic gain due to channel conductance is accounted for through the intrinsic device model. Nonetheless, scaling of CMOS devices tends to offer primarily improvements in the f_T while f_{MAX} comes from more significant changes to the processing of silicon devices.

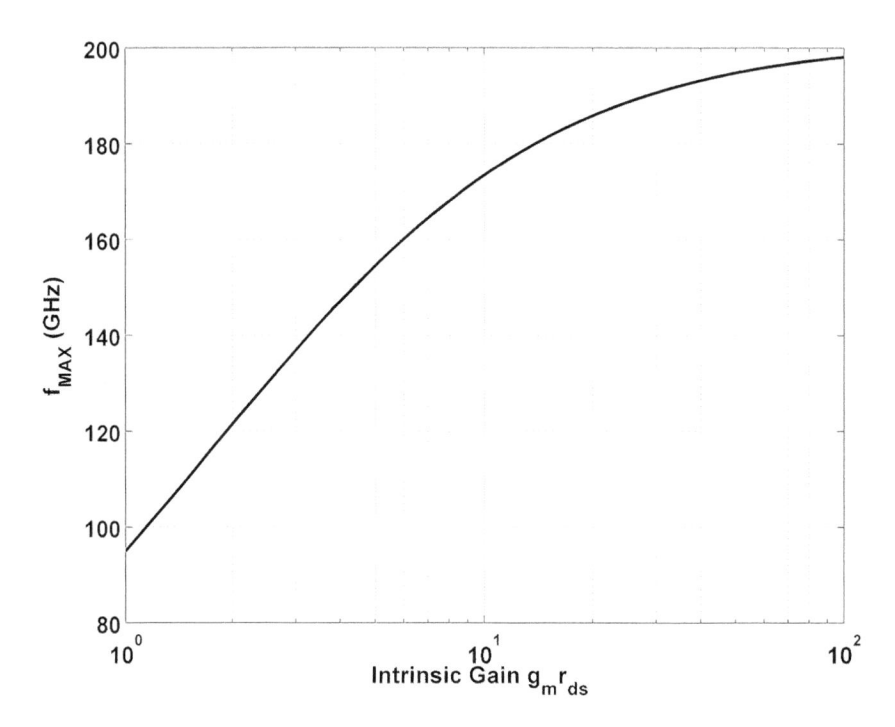

Fig. 4.8 Degradation in MAG cut-off frequency as a function of intrinsic device gain ($g_m = 1.7\,\text{S}$, $C_{gs} = 1\,\text{pF}$, $C_{gd} = 400\,\text{fF}$, $r_g = 0.5\,\text{Ohms}$).

4.3.2 mm-Wave transistor layout

While the native device might offer extremely high f_T/f_{MAX}, the mm-wave power transistor is a large transistor which typically requires interconnection to higher metal layers where thicker – higher Q – passives can be realized. For this reason, the circuit designer should characterize a transistor which includes relevant interconnects. The consequence of these interconnects is the introduction of additional sources of capacitance, resistance, and – perhaps most notably – inductance in the device model. Even a few picohenry of inductance introduces a significant impedance at 60 GHz. There are a number of approaches to making connections to the gate, drain, and source of the transistor and hence trade-offs are encountered as one attempts to minimize the parasitic components. As is evident in (4.12), the gate resistance and gate–drain capacitance are two which directly effect the f_{MAX}. However, source parasitics will also be shown to contribute an unfortunate degradation to the device performance. Exact expressions for the MAG in the presence of various extrinsic parasitic components are complicated to derive. As an alternative, the impact of each extrinsic parasitic component is analyzed for an intuitive approach. By characterizing the reduction in f_{MAX}, the reduction in the MAG can be determined. As a rule of thumb, the frequency of f_T/f_{MAX} for the extrinsic device is commonly around 60%–70% that of the original intrinsic device.

Gate resistance

The extrinsic gate resistance occurs due the metal loss and additionally via losses required to distribute a signal to the gate of the entire transistor. While the double-contact transistor might reduce the gate resistance for the intrinsic device, additional wiring might be required to make contact with both sides of the gate. For straightforward analysis, the extrinsic and intrinsic gate resistances can be lumped according to $r'_g = r_g + r_{ge}$.

Gate inductance

To minimize the gate–drain capacitance, single-contact transistor configurations are sometimes preferable. The gate connection is penalized at the expense of the drain and source routing and introduces additional gate inductance as well as potential mutual coupling between the gate and drain nodes. As shown in Fig. 4.7, any inductance in series with the gate of the transistor can be appropriately resonated through proper input impedance matching with no degradation in f_{MAX}. However, exact formulas that account for the self- and mutual inductance of the gate interconnection are difficult to derive and require electromagnetic simulation to estimate.

Gate–drain capacitance

In the absence of an extrinsic gate resistance, additional parasitic gate–drain capacitance should not explicitly change f_{MAX} since the additional reactance could be canceled through appropriate choice of reactance at the input of the PA. Thus, the extrinsic gate–drain capacitance does not greatly impact MAG but impacts the stability conditions for the amplifier. However, the combination of extrinsic gate–drain capacitance and the gate resistance mentioned previously does reduce f_{MAX}. Considering the total gate–drain capacitance as the shunt combination of intrinsic and extrinsic elements suggests $C'_{gd} = C_{gd} + C_{gde}$. The new expression for f_{MAX} is determined by substituting r'_g and C'_{gd} into (4.12).

Source resistance

Feeding the source from each transistor to a common ground plane introduces a path of resistance to the transistor. Resistive degeneration at the source of a transistor reduces the power gain and directly reduces f_{MAX}. An expression for f_{MAX} in the presence of source resistance takes into account degeneration through substitution of the degenerated transconductance and increased C_{gs} of the transistor. While the source resistance does not change f_T, f_{MAX} becomes

$$f_{MAX} = \sqrt{\frac{1}{4C_{gd}r_g} \frac{g_m}{C_{gs} + C_{gd}(1+g_m r_{se})}} \approx \sqrt{\frac{1}{8\pi} \frac{f_T}{C_{gd}r_g} \frac{1}{1+\frac{C_{gd}g_m r_{se}}{C_{gs}}}}. \quad (4.14)$$

Here, the impact of the source resistance depends strongly on the transconductance. The left-hand plot of Fig. 4.9 compares the degradation of a large transistor relative to a small transistor (one-tenth the size of the larger device). For additional source resistance, the larger device suffers far more dramatically from the source resistance. The right-hand plot of Fig. 4.9 normalizes the impact of the transconductance with respect to the source degeneration as a feedback factor $g_m r_{se}$. Re-examining (4.14), the feedback factor should be minimized to the point that the degradation in the f_{MAX} is only 10%. In this case,

$$g_m r_{se} \leq 0.2 \left(1 + \frac{C_{gs}}{C_{gd}}\right). \quad (4.15)$$

Clearly, the acceptable feedback factor is limited by the ratio C_{gd}/C_{gs}. As mentioned earlier, this parameter ratio tends to worsen device scaling and, for processes lower than 22 nm, this ratio may approach one. In this case, the feedback factor must be less

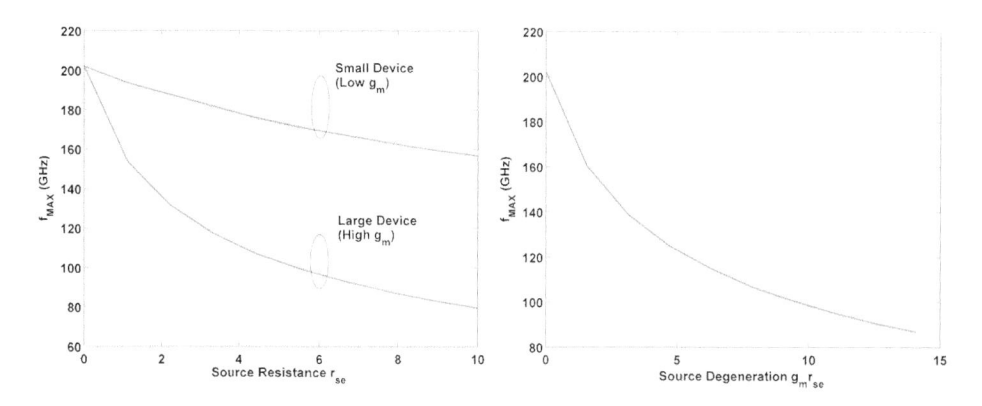

Fig. 4.9 Degradation in MAG cut-off frequency as a function of source resistance (left) and source degeneration feedback factor (right) ($g_m = 1.7$ S, $C_{gs} = 1$ pF, $C_{gd} = 400$ fF, $r_g = 0.5$ Ohms).

than 0.5, limiting the tolerable transconductance and hence tolerable transistor size and output power.

The study of the source resistance in the CMOS transistor model leads us to make a fundamental observation about mm-wave PA design. Amplifier circuits that are not required to deliver high power levels are not as difficult to design as mm-wave PAs because the smaller devices are less sensitive to the effects of these device parasitics.

Source inductance

Source inductance provides another significant source of degeneration to the transistor at mm-wave bands since it produces a non-negligible impedance at high frequency. Analysis of the MAG and the corresponding f_{MAX} cannot be expressed in simplified formulas. Alternatively, the impact of source inductance can be included with the intrinsic device model using computational techniques that can be corroborated with intuition. Here, the source resistance will be assumed to be zero. A plot of the G_{MAG} for our simplified device model is shown in the left-hand plot of Fig. 4.10. It becomes clear that increasing source inductance does *not* change the f_{MAX} since all curves intersect for 0 dB. This agrees with intuition that the choice of source inductance can reactively cancel with the input reactance at one frequency.

However, the source inductance suppresses the maximum available gain below f_{MAX}. The "cliff" in the MAG behavior occurs where the device becomes unconditionally stable. This cliff moves to a lower frequency, which substantially reduces the gain at frequencies below f_{MAX}. Second, the roll-off in the MAG curve becomes more gradual. Comparing a large device and a small device, the device with lower transconductance does not exhibit reduced f_{MAX}. In fact, it may be slightly enhanced – by the presence of small amounts of source inductance. However, a large device is extremely sensitive to small amounts of source inductance due to the shift in the "cliff" in the MAG curve. In this example, a 1 pH source inductance drops the gain at 60 GHz by 3 dB.

For mm-wave PA designs, the MAG can be approximated to account for the effect of source resistance and inductance [8]:

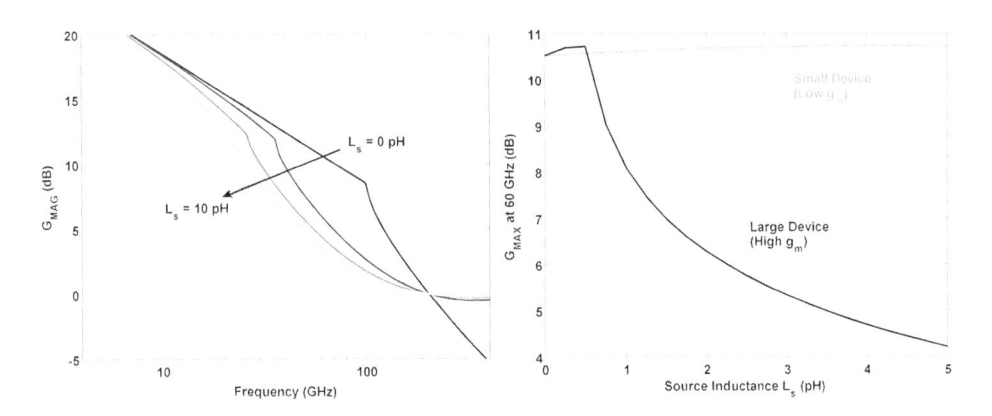

Fig. 4.10 Degradation in MAG cut-off frequency as a function of intrinsic device gain ($g_m = 1.7$ S, $C_{gs} = 1$ pF, $C_{gd} = 400$ fF, $r_g = 0.5$ Ohms).

154 Linear-mode mm-wave silicon power amplifiers

$$MAG = \left(\frac{1}{f_o}\right)^2 \frac{f_T}{4\pi C_{gd} r_g \left(\frac{R_s}{r_g} + 1 + \omega_T \frac{L_s}{2r_g}\right)} \approx \left(\frac{f_{MAX}}{f_o}\right)^2 \frac{1}{1 + \frac{R_s}{r_g} + \frac{\omega_T L_s}{2r_g}}. \quad (4.16)$$

This expresssion is only approximate due to the complex behavior of the MAG below f_{MAX}.

Drain resistance

Drain resistance does not appreciably limit the gain of the transistor as long as the effect of channel modulation is relatively small. However, drain resistance may substantially impact the power available to the load since even a small amount of drain resistance might be substantial compared with the loadline resistance of the device. Consequently, the drain resistance can play a role in the efficiency and peak output power but to first order might be ignored for calculating gain.

Drain–source capacitance

Most designs tend to be more tolerant of drain–source capacitance since this capacitance can be incorporated into the output matching network. Consequently, the layout of the transistor tends to prefer capacitance between the drain and source to capacitance between the drain and gate, which has been suggested to impact the stability and speed of the transisor. However, in switched-mode amplifiers, this capacitance plays a critical role in the waveform at the drain node and limits the power and frequency of operation.

4.3.3 Example: effect of layout on MAG for a CMOS SOI FET

The layout of a power transistor is shown in Fig. 4.11. This layout is sometimes referred to as a zipper layout because of the interleaved connections between the source and drain. Note that the gate connection is a single contact to prevent excessive overlap between the gate and drain interconnects. Furthermore, the source and drain

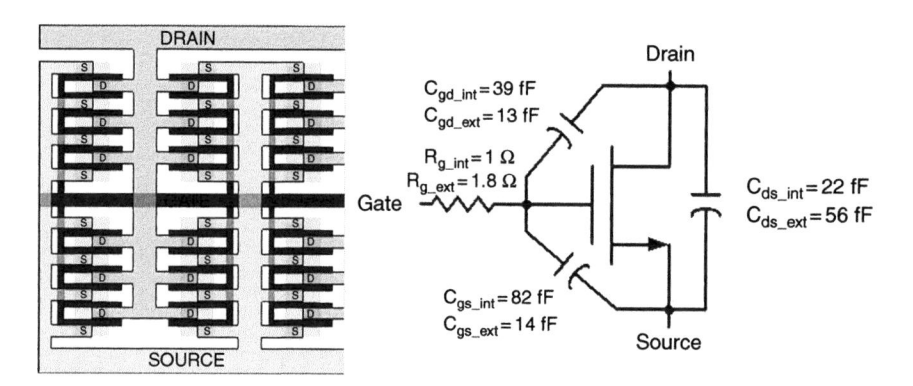

Fig. 4.11 Zipper layout of a mm-wave power device. Schematic indicates extracted RC device model for a 160-μm transistor associated with the intrinsic and extrinsic capacitances. Adapted from [2].

metallization avoids direct overlap to limit the contribution of the fringing capacitance. Characterization of f_{MAX} is highly dependent on the device layout and the parasitics associated with the interconnect stack. A top aluminum interconnect layer ($t = 2.2$ μm) is 9 μm above the substrate. The capacitances of the floating source and drain nodes are reduced by a 145-nm-thick buried-oxide layer (BOX), which reduces the losses in the relatively low-resistivity (13 $\Omega \cdot$ cm) silicon substrate.

A strategy for estimating the parasitics of the interconnects to the transistor involves two separate simulation techniques. First, the *RC* parasitics are extracted through the parameters provided with the process design kit. For physical connections between the device diffusions and the second metal layer, *RC* extraction of the layout of the device is used. Second, the interconnects above the lowest metal layers are simulated in an electromagnetic simulator to properly account for the inductive and capacitive coupling between the different device terminals. The drain, gate, and source interconnections are simulated with an electromagnetic field solver from the top metal layer to the second metal layer.

A small-signal model is shown in Fig. 4.11, which accounts for the parasitic capacitance and resistance introduced by the lowest two metallization layers. This breaks down the contributions due to intrinsic and extrinsic device capacitances for a 160-μm transistor. The extrinsic capacitance adds only about 25% and 33% more capacitance to the gate–source and gate–drain capacitances, respectively. However, the extrinsic capacitance dominates the drain–source capacitance. This capacitance can be tuned out in the case of a common-source amplifier. Additionally, the extrinsic capacitance is found to be about the same size as the intrinsic capacitance, suggesting substantial power loss in the input of the transistor.

A simulation of the f_T and f_{MAX} of the device with S-parameters extracted for the interconnect structure between the second metallization and the top metal layer is shown in Fig. 4.12. Whereas the f_T of the native device model is roughly 380 GHz, accounting for the interconnect to the transistor reduces f_T/f_{MAX} by around 40%. Above a current density of 0.27 mA/μm, f_T and f_{MAX} are greater than 200 GHz.

4.4 Linear classes of operation

Power amplifiers are categorized based on bias point and harmonic control. Linear classes of operation are generally divided into classes A, AB, B, and C, which are based on the bias point of the gate and, therefore, the DC drain current levels. Generally, class-A operation is desirable for high linearity and high gain but the drain efficiency is limited. Class-B operation increases the drain efficiency of the amplifier at the expense of linearity in the output waveform. In this section, we will briefly review the dependence of the loadline resistance, shape factor, output power, and gain of the power amplifier at different bias points. By relating these factors to the conduction angle, we can demonstrate how to optimize efficiency for the device technology, circuit, and power specifications.

Table 4.1 Conduction angle for linear modes of operation.

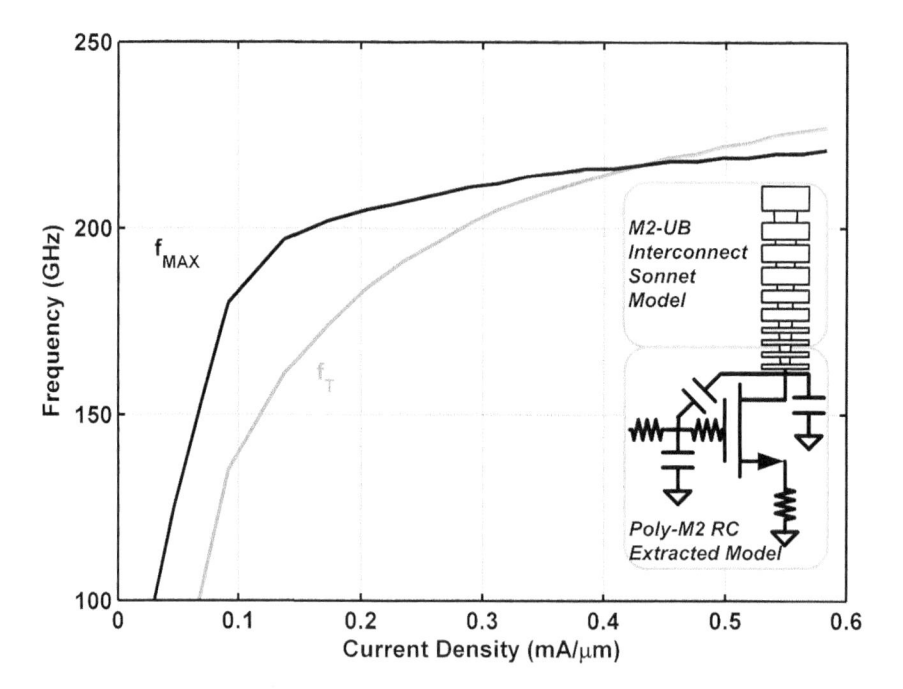

Class	A	AB	B	C
Conduction angle	2π	$2\pi > \Phi > \pi$	π	$\pi > \Phi$

Fig. 4.12 Simulated f_T/f_{MAX} for a 160-μm 45-nm n-FET. The inset illustration shows the combination of extracted RC device and SONNET modeling for metal interconnects. After [4]. Reprinted with permission from the IEEE.

4.4.1 Conduction angle

The conduction angle defines the phase interval during which the transistor drain current is conducting. For power amplifiers defined by the bias-point operation, the dependences of the shape factor, gain, and output power are related through the conduction angle Φ of the transistor, which occurs at peak power. The quiescent gate biasing voltage determines the conduction angle. The conduction angle is defined for each mode of linear operation in Table 4.1.

To describe the drain-current waveform at an arbitrary power level, the drain-current waveform can be expressed as a function of the peak conduction angle Φ and the normalized input power level x where the value x ranges from zero to unity. Normalizing the RF period to a phase θ over 0 to 2π, the drain current is

$$i_D(\theta) = \begin{cases} \frac{I_{MAX}}{1-\cos\left(\frac{\Phi}{2}\right)} \left(x\cos(\theta) - \cos\left(\frac{\Phi}{2}\right)\right) & \frac{-\theta_x}{2} \leq \theta < \frac{\theta_x}{2} \\ 0 & \theta > \frac{\theta_x}{2}, \theta \leq \frac{-\theta_x}{2} \end{cases}, \quad (4.17)$$

4.4 Linear classes of operation

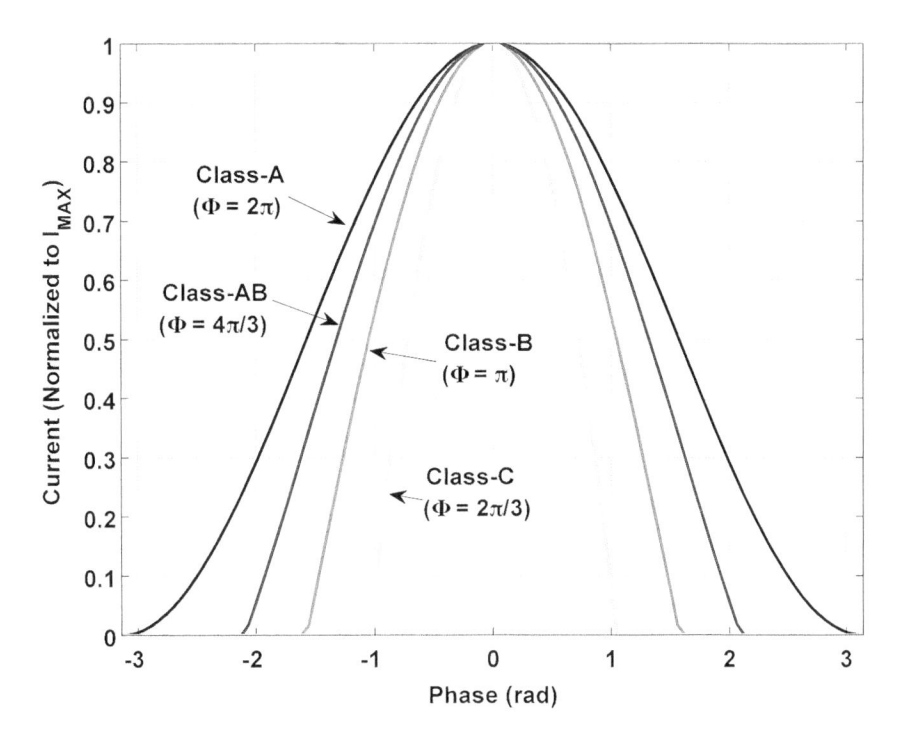

Fig. 4.13 Drain current for different classes of operation ($x = 1$).

where θ_x is the phase interval over which the device conducts at a normalized input power level x and is found from $\theta_x = 2 \arccos\left(\frac{1}{x} \cos\left(\frac{\Phi}{2}\right)\right)$ [9]. These waveforms have been plotted in Fig. 4.13 for operation in classes A, AB, B, and C. When the transistor gate is biased such that the RF swing keeps the transistor drain current in the saturated mode of operation, the transistor conducts throughout the entire RF cycle. As the gate bias is reduced towards the threshold voltage of the transistor, the RF swing at the input of the transistor pushes the gate voltage below the threshold voltage of the transistor and the device turns off for some period.

The waveforms in Fig. 4.13 clearly exhibit DC and RF current components that change under different conduction angles and at different input power levels. The drain current can be decomposed into the DC and RF drain current components that depend on the input power x. The DC drain current is

$$I_D(x) = \begin{cases} I_{D,Q} & x < \left|\cos\left(\frac{\Phi}{2}\right) 1\right| \\ x\frac{I_{MAX}}{2\pi} \frac{2\sin\left(\frac{\theta_x}{2}\right) - \theta_x \cos\left(\frac{\theta_x}{2}\right)}{1 - \cos\left(\frac{\Phi}{2}\right)} & x \geq \left|\cos\left(\frac{\Phi}{2}\right)\right| \end{cases}, \quad (4.18)$$

where $I_{D,Q}$ is the drain current under quiescent conditions, e.g. $x = 0$, and

$$I_{MAX} = \begin{cases} 0 & \Phi \leq \pi \\ I_{D,Q}\left(1 - \sec\left(\frac{\Phi}{2}\right)\right) & \Phi > \pi \end{cases}. \quad (4.19)$$

If we assume class-A operation ($\Phi = 2\pi$), $I_{MAX} = I_{D,Q}$ and $I_D(x) = x\frac{I_{D,Q}}{\pi}$. In other words, the maximum current swing is twice the DC swing for a class-A amplifier as found in the loadline discussion earlier. As the conduction angle is reduced, the ratio of the maximum current swing relative to the DC value increases. Therefore, a normalized DC drain-current expression can relate the DC drain current at an arbitrary conduction angle to class-A operation:

$$\frac{I_{D,\Phi}}{I_{D,A}} = \frac{2\sin\left(\frac{\Phi}{2}\right) - \Phi\cos\left(\frac{\Phi}{2}\right)}{\pi\left(1 - \cos\left(\frac{\Phi}{2}\right)\right)}. \tag{4.20}$$

The drain current generated at the fundamental RF frequency is

$$i_d(x) = \begin{cases} 0 & x < \left|\cos\left(\frac{\Phi}{2}\right)\right| \\ x\frac{I_{MAX}}{2\pi}\frac{\theta_x - \sin(\theta_x)}{1 - \cos\left(\frac{\Phi}{2}\right)} & x \geq \left|\cos\left(\frac{\Phi}{2}\right)\right| \end{cases}. \tag{4.21}$$

The condition on the RF current is meaningful for amplifiers biased in the class-C region where the RF current is off until the input power forces the transistor to begin to conduct. The dependence of DC and RF currents on the normalized input power levels is plotted in Fig. 4.14 for conduction angles representing class-AB and class-C operation. In class AB, the amplifier remains in class A until the input power level defined by $x = \left|\cos\left(\frac{\Phi}{2}\right)\right|$. In this region, the DC current remains constant while the RF current increases. Above this power level, the amplifier behaves similarly to class B where the

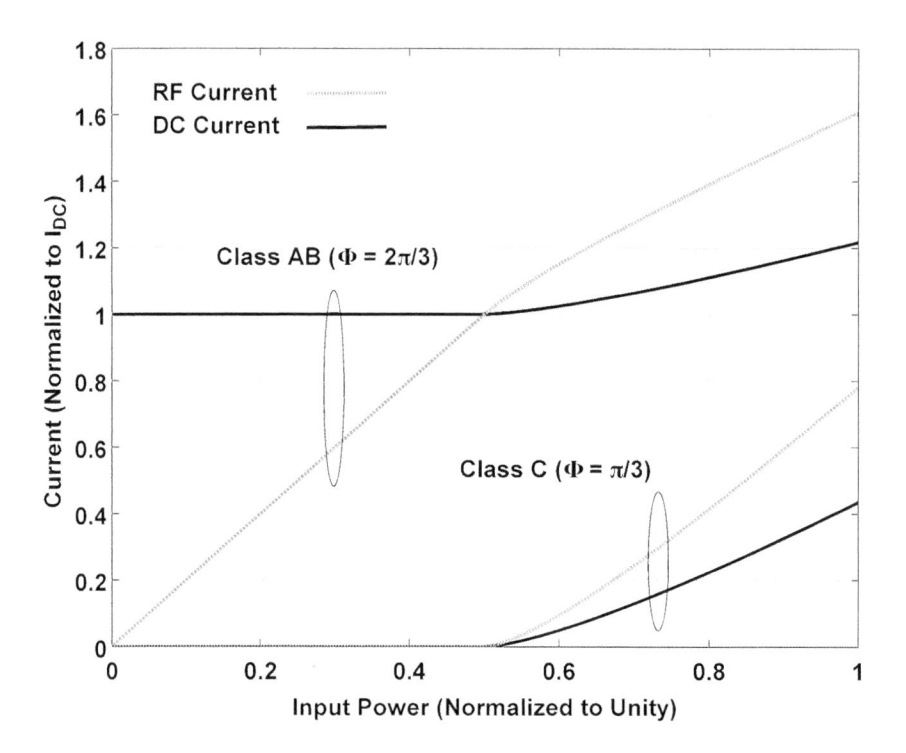

Fig. 4.14 Drain current DC and RF components as a function of input power.

DC current and RF current increase. In class C, the amplifier remains off until the power level defined by $x = \left|\cos\left(\frac{\Phi}{2}\right)\right|$ and begins to conduct above this power level. Notably, the same normalized power level defines the "activation" power in this example for both the class-AB and class-C examples since $\left|\cos\left(\frac{2\pi}{3}\right)\right| = \left|\cos\left(\frac{4\pi}{3}\right)\right|$.

Finally, we should consider the change in the peak RF current for different conduction angles. When $x = 1$, $\theta_x = \Phi$ and (4.21) reduces to

$$i_d = \frac{I_{MAX}}{2\pi} \frac{\Phi - \sin(\Phi)}{1 - \cos\left(\frac{\Phi}{2}\right)}. \tag{4.22}$$

For example, class-A operation is defined by $\Phi = 2\pi$ and the amplitude of the RF drain current is

$$i_{d,A} = \frac{I_{MAX}}{2\pi} \frac{2\pi}{2} = \frac{I_{MAX}}{2} = I_{D,Q}. \tag{4.23}$$

In other words, the RF amplitude increases until it is equal to the DC bias current. More generally, (4.22) shows that the RF current is highest in class-AB operation and drops as the amplifier is pushed into class-C operation. We can express the ratio of the RF drain current at any conduction angle with respect to class A as

$$\frac{i_{d,\Phi}}{i_{d,A}} = \frac{\Phi - \sin(\Phi)}{\pi \left(1 - \cos\left(\frac{\Phi}{2}\right)\right)}. \tag{4.24}$$

This ratio is helpful in investigating the dependence of the loadline, output power, and gain of the amplifier on the conduction angle.

4.4.2 Shape factor

The shape factor of the waveform is defined from the ratio of the RF drain current and the DC drain current,

$$S = \frac{P_o}{P_{DC}} = \frac{\frac{1}{2}i_D^2 R_L}{I_D V_{DD}} = \frac{1}{2}\frac{i_d}{I_D}, \tag{4.25}$$

where the final equality holds if $i_D R_L = V_{DD}$. Notably, the shape factor is important in determining the peak drain efficiency of the power amplifier since the RF and DC currents are related to the RF power and DC power consumption of the amplifier. Because both the RF and DC currents change with the conduction angle, the shape factor changes. The numerator of (4.15) has been expressed as a function of the conduction angle in (4.24). From (4.24) and (4.20), the shape factor can be expressed as a function of Φ:

$$S = \frac{1}{2}\frac{\Phi - \sin\Phi}{2\sin\left(\frac{\Phi}{2}\right) - \Phi\cos\left(\frac{\Phi}{2}\right)}. \tag{4.26}$$

The dependence of the shape factor on the conduction angle is plotted in Fig. 4.15. At $\Phi = 2\pi$, the amplifier is in class-A operation and the maximum shape factor is 50%. As the conduction angle decreases from 2π, the shape factor increases in the class-AB

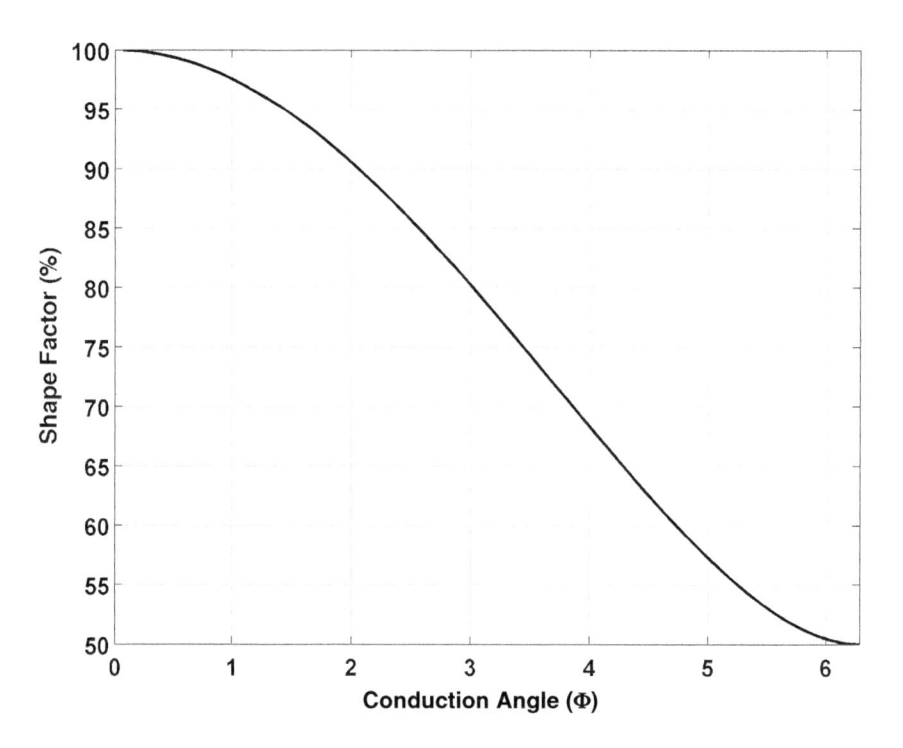

Fig. 4.15 Shape factor vs conduction angle.

region. When $\Phi = \pi$, the amplifier is operated in class-B operation and turns on for exactly one-half of the cycle. As the Φ approaches zero, the amplifier conducts over a short period, which further increases the shape factor towards 100%. High shape factor is promising for efficient operation. However, the loadline, output power, and gain of the amplifier also change as a function of the conduction angle.

4.4.3 Output power

The penalty of pushing the bias point of the device deep into class-C operation where efficiency is high is a reduction in output power. From (4.24), we anticipate the decrease in the RF current for reduced conduction angles, which translates into lower output power of the amplifier:

$$P_\Phi = \frac{P_A}{\pi} \frac{\Phi - \sin \Phi}{1 - \cos\left(\frac{\Phi}{2}\right)}, \qquad (4.27)$$

where P_A is the output power in class-A operation. The normalized output power is plotted in Fig. 4.16. The output power improves slightly as the conduction angle is pushed into class-AB mode and the RF current amplitude increases. The output power then returns to the class-A level once we reach the class-B bias. As the conduction angle reduces below π, the output power drops precipitously.

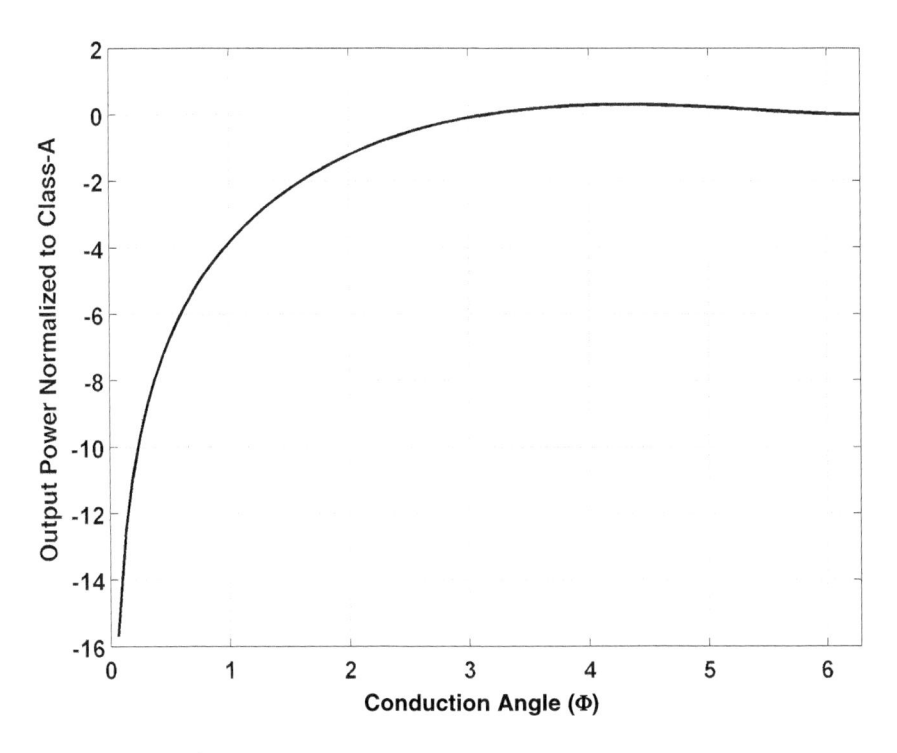

Fig. 4.16 Output power vs conduction angle.

4.4.4 Loadline resistance

The loadline resistance previously determined in (4.5) assumed class-A operation during which the device is always conducting. Using the dependence of the peak RF current on conduction angle in (4.22), the loadline impedance for an arbitrary conduction angle is

$$R_L(\Phi) = R_{L,A} \pi \frac{1 - \cos\left(\frac{\Phi}{2}\right)}{\Phi - \sin \Phi}, \qquad (4.28)$$

where $R_{L,A}$ is the loadline for a class-A amplifier. The normalized loadline impedance is plotted in Fig. 4.17. As the conduction angle reduces, the loadline resistance drops slightly before increasing for small conduction angles. Notably, the loadline resistance is the same in class-A and class-B modes of operation. This is generally useful for testing the power amplifier at different bias points without causing a significant change in the loadline condition and hence the amplifier should remain matched regardless of operation in class-A, class-AB, or class-B regions.

4.4.5 Gain

Finally, the gain of the power amplifier can be expressed when normalized to class-A operation. Specifically, the gain of the amplifier depends both on the loadline matching and on the RF power generated by the amplifier:

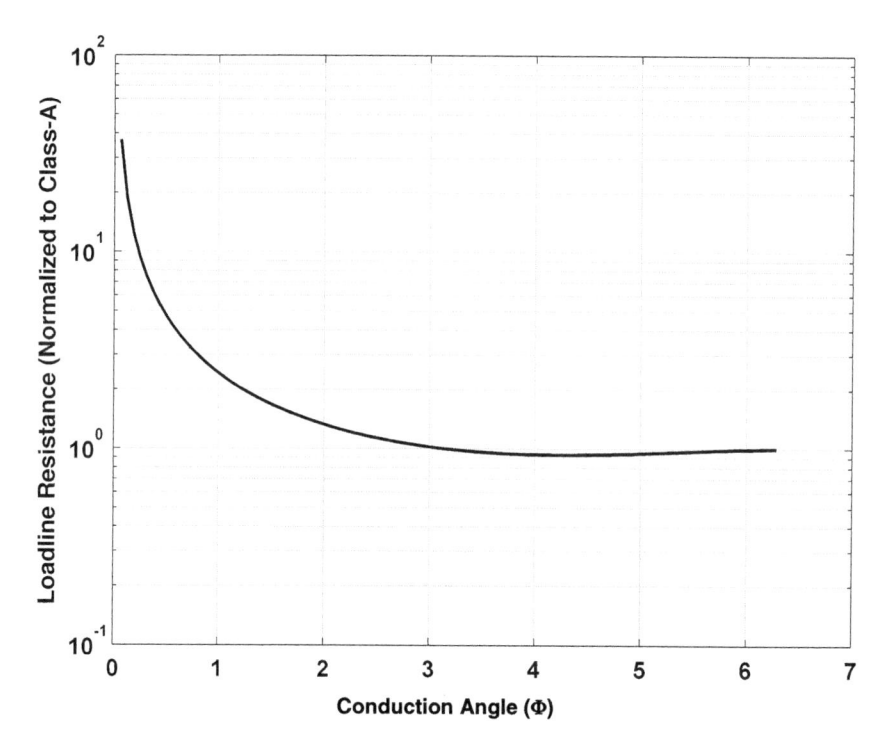

Fig. 4.17 Loadline resistance vs conduction angle normalized to class-A operation.

$$G = G_A \frac{R_L(\Phi)}{R_{LA}} \frac{\Phi - \sin \Phi}{2\sin\left(\frac{\Phi}{2}\right) - \Phi\cos\left(\frac{\Phi}{2}\right)}, \quad (4.29)$$

where G_A is the gain in class-A operation. The gain is plotted as a function of conduction angle in Fig. 4.18. As small conduction angles are desired for better efficiency, it is clear that the penalty is even greater reduction in the gain of the amplifier

In the following section, these different factors will be coordinated to make some conclusions about the optimal biasing of a linear amplifier in the mm-wave frequency regime.

4.5 Optimization of mm-wave amplifiers: why linear?

With a framework to assess PAE against device and circuit factors as well as conduction angle, a bound on the PAE presented in (4.2) can be refined for linear operation. Unifying previous factors related to the device technology, the biasing of the transistor, and circuit topology leads to understanding how to optimally bias and what PAE bounds are expected. To establish this bound, two design specifications are chosen – the design frequency f_o and the desired output power P_o – subject to constraints imposed by the device technology, e.g. V_{DD}, V_K, J_{MAX}.

Based on the desired output power and maximum supply that can be applied to the device, (4.4) determines the transistor width. In other words, $W =$

4.5 Optimization of mm-wave amplifiers

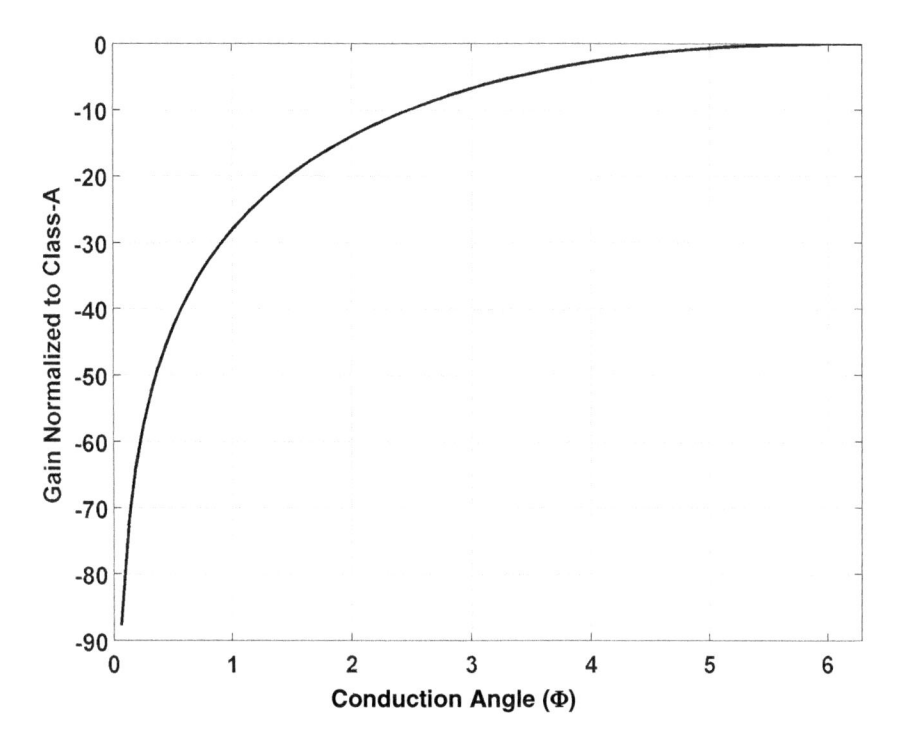

Fig. 4.18 Gain vs conduction angle.

$f_1(P_o, V_{DD} - V_K, J_{MAX})$. The transistor geometry can be used to account for the parasitics that must be incorporated into the device model at high frequency to accurately predict the gain, i.e. $G = f_2(f_o, f_T, W, \Phi)$. The design frequency impacts the MAG available from the transistor – particularly when considering parasitic elements. The parasitics of the transistor that strongly impact the predicted f_{MAX} of the transistor include R_S, L_S, R_G, L_G, and C_{GD} and are strongly dependent on the width of the transistor and precisely how the transistor interconnections are made. Additionally, the gain depends on the biasing of the transistor.

Next, the classical analysis of the shape factor has predicted the dependence on conduction angle. Therefore, the drain efficiency of the amplifier can be expressed from the product of the shape factor and the knee voltage factor, i.e. $\eta = S(\Phi)\left(1 - \frac{V_K}{V_{DD}}\right)$. Finally, the losses in the output matching network depend on the frequency of operation, the loadline matching, the quality factor of passives that realize the device technology, and – to some degree – the conduction angle of operation, i.e. $L = f_3(f_o, W, Q, \Phi)$.

Putting these different elements together, the best PAE can be predicted by examining the contribution of these different factors as defined from (4.2):

$$PAE = h(W, G, \eta, L). \tag{4.30}$$

To provide some insight into the bounds that can be developed, Table 4.2 summarizes typical parameters for an advanced CMOS device. The table supplies key device

Table 4.2 Table of device parameters for Figures 4.19–4.21.

Parameter	Value
V_{DD}	1.2 V
V_K	0.4 V
f_T/f_{MAX}	250/200 GHz
J_P	0.4 mA/μm
Q	10
C_{GSW}	1.5 fF/μm
C_{GDW}	0.6 fF/μm

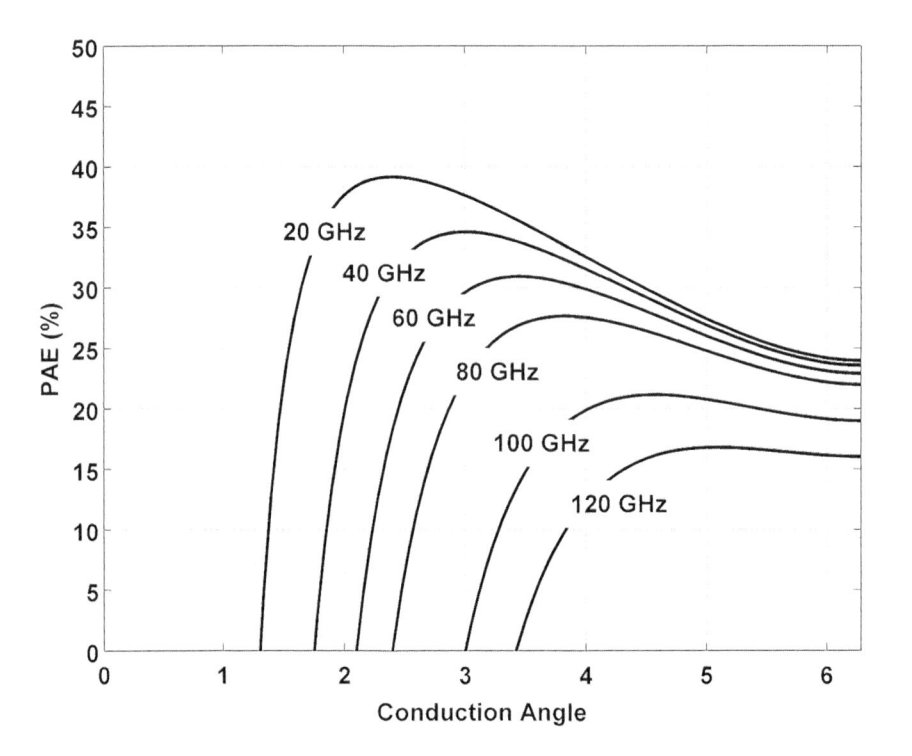

Fig. 4.19 Optimal PAE vs conduction angle for different frequencies of operation ($P_o = 20$ dBm).

parameters including the DC and knee voltages, the peak current density, and the per-unit-length capacitance seen from gate to source and gate to drain. Other device parameters can be extracted from these given parameters. For instance, the value of r_g can be found, given the f_{MAX}.

The maximum PAE for different frequencies of operation is plotted in Fig. 4.19, given a desired output power of 20 dBm. For this baseline simulation, the transistor is assumed to have no substantial parasitic resistance, capacitance, or inductance that further degrades the transistor gain. Several notable conclusions can be reached. First, the peak PAE is expected to be bounded to 40% at 20 GHz, the lowest frequency plotted.

4.5 Optimization of mm-wave amplifiers

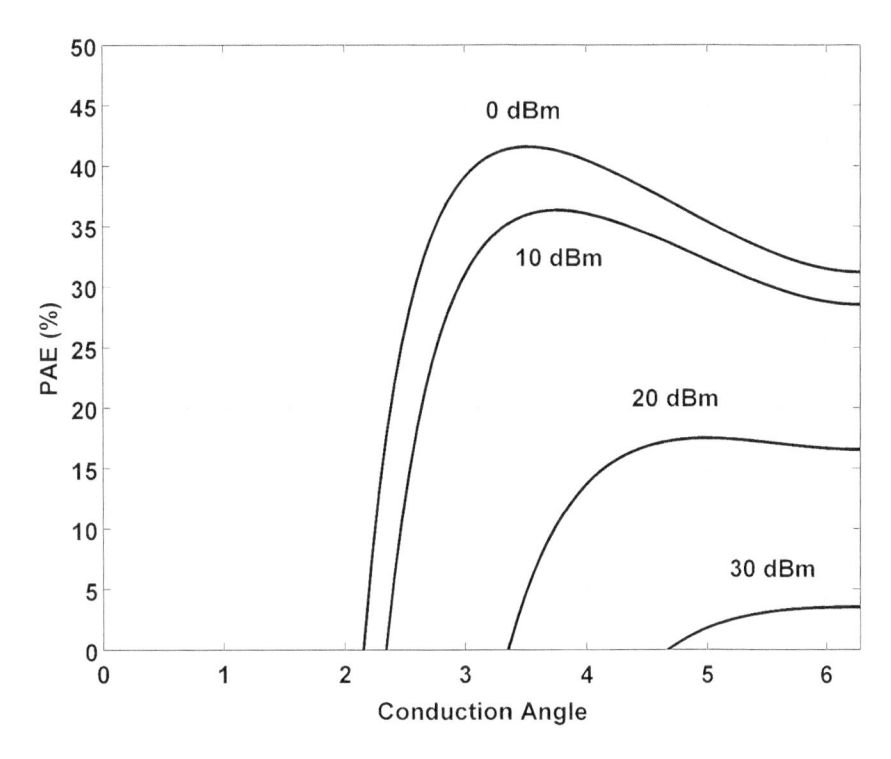

Fig. 4.20 Output power vs conduction angle (60 GHz).

This peak PAE decreases as the frequency of operation increases because the transistor gain decreases, which reduces the PAE. Second, the optimal conduction angle for the peak PAE also increases with frequency. At 20 GHz, the amplifier achieves the peak PAE in class C. However, the same device operated at 120 GHz achieves the peak PAE in class AB. As the gain becomes more limited at higher frequency, the limiting factor in the PAE becomes the gain.

Another perspective on this trend is to examine the PAE at one frequency for different desired output powers. Figure 4.20 demonstrates the reduction in the PAE at 60 GHz on increasing the transistor size. Again, no layout-dependent parasitic elements are assumed at this point. At 0 dBm, a peak PAE of 40% is found as the amplifier operates close to class B. As the output power is increased to 30 dBm, the efficiency drops precipitously to 3% and the amplifier is optimally biased closer to class A. The explanation for this behavior is that the loss factor of the output matching network increasingly dominates the overall PAE.

Finally, the impact of layout-dependent parasitics can be examined by plotting the peak PAE as a function of operating frequency for a given output power. Figure 4.21 is a composite plot that illustrates the peak PAE found at different operating frequencies from Fig. 4.19. By implicitly choosing the best conduction angle for operation, the PAE can be plotted as a function of the operating frequency. The upper curve is the baseline device performance when the influence of the layout-dependent parasitics is ignored.

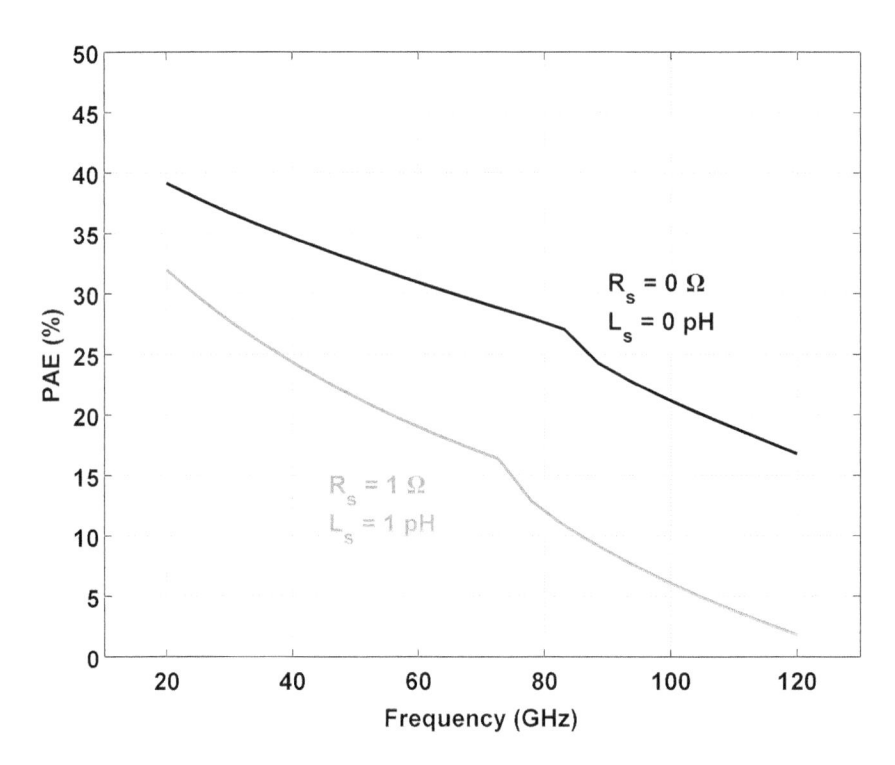

Fig. 4.21 Maximum PAE vs operating frequency for transistors analyzed with and without parasitic source resistance and inductance ($P_o = 20$ dBm).

This provides an optimistic estimation of the PAE. When a nominal source resistance of 1 Ω and source inductance of 1 pH are included in the device model (lower curve), the PAE falls off more steeply with frequency, indicating that the notional PA is not feasible above 120 GHz. At a nominal frequency of 60 GHz, the PAE drops from just over 30% to around 18%.

Several general conclusions can be reached about the operation of linear mm-wave PAs from this optimization approach. First, pushing deeper into the mm-wave regime greatly limits the PAE that a PA can achieve. Furthermore, the dependence of the layout of the device becomes extremely critical since the parasitic elements create feedback which further reduces the already limited gain. As the gain is reduced, the optimal conduction angle always approaches class-A operation. Finally, higher power requirements imply larger devices with more significant layout-dependent parasitic elements. Therefore, large mm-wave PAs are difficult to design and operate efficiently.

4.6 Case study: Q-band SiGe power amplifier

In this case study, a 45-GHz PA is demonstrated using a 0.12-μm SiGe BiCMOS process and has cut-off frequencies of f_{MAX}/f_T of 240/200 GHz. The output power is specified to deliver 15 dBm.

4.6 Case study: Q-band SiGe power amplifier

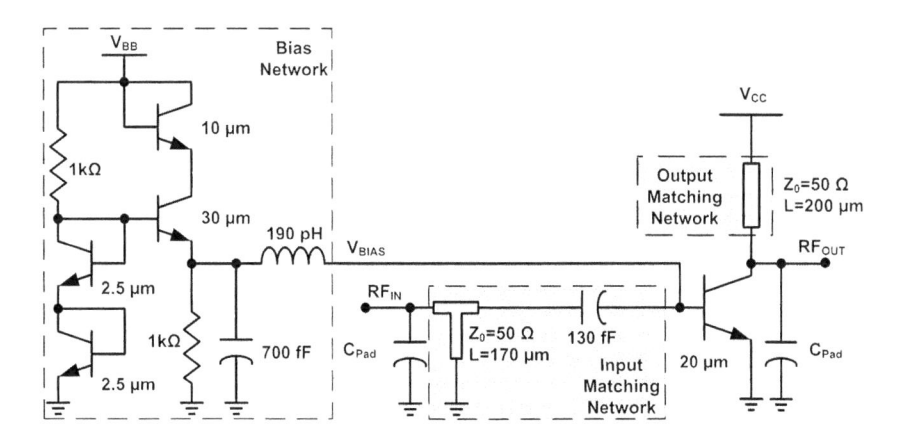

Fig. 4.22 Schematic of a SiGe PA for operation at 45 GHz. After [12]. Reprinted with permission from the IEEE.

To achieve a breakdown voltage close to the BV_{CBO}, the impedance seen looking from the base should be low-impedance to mitigate the effect due to impact ionization. Furthermore, the bias circuit must be capable of sweeping electrons out of the base region and a low-impedance base biasing circuit is required, which can be both sink and source current [10, 11]. This is particularly effective in the SiGe HBT technology, where BV_{CBO} is significantly higher than BV_{CEO}. A biasing circuit for a mm-wave PA is shown in Fig. 4.22 [12]. The impedance seen looking from the base at low frequency is defined by $1/g_m$ into the emitter of the HBT. Under low-power conditions, the bias circuit produces a base current flowing into the base. Simulation of the large-signal I_C vs V_{CE} demonstrates that if the impedance presented to the base by the biasing network is approximatly 40 Ω, the peak collector voltage can be more than 4 V. Therefore, the DC collector voltage is set to 2.4 V.

To reach an output power level of 15 dBm, the supply voltage is determined based on the peak swing of 4.4V and knee voltage of 0.8 V. Hence, the DC collector supply voltage is 2.4 V and the maximum collector current is determined:

$$I_{MAX} = \frac{4P_o}{V_{CC} - V_K} = \frac{4 \cdot 32\,\text{mW}}{2.4 - 0.8} = 80\,\text{mA}. \tag{4.31}$$

The peak current swing should be 80 mA with a DC collector current of 40 mA, assuming class-A operation. The peak current determines the size of the required HBT device for high-speed operation. Figure 4.23 plots the f_{MAX} as a function of collector current for a 20-μm transistor. The peak f_{MAX} drops from 243 GHz to 196 GHz when the layout parasitics are included. From this plot, the current density may vary between 1 and 3 mA/μm. The DC and peak RF currents are readily handled through this choice of emitter length while maintaining high-speed operation. Careful layout aids in minimizing the additional parasitics, in particular the ground inductance.

Finally, the loadline impedance for the proposed power level would be

$$R_{L,A} = \frac{2(V_{CC} - V_K)}{I_{MAX}} = \frac{3.2\,\text{V}}{80\,\text{mA}} = 40\,\Omega. \tag{4.32}$$

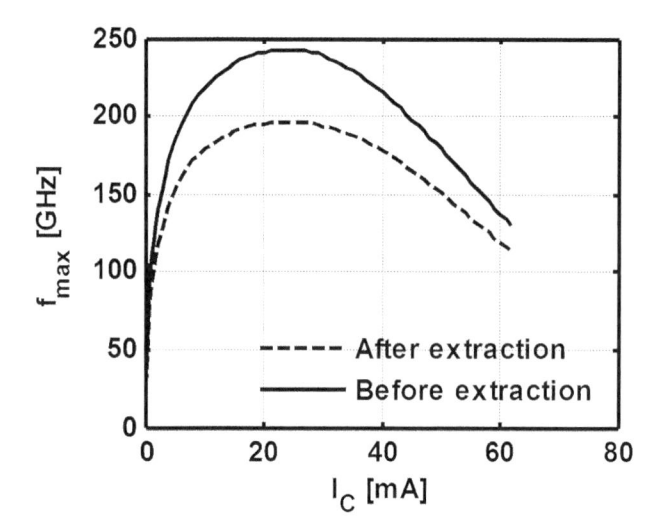

Fig. 4.23 Simulated f_{MAX} for pre- and post-layout of 20-μm SiGe HBT device. After [12]. Reprinted with permission from the IEEE.

The loadline impedance at the fundamental frequency of 45 GHz for the 20-μm HBT is simulated in Fig. 4.24 and is approximately $50 + j25$ Ω for optimal efficiency. Therefore, the reactive component can be implemented with a single element output match using a 50-Ω shunt CPW, which also connects the transistor's collector to the supply. This technique avoids losses in a separate matching network. Additional desired harmonic impedances are shown in Fig. 4.24, but intentional harmonic matching has not been applied, in order to avoid additional loss. The input is matched to 50 Ω with a shunt stub implemented with a shielded coplanar waveguide and a series metal–insulator–metal capacitor. The real part of the fundamental frequency load impedance Z_L should roughly correspond to the loadline resistance and the imaginary part of Z_L should tune out the output capacitance of the HBT.

The simulated and measured S-parameters are plotted in Fig. 4.25. The peak amplifier gain of 11.3 dB occurs at 37.5 GHz; the 3-dB bandwidth is 15.8 GHz (from 31.7 to 47.5 GHz). The input return loss is less than 20 dB over 36 GHz to 43 GHz. The output is well matched from 34 to 47 GHz. The return isolation is below 20 dB through the measured frequency range.

The large-signal behavior at 45 GHz for class-A operation is shown in Fig. 4.26. The gain for low input powers of the PA in this measurement is 9.2 dB and matches the results in the S-parameters measurement. The output referred to the 1-dB compression point is 12.4 dBm. The peak PAE is 25% at 13 dBm output power and the P_{sat} is 13.6 dBm.

By adjusting the base bias, the amplifier can be operated close to class B. The gain and PAE for class-B operation are shown in Fig. 4.27. In this case, the peak gain is 7.8 dB and the peak PAE is 30.8% at 13.3 dBm output power. The P_{sat} reaches 14.75 dBm.

We can conclude two things from this case study. First, the PA demonstrates higher PAE and output power in class B rather than class A. Second, the peak PAE is around

4.6 Case study: Q-band SiGe power amplifier

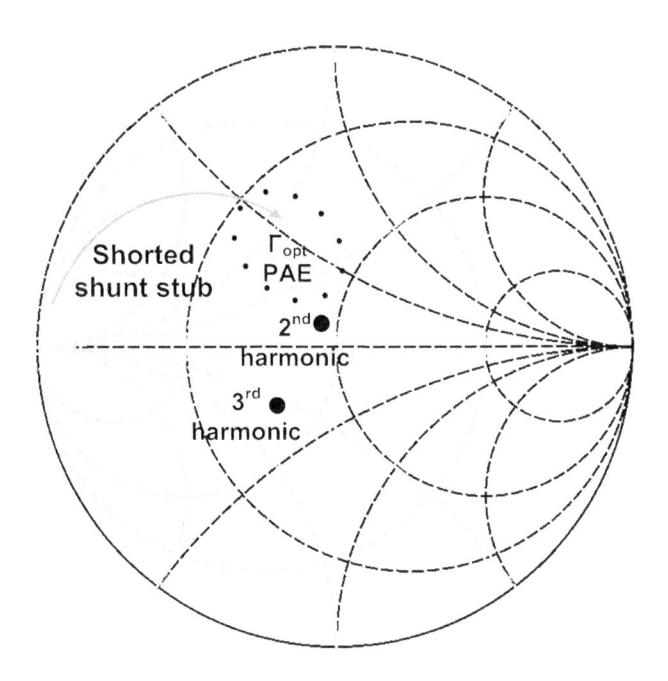

Fig. 4.24 The 50 Ω load and pad capacitance are transformed by a shunt stub (solid line) to a load impedance for optimal PAE inside the region enclosed by the dotted line. The impedances seen at the second and third harmonic from the matching network are as shown. After [12]. Reprinted with permission from the IEEE.

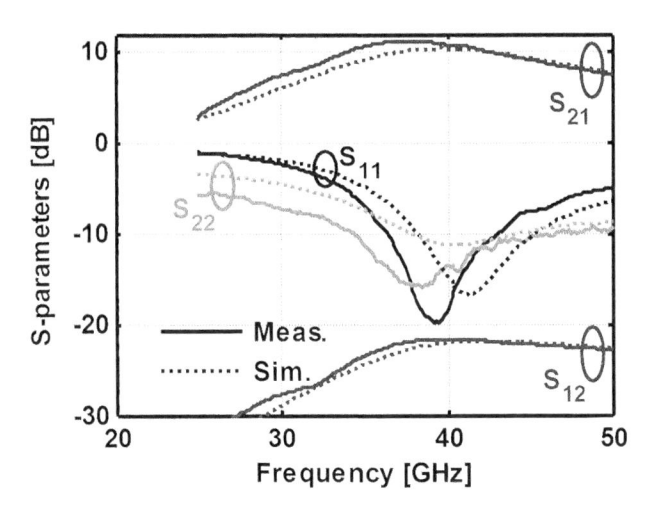

Fig. 4.25 Measured and simulated S-parameters of the PA in class-A mode ($V_{CC} = 2.4$ V). After [12]. Reprinted with permission from the IEEE.

30% and drops around 5% for class-A operation. These results help to put the predictions presented in Section 4.5 into context. Comparing these results with Fig. 4.19 at 45 GHz, we find that a peak PAE of roughly 35% is achieved near class-B conduction angle. Second, we find that the PAE for the corresponding class-A operation is around 6%

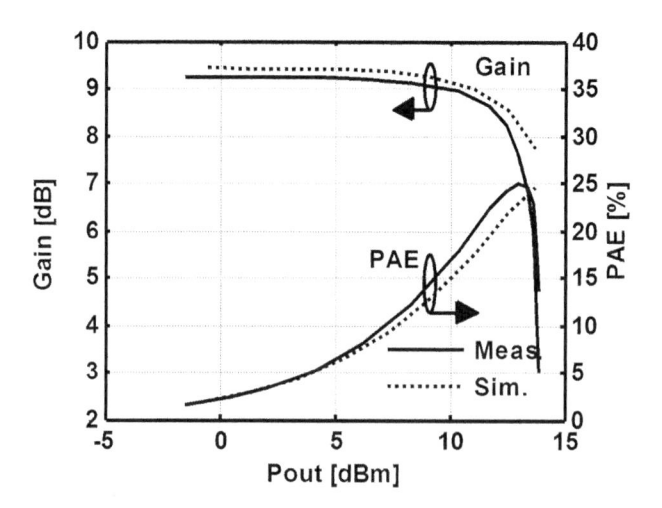

Fig. 4.26 Measured and simulated large-signal gain and PAE in class-A mode at 45 GHz. After [12]. Reprinted with permission from the IEEE.

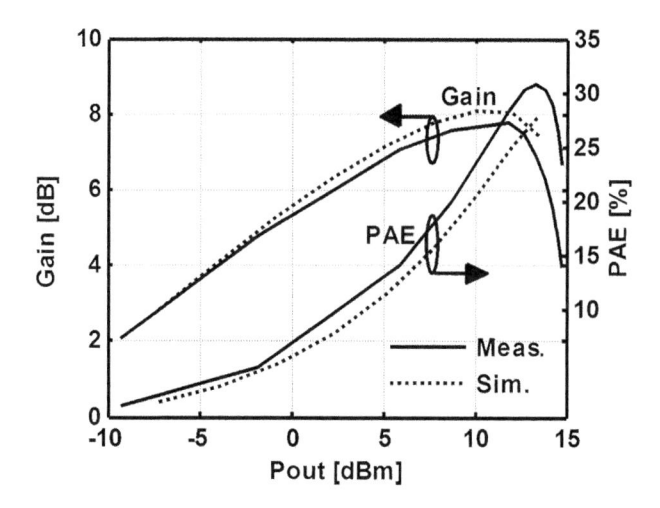

Fig. 4.27 Measured and simulated large-signal gain and PAE in class-B mode at 45 GHz. After [12]. Reprinted with permission from the IEEE.

lower than at the peak PAE. This SiGe PA study demonstrates good agreement between the analysis, simulated, and measured mm-wave PAs in terms of S-parameters and large-signal parameters.

4.7 Doherty amplifiers

Linear amplifiers exhibit the best PAE when driven near the saturated output power. To compensate for the reduction in efficiency under back-off conditions, active load modulation is introduced to allow a *pair* of linear amplifiers to operate at high *average*

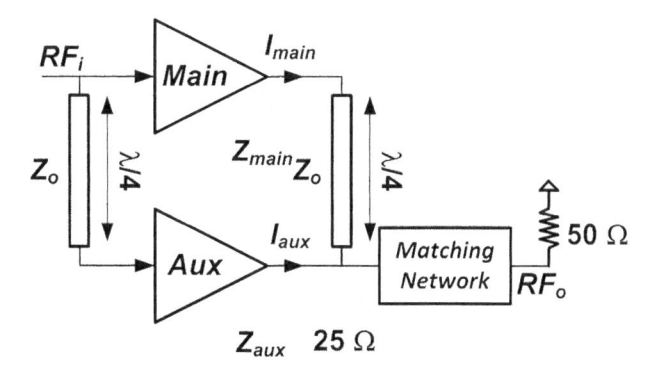

Fig. 4.28 Schematic of a mm-wave Doherty amplifier.

efficiency. Referring back to the example shown in Fig. 4.1, complex signals such as QAM require the power amplifier to operate at an average power that is lower than the peak power. Operation at high peak-to-average ratio causes the average efficiency to degrade. Section 4.5 demonstrated that mm-wave PAs would presumably operate best between class-B and class-A operation. We examine this in the context of the Doherty amplifier design for mm-wave bands.

The conventional Doherty amplifier is shown in Fig. 4.28 and consists of a main amplifier and an auxiliary amplifier [13]. At low power levels, the auxiliary amplifier is off and the main amplifier provides all of the output power. The main amplifier is designed to operate at peak PAE at this back-off power level. At high power levels, the auxiliary amplifier begins to conduct and both amplifiers supply power to the output. To keep the main amplifier operating at peak efficiency, the Doherty amplifier relies on active load modulation to change the load seen at the main amplifier.

4.7.1 Active load modulation

Active load modulation occurs when two current sources independently source current into the load resistance to generate a voltage v_l as shown in Fig. 4.29:

$$Z_1 = R_L \left(1 + \frac{i_2}{i_1}\right)$$
$$Z_2 = R_L \left(1 + \frac{i_1}{i_2}\right). \tag{4.33}$$

If the current provided from both current generators is equal, the impedance from either side of the load is twice the load resistance seen if either of the currents is turned off. This impedance change is exploited to achieve load modulation over a range of output power levels. While the current i_2 can be associated with the auxiliary amplifier current, the current i_1 is produced through an impedance transformer from the main amplifier. To understand the effect of a lossy, mm-wave impedance transformer on the voltage seen at the main device, the following section considers the lossy quarter-wave, mm-wave transformer.

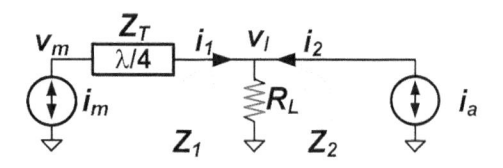

Fig. 4.29 Illustration of load modulation in Doherty amplifiers.

4.7.2 Impedance inversion

The quarter-wave ($\lambda/4$) transmission line at the output plays a critical role in inverting the impedance seen at the main amplifier. Using an *ABCD* matrix to relate the current and voltage on either side of the impedance transformer, the transmission line inverts the voltage and current according to

$$\begin{bmatrix} v_m \\ i_m \end{bmatrix} = \begin{bmatrix} \cosh(\gamma l) & Z_T \sinh(\gamma l) \\ \frac{1}{Z_T} \sinh(\gamma l) & \cosh(\gamma l) \end{bmatrix} \begin{bmatrix} v_l \\ i_1 \end{bmatrix}, \qquad (4.34)$$

where $\gamma = \alpha + j\beta$ is the propagation constant of the transmission line and l is the length of the line. Assuming that the line length is $\lambda/4$ and that the losses are small,

$$\begin{bmatrix} v_m \\ i_m \end{bmatrix} \approx \begin{bmatrix} 0 & jZ_T \\ \frac{j}{Z_T} & 0 \end{bmatrix} \begin{bmatrix} v_l \\ i_1 \end{bmatrix}. \qquad (4.35)$$

To understand the operation of active load modulation, consider the impedance seen at the output of the main device from the ratio of v_m and i_m. Ignoring the losses of the impedance inverter, the load at the main amplifier is

$$Z_m = \frac{v_m}{i_m} = \frac{i_1}{v_l} Z_T^2 = \frac{Z_T^2}{Z_1} = \frac{Z_T^2}{R_L \left(1 + \frac{i_2}{i_1}\right)}, \qquad (4.36)$$

where the final equality substitutes the load modulation from (4.33). Two features are evident. First, the load resistance is reduced when transformed to the output of the main amplifier. Second, the impedance seen at the main amplifier is modulated by the current provided from the auxiliary amplifier. The voltage maintained at the output of the main amplifier is found from the previous expression since $v_m = i_1 Z_T$ in (4.35):

$$v_m = i_m Z_m = \frac{i_m Z_T^2}{R_L \left(1 + \frac{i_2 Z_T}{v_m}\right)}. \qquad (4.37)$$

Rewriting this in terms of v_m,

$$v_m = \left(\frac{i_m Z_T}{R_L} - i_2\right) Z_T. \qquad (4.38)$$

This expression demonstrates that while the auxiliary amplifier and the main amplifier increase the current supplied to the load of the amplifier, the voltage swing does not need to change. Conditions for Doherty operation are analyzed assuming different biasing conditions on the main and auxiliary amplifiers in the following sections.

4.7.3 Class-B/Class-B biasing

A classic example analyzed by Raab considers using class-B biasing for both the main amplifier and the auxiliary amplifiers and illustrates ideal Doherty behavior [14]. This will be shown to be difficult to implement for mm-wave power amplifiers but serves as a useful illustration of the ideal behavior of a Doherty amplifier. From (4.21), the RF drain current amplitude when the amplifier is biased in class B is

$$i_{d,B}(x) = x \frac{I_{MAX}}{2}. \tag{4.39}$$

The RF current in class-B operation is directly proportional to the input power. The conventional arrangement assumes that the current supplied by the main amplifier and auxiliary amplifier is equal at peak output power and that the auxiliary amplifier begins to conduct at one-quarter of the maximum output power. Under these conditions, the current provided by the main and auxiliary amplifiers is

$$\begin{aligned} i_m &= \frac{I_{MAX}}{4} x \\ i_a &= \frac{I_{MAX}}{2} \left(x - \frac{1}{2} \right). \end{aligned} \tag{4.40}$$

Substituting (4.40) into (4.38), the voltage swing at the main amplifier is

$$v_m = \left(\frac{x I_{MAX}}{2} \left(\frac{Z_T}{2 R_L} - 1 \right) + \frac{I_{MAX}}{4} \right) Z_T. \tag{4.41}$$

The voltage across the main amplifier can be independent of the power level in the Doherty region if the characteristic impedance of the amplifier is chosen to be

$$Z_T = 2 R_L. \tag{4.42}$$

This is an important feature for Doherty amplifiers since the constant voltage swing enables the main amplifier to remain at peak efficiency as the input power level increases. In other words, the loadline resistance that the main amplifier sees when the auxiliary amplifier turns on is

$$R_m = \frac{v_m}{i_m} = \frac{2(V_{DD} - V_K)}{I_{MAX}/4}. \tag{4.43}$$

Note that this loadline resistance is four times the value predicted from (4.5).

The ideal class-B behavior is plotted in Fig. 4.30 and shows the voltage and the output of the main amplifier as well as the RF current contributions assumed under class-B conditions for the main and auxiliary amplifiers.

The RF output powers of the main and auxiliary amplifiers are given as

$$\begin{aligned} P_{M,RF} &= i_m^2 R_m = x^2 \frac{I_{MAX}}{2} (V_{DD} - V_K) \\ P_{A,RF} &= i_a^2 R_a = 2 \left(x - \frac{1}{2} \right)^2 I_{MAX} (V_{DD} - V_K). \end{aligned} \tag{4.44}$$

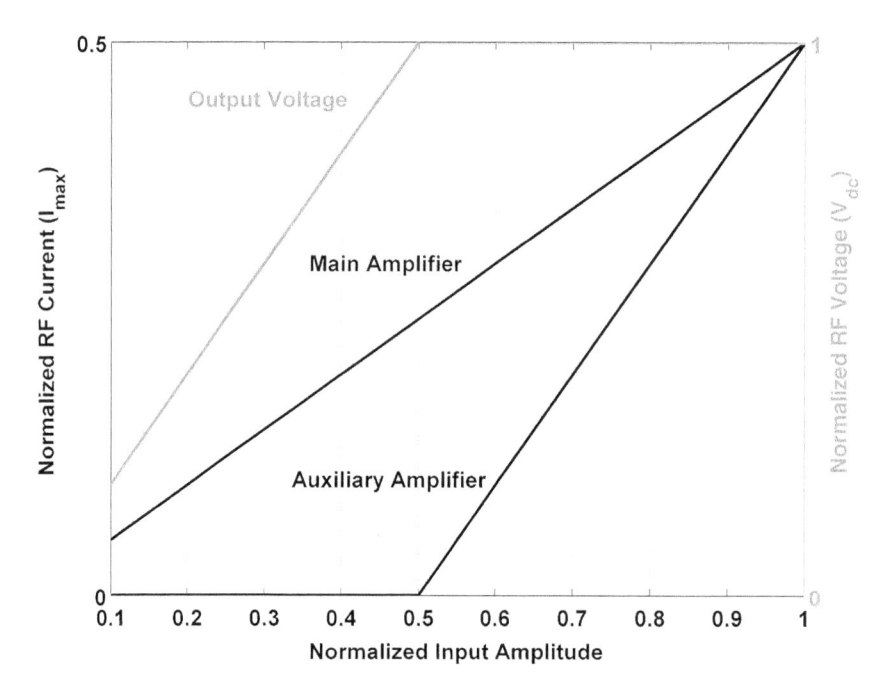

Fig. 4.30 Illustration of class-B/class-B Doherty PA currents and voltages.

The DC power consumption of the main and auxiliary amplifiers is given as

$$P_{M,DC} = x \frac{I_{MAX}}{\pi} V_{DD}$$
$$P_{A,DC} = 2 \left(x - \frac{1}{2} \right) \frac{I_{MAX}}{\pi} V_{DD}. \tag{4.45}$$

Therefore, the efficiency of the class-B/class-B Doherty amplifier is

$$\eta(x) = \frac{P_{M,RF} + P_{A,RF}}{P_{M,DC} + P_{A,DC}} = \begin{cases} 2x\frac{\pi}{4} \left(1 - \frac{V_K}{V_{DD}}\right) & x < 0.5 \\ \frac{\pi}{2} \frac{x^2}{3x-1} \left(1 - \frac{V_K}{V_{DD}}\right) & x \geq 0.5 \end{cases}. \tag{4.46}$$

This function is illustrated in Fig. 4.31 and demonstrates the efficiency as a function of the normalized input power. The peak efficiency of 78% is achieved at the maximum output power and at the 6-dB back-off point. Notably, we can re-evaluate the Doherty results for lossy transmission lines that are inevitable for mm-wave impedance transformation. By returning to (4.34) and re-formulating the current and voltage relationships due to the lossy transformer, the drain efficiency in the presence of the lossy transformer can also be calculated. These are compared in Fig. 4.31. Notably, the results indicate that high loss diminishes the advantages of a Doherty amplifier compared with a single class-B amplifier operated under back-off conditions.

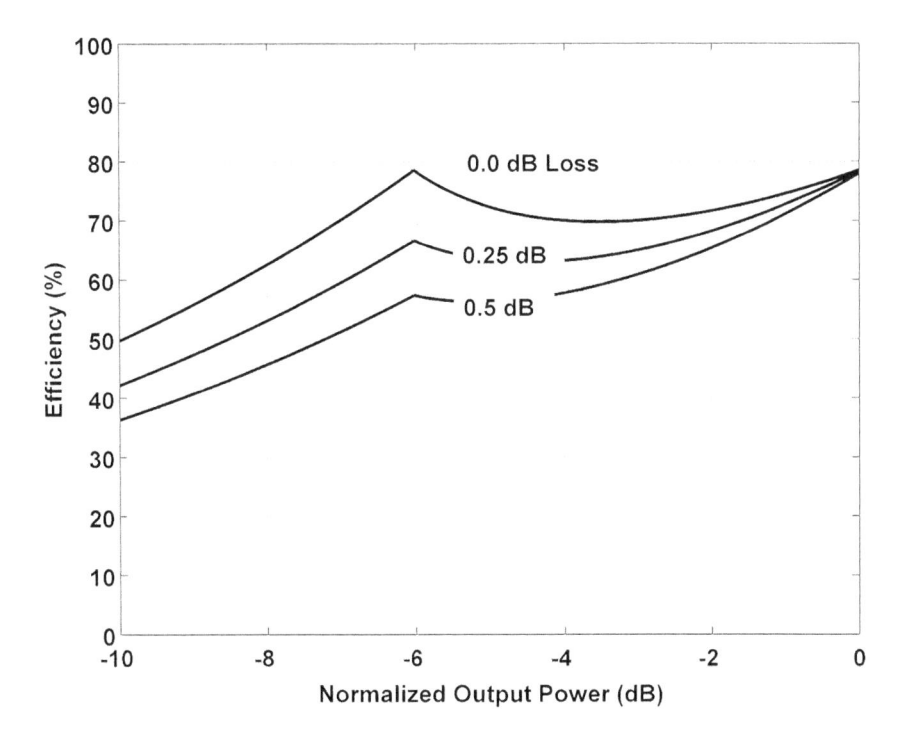

Fig. 4.31 Illustration of class-B/class-B Doherty PA efficiency for lossless and lossy quarter-wave transmission lines.

4.8 Case study: a Q-band Doherty power amplifier

A Doherty amplifier designed to operate at 45 GHz is realized in a 45-nm CMOS SOI process [15]. The schematic of the Doherty amplifier is shown in Fig. 4.32 and consists of a main amplifier and auxiliary amplifier which are connected through $\lambda/4$ lines at the input and the output.

A two-stack FET power-amplifier stage is used for the main and auxiliary PAs to overcome difficulties associated with the low breakdown voltage of scaled CMOS FETs [16]. The stacking technique described in a subsequent chapter of this book provides for the series connection of N FETs such that the voltage supply across the series devices is $N \cdot BV_{DS}$ and the RF output power increases by a factor of N relative to the single-transistor operation described in this chapter. The transistors in the auxiliary amplifier are twice the width of the main amplifier to provide twice the gain of the main amplifier in the high-power region of operation. Finally, an additional impedance matching network is provided to transform the 50 Ohm load to 25 Ohms at the output connection of the auxiliary amplifier and main amplifier. This is necessary because of the characteristic impedance condition demonstrated in (4.42), which suggests that the transmission line characteristic impedance should be twice the load resistance. For on-chip transmission lines, realizing transmission lines with

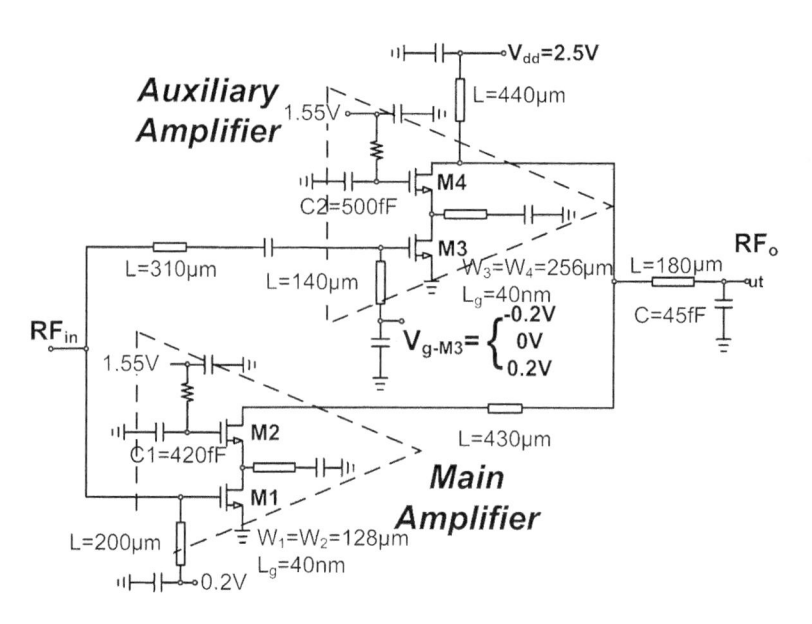

Fig. 4.32 Schematic of a Doherty amplifier implemented in 45-nm CMOS SOI. Adapted from [15].

characteristic impedance greater than 70 Ohms is not easily realized. Therefore, the output impedance transformation is necessary for integrated circuit implementations of a Doherty amplifier.

Slow-wave transmission lines were used for tuning elements in the amplifier as well as for the input and output quarter-wave ($\lambda/4$) transmission lines. Since the $\lambda/4$ line is relatively long (800 μm) at 45 GHz, the overall die area of the Doherty amplifier is determined by the input and output networks which both contain a $\lambda/4$ line, and their losses reduce the overall PAE. Therefore, reducing the length and loss of these transmission lines is highly desirable. The slow-wave transmission line includes floating metal lines under the signal line to increase the distributed capacitance without reducing the distributed inductance. Therefore, a higher delay $\sqrt{LC}$ per unit length is achieved with designs that maintain the same characteristic impedance of $\sqrt{L/C}$. This translates to higher phase shift per unit length [17]. From electromagnetic simulation, the conventional $\lambda/4$ transmission line with a ground shield has a length of 800 μm while the slow-wave $\lambda/4$ transmission line has a length of 620 μm. Further reduction in the physical length of the $\lambda/4$ lines occurs due to capacitive loading on the transmission lines. S-parameter simulations predict 0.25 dB lower loss with the slow-wave structure at 45 GHz compared with the conventional transmission line.

The Doherty amplifier is measured at 42 GHz with a 2.5 V supply. The use of a two-stack amplifier allows the DC voltage (and RF voltage swing) to be twice as large as what might be applied to an individual device. The measured and simulated gain are plotted in Fig. 4.33. For the simulation and measurements, the bias of the auxiliary amplifier is adapted for the peak gain at different output power levels. This adaptive biasing allows the Doherty amplifier to maintain high gain and therefore the highest PAE over the power range. The bias voltage at the gate of M3 in Fig. 4.32 is held to −0.2 V for deep class-C operation under low-power conditions. At high power, the gate

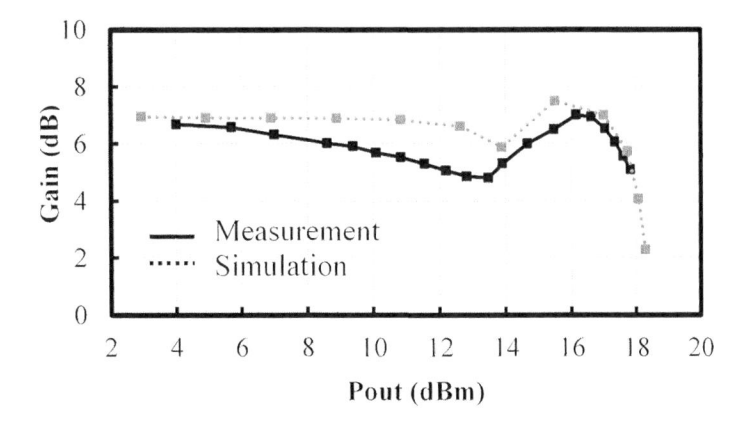

Fig. 4.33 Simulated and measured gain at 45 GHz. Adaptive biasing of the auxiliary device is implemented to maintain peak gain. Adapted from [15].

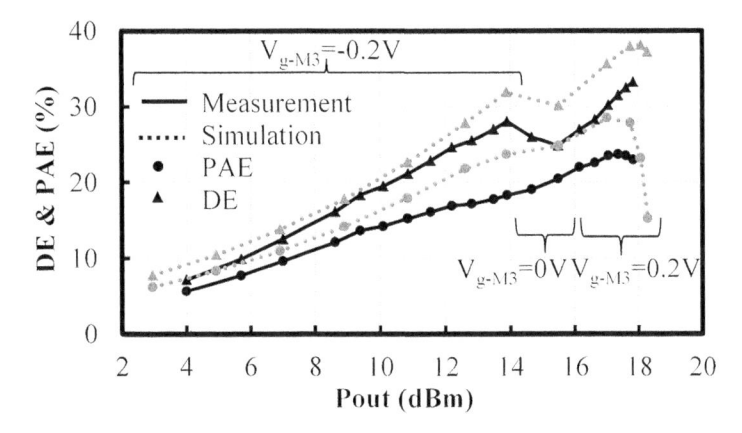

Fig. 4.34 Simulated and measured drain and power-added efficiency for a Doherty amplifier at 42 GHz. Adapted from [15].

voltage is increased to 0.2 V to keep the amplifier closer to class-B bias. While the simulated gain is less than 0.7 dB, the measured gain deviates by more than 2 dB.

The peak power in both measurement and simulation is just over 18 dBm. Large-signal simulations and measurements demonstrate the drain efficiency and power-added efficiency of the amplifier in back-off. While the simulated drain efficiency remains above 30% at 4 dB back-off, the measured drain efficiency is degraded by the somewhat lower gain of the measured circuit. However, the drain efficiency of the Doherty amplifier with the $\lambda/4$ transmission line is approximately 3% higher at back-off power than that of a Doherty amplifier that was designed using conventional $\lambda/4$ transmission lines (see Fig. 4.34).

4.9 Summary

This chapter has determined the fundamental bounds for linear-mode, mm-wave power amplifiers implemented in silicon integrated circuit technologies based on device and circuit limitations. These device and circuit factors have been put in the context of classical linear-mode operation and related to the conduction angle of a power amplifier to show the biasing that achieves peak efficiency. For mm-wave circuits using fine-line CMOS, the optimum linear mode of operation is between class A and class B, depending on the frequency and power level that is desired. Furthermore, this analysis has given insight into the importance of layout parasitics in mm-wave PA design and the severe performance penalties imposed by parasitic source resistance and inductance. Finally, this chapter discussed the use of Doherty PAs to maintain high average efficiency for applications that require operation under back-off conditions. Case studies have been provided for a SiGe class-B PA at 45 GHz and a CMOS SOI Doherty PA at 45 GHz.

References

[1] S. Pornpromlikit, J. Jeong, C. D. Presti, A. Scuderi, and P. M. Asbeck, "A watt-level stacked-FET linear power amplifier in silicon-on-insulator CMOS," *IEEE Transactions on Microwave Theory and Techniques*, vol. **58**, no. 1, pp. 57–64, 2010. [Online]. Available: http://ieeexplore.ieee.org/stamp/stamp.jsp?arnumber=5352239

[2] S. Pornpromlikit, H.-T. Dabag, B. Hanafi, *et al.*, "A Q-band amplifier implemented with stacked 45-nm CMOS FETs," in *Compound Semiconductor Integrated Circuit Symposium (CSICS), 2011 IEEE*, 2011, pp. 1–4. [Online]. Available: http://ieeexplore.ieee.org/stamp/stamp.jsp?arnumber=6062465

[3] H. Dabag, B. Hanafi, F. Golcuk, *et al.*, "Analysis and design of stacked-FET millimeter-wave power amplifiers," *IEEE Transactions on Microwave Theory and Techniques*, vol. **61**, no. 4, pp. 1543–1556, 2013. [Online]. Available: http://ieeexplore.ieee.org/stamp/stamp.jsp?arnumber=6475208

[4] J. Kim, H. Dabag, P. Asbeck, and J. F. Buckwalter, "Q-band and W-band power amplifiers in 45-nm CMOS SOI," *IEEE Transactions on Microwave Theory and Techniques*, vol. **60**, no. 6, pp. 1870–1877, 2012. [Online]. Available: http://ieeexplore.ieee.org/stamp/stamp.jsp?arnumber=6197245

[5] P. Habas, "Hot-carrier NMOST degradation at periodic drain signal," in *Microelectronics, 2002. 23rd International Conference on*, vol. 2, 2002, pp. 731–734. [Online]. Available: http://ieeexplore.ieee.org/stamp/stamp.jsp?arnumber=1003361

[6] C. D. Presti, F. Carrara, A. Scuderi, S. Lombardo, and G. Palmisano, "Degradation mechanisms in CMOS power amplifiers subject to radio-frequency stress and comparison to the DC case," in *Reliability Physics Symposium, 2007. Proceedings. 45th Annual. IEEE International*, 2007, pp. 86–92. [Online]. Available: http://ieeexplore.ieee.org/stamp/stamp.jsp?arnumber=4227614

[7] A. Sayag, D. Ritter, and D. Goren, "Compact modeling and comparative analysis of silicon-chip slow-wave transmission lines with slotted bottom metal ground planes," *IEEE Transactions on Microwave Theory and Techniques*, vol. **57**, no. 4, pp. 840–847, 2009. [Online]. Available: http://ieeexplore.ieee.org/stamp/stamp.jsp?arnumber=4803741

[8] I. Bahl, *Fundamentals of RF and Microwave Transistor Amplifiers*. Wiley, 2009.

[9] P. Colantonio, F. Giannini, and E. Limiti, *High Efficiency RF and Microwave Solid State Power Amplifiers*. Wiley, 2009.

[10] M. Rickelt, H.-M. Rein, and E. Rose, "Influence of impact-ionization-induced instabilities on the maximum usable output voltage of Si-bipolar transistors," *IEEE Transactions on Electron Devices*, vol. **48**, no. 4, pp. 774–783, 2001.

[11] U. R. Pfeiffer and A. Valdes-Garcia, "Millimeter-wave design considerations for power amplifiers in an SiGe process technology," *IEEE Transactions on Microwave Theory and Techniques*, vol. **54**, no. 1, pp. 57–64, 2006.

[12] H.-T. Dabag, J. Kim, L. E. Larson, J. F. Buckwalter, and P. M. Asbeck, "A 45-GHz SiGe HBT amplifier at greater than 25% efficiency and 30-mW output power," in *Bipolar/BiCMOS Circuits and Technology Meeting (BCTM), 2011 IEEE*, 2011, pp. 25–28. [Online]. Available: http://ieeexplore.ieee.org/stamp/stamp.jsp?arnumber=6082742

[13] W. H. Doherty, "A new high efficiency power amplifier for modulated waves," *Proceedings of the Institute of Radio Engineers*, vol. **24**, no. 9, pp. 1163–1182, 1936.

[14] S. C. Cripps, *RF Power Amplifiers for Wireless Communications*. Artech House, 2006.

[15] A. Agah, B. Hanafi, H. Dabag, *et al.*, "A 45 GHz Doherty power amplifier with 23% PAE and 18 dBm output power in 45 nm SOI CMOS," in *Microwave Symposium Digest (MTT), 2012 IEEE MTT-S International*, 2012, pp. 1–3. [Online]. Available: http://ieeexplore.ieee.org/stamp/stamp.jsp?arnumber=6259632

[16] A. Agah, H. Dabag, B. Hanafi, *et al.*, "A 34% PAE, 18.6 dBm 42–45 GHz stacked power amplifier in 45-nm SOI CMOS," in *Radio Frequency Integrated Circuits Symposium (RFIC), 2012 IEEE*, 2012, pp. 57–60. [Online]. Available: http://ieeexplore.ieee.org/stamp/stamp.jsp?arnumber=6242231

[17] T. S. D. Cheung and J. R. Long, "Shielded passive devices for silicon-based monolithic microwave and millimeter-wave integrated circuits," *IEEE Journal of Solid-State Circuits*, vol. **41**, no. 5, pp. 1183–1200, 2006. [Online]. Available: http://ieeexplore.ieee.org/stamp/stamp.jsp?arnumber=1624408

5 Switch-mode mm-wave silicon power amplifiers

Harish Krishnaswamy, Hossein Hashemi, Anandaroop Chakrabarti, and Kunal Datta

5.1 Introduction to switching power amplifiers

Switch-mode power amplifiers (PAs) are motivated by the insight that power-amplifier efficiency is maximized by minimizing the amount of overlap between device current and device voltage. In other words, it is desirable to minimize the amount of time spent by the device supporting a non-zero current and non-zero voltage simultaneously. Switch-mode power amplifiers accomplish this by employing the active device as a switch that transitions between two states – an ON state, where the resistance of the device is ideally zero and in practice small compared with the other impedances in the circuit, and an OFF state, where the resistance of the device is ideally infinite and in practice high compared with the other circuit impedances. Consequently, the active device forces the current and voltage waveforms to be non-overlapping – during the ON state, the switch supports non-zero current but the voltage across it is close to zero, and during the OFF state, the switch supports non-zero voltage but the current through it is zero. Under idealized conditions, 100% efficiency can be achieved assuming switching losses are eliminated. This typically requires the voltage across the switch to be shaped to zero at the end of the OFF state as the switch is turning ON, so that the switch is not turned ON with non-zero charge stored on its output capacitance. Assuming such a "zero-voltage switching" (ZVS) condition is met, 100% efficiency can be achieved, unlike for current-source-based power amplifiers, where 100% efficiency can be achieved only in class-C operation as the output power approaches zero or by using class-F tuning, where 100% efficiency is achieved asymptotically with infinite harmonic tuning. In practice, however, the efficiency is limited by parasitic effects, such as conduction loss in the non-zero switch resistance and loss in the passive output matching network.

Switch-mode power amplifiers are distinct from their current-source counterparts (Fig. 5.1) in that they are (at least partially) voltage-forcing amplifiers. During the ON portion of the RF cycle, the device voltage is forced to ground through the switch, and during the OFF portion, the voltage is determined either by the load network or perhaps by the presence of a complementary switch (as is the case in inverter-like class-D power amplifiers). Current-source-based power amplifiers on the other hand use the active device as a transconductance element, and the current of the active device is determined by the nature and shape of the input drive voltage and device bias point. The

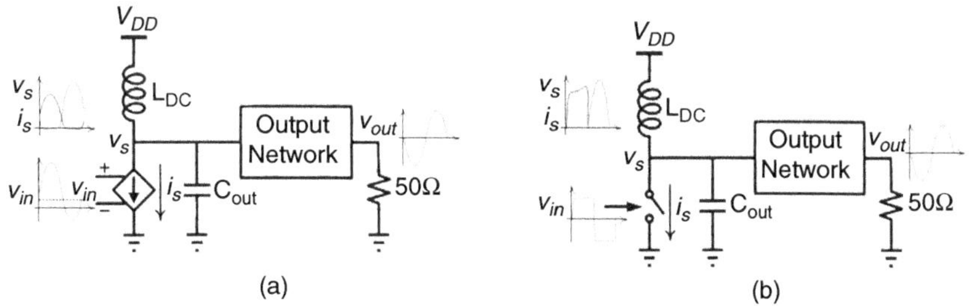

Fig. 5.1 (a) Generic current-source-based power amplifier. (b) Generic switch-mode power amplifier.

device voltage on the other hand is purely determined by the load network tuning that selects one or more of the device current harmonics for voltage shaping.

Switch-mode power amplifiers have been extensively investigated at RF frequencies below 10 GHz in both CMOS and SiGe technologies [1–9]. However, technology scaling has afforded us with CMOS and SiGe technology nodes whose f_Ts approach or even exceed 200 GHz. This has enabled research into the implementation of switch-mode power amplifiers at millimeter-wave (mm-wave) frequencies. This chapter explores design considerations for CMOS and SiGe switch-mode power amplifiers and describes several reported designs in the literature as case studies. The chapter finally concludes with a description of linearizing architectures that can be used to enable switch-mode power amplifiers to support complex modulations that include amplitude components.

5.2 Design issues for CMOS mm-wave switching power amplifiers

5.2.1 Parasitics that become significant at mm-wave frequencies

Switch-mode CMOS power amplifiers such as the class-E PA [10] have been the subject of extensive research at RF frequencies. However, efficient switch-mode operation at mm-wave frequencies in CMOS is rendered challenging due to the effects of several active and passive device parasitics. In this section, we focus on parasitic effects that necessitate a revisitation of the design methodology of the class-E PA. The classical class-E design methodology uses a switch with an infinite DC-feed inductance, both assumed to be lossless, and derives the optimal load impedance that results in 100% efficiency through the achievement of the "class-E switching conditions," namely ZVS and ZDVS (zero-derivative-of-voltage at switching). However, at mm-wave frequencies in CMOS, the following parasitic effects become significant.

(1) The use of an infinite DC-feed inductance or choke is impractical due to the associated loss and the finite self-resonant frequency of on-chip inductors.
(2) Switch ON-resistance can play a significant role at mm-wave frequencies and must be considered.
(3) The loss of passive components, namely the finite DC-feed inductance as well as the load inductance, must be taken into account.
(4) The input power that must be supplied to drive the active device into switch-mode operation can be significant, impacting power-added efficiency (PAE).

5.2.2 A mm-wave loss-aware class-E PA design methodology

In this subsection, we focus on a detailed analysis of the class-E PA including all parasitic effects that are significant at mm-wave frequencies. The presence of a finite DC-feed inductance, switch ON-resistance and passive loss contribute significant complexity to the mathematical analysis of the circuit. However, integrated solutions at mm-wave frequencies necessitate that these non-idealities be taken into account so as to avoid sub-optimal designs. The different sources of loss can be accounted for in two ways: (1) perturbation analysis, which assumes that losses are small enough that currents and voltages remain unchanged; or (2) comprehensive circuit analysis with all parameters derived in the presence of loss.

Some workers have performed a comprehensive analysis with switch ON-resistance (R_{on}) and/or finite DC-feed inductance [11–15], but impose one or both of the "class-E switching conditions," namely ZVS and ZDVS. It must be emphasized that these conditions are no longer optimal at mm-wave frequencies for achieving high-efficiency operation in the presence of high loss levels in the switching device and/or passive components. In the presence of appreciable R_{on}, it might be beneficial to sustain some ZVS loss in order to reduce conduction loss.

To circumvent this issue, an improved loss-aware class-E design methodology is proposed [16], which formally takes switch loss and passive loss into account. The methodology also incorporates the input power required to drive the switch and enables optimization of PAE rather than drain efficiency. In essence, the methodology is an analytical load-pull for optimizing PAE in the presence of high loss levels and input power requirements, thereby making it suitable for mm-wave PA design.

Circuit model and assumptions

The circuit diagram of the class-E CMOS PA is shown in Fig. 5.2. In the absence of R_{on} and passive loss, the switch voltage and switch current resemble those depicted in the inset in Fig. 5.2 when class-E switching conditions are satisfied. For the ensuing derivations, we make the following assumptions.

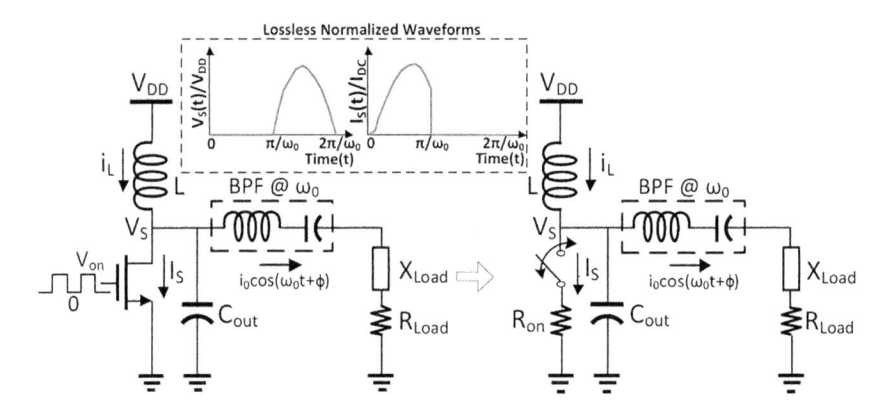

Fig. 5.2 Class-E PA with finite DC-feed inductance and non-zero switch ON-resistance.

(1) The MOSFET can be represented by a switch with finite series ON-resistance R_{on} in parallel with a linear capacitor C_{out}.
(2) $R_{on} \ll \frac{1}{\omega_0 C_{out}}$.
(3) The loaded quality factor (Q_L) of the series resonant filter in the output network is large.
(4) The duty cycle of the switch is 50%, though the analysis can be extended to any arbitrary duty cycle.
(5) The filter is assumed to be lossless.

Loss-aware class-E analysis

Let us assume that the switch is open ("OFF") for $0 \leq t < \frac{T}{2}$ and closed ("ON") for $\frac{T}{2} \leq t < T$, where $T = \frac{2\pi}{\omega_0}$ is the switching period. We use the subscripts *ON* and *OFF* for voltages and currents to indicate the respective half-cycles. Using assumption (3), the load current can be represented as

$$i_{load} = i_0 \cos(\omega_0 t + \phi). \quad (5.1)$$

During the "ON" half-cycle $\frac{T}{2} \leq t < T$, we have the following relations:

$$V_{DD} - V_{S,ON} = L \frac{di_{L,ON}}{dt} \quad (5.2)$$

and

$$V_{S,ON} = \left(i_{L,ON} - i_0 \cos(\omega_0 t + \phi)\right) R_{on}. \quad (5.3)$$

The current through C_{out} is neglected in view of assumption (2). Using Eqn (5.3), we can rewrite Eqn (5.2) as

$$\frac{dV_{S,ON}}{dt} + \left(\frac{R_{on}}{L}\right) V_{S,ON} - i_0 \omega_0 R_{on} \sin(\omega_0 t + \phi) - \left(\frac{V_{DD} R_{on}}{L}\right) = 0. \quad (5.4)$$

The solution to this linear differential equation is of the form

$$V_{S,ON}(t) = V_{DD} + a_1 e^{\beta t} + a_2 \cos(\omega_0 t + \phi) + a_3 \sin(\omega_0 t + \phi), \quad (5.5)$$

where

$$a_1 = V_{S,ON}\left(\frac{T}{2}\right) - V_{DD} - \frac{R_{on} i_0 e^{-\frac{\beta T}{2}} + \frac{\beta}{\omega_0} R_{on} i_0 e^{-\frac{\beta T}{2}} \sin(\phi)}{\left(1 + \frac{\beta^2}{\omega_0^2}\right)}$$

$$a_2 = \frac{-R_{on} i_0}{1 + \frac{\beta^2}{\omega_0^2}}, \quad a_3 = \frac{-R_{on} i_0 \beta}{\omega_0 \left(1 + \frac{\beta^2}{\omega_0^2}\right)}, \quad \beta = \frac{-R_{on}}{L}, \quad (5.6)$$

and $V_{S,ON}(\frac{T}{2})$ is a constant to be evaluated.

For the "OFF" half-cycle $0 \leq t < \frac{T}{2}$, when the switch is open, we can write equations identical to (5.2) and (5.3) and arrive at

$$\frac{d^2 V_{S,OFF}}{dt^2} + \frac{V_{S,OFF}}{LC_{out}} - \frac{i_0 \omega_0}{C_{out}} \sin(\omega_0 t + \phi) - \frac{V_{DD}}{LC_{out}} = 0. \tag{5.7}$$

The solution to this second-order linear differential equation is given by

$$\begin{aligned} V_{S,OFF}(t) = {} & V_{DD}\left[1 - \cos(\omega_s t)\right] + V_{S,OFF}(0)\cos(\omega_s t) \\ & + \frac{V'_{S,OFF}(0)}{\omega_s} \sin(\omega_s t) \\ & + \frac{i_0 \omega_0 \sin(\phi)}{C_{out}\left(\omega_s^2 - \omega_0^2\right)} \left[\cos(\omega_0 t) - \cos(\omega_s t)\right] \\ & + \frac{i_0 \omega_0^2 \cos(\phi)}{C_{out}\left(\omega_s^2 - \omega_0^2\right)} \left[\frac{\sin(\omega_0 t)}{\omega_0} - \frac{\sin(\omega_s t)}{\omega_s}\right], \end{aligned} \tag{5.8}$$

where $\omega_s = \frac{1}{\sqrt{LC_{out}}} = n\omega_0$, while $V_{S,OFF}(0)$ and $V'_{S,OFF}(0)$ are constants to be evaluated. The values for $V_{S,ON}(\frac{T}{2})$, $V_{S,OFF}(0)$, and $V'_{S,OFF}(0)$ can be arrived at by imposing the following continuity conditions:

$$i_{L,OFF}(0^+) = i_{L,ON}(T^-), \quad V_{S,OFF}\left(0^+\right) = V_{S,ON}\left(T^-\right) \tag{5.9}$$

and

$$i_{L,OFF}\left(\frac{T^+}{2}\right) = i_{L,ON}\left(\frac{T^-}{2}\right). \tag{5.10}$$

The load impedance Z_{load} is computed as the ratio of the fundamental component of the switch voltage to that of the load current. Since no constraints have been imposed on either the switch voltage or its derivative at switch turn-on, we need to account for possible capacitive discharge loss. Under the assumption $R_{on} \ll \frac{1}{\omega_0 C_{out}}$, this loss can be estimated as

$$P_{loss,cap} = 0.5 f_0 C_{out} \left[V_{S,OFF}^2\left(\frac{T^-}{2}\right) - V_{S,ON}^2\left(\frac{T^+}{2}\right)\right]. \tag{5.11}$$

The loss in the switch is given by

$$P_{loss,switch} = R_{on} * \frac{1}{T}\int_{\frac{T}{2}}^{T}\left(\frac{V_{S,ON}}{R_{on}}\right)^2 dt. \tag{5.12}$$

In order to incorporate input power into the formulation, the input power (P_{in}) is approximated as

$$P_{in} = k f_0 C_{in} V_{on}^2, \tag{5.13}$$

where $C_{in} = C_{gs} + C_{gd}$ in the triode region, V_{on} is the input drive level in the "ON" half-cycle, and k is a fitting parameter determined from schematic simulations [17]. Finite reverse isolation (i.e. $C_{gd} \neq 0$) causes the value of parameter k to vary with the parameter n (since output network component values change), but for preliminary

5.2 Design issues for CMOS mm-wave switching power amplifiers

analysis, this dependence is ignored. Finally, if $R_{choke} = \frac{\omega_0 L}{Q_{choke}}$ is the series resistance in the DC-feed inductance, its loss can be calculated using perturbation analysis as

$$P_{loss,choke} = R_{choke} * \frac{1}{T} \left(\int_0^{\frac{T}{2}} i_{L,OFF}^2 \, dt + \int_{\frac{T}{2}}^{T} i_{L,ON}^2 \, dt \right). \tag{5.14}$$

The circuit may now be optimized for PAE by choosing the appropriate load impedance. This is achieved by means of a MATLAB code, which sweeps the magnitude i_0 and phase ϕ of the load current to arrive at a design point with optimal PAE for a given device size, input drive level V_{on}, and the parameter n. A global optimization is performed subsequently by varying V_{on} and n to select the design point with highest PAE for a fixed device size. If the load impedance is different from 50 Ω, then a matching network needs to be designed to perform impedance transformation. Alternatively, subsequent to PAE optimization, the device size (and all other circuit components) can be scaled so that $R_{load} = 50\,\Omega$ to determine the power that can be delivered to a 50-Ω load and to eliminate the loss in a matching network. Based on the theoretical design points provided by this design methodology, a single-device class-E PA driving a 50-Ω load (Fig. 5.3a) is designed in IBM 45-nm SOI CMOS using floating-body devices. Figure 5.3b compares normalized theoretical and device-based waveforms for tuning $n = 0.8$, while Figs. 5.3c, d compare the theoretical output power and PAE as a function of the tuning parameter n. The characteristics of the PA for $n = 0.8$ and for three different frequencies are summarized in Fig. 5.4.

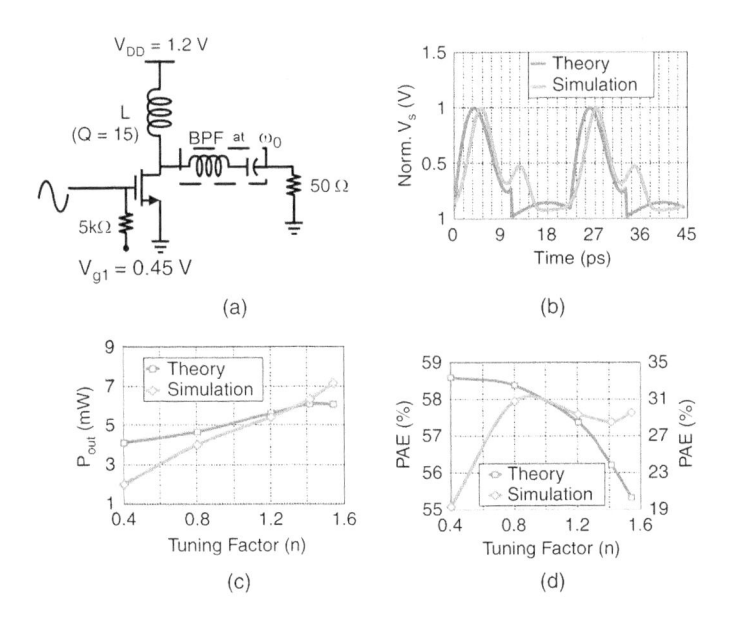

Fig. 5.3 (a) Single-device class-E PA with finite DC-feed inductance driving a 50-Ω load, (b) comparison of theoretical and device-based waveforms for tuning parameter $n = 0.8$ and simulation results (all at 45 GHz) for (c) output power and (d) PAE as a funtion of n in IBM 45-nm SOI CMOS using floating-body device.

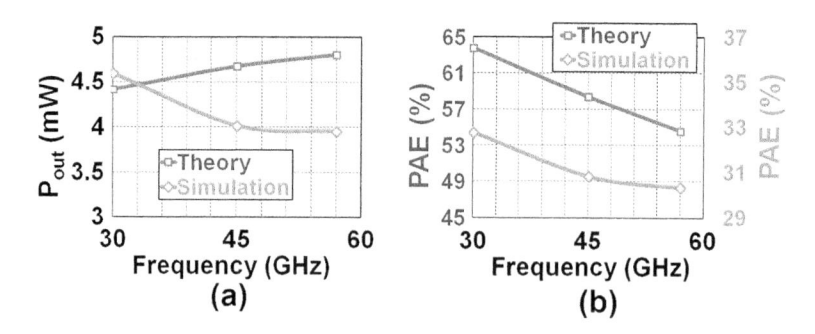

Fig. 5.4 Simulated (a) output power and (b) PAE for single-device class-E PA in IBM 45-nm SOI CMOS for tuning parameter $n = 0.8$ across frequency.

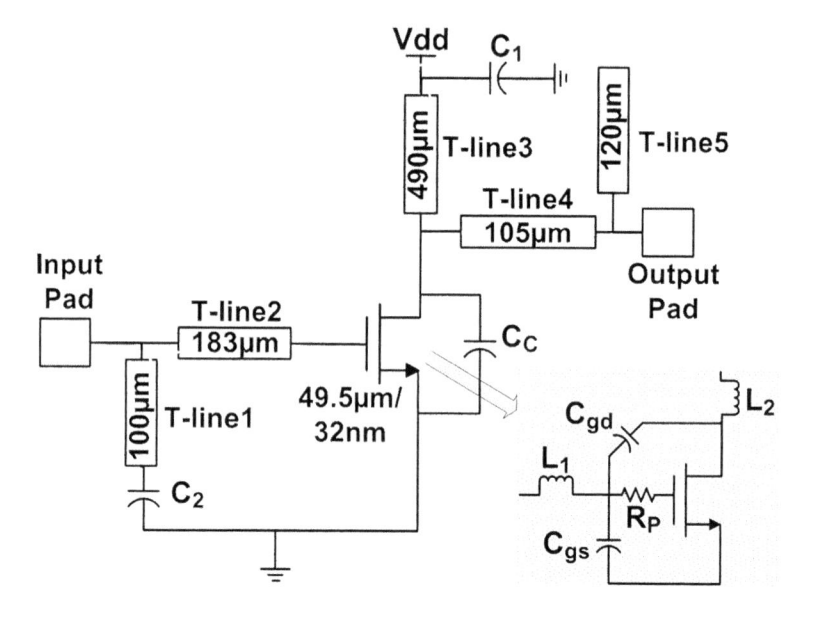

Fig. 5.5 Circuit diagram of a single-ended 60-GHz class-E PA implemented in 32-nm SOI CMOS.

5.2.3 Case study – a 60-GHz class-E PA in 32-nm SOI CMOS

Researchers at IBM recently reported 60GHz class-E power amplifiers in 32-nm SOI CMOS. Two versions were implemented – one single-ended and the other differential. The circuit diagram of the single-ended version is shown in Fig. 5.5.

The design methodology for the load impedance follows the original Sokal class-E design equations, followed by large-signal simulation-based design optimization. The device size was chosen so that 175 mV of input voltage change around the bias point (chosen to be the threshold voltage) produces the required maximum drain current of 30 mA. This maximum drain current is limited by the AC breakdown voltage of the process technology. The DC-feed transmission line was chosen to cancel out the excess output capacitance of the chosen device to achieve the class-E output shunt

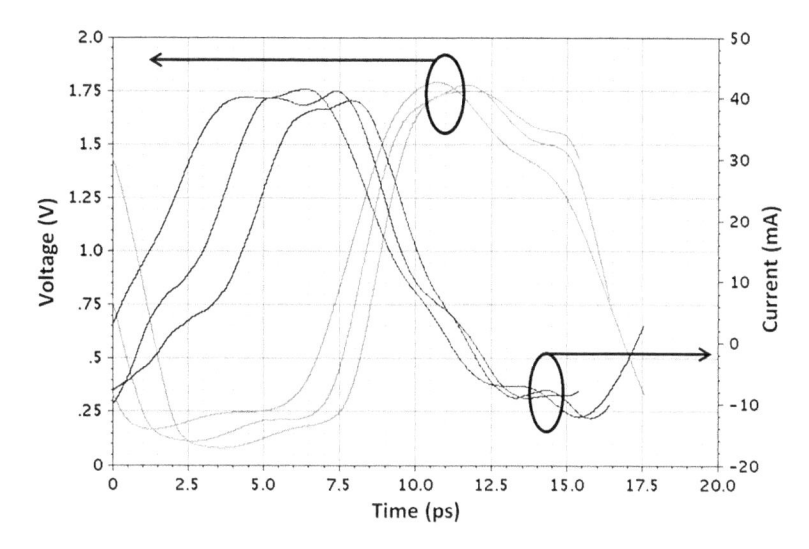

Fig. 5.6 Switch current and voltage waveforms at various frequencies of operation for the 60-GHz 32-nm SOI CMOS class-E PA.

capacitance. It is expected that such a simulation-based design approach yields a design point that is close to the loss-aware class-E design methodology described earlier. The switch current and voltage waveforms are shown in Fig. 5.6 and their harmonic-enriched non-overlapping nature confirms the class-E tuning of the PA.

The single-ended PA achieved a power gain of 8.8 dB, saturated output power >9 dBm and peak PAE of 27% when running off a 0.9-V supply at 60 GHz. The differential PA achieved a power gain of 10 dB, saturated output power >12.5 dBm and peak PAE of 30% with a 0.9-V supply at 60 GHz.

5.3 Design issues for SiGe HBT mm-wave switching power amplifiers

5.3.1 Challenges and opportunities in SiGe HBT switching amplifiers at mm-wave

The fundamental trade-off between the breakdown voltage and the maximum frequency of operation of semiconductor devices [18] is a mixed blessing for mm-wave power-amplifier designers. On the one hand, the higher f_{max} and f_T of advanced technology nodes allow power generation at higher frequencies, while on the other hand, the reduced maximum voltage swing across the devices directly translates to lower power being generated from unit power-amplifier cells. mm-Wave power-amplifier designs generating a moderate amount of power (>20 dBm) thus typically employ a variety of passive power-combining schemes, with a lower overall collector efficiency and PAE [19]. For mm-wave switching-amplifier designs, this breakdown voltage versus transistor speed trade-off is even more critical. Although switching power amplifiers are nominally more efficient than their linear counterparts due to the non-overlapping of voltage and current waveforms, at mm-waves this efficiency can be achieved only by properly manipulating and managing the harmonics of the fundamental frequency of operation, which becomes more and more difficult as the operational frequency becomes closer to the f_{max} of the device. The transistors in the switching amplifiers are expected to generate not only the fundamental frequency of operation but also its harmonics that, when properly weighted in conjunction with passive networks, create non-overlapping voltage and current waveforms across the transistors. Given the engineering effort required for a proper switching-amplifier design, it is critical that mm-wave switching-amplifier architectures are not only more efficient, but also able to maintain this high efficiency while generating significantly higher output power at mm-waves than the competing linear class of amplifiers. In the next few paragraphs, the design of mm-wave switching power amplifiers using SiGe HBT transistors is discussed, where the breakdown mechanism of the HBTs has been fully harnessed, in an effort to generate high power simultaneously with high PAE at mm-waves.

5.3.2 Beyond-BV_{CEO} operation of SiGe HBTs

SiGe HBTs in general have a higher breakdown voltage than the silicon CMOS FETs for the same small-signal f_{max}. Generating higher harmonics at mm-waves is feasible using modern SiGe HBT with f_{max} reaching or even exceeding 300 GHz [20, 21], enabling design of switching amplifiers with high efficiency. In addition, the non-overlapping voltage and current waveforms of switching amplifiers also enhance the maximum voltage swing achievable across the HBT transistor without device breakdown. In a SiGe HBT, the BV_{CEO} is the breakdown voltage of the device only when it is used in an open-base, base current source driven configuration. This is usually true in linear amplifiers, where the transistors are modeled as current driven current sources to maintain linearity, but at the cost of reduced voltage swing and lower output power. In a low base impedance, fixed voltage driven configuration, the breakdown voltage of SiGe HBTs increases and it can be as large as BV_{CBO} of 5.9 V [22]. Under high electric field, the minority carriers in the n-p-n transistors, the electrons transiting from the emitter to the collector, reach critical velocity and collide with the collector–base junction lattice creating electron–hole pairs (EHPs) in a process known as "impact ionization." For open-base transistor configurations, the hole of this newly created EHP can reach the emitter and create new EHPs, accelerating the process, which eventually leads to transistor breakdown due to heating effects. In contrast, a low base impedance termination allows the majority of these holes to exit through the base terminal, allowing a larger range of collector voltage over which the HBT transistors can operate. This property of HBT transistors can be particularly useful for mm-wave switching-amplifier configurations. In a properly designed switching amplifier, the collector voltage peaks when the collector current is close to zero and vice versa. Hence, for low collector current density, the collector voltage in switching amplifiers can be as high as the BV_{CBO} of 5.9 V, as shown in Fig. 5.7, resulting in high output power generation even at mm-waves while maintaining high efficiency. In the following case studies, this beyond-BV_{CEO} operation of SiGe HBTs has been demonstrated at mm-waves in some Q-band class-E switching amplifier designs [23].

5.3 Design issues for SiGe HBT mm-wave switching power amplifiers

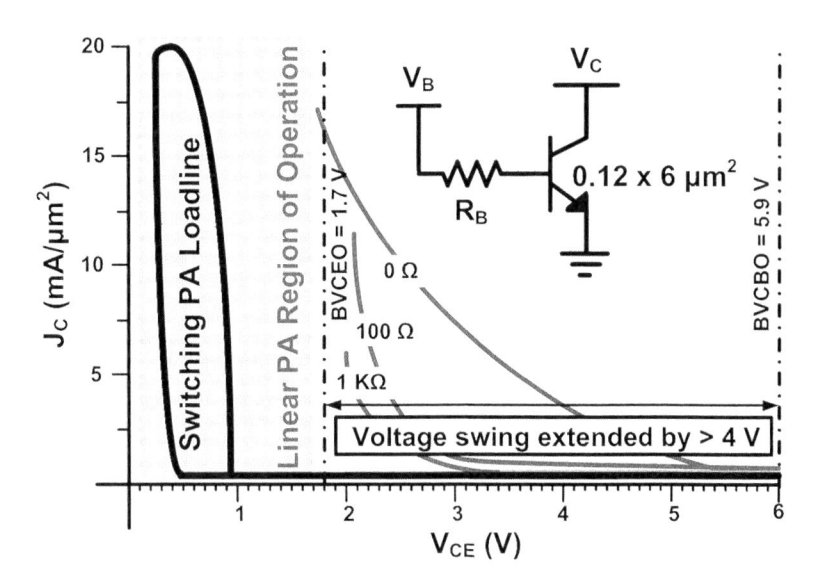

Fig. 5.7 Maximum allowable collector–emitter voltage (V_{CE}) of a SiGe HBT vs collector current density (J_C) for different base impedance terminations (R_B).

Using SiGe HBTs in a switching-amplifier configuration presents several challenges. Firstly HBT transistors, in general, act as better current sources than as ON–OFF switches compared with the CMOS devices. Although the saturation voltage (or knee voltage) of HBTs is low compared with the maximum collector voltage, it is still significant in a switching amplifier configuration, where minimizing the voltage–current overlap is critical. This translates into a higher dynamic ON-resistance of the device, leading to efficiency degradation due to conduction loss. The conduction loss of the device can be mitigated, by pushing the transistors deeper into saturation, but at the cost of significant base voltage overdrive which at mm-waves translates to high input power consumption and lower intrinsic gain from the devices. Low power gain of switching amplifiers is an important demerit as it can lead to lower PAE even with high collector efficiency. However, as shown in some of the case studies later, the power gain of switching amplifier stages can be successfully overcome by using stacking of transistors in switching amplifiers as well [24]. In such designs, all of the series stacked HBTs act as synchronous switches, turning ON and OFF simultaneously but with a larger overall voltage swing. The larger output power resulting from series stacking of devices is achieved at the cost of the same input power as before, thereby increasing the power gain of the system.

Using SiGe HBTs in the beyond-BV_{CEO} mode presents some additional challenges that must be addressed in the design stage [25]. Firstly, the holes coming out of the base junction, during high-voltage swings across the transistor, reduce the base current of the device until it becomes zero in a phenomenon known as base-current reversal (BCR). Beyond BCR, the base current becomes negative, which can lead to several types of transistor instabilties that are well documented. In particular, the real part of the input impedance looking into the device becomes negative, making the device susceptible to

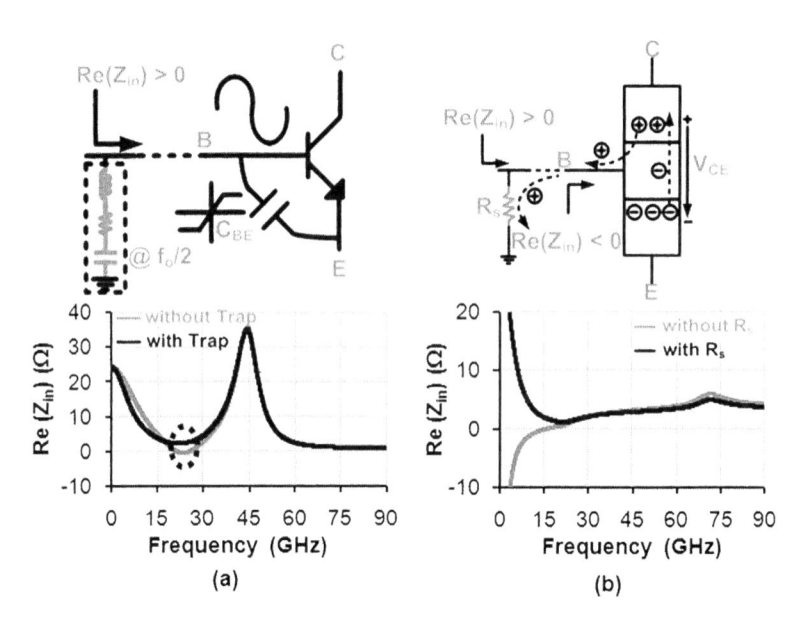

Fig. 5.8 Instabilities due to beyond-BV_{CEO} class-E operation: (a) half-harmonic oscillation; (b) base-current reversal.

low-frequency oscillation. Putting a low-impedance resistance in parallel to the base seems to mitigate this issue, as it prevents the quiescent base voltage from changing due to BCR. In addition, the hard switching of the HBTs to ensure non-overlapping class-E waveforms can cause the generation of sub-harmonic tones, parametrically, due to hard switching of the device's large nonlinear base–emitter capacitance. Half-harmonic traps added to the base of each stage in a multi-stage design are critical in overcoming the possibility of spurious oscillations. These effects and the ways to mitigate them are highlighted in Fig. 5.8. Finally, as in all large-signal power-amplifier designs, periodic steady-state simulations followed by large-signal S-parameter simulation of the two-port networks are critical in detecting and mitigating both small-signal and large-signal stability issues.

5.3.3 Case studies at Q-band

The beyond-BV_{CEO} operation of SiGe HBTs in mm-wave switching amplifiers has been demonstrated in the following case studies. In particular, a mm-wave class-E architecture was chosen as the transistor output capacitance can be especially incorporated into the amplifier-design procedure thus ensuring non-overlapping voltage and current waveforms, even at mm-waves.

Q-band two-stage class-E amplifier

Low-frequency class-E designs based on an ideal/non-ideal switch model have been extensively studied [26] (Fig. 5.9). However, such design methodologies are not necessarily applicable to mm-wave SiGe HBT designs due to large nonlinear device collector

5.3 Design issues for SiGe HBT mm-wave switching power amplifiers

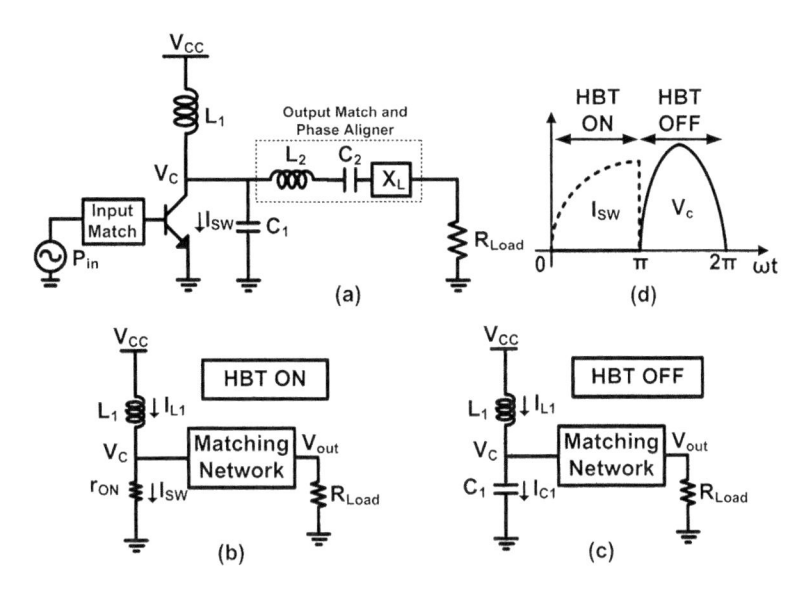

Fig. 5.9 A switch-model-based class-E power-amplifier analysis: (a) generic class-E architecture; (b) during ON cycle; (c) during OFF cycle; and (d) ideal class-E transient waveforms.

capacitance, base-to-collector capacitance, and the inability of the HBTs to perform as an ideal ON/OFF switch. In low-frequency class-E designs, the supply voltage V_{CC} is chosen to ensure that the peak collector voltage V_{CMax} for different values of the load resistance R_{Load} is lower than the transistor breakdown voltage. R_{Load} also determines the maximum switch current needed to be supplied by the transistor to enable the collector voltage to reach its peak value. At low radio frequencies, the effect of the output capacitance of the transistors is negligible, and hence the transistors are chosen as large as possible to supply the required switch current and lower conduction loss simultaneously.

At mm-waves, however, the design methodology has to be completely different. In particular, the class-E charging capacitor C_1 is primarily realized by the device intrinsic collector capacitance. Thus, for a fixed R_{Load}, the transistor of the chosen size must not only satisfy the class-E capacitance budget but at the same time must be able to satisfy the peak switch current requirement, to ensure a large voltage swing and high output power from the transistor. It can be shown that at mm-waves for the 0.13-μm process, it is not possible to meet both requirements simultaneously. One way of solving this dichotomy is to choose the device size according to the capacitance budget requirement and then to increase the supply voltage V_{CC} to increase the current density in the transistors to meet the peak switch-current requirement. Needless to say this comes at the cost of lower V_{CMax}-to-V_{CC} ratio or the "voltage utilization ratio," which is ≈3.56 for low-frequency class-E designs but degrades to ≈2.5 at mm-waves. Based on this design strategy, the maximum output power with best achievable PAE at 45 GHz is shown in a class-E performance chart for different load resistances (Fig. 5.10). The theory behind predicting the best achievable output power and PAE in SiGe HBT class-E designs has

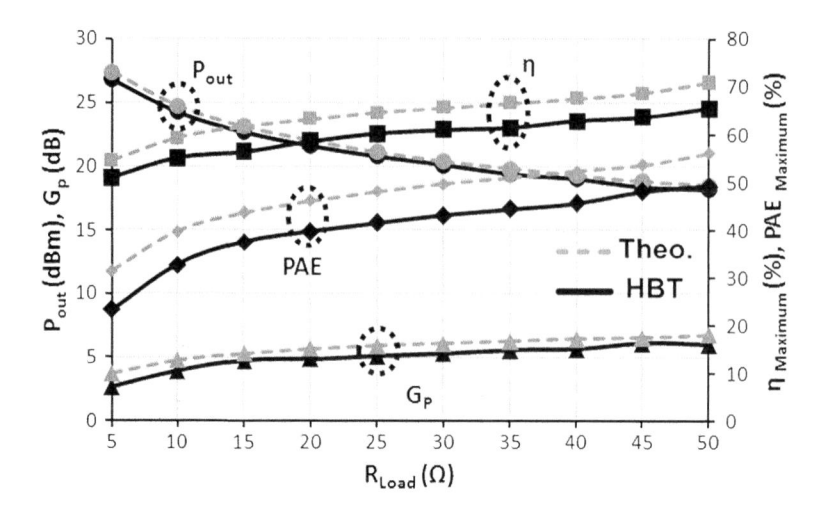

Fig. 5.10 Maximum achievable output power, power gain, collector efficiency, and PAE vs the load resistance for a 45-GHz, 130-nm SiGe HBT, class-E amplifier ($Q_{Passive} = 20$).

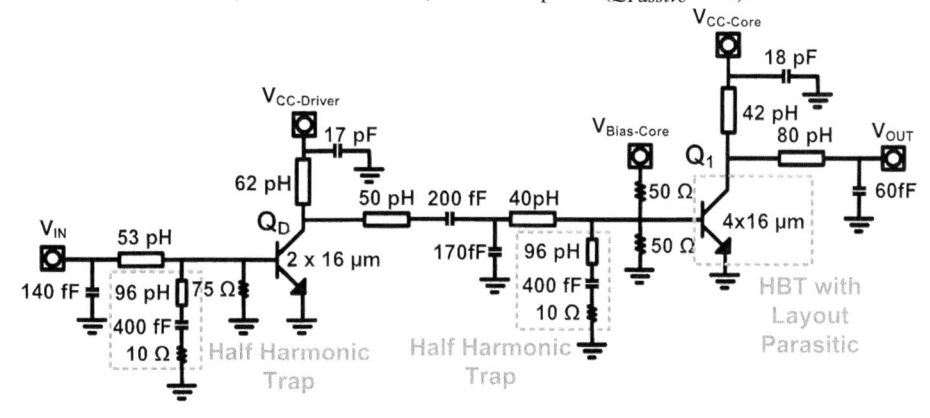

Fig. 5.11 Q-band two-stage class-E amplifier schematic.

been discussed in detail in [27]. Based on the above-mentioned mm-wave class-E design methodology, a two-stage Q-band class-E power amplifier (Fig. 5.11) was implemented in the 0.13-μm SiGe BiCMOS process [23]. In this design, the inductors were realized as low-loss microstrip transmission lines and the capacitors as metal–insulator–metal (MIM) capacitors. The beyond-BV_{CEO} operation of the SiGe HBTs at mm-waves can be observed from Fig. 5.12a where around 20 dBm output power was generated at 45 GHz from a single unit cell without recourse to any power-combining techniques. The benefit of using switching class-E amplifier topology can be seen in Figs. 5.12b, c where an overall peak PAE of 31.5% is achieved at the maximum output power levels.

Q-band double-stacked class-E amplifier

Stacking of active devices has been used in low-frequency linear and switching amplifiers and has also been demonstrated recently at mm-waves [24]. In the following example, a double-stacked Q-band class-E amplifier example is discussed, where

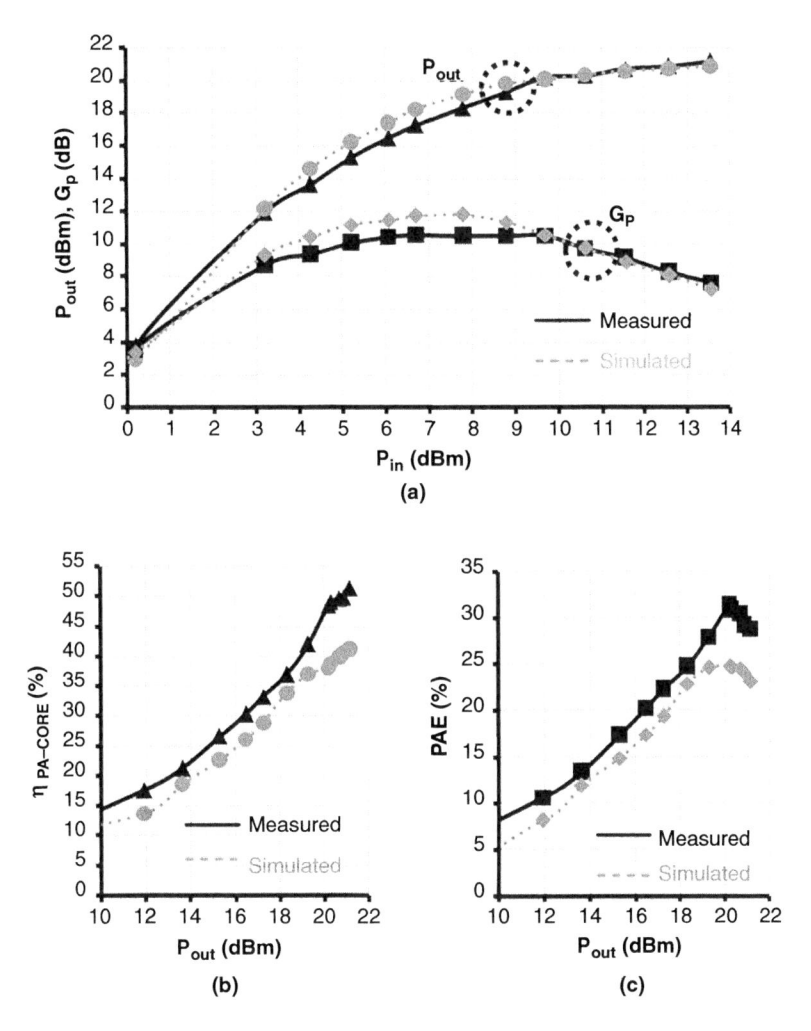

5.3 Design issues for SiGe HBT mm-wave switching power amplifiers

Fig. 5.12 Measured Q-band two-stage class-E performance: (a) output power and power gain vs input power; (b) collector efficiency vs output power; and (c) PAE vs output power.

two SiGe HBTs are stacked in series where each of the stacked devices operates in beyond-BV_{CEO} switching class-E mode. This is to demonstrate that the beyond-BV_{CEO} SiGe HBT operation discussed in the previous section can be extended to stacked architecture as well, leading to large output power levels ($\approx$23 dBm) while maintaining high efficiency.

The series-stacked transistors (Q_1 and Q_2) in the double-stacked class-E amplifier are designed to turn ON and OFF simultaneously (Fig. 5.13). This ensures that the voltage swing across the transistors adds up in-phase while still being less than the BV_{CBO} breakdown limit, leading to larger overall output voltage swing and power delivered to a fixed load. Ideally, the dynamic voltage swing must be equally divided amongst all the series-stacked HBTs to avoid stressing any single HBT. When the switching HBTs are

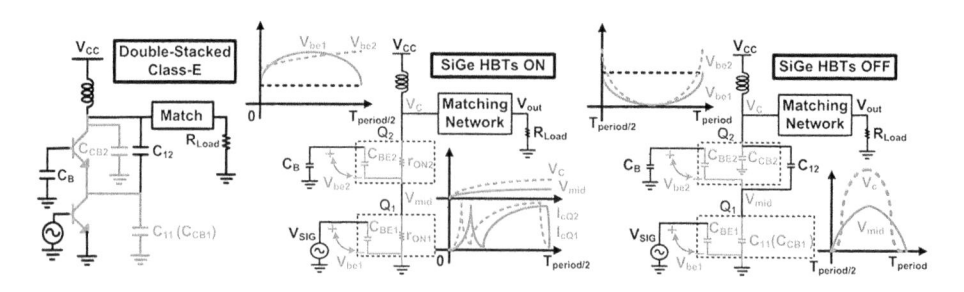

Fig. 5.13 Series stacking of SiGe HBTs in a class-E amplifier architecture when (a) HBTs are ON, and (b) HBTs are OFF.

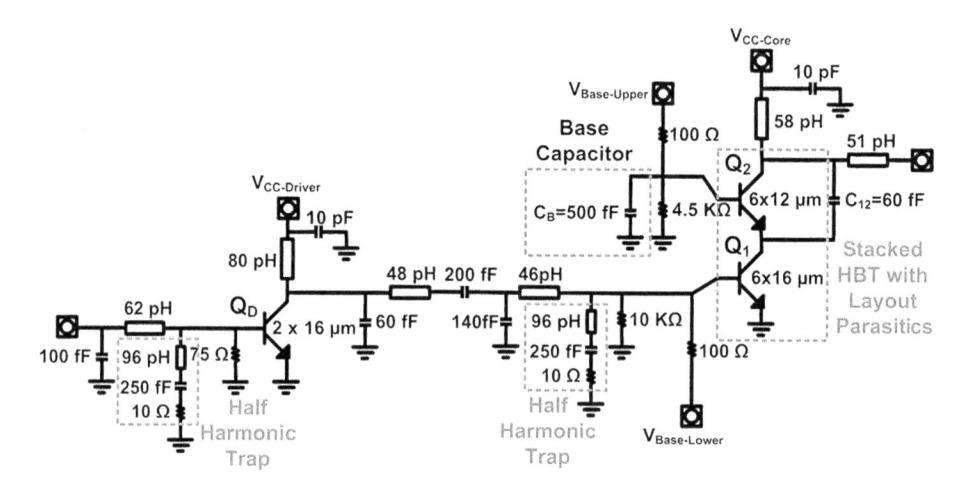

Fig. 5.14 Double-stacked Q-band class-E amplifier schematic.

ON, the ratio of the device ON-resistances, r_{ON1}/r_{ON2}, determines the voltage division across the devices. In the OFF state, a capacitive ladder network comprising C_{11} and C_{12} is used for voltage division in the intermediate node. In a mm-wave class-E design, the bottom capacitor of the ladder network (C_{11}) is realized mostly by the intrinsic collector to bulk capacitance of Q_1, C_{CB1}, which makes sizing of Q_1 a design parameter, while the capacitor C_{12} is realized by an additional explicit capacitor. The base terminal of Q_2 is terminated with the capacitance C_B, whose value depends on the base–emitter capacitance C_{BE2} of the stacked transistor Q_2. The $C_{BE2}-C_B$ capacitive divider ensures that the base terminal of Q_2 swings up along with its collector in the dynamic operation, thus preventing Q_2 collector–base junction breakdown, and also achieves the synchronous switching of the stacked devices.

The Q-band double-stacked class-E amplifier shown in Fig. 5.14 was implemented in the 0.13-μm SiGe BiCMOS process. Careful consideration must be employed in the layout and modeling of the stacked device core to minimize the effect of interconnect parasitics. The output power and power gain of the fabricated chip are shown in Fig. 5.15a while plots of the collector efficiency and PAE versus output power are shown in Figs. 5.15b, c. A peak output power of 23.4 dBm is reported for a peak PAE of 34.9% with 14.5 dB of power gain at 41 GHz. The benefit of stacking SiGe HBTs even

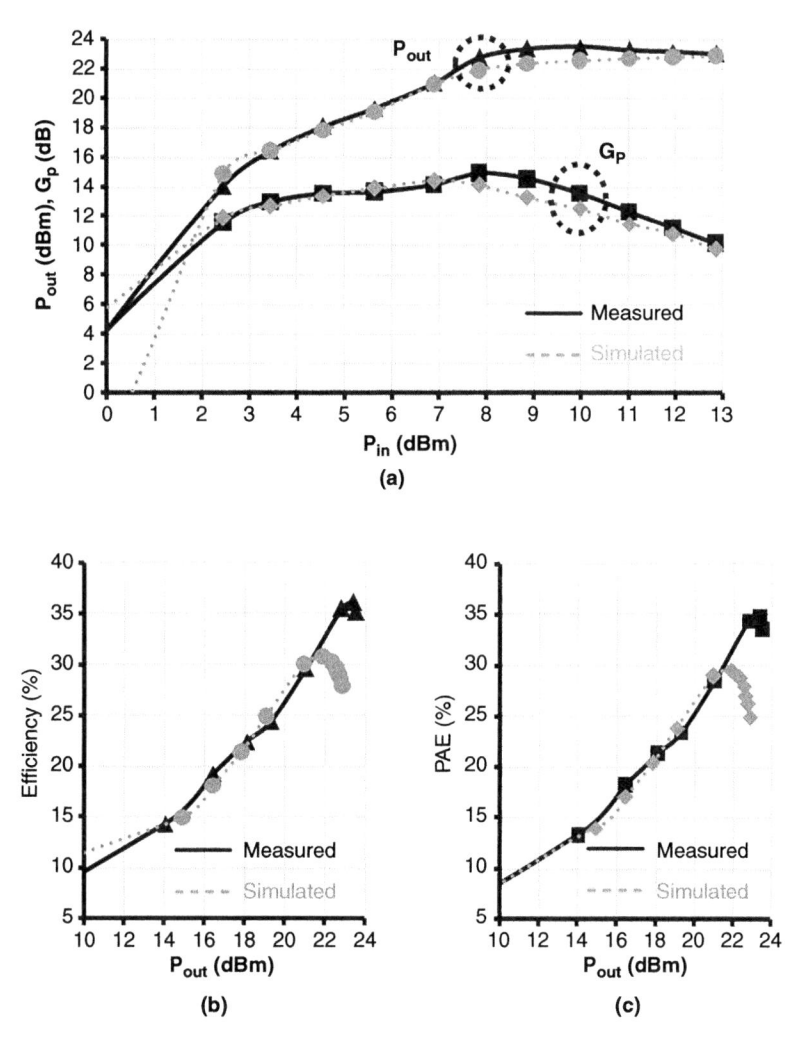

5.3 Design issues for SiGe HBT mm-wave switching power amplifiers

Fig. 5.15 Measured Q-band double-stacked class-E performance: (a) output power and power gain vs input power; (b) collector efficiency vs output power; and (c) PAE vs output power.

at mm-waves is evident from the improved output power and efficiency achievable in stacked architectures compared with conventional power-combining schemes.

Which is more important: f_{max} or BV_{CBO}? A design example

Scaling of technologies, while undoubtedly beneficial for digital circuits, is not necessarily advantageous for power-amplifier designs. This is primarily because the increase of f_T and f_{max} of scaled processes comes at the cost of lower breakdown voltages. Thus choosing between f_{max} and BV_{CBO} is often a real dilemma faced by mm-wave power-amplifier designers. In this section, a 45-GHz class-E amplifier design has been studied under two different SiGe HBT processes. The 0.13-μm 8HP SiGe HBT BiCMOS process offered by the IBM foundry has f_T/f_{max} of 200/280 GHz with a BV_{CBO} of 5.9 V.

The new-generation 90-nm 9HP SiGe HBT BiCMOS process also offered by the IBM foundry has f_T/f_{max} of 300/350 GHz with a BV_{CBO} of 5.2 V. The frequency of operation is chosen to be 45 GHz, since it is low enough compared with the f_{max} of the devices to enable generation of proper class-E waveforms in both the technologies, while being high enough to demonstrate all the complexities of mm-wave power-amplifier design.

For a fair comparison between the two technologies, an output power of 20 dBm is targeted, assuming the same quality factor of 20 for the passives (inductors and capacitors) at 45 GHz. It has been shown in the literature [27] that the maximum output power ($P_{out\ Maximum}$) and the maximum achievable collector efficiency ($\eta_{Maximum}$) in a class-E amplifier architecture can be defined in terms of transistor technology parameters,

$$\begin{cases} P_{out,max} = \frac{1}{2} \frac{BV_{CBO}^2}{R_{Load}} \psi^2 \approx \frac{BV_{CBO}^2}{9.75 R_{Load}}, \\ \eta_{max} = \frac{1}{1+\omega \tau_{out} \chi} \approx \frac{1}{1+0.13 \omega \tau_{out}}, \\ G_{P,max} = \frac{1}{\omega} \frac{BV_{CBO}^2}{\tau_{in}\alpha} \frac{\psi^2}{K_C(q_{max}) V_T^2} \left[\ln(BV_{CBO}) + \ln(\omega) + \ln(K_{ClassE}) \right. \\ \left. + \ln\left(\frac{C_1/\mu m^2}{I_{DC,Q}/\mu m^2}\right) \right]^{-2}, \\ PAE_{max} = \left(1 - \frac{1}{G_{P,max}}\right) * \eta_{max}, \end{cases} \quad (5.15)$$

where $\psi \approx 0.453$, $\chi \approx 0.1267$, $K_C(q_{max}) \approx 0.6765$, and $K_{ClassE} \approx 1.484$ are constants that are introduced in [27] and the remaining terms are technology parameters shown in Table 5.1. Similar results for maximum output power and maximum achievable PAE for double-stacked and triple-stacked SiGe HBT class-E power amplifiers have also been derived [27]. Figures 5.16 and 5.17 show the maximum achievable PAE in a class-E

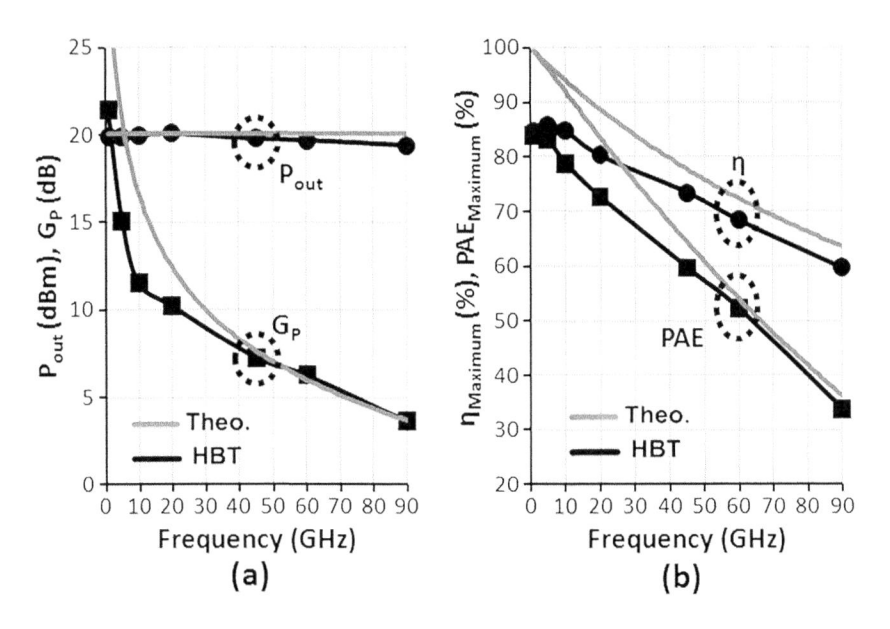

Fig. 5.16 Maximum achievable output power, power gain, collector efficiency, and PAE vs frequency in a 130-nm SiGe HBT class-E amplifier.

5.3 Design issues for SiGe HBT mm-wave switching power amplifiers

Table 5.1 Technology parameters for representative SiGe HBT processes.

Technology	HBT	BV_{CEO} (V)	BV_{CBO} (V)	f_T (GHz)	f_{max} (GHz)	T_{in} (ps)	T_{out} (ps)	C_{in} (pF/μm²)	C_{out} (fF/μm²)	α (C_{in}/C_{out})	$I_{DC,Q}$ (mA/μm²) ($V_B = 0.75$ V)
0.25-μm 7WL	High f_T	3.3	11	60	85	22	0.53	0.68	4	172	0.035
0.13-μm 8HP	High f_T	1.7	5.9	200	280	8	0.22	0.325	9.76	33.3	0.07
	High BV	3.55	12	60	120	18	1.6	0.68	9.76	69	0.035
90-nm 9HP	High f_T	1.5	5.3	300	350	5	0.1	0.4	12	33.3	0.22

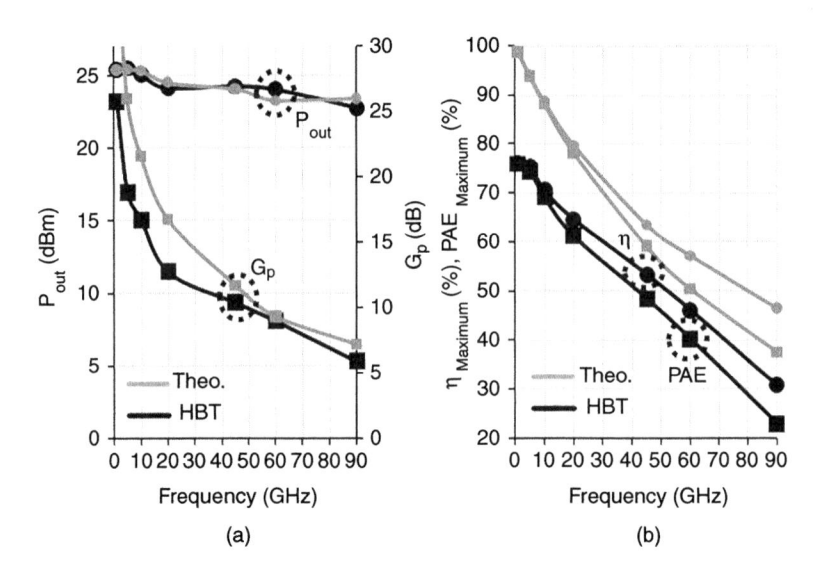

Fig. 5.17 Maximum achievable output power, power gain, collector efficiency, and PAE vs frequency in a 130-nm SiGe HBT double-stacked class-E amplifier.

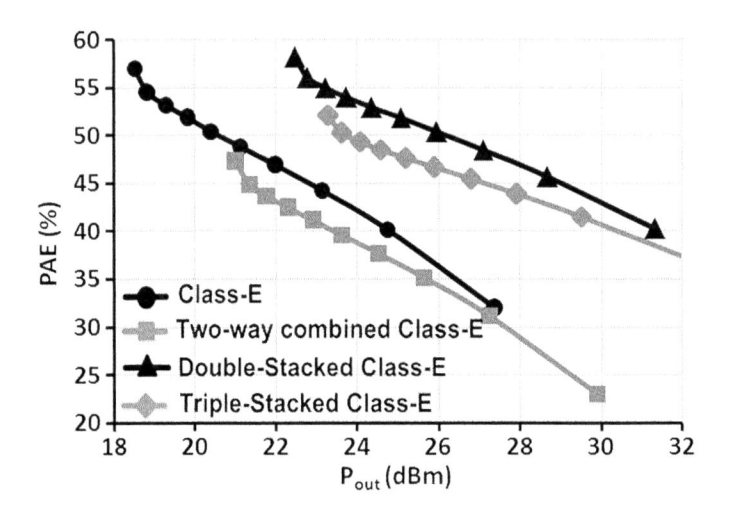

Fig. 5.18 Maximum achievable PAE vs output power at 45 GHz for different class-E architectures assuming $Q_{Passive} = 20$.

and a double-stacked class-E SiGe HBT PA over frequency. Figure 5.18 shows the maximum achievable PAE for different output power levels at 45 GHz using different class-E architectures for the same 0.13-μm SiGe BiCMOS technology. It can be observed that, at mm-wave frequencies, accounting for the high power-combining loss in the passives, a stacked class-E architecture achieves higher PAE at higher output power levels than do conventional power-combining schemes [27].

5.3 Design issues for SiGe HBT mm-wave switching power amplifiers

In the 0.13-μm process, for a conventional class-E design, a load resistance of ≈32 Ω is required to generate 20 dBm with 5.9-V breakdown voltage, while the 90-nm SiGe process requires a lower load resistance of ≈25 Ω for the same output power due to its lower breakdown voltage of 5.2 V. The transient waveforms normalized to the BV_{CBO} and the peak current are shown in Fig. 5.19. It is needless to say that, for ideal passives with infinite quality factor, the 90-nm SiGe process would lead to higher collector efficiency, due to its higher f_{max} which also results in a lower τ_{out}. In fact, as seen from Eqn (5.15), since the maximum achievable collector efficiency is independent of the breakdown voltage, using faster devices with lower breakdown voltages, we can always generate higher output power by choosing low load resistances with higher efficiency. More importantly, the power gain of the 90-nm 9HP process is 2 dB higher due to τ_{in} being 60% lower. This plays a key role in maintaining high PAE in mm-wave frequencies. Now assuming the output network has to be matched to a standard 50-Ω interface using a simple finite-quality-factor LC matching network, the loss in the output matching network in the 8HP process comes to be about 6%, compared with ≈10% for the 9HP design. Despite a higher impedance transformation ratio, the 9HP process gives better results. However, power amplifiers need to be driven by a signal source and hence have to be matched to 50 Ω also at their input. Since the higher f_{max} of the 90-nm process is obtained by minimizing the gate resistance, the matching ratio for the 90-nm 9HP process is larger than the transformation ratio for the 0.13-μm 8HP process. Even so, for $Q_{Passive} = 20$, the overall power gain of the 9HP design is higher. In conjunction with higher collector efficiency, the higher G_P leads to an overall higher PAE for the more advanced node for the same output power levels at mm-waves. Such designs can prove the benefit of scaling technologies even for mm-wave designs. The different steps of this design are highlighted in Fig. 5.20. We must point out that this case study is only indicative of the complexities of mm-wave power-amplifier design where both the active device and passive loss play an equal role in determining the overall system performance.

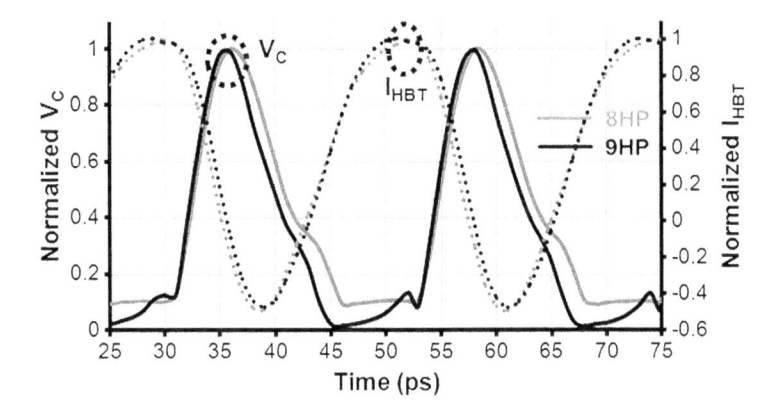

Fig. 5.19 Normalized transient voltage and current waveforms of a 45-GHz, 20-dBm class-E amplifier in the 0.13-μm 8HP and the 90-nm 9HP SiGe BiCMOS process.

	1. Ideal Passives + HBT		2. Output Matching		3. Input Matching	

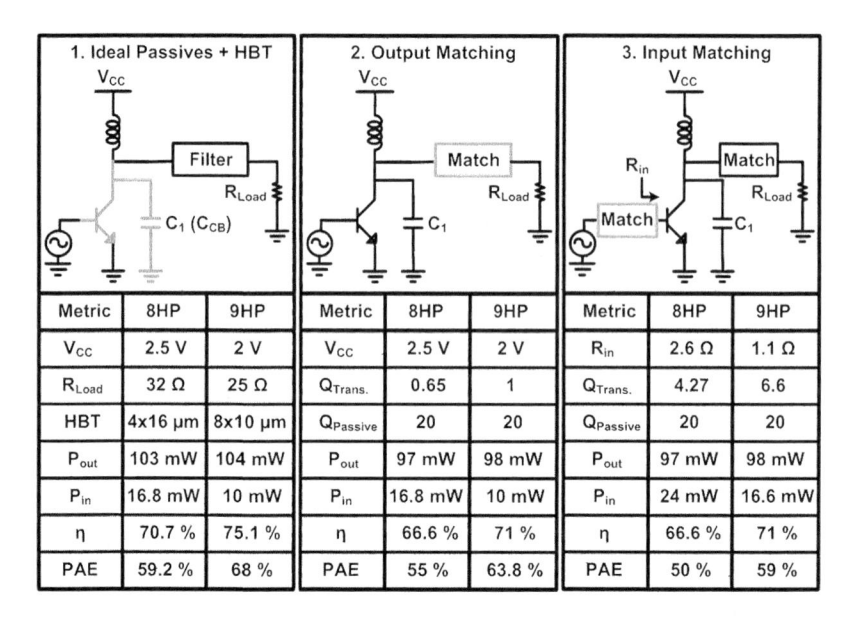

Metric	8HP	9HP
V_{CC}	2.5 V	2 V
R_{Load}	32 Ω	25 Ω
HBT	4x16 μm	8x10 μm
P_{out}	103 mW	104 mW
P_{in}	16.8 mW	10 mW
η	70.7 %	75.1 %
PAE	59.2 %	68 %

Metric	8HP	9HP
V_{CC}	2.5 V	2 V
$Q_{Trans.}$	0.65	1
$Q_{Passive}$	20	20
P_{out}	97 mW	98 mW
P_{in}	16.8 mW	10 mW
η	66.6 %	71 %
PAE	55 %	63.8 %

Metric	8HP	9HP
R_{in}	2.6 Ω	1.1 Ω
$Q_{Trans.}$	4.27	6.6
$Q_{Passive}$	20	20
P_{out}	97 mW	98 mW
P_{in}	24 mW	16.6 mW
η	66.6 %	71 %
PAE	50 %	59 %

Fig. 5.20 Performance comparison of a 20-dBm 45-GHz class-E PA design in the 0.13-μm 8HP and the 90-nm 9HP SiGe BiCMOS process.

5.4 Linearizing architectures for switch-mode power amplifiers

While switch-mode power amplifiers are advantageous in their ability to achieve high peak efficiencies, they exhibit poor linearity to input amplitude modulation as they are designed for hard-switching operation. Consequently, they are most readily used with constant-envelope phase-modulated signals. However, the desire for high data rates, especially at mm-wave frequencies, demands efficient utilization of the spectrum and dictates complex modulations that combine both amplitude and phase, such as quadrature amplitude modulation (QAM) or orthogonal frequency division multiplexing (OFDM). Consequently, architectural solutions at the transmitter level are commonly pursued at RF frequencies to enable the use of switching PAs with complex modulations.

Polar transmitters

The polar transmitter, also called the envelope elimination and restoration (or EER) and the Kahn transmitter [28], is one such architecture that utilizes the fact that switching PAs are linear from the drain side. In other words, a linear change in the supply voltage of a switching PA results in a linear change in the output amplitude. As a result, complex modulations are supported by splitting the modulation into amplitude and phase components, and applying the phase modulation to the input of the switching PA and the amplitude modulation to the supply, typically by means of a supply modulator (Fig. 5.21).

The polar transmitter has been extensively explored at RF frequencies. Two main challenges arise in the implementation of a polar transmitter. The first is linked to the

5.4 Linearizing architectures for switch-mode power amplifiers

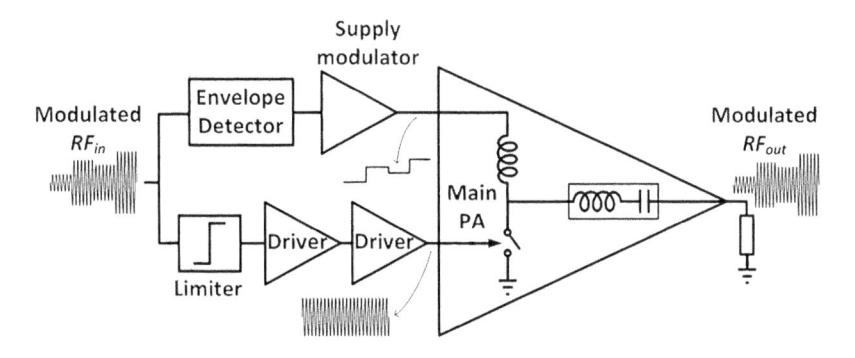

Fig. 5.21 Conventional polar or Kahn transmitter.

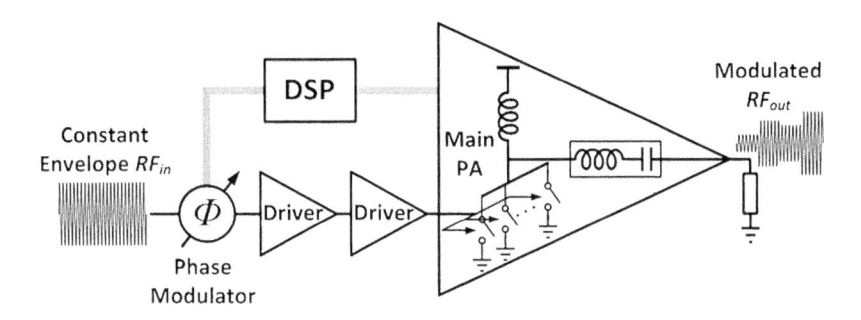

Fig. 5.22 A digital polar transmitter architecture.

expanded bandwidth of the amplitude and phase components. Typically, the bandwidths of the amplitude and phase are 3–5× larger than that of the original modulation due to the nonlinear transformation from the in-phase (I) and quadrature (Q) components of the original modulation to the amplitude and phase. This necessitates high-speed phase and supply modulators, which can be very challenging to design, especially when high levels of precision are required. This challenge is exacerbated at mm-wave frequencies, where wide signal bandwidths are typically used to achieve high data rates. The second challenge arises from delay mismatches between the amplitude and phase paths. Since these paths are not similar and hence not matched with each other, inevitable delay differences result in distortion in the output signal, manifesting itself in reduced EVM and spectral regrowth. Since the allowable delay difference is typically a fraction of the symbol rate (and determined by the required precision) [29], mm-wave operation with wide bandwidths exacerbates this challenge as well and demands tighter delay matching between the amplitude and phase paths.

Recently, digital polar transmitters have gained momentum at RF frequencies (Fig. 5.22), where the amplitude and phase modulation are accomplished digitally [30, 31]. Digital-to-phase converters at RF may be implemented in several ways, including digitally controlled vector modulators [30], tapped inverter-based delay lines [32], and digitally controlled inverter-based delay lines [32]. Digital amplitude modulation is commonly achieved through the implementation of a DAC-like switching power

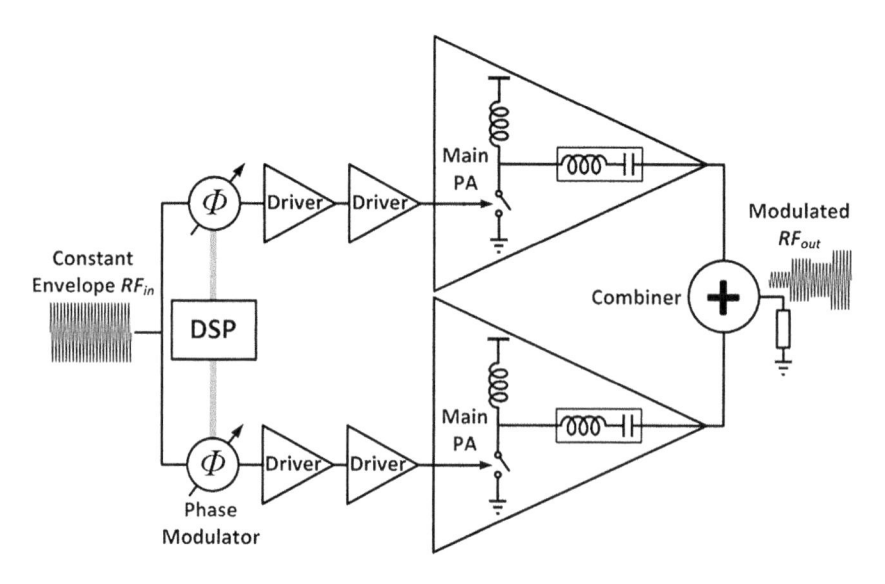

Fig. 5.23 Outphasing transmitter.

amplifier – the switch is segmented into a sea of binary-/thermometer-coded switches, each of which can be enabled by means of a digital control signal [30, 31, 33, 34]. The output amplitude is regulating by controlling the number of switches that are enabled. In other words, the conductance of the switching power amplifier is digitally modulated to modulate the amplitude. Digital modulation enables retiming and hence tighter control of the delay difference between the amplitude and phase bits. There is significant AM–AM and AM–PM distortion associated with the digital conductance modulation, necessitating digital pre-distortion.

Such techniques have not been extensively explored at mm-wave frequencies. In particular, high-resolution mm-wave DAC-like switching power-amplifier implementations have only recently been investigated [35] due to the challenges associated with layout parasitics and digital control of switch gate drive.

Outphasing transmitters

The outphasing transmitter combines two constant amplitude signals with time-varying phase difference to create a signal with modulated amplitude. This architecture was invented in the 1930s by Chireix to obtain high-quality amplitude-modulated signals from vacuum tubes exhibiting poor linearity [36]. It was used until about 1970 in RCA ampliphase AM-broadcast transmitters. In 1974, Cox proposed the use of outphasing at microwave frequencies under the name LINC (linear amplification using nonlinear components) [37]. Such an architecture allows switching power amplifiers to be used at their peak efficiency point for the two constant amplitude signals.

Two variants of the outphasing transmitter exist. In what is commonly called the LINC transmitter, an isolating combiner is used to combine the two constant-envelope signals. Consequently, as the outphasing angle varies, the load impedance seen by the switching

PAs remains constant. This results in good linearity and peak efficiency, but the efficiency under back-off is no better than a class-A PA, as the DC power consumption of the switching PAs remains unchanged as the output power reduces with an increase in the outphasing angle. Essentially, the excess RF power being delivered to the isolating combiner is dissipated within the combiner.

The Chireix technique solves this problem by using a non-isolating combiner to combine the outphased signals, typically implemented as a transformer–combiner. As the outphasing angle increases, the reactive loads seen by the switching PAs change, which reduces the RF power produced by them as well as the DC power consumed by them, improving efficiency under back-off. Shunt reactances are typically used at the inputs of the combiner to tune out the input reactances for a particular output amplitude, maximizing the efficiency in the vicinity of that amplitude. The most common approach is to maximize efficiency at the peak output amplitude, and the efficiency typically stays high to around 6-dB back-off. It is possible to choose the reactances to maximize average efficiency for a specific signal probability density function (PDF) as well.

Outphasing has recently been explored at 60 GHz using scaled CMOS technology [38]. The main challenges associated with outphasing are the bandwidth expansion from the I and Q components of the signal to the outphasing angle, and the digital signal processing associated with generating the outphasing modulation as well as the necessary pre-distortion.

5.5 Conclusions

The increase of f_T and f_{max} of modern transistors, coupled with innovative circuit solutions, has led to the realization of switching amplifiers at mm-wave frequencies. The non-overlapping nature of transistor voltage and current waveforms in switching amplifiers leads to higher efficiency and ability to handle higher voltage and current levels without transistor breakdown. To ensure non-overlapping voltage and current waveforms across the switching transistors at mm-wave frequencies, proper passive networks must be added. A special case of such waveform engineering applied to switching power amplifiers is in the context of transistor stacking leading to higher voltage swing and output power levels.

Switching amplifiers enable realization of digital transmitter architectures such as digital Cartesian transmitters, digital polar transmitters, and outphasing transmitters. In these schemes, low-loss signal combining is needed to achieve a high overall efficiency at mm-wave frequencies. Fortunately, short wavelength and high-quality metallization of modern silicon processes facilitate compact efficient passive power combiners. Research towards realization of extremely high-speed efficient digital transmitters at mm-wave frequencies is ongoing.

The ultimate achievable performance of linear amplifiers and circuits can be represented as a function of passives and transistor small-signal parameters such as *MAG*, *MSG*, and f_{max}. For a given desired output power level, the highest achievable power gain and efficiency of switching amplifiers can be related to a few transistor parameters and the quality of waveform-shaping passive components. The basic principles introduced in this chapter are independent of semiconductor technology and can be applied to silicon CMOS FETs, BJTs, and HBTs as well as non-silicon transistors. The preferred technology choice is a function of the specific application. In the silicon domain, at a given technology node, SiGe HBTs have a higher f_{max} than CMOS transistors. Typically, at a given f_{max}, SiGe HBTs have a higher breakdown voltage than CMOS transistors. On the other hand, at any given time, the CMOS transistors of a SiGe BiCMOS process are inferior to those in a CMOS-only process. High-performance mm-wave transmitters supporting complex modulations require not only efficient power amplifier unit cells and power combiners, but also high-speed low-power mixed-signal and digital circuitry. The desire to integrate the transmitter with the receiver and possibly realize a mm-wave SOC also affects the choice of technology.

References

[1] O. Lee, J. Han, K. H. An, et al., "A charging acceleration technique for highly efficient cascode class-E CMOS power amplifiers," *Solid-State Circuits, IEEE Journal of*, vol. **45**, no. 10, pp. 2184–2197, Oct. 2010.

[2] Y. Song, S. Lee, E. Cho, J. Lee, and S. Nam, "A CMOS class-E power amplifier with voltage stress relief and enhanced efficiency," *Microwave Theory and Techniques, IEEE Transactions on*, vol. **58**, no. 2, pp. 310–317, Feb. 2010.

[3] A. Mazzanti, L. Larcher, R. Brama, and F. Svelto, "Analysis of reliability and power efficiency in cascode class-E PAs," *Solid-State Circuits, IEEE Journal of*, vol. **41**, no. 5, pp. 1222–1229, May 2006.

[4] R. Brama, L. Larcher, A. Mazzanti, and F. Svelto, "A 30.5 dBm 48class-E PA with integrated balun for RF applications," *Solid-State Circuits, IEEE Journal of*, vol. **43**, no. 8, pp. 1755–1762, Aug. 2008.

[5] M. Apostolidou, M. van der Heijden, D. Leenaerts, et al., "A 65 nm CMOS 30 dBm class-E RF power amplifier with 60% PAE and 40% PAE at 16 dB back-off," *Solid State Circuits, IEEE Journal of*, vol. **44**, no. 5, pp. 1372–1379, May 2009.

[6] J. Popp, D.-C. Lie, F. Wang, D. Kimball, and L. Larson, "Fully-integrated highly-efficient RF class E SiGe power amplifier with an envelope-tracking technique for edge applications," in *Radio and Wireless Symposium, 2006 IEEE*, Jan. 2006, pp. 231–234.

[7] D.-C. Lie, J. Lopez, J. Popp, et al., "Highly efficient monolithic class E SiGe power amplifier design at 900 and 2400 MHz," *Circuits and Systems I: Regular Papers, IEEE Transactions on*, vol. **56**, no. 7, pp. 1455–1466, July 2009.

[8] D. Chowdhury, S. Thyagarajan, L. Ye, E. Alon, and A. Niknejad, "A fully-integrated efficient CMOS inverse class-D power amplifier for digital polar transmitters," *Solid-State Circuits, IEEE Journal of*, vol. **47**, no. 5, pp. 1113–1122, May 2012.

[9] J. Fritzin, C. Svensson, and A. Alvandpour, "Analysis of a 5.5-V class-D stage used in 30-dBm outphasing RF PAs in 130- and 65-nm CMOS," *Circuits and Systems II: Express Briefs, IEEE Transactions on*, vol. **59**, no. 11, pp. 726–730, Nov. 2012.

[10] N. Sokal and A. Sokal, "Class E – a new class of high-efficiency tuned single-ended switching power amplifiers," *Solid-State Circuits, IEEE Journal of*, vol. **10**, no. 3, pp. 168–176, June 1975.

[11] J. Hasani and M. Kamarei, "Analysis and optimum design of a class E RF power amplifier," *Circuits and Systems I: Regular Papers, IEEE Transactions on*, vol. **55**, no. 6, pp. 1759–1768, July 2008.

[12] C. Avratoglou, N. Voulgaris, and F. Ioannidou, "Analysis and design of a generalized class E tuned power amplifier," *Circuits and Systems, IEEE Transactions on*, vol. **36**, no. 8, pp. 1068–1079, Aug. 1989.

[13] M. Acar, A. Annema, and B. Nauta, "Analytical design equations for class-E power amplifiers with finite DC-feed inductance and switch on-resistance," in *Circuits and Systems, 2007. ISCAS 2007. IEEE International Symposium on*, May 2007, pp. 2818–2821.

[14] C. Wang, L. Larson, and P. Asbeck, "Improved design technique of a microwave class-E power amplifier with finite switching-on resistance," in *Radio and Wireless Conference, 2002. RAWCON 2002. IEEE*, 2002, pp. 241–244.

[15] M. Acar, A. Annema, and B. Nauta, "Generalized analytical design equations for variable slope class-E power amplifiers," in *Electronics, Circuits and Systems, 2006. ICECS '06. 13th IEEE International Conference on*, Dec. 2006, pp. 431–434.

[16] A. Chakrabarti and H. Krishnaswamy, "An improved analysis and design methodology for RF class-E power amplifiers with finite DC-feed inductance and switch on-resistance," in *Circuits and Systems (ISCAS), 2012 IEEE International Symposium on*, May 2012, pp. 1763–1766.

[17] S. Kee, I. Aoki, A. Hajimiri, and D. Rutledge, "The class-E/F family of ZVS switching amplifiers," *Microwave Theory and Techniques, IEEE Transactions on*, vol. **51**, no. 6, pp. 1677–1690, June 2003.

[18] E. Johnson, "Physical limitations on frequency and power parameters of transistors," in *IRE International Convention Record*, vol. **13**, Mar. 1965, pp. 27–34.

[19] W. Tai, L. Carley, and D. Ricketts, "A 0.7 W fully integrated 42 GHz power amplifier with 10% PAE in 0.13 μm SiGe BiCMOS," in *Solid-State Circuits Conference Digest of Technical Papers (ISSCC), 2013 IEEE International*, Feb. 2013, pp. 142–143.

[20] *BiCMOS8HP Technology Design Manual*, Semiconductor Research & Development, IBM.

[21] *BiCMOS9HP Technology Design Manual*, Semiconductor Research & Development, IBM.

[22] C. Grens, "A comprehensive study of safe-operating-area, biasing constraints, and breakdown in advanced SiGe HBTs," Master's thesis, School of Electrical and Computer Engineering, Georgia Institute of Technology, Atlanta, Georgia, Aug. 2005. [Online]. Available: http://hdl.handle.net/1853/7124

[23] K. Datta, J. Roderick, and H. Hashemi, "A 20 dBm Q-band SiGe class-E power amplifier with 31% peak PAE," in *Custom Integrated Circuits Conference (CICC), 2012 IEEE*, Sept. 2012, pp. 1–4.

[24] K. Datta, J. Roderick, and H. Hashemi, "Analysis, design and implementation of mm-wave SiGe stacked class-E power amplifiers," in *Radio Frequency Integrated Circuits Symposium (RFIC), 2013 IEEE*, June 2013, pp. 275–278.

[25] C. Grens, J. Cressler, and A. Joseph, "On common-base avalanche instabilities in SiGe HBTs," *Electron Devices, IEEE Transactions on*, vol. **55**, no. 6, pp. 1276–1285, June 2008.

[26] M. Acar, A. Annema, and B. Nauta, "Analytical design equations for class-E power amplifiers," *Circuits and Systems I: Regular Papers, IEEE Transactions on*, vol. **54**, no. 12, pp. 2706–2717, Dec. 2007.

[27] K. Datta and H. Hashemi, "Performance limits, design and implementation of mm-wave SiGe HBT class-E and stacked class-E power amplifiers," accepted in *Solid-State Circuits, IEEE Journal of*, Sept. 2014.

[28] L. Kahn, "Single-sideband transmission by envelope elimination and restoration," *Proceedings of the IRE*, vol. **40**, no. 7, pp. 803–806, July 1952.

[29] F. Raab, "Intermodulation distortion in Kahn-technique transmitters," *Microwave Theory and Techniques, IEEE Transactions on*, vol. **44**, no. 12, pp. 2273–2278, Dec. 1996.

[30] L. Ye, J. Chen, L. Kong, E. Alon, and A. Niknejad, "Design considerations for a direct digitally modulated WLAN transmitter with integrated phase path and dynamic impedance modulation," *Solid-State Circuits, IEEE Journal of*, vol. **48**, no. 12, pp. 3160–3177, Dec. 2013.

[31] L. Ye, J. Chen, L. Kong, et al., "A digitally modulated 2.4 GHz WLAN transmitter with integrated phase path and dynamic load modulation in 65 nm CMOS," in *Solid-State Circuits Conference Digest of Technical Papers (ISSCC), 2013 IEEE International*, Feb. 2013, pp. 330–331.

[32] A. Ravi, P. Madoglio, M. Verhelst, et al., "A 2.5 GHz delay-based wideband OFDM outphasing modulator in 45 nm-lp CMOS," in *VLSI Circuits (VLSIC), 2011 Symposium on*, June 2011, pp. 26–27.

[33] C. Presti, F. Carrara, A. Scuderi, P. Asbeck, and G. Palmisano, "A 25 dBm digitally modulated CMOS power amplifier for WCDMA/EDGE/OFDM with adaptive digital predistortion and efficient power control," *Solid-State Circuits, IEEE Journal of*, vol. **44**, no. 7, pp. 1883–1896, July 2009.

[34] D. Chowdhury, L. Ye, E. Alon, and A. Niknejad, "An efficient mixed-signal 2.4-GHz polar power amplifier in 65-nm CMOS technology," *Solid-State Circuits, IEEE Journal of*, vol. **46**, no. 8, pp. 1796–1809, Aug. 2011.

[35] A. Balteanu, I. Sarkas, E. Dacquay, et al., "A 2-bit, 24 dBm, millimeter-wave SOI CMOS power-DAC cell for watt-level high-efficiency, fully digital m-ary QAM transmitters," *Solid-State Circuits, IEEE Journal of*, vol. **48**, no. 5, pp. 1126–1137, May 2013.

[36] H. Chireix, "High power outphasing modulation," *Radio Engineers, Proceedings of the Institute of*, vol. **23**, no. 11, pp. 1370–1392, Nov. 1935.

[37] D. Cox, "Linear amplification with nonlinear components," *Communications, IEEE Transactions on*, vol. **22**, no. 12, pp. 1942–1945, Dec. 1974.

[38] D. Zhao, S. Kulkarni, and P. Reynaert, "A 60-GHz outphasing transmitter in 40-nm CMOS," *Solid-State Circuits, IEEE Journal of*, vol. **47**, no. 12, pp. 3172–3183, Dec. 2012.

6 Stacked-transistor mm-wave power amplifiers

Peter Asbeck and Harish Krishnaswamy

6.1 Introduction

A dominant theme in the evolution of integrated electronics over past decades has been Moore's law, which has led to development of silicon-based transistors with dramatically improved operating speeds and levels of integration, as well as reduced power dissipation. Although the driving motivation for the technology development has been primarily application to digital circuits, silicon-based transistors are poised, by virtue of their very high frequency response, to provide mm-wave and THz circuits with also dramatically improved performance as well as low cost. One of the tradeoffs that has been necessary in the development of scaled silicon devices, however, has been a reduction in voltage handling capability. As a result, conventional scaled silicon transistors are not very well suited to power amplifiers, which perform best with high voltage swings along with high current swings. Recent research, however, has pointed to a circuit design approach that can overcome to a considerable degree the limitations of the basic transistors: the technique of series-connection, or stackingĺ of multiple devices to generate a composite structure that can handle higher voltages. The stacking technique has opened the door to higher output power as well as higher efficiency and greater bandwidth, since the resulting impedance transformations needed to generate amplifier outputs at 50 ohm levels become easier to implement with low loss and high bandwidth. This chapter describes in detail the background, design considerations and benefits of stacked Si devices in mm-wave power amplifiers. Emphasis is given to stacked CMOS devices, although bipolar transistors and HBTs can also be stacked as described in Chapter 5. A review of Si FET characteristics is first given, along with a description of the impact of voltage handling on power amplifier characteristics. Stacked FET amplifiers at low microwave frequencies, which are emerging as important contenders for cell phone handsets, is then reviewed. Key design considerations for stacking at mm-wave frequencies are then presented. The chapter also describes a number of mm-wave power amplifier example designs in detail.

6.2 Motivation for stacking

As a result of Moore's law, transistor current gain cut-off frequencies (f_T) have been increasing rapidly as the transistor dimensions have been reduced. Figure 6.1 illustrates

208 Stacked-transistor mm-wave power amplifiers

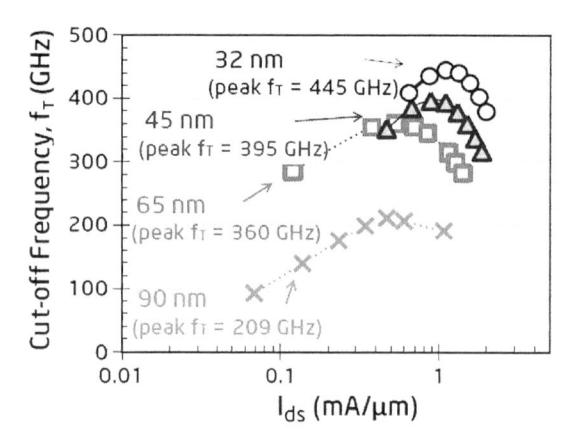

Fig. 6.1 Recently reported nMOS f_T values vs. current density for different gate lengths L_g [1].

recent f_T measurement results for n-MOS devices at the 90nm node and below, as a function of gate length (which typically is smaller than the metallization half-pitch which defines the technology node). For the 32nm node, f_T values reported by Intel for the metal gate, high-K dielectric transistors reached 445GHz [1]. Although reported values are de-embedded to the transistor terminals, and in practice f_T is degraded from the addition of interconnect parasitics, these devices can provide copious levels of current gain at frequencies reaching up to 100s of GHz. The maximum frequency of oscillation, f_{max}, another critical metric for the transistors, describing the frequency for which power gain above unity can be achieved. It is strongly dependent on transistor layout and gate feed resistance, but is typically also above 300GHz for advanced technologies. Digital circuit technology development has also led to numerous thick metal layers (largely of copper) and thick dielectrics which are very favorable for mm-wave applications. As the device speeds have soared, however, the voltage handling capability of Si transistors has been trending to lower values over a considerable period. Figure 6.2 illustrates, for example, the power supply voltage V_{DD} values used in high end digital CMOS circuits in the recent past and in current projections for the future [2]. For digital circuits, the steady decrease in V_{DD} is a critical consideration in order to avoid excessive increases in power density in the digital circuits (which follows $P_{diss} \approx \alpha \times f \times C \times V_{DD}^2$, where C is the overall average load capacitance, f is the switching frequency and α is an activity coefficient). Reductions of V_{DD} have been relatively successful in maintaining digital IC power dissipation under control over many years, until recently (when further scaling of V_{DD} has encountered problems associated with sub-threshold current swings and transistor leakage current in the OFF-state).

While V_{DD} expresses the maximum operating drain voltage in digital circuits, the devices can typically handle greater values of V_{DS}. There are limits to V_{DS}, and related limits to V_{GS} and V_{GD}, nonetheless, imposed by a variety of physical mechanisms: (1) gate dielectric breakdown, and onset of large gate dielectric leakage associated with Fowler-Nordheim tunneling; (2) gate dielectric reliability problems associated with time-dependent dielectric breakdown; (3) impact ionization at the drain edge of the

6.2 Motivation for stacking

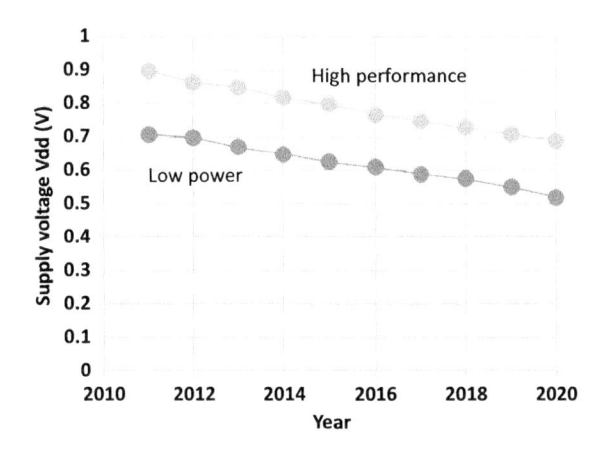

Fig. 6.2 Power supply voltage trends for digital CMOS circuits (after [2]).

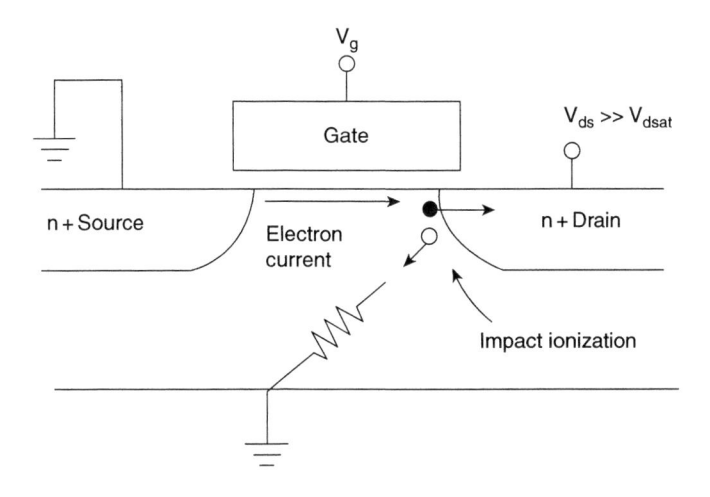

Fig. 6.3 Impact ionization process near drain-channel junction.

channel, leading to excess current, and often to snap-back behavior; (4) hot carrier injection, leading to generation of defects at the channel / dielectric interface and progressive degradation; (5) development of tunnel currents at the body–drain junction. Mechanism (4), illustrated schematically in Fig. 6.3, is a particularly strong limit in power amplifiers, although its onset is dependent on the presence of both voltage and current in the channel, and thus is dependent on specific amplifier design and dynamic loadline. Although at present there is not a strong theoretical basis for calculating the maximum value of VDS that can be sustained reliably in mm-wave circuits, empirical results show that the values have been decreasing with transistor scaling.

For design of high performance power amplifier ICs, the maximum output voltage that can be handled, V_{max}, generally emerges as a key consideration for both maximum output power and efficiency. The impact of V_{max} can be appreciated from a representative load line, shown schematically in Figure 6.4 for a Class B amplifier. As

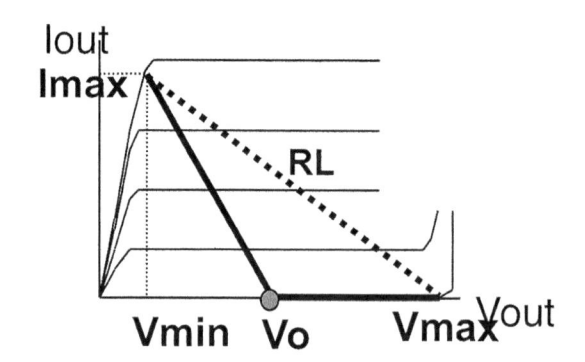

Fig. 6.4 Schematic output transistor characteristics and dynamic load line for class-B operation.

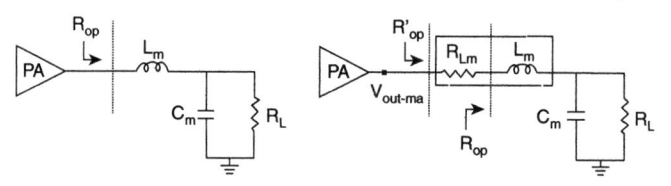

Fig. 6.5 Representative impedance matching network at PA output, showing contribution to loss from parasitic inductor resistance.

V_{max} is increased, the maximum output power (given by the area of the power triangle $P_{out} \approx 1/8 \times V_{max} \times I_{max}$) increases. Of equal importance, the load resistance that should be presented to the drain, $R_{L,opt} \approx \frac{V_{max} - V_{min}}{I_{max}}$ increases. It is desirable to keep $R_{L,opt}$ as close to 50 ohms as possible in order to minimize the difficulty of impedance matching to an overall amplifier output load of 50 ohms. $R_{L,opt}$ tends to be very low for Si-based PAs (for example, if $V_{max} - V_{min}$ is 3V, then $R_{L,opt}$ is only 1.1 ohms for $P_{out} = 1W$). If $R_{L,opt}$ cannot be increased, then the impedance matching to 50 ohms becomes problematic from the standpoint of both loss and bandwidth. Figure 6.5 shows a representative single-stage matching network used to transform $R_{L,out} = 50$ ohms to the $R_{L,opt}$ at the drain of the PA transistor. The efficiency of the impedance transformation is limited due to the resistive losses (finite Q) of the passive components L_m and C_m. For the case where the inductor Q_{ind} is limiting, the efficiency can be calculated as

$$\eta = \frac{Q_{ind}}{Q_{ind} + Q_{trans}} \qquad (6.1)$$

$$Q_{trans} = \left(\frac{R_{L,out}}{R_{L,opt}} - 1\right)^{1/2} \qquad (6.2)$$

Since the Q values for inductors, capacitors, transmission lines and related passives are very limited at mm-wave frequencies, particularly in an integrated circuit embodiment, it is very important to keep Q_{trans} as low as possible. This can be achieved by increasing V_{max}, with the use of transistor stacking.

It should be noted that with design rules in use in today's CMOS, it is in principle possible to engineer transistors with higher voltage handling capabilities than those in use in digital circuits, by properly designing the gap between the gate and the drain of the device [3]. Such transistors, drain extensionı̇ devices and their close relatives, laterally diffused MOSFETs or LDMOS, can allow higher-voltage operation, although with some sacrifice in f_T, and, importantly, they typically utilize a process technology that is not the standard digitalı̇ variety. The stacking technique allows amplifiers to be implemented with no process modifications from what is used in digital ICs.

6.3 Principles of transistor stacking

The generic configuration of a stacked FET amplifier is shown in Fig. 6.6. A sequence of K devices are series-connected, in order to share the output voltage appearing at the drain of the topmost device (with $K=4$ in the example shown). The RF input is applied to the bottom device in the stack, which operates in common-source configuration, while the upper devices are operated with no specific terminal grounded. In the most straightforward designs, equal sharing of the voltage is desired, leading to a configuration with $V_{DS,max,k} = V_{max}/K$, where $V_{DS,max,k}$ is the maximum V_{DS} swing to be experienced by the kth transistor in the stack. The assumption of equal sharing will be revisited in the discussion below, particularly if the technology permits a mix of thin oxide and thicker oxide devices which have different voltage handling limits. IMN and OMN are impedance matching networks at input and output; the latter is configured so that the impedance seen at the top drain has real part R_L.

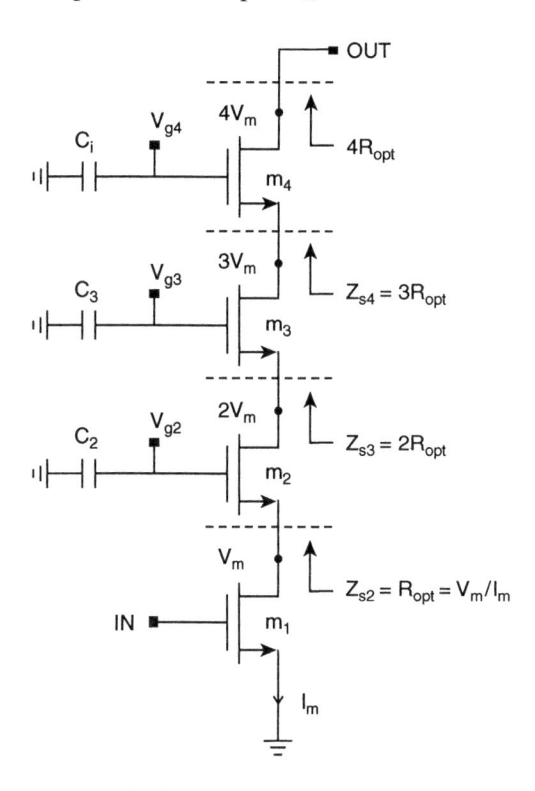

Fig. 6.6 Representative schematic stacked FET amplifier.

In order to set the quiescent drain voltages of the transistors in the stack, external bias voltages $V_{G,k}$ are applied to the gates of the upper transistors. When these transistors are active, their corresponding source voltages $V_{S,k}$ (which coincide with the drain voltages of the transistors below them, $V_{S,k} = V_{D,k-1}$) are given by $V_{S,k} = V_{G,k} - V_{GS,k}$, where $V_{GS,k}$ is controlled by the current flowing in the kth transistor. Resistive dividers can be used to establish the proper $V_{G,k}$ gate bias voltages (although, since $V_{GS,k}$ is current dependent, the sequence of drain voltages $V_{D,k}$ is not necessarily accurately pinned).

In order to maintain equal division of the drain voltages along the stack in large signal operation, it is important to establish a sequence of impedances seen at the drains of the different devices that scale appropriately. As shown in Fig. 6.6, the impedance seen at the drain of the kth device should increase with k. If we assume that the drain currents of all the devices are equal to I_d (a reasonable assumption at low frequency) and neglect reactive components, then the impedances seen at the drains should be $Z_{d,k} = k \times R_{opt}$ [4]. The load resistance applied to the overall stack should be $R_L = K \times R_{opt}$.

An additional important constraint in the stack design is the distribution of gate voltages. As the drain voltage swings increase along the stack, the gate voltages must also have a voltage swing that increases along the stack, to limit $V_{GD,k}$ under large signal conditions. Ideally $V_{G,k}$ should be incremented going up the stack by the voltage swing $I_d \times R_{opt}$, just as the drain swing increases by this value.

In the simplest approximation, the constraints on the drain voltage swing and the gate voltage swing can both be met at the same time. Both can be achieved in straightforward fashion by connection of an appropriate passive network to the gate of the kth transistor (although a variety of other techniques can also be used). Figure 6.6 shows external capacitors C_{xk} of an appropriate size attached to the different gates of the upper transistors. This is a simple and often used approach. Transformer drive circuits or transistor-based active drive circuits are useful alternatives that will be discussed in later sections.

Design of the passive network placed at the gate of the kth transistor in the stack is straightforward using a simple approximation to the transistor model. With the small signal model given in Fig. 6.7, and neglecting C_{gd} for the moment, it can be readily shown that the voltage swing at the gate of the kth transistor is given by $V_{g,k} = I_{g,k}/j\omega C_{xk}$, where $I_{g,k}$ is the gate current drawn by the transistor and C_{xk} is an external capacitor placed at its gate. $I_{g,k}$ is given by $j\omega C_{gs,k} V_{gs,k} = j\omega C_{gs,k} \times I_{d,k}/g_m$. From these,

$$Z_i = \left(1 + \frac{C_{gs}}{C_i}\right) \times \left(\frac{1}{g_m} || \frac{1}{sC_{gs}}\right) \qquad (6.3)$$

$$Z_i \approx \left(1 + \frac{C_{gs}}{C_i}\right) \times \frac{1}{g_m}, \; f_0 \ll f_T \qquad (6.4)$$

The overall voltage gain of the stack, A_v, is given (for the case of a 3-stack amplifier) by

$$A_v = \frac{g_{m1} R_L}{\left(1 + \frac{sC_{gs,2}}{g_{m,2}}\right)\left(1 + \frac{sC_{gs,3}}{g_{m,3}}\right)} \qquad (6.5)$$

$$A_v \approx g_{m1} R_L, \; f_0 \ll f_T \qquad (6.6)$$

6.3 Principles of transistor stacking

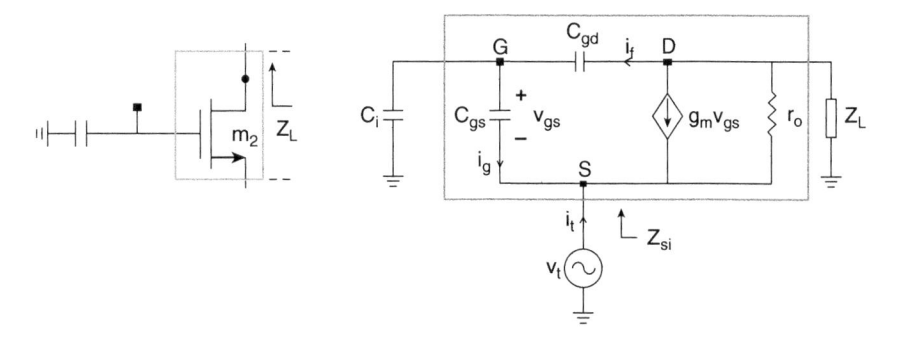

Fig. 6.7 Circuit configuration for upper transistors in stack and small-signal model.

To attain equal voltage swings along the stack, the impedance (assumed to be real) should be set to $k \times R_{opt}$, the optimal load line impedance for an individual transistor (and the overall stack of K transistors should have a load impedance of $K \times R_{opt}$). To set the drain impedances and the gate voltage swings appropriately, the selection of C_{xk} should be according to:

$$C_{xk} = \frac{C_{gs,k}}{(k-1)g_m R_{opt} - 1} \qquad (6.7)$$

The design principle can be viewed intuitively in various ways. According to one point of view, due to the finite current gain of the FETs, a sample of the drain current of each transistor appears at its gate. A gate voltage swing is produced by the impedance of the externally placed capacitor together with this gate current. Accordingly, the capacitor values are large at the bottom of the stack, and smaller at the top of the stack where the voltage swings must be larger. According to a different point of view, the voltage at the gate is the result of a voltage divider among the capacitors C_{xk} and $C_{gs,k}$, where the voltage across the $C_{gs,k}$ capacitor is I_{dk}/g_m. The voltage appearing at the source of the transistor follows the voltage at its gate if we assume that the g_m of the transistor is very high so that $V_{gs,k}$ can be neglected.

Within the approximations derived above, $V_{ds,k}$ and $V_{gs,k}$ values of each of the transistors can be made equal by proper selection of the external capacitors and bias points, for class-A operation. The ideal situation is shown schematically in Fig. 6.8 for a 3 stack. The figure shows clearly that there are large RF voltage swings on the gates of the upper transistors in the stack. This contrasts with what would be expected in the case of common-gate operation of the upper transistors (as would occur in a cascade configuration for a 2 FET combination, or a multiply cascaded structure). In the cascade case, large capacitors would be placed at the gates in order to short the RF signal to ground. The cascade configuration leads to increasing gate-drain voltages as one goes up the stack, so it is not capable of significant increases in output voltage handling for the FETs.

The approximate analysis given above must be supplemented for more realistic models of the transistor and more general regimes of operation. If we take into account C_{gd} of the transistors, a more complete model can be derived as follows [5]. The impedance $Z_{d,k-1}$ presented to the drain of the $(k-1)$ th transistor can be found to be:

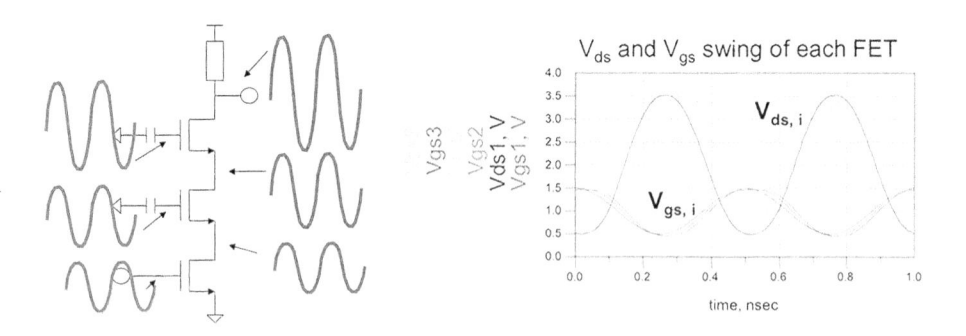

Fig. 6.8 Schematic 3 stack structure illustrating voltage waveforms for gates and drains. Equal swings for all devices can be ideally achieved.

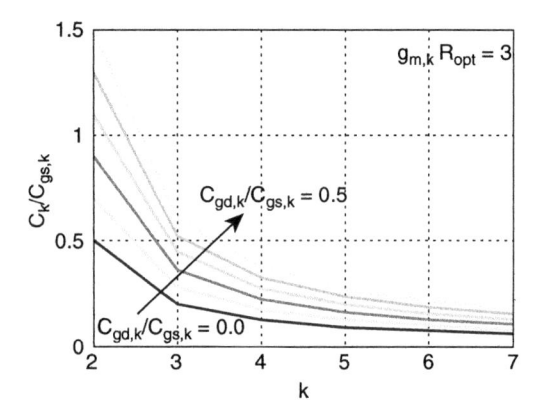

Fig. 6.9 Values of C_{xk} (normalized to $C_{gs,k}$) computed for various values of C_{gd} relative to C_{gs}.

$$Z_{d,k-1} = \frac{C_{gs,k} + C_k + C_{gd,k}(1 + g_{m,k}Z_{d,k})}{(g_{m,k} + sC_{gs,k})(C_{gd,k} + C_k)} \tag{6.8}$$

For equal voltage swings as detailed above, this constraint leads to the selection of C_{xk} according to

$$C_k = \frac{C_{gs,k} + C_{gd,k}(1 + g_{m,k}R_{opt})}{(k-1)g_{m,k}R_{opt} - 1} \tag{6.9}$$

Figure 6.9 shows representative values for the choice of C_{xk}, normalized to $C_{gs,k}$, for different cases of C_{gd}/C_{gs}. A voltage gain per device of $g_m R_{opt} = 3$ is used in this example. For large stacks, the values of C_{xk} become small relative to other device capacitances.

The voltage handling characteristic of stacked FETs is increased by K, according to the above design principles, with the introduction of appropriate external capacitors C_{xk}. It should be noted that the voltage swings on the gates maintain their proper design values over a large frequency range, but near dc the gate voltages will revert to their dc bias values. If these gate bias voltages are set to be constant values, then the stack may not be able to handle voltages as large under static conditions as it can under ac operation. In particular, if the transistor stack is ON on a dc basis, there may be gate-channel breakdown of the topmost transistors. To avoid this problem, the bias voltages on the gates are frequently set by a resistive divider whose positive side is connected to the top drain voltage rather than to a power supply.

The analysis of transistor stacking outlined here still relies on a number of assumptions that limit its accuracy. The impedances at the critical nodes have been assumed to be real. While this can be approximately true at low frequencies, at mm-waves it is not a proper assumption. Effects of C_{gs}, C_{ds} and of capacitance to ground loading intermediate nodes have not been taken into account. Approaches to deal with this complication are detailed below. The transistors have been assumed to operate in the active region, under small signal conditions. In many power amplifier scenarios, however, the transistors operate with large swings that cause capacitances to vary, and further, as devices go from cutoff to saturation, g_m and V_{gs} depart dramatically from their small signal values. Importantly, when the upper transistors go into the triode region, the gate voltage loses control of the source voltage, and the impedances seen at transistor drains are no longer as specified by the above equations. Furthermore, for optimal power amplifier operation in general it is desirable to carry out waveform engineeringĺ by providing appropriate harmonic terminations at the transistor drain. The methodology outlined here provides a voltage waveform at the intermediate drain nodes that is derived from the drain current in a simple way that precludes direct harmonic matching.

The advantages of stacking transistors in this fashion for power-amplifier applications are numerous:

(1) Overall voltage handling (V_{max}) is increased.
(2) Output power is increased; the structure thus acts as a power combiner.
(3) Load impedance for the overall structure is increased.
(4) Efficiency is increased in general because the impedance transformation between overall $R_{L,opt}$ and 50 ohms is less challenging. The efficiency as limited by the device ON-resistance (or knee voltage), on the other hand, is unchanged, because the overall ON-resistance increases linearly with the number of transistors; each transistor ideally experiences the same load line as it would in a single transistor amplifier, so that $(V_{max} - V_{min})/V_{max}$ is unchanged.
(5) The voltage gain of the amplifier goes up by a factor of K, where K is the number of stacked devices.
(6) The input impedance remains the same as that of an individual transistor, and thus it is larger (more favorable for matching) than the impedance of an equivalent current combined amplifier of the same output power.
(7) The drain current for the stack is reduced compared with that of an equivalent current combined amplifier of the same output power. One important consequence is that the effect of parasitic source inductance in degeneration of the gain of the amplifier is correspondingly reduced.

When stacked transistor designs are compared with alternatives, the quantitative benefits of power combining and load impedance increase can be obtained in different amounts according to the strategy of how the transistor sizes are set when using the stacked strategy is introduced. In one scenario, the transistor size (gate width) is kept constant while stacking K transistors. For such a case, the output voltage increases in proportion to K, and the output load impedance and output power also increase proportionally with K. In another scenario, as the transistors are stacked their width (and current carrying capability) is also increased by a factor K, so that the overall load impedance of the amplifier is kept constant (for example, $R_{L,opt}$ is kept close to 50 ohms for any value of K chosen). For this case, both the current I_d and the output voltage increase in proportion to K, and overall output power increases in proportion to K^2. There is a variety of strategies that range between these two limits of the scaling of output power and output load impedance with K.

6.4 Transistor stacking for switch-mode operation

As was discussed in the previous chapter, switching power amplifiers are extensively utilized at RF frequencies owing to their (ideally) lossless operation. The class-E PA [6] has been of particular interest because of its relatively simple output network. The design of switching PAs in CMOS at mmWave frequencies is challenging due to the lack of ideal square-wave drives (resulting in soft switching), low PAE due to the high input drive levels required to switch the devices and high loss levels in the device/switch. Nevertheless, as was discussed in the previous chapter, recent research coupled with technology scaling has demonstrated that "switch-like" PAs may be implemented at millimeter-wave frequencies and exhibit the benefits expected from switch-mode operation [7].

Device stacking may be applied to switch-mode power amplifiers, affording benefits similar to those afforded to linear power amplifiers. The analysis presented in the previous section for linear PAs assumed linear operation of the device, enabling the device to be replaced with a small-signal model. While the resulting insights regarding the increase in overall voltage handling, output power, load impedance and efficiency generally hold true for stacked switch-mode power amplifiers as well, the analysis and resulting design methodogy are distinct, as would be expected based on the switch-mode operation.

Fig. 6.10(a) depicts the concept of a stacked CMOS class-E-like PA. The stacked configuration consists of multiple series devices, which might be of equal or different size. In order to preserve input power and improve PAE, only the bottom device is driven by the input signal. The devices higher up in the stack turn on and off due to the swing of the intermediary nodes. The topmost drain is loaded with an output network that is designed based on class-E principles, and consequently sustains a class-E-like voltage waveform. The intermediary drain nodes must also sustain class-E-like voltage swings with appropriately scaled amplitudes so that the voltage stress is shared equally among all devices. For long-term reliability, the maximum swing across any two transistor junctions under large signal operation must be limited to $2 \times V_{DD}$ [8]. Consequently, for a PA with n stacked devices, the peak output swing is $2n \times V_{DD}$ as marked on Fig. 6.10(a), and the appropriate intermediary node swings are also noted. Appropriate voltage swing

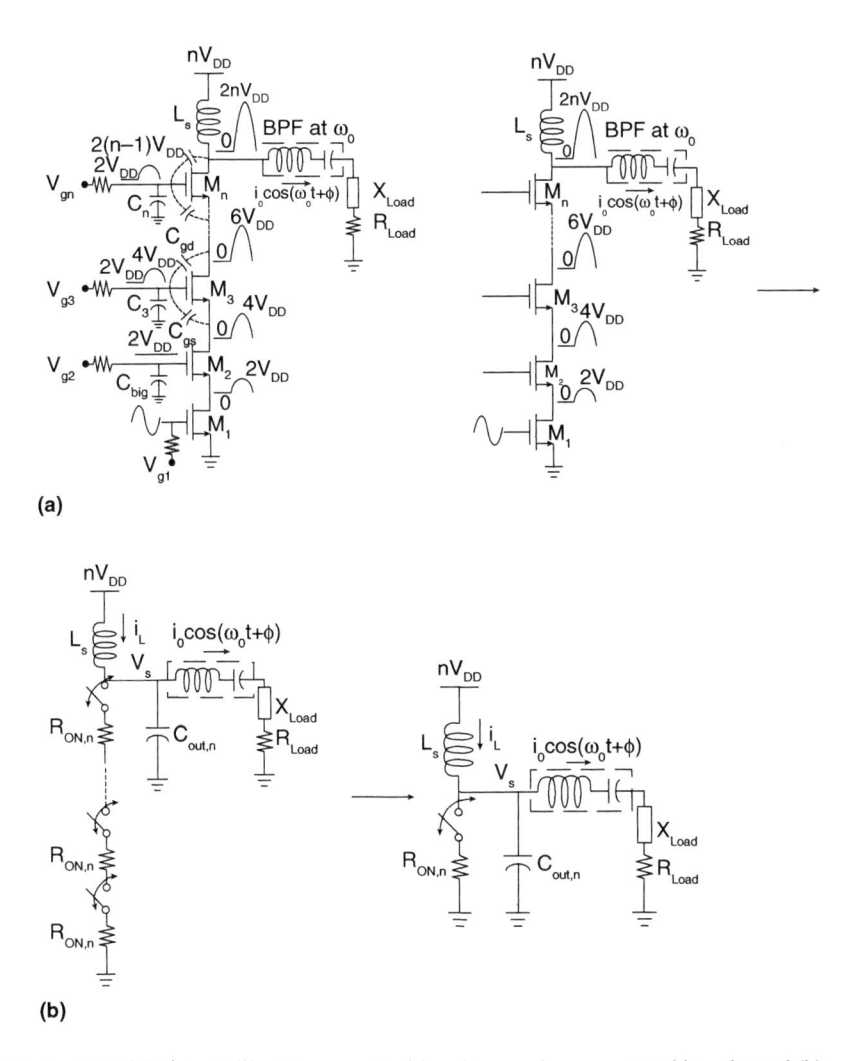

Fig. 6.10 (a) Stacked CMOS class-E-like PA concept with voltage swings annotated in volts and (b) loss-aware class-E design methodology for stacked CMOS PAs.

may be induced at the intermediary nodes through various techniques such as inductive tuning [8], capacitive charging acceleration [9], and placement of class-E load networks at intermediary nodes [10]. These techniques will be discussed in detail later in this chapter and are not shown in Fig. 6.10(a) for simplicity. In order to conform to the peak AC swing limit across the gate-source junction in the on half-cycle and the gate-drain junction in the off half-cycle, the gates of the devices in the stack must swing as shown in Fig. 6.10(a). The swing at each gate is induced through capacitive coupling from the corresponding source and drain node via C_{gs} and C_{gd} respectively and is controlled through the gate capacitor C_n. The DC biases of all gates may be applied through large resistors.

In order to model, analyze and design a stacked Class-E-like amplifier, the devices (taken to be equal in size) in a stacked switching PA are assumed to behave as a single switch with linearly increased breakdown voltage and ON-resistance (Fig. 6.10(b)). To simplify the theoretical analysis, only the total ON-resistance of the stacked configuration and the output capacitance of the top device are considered. The output capacitance of a stacked configuration should ideally scale down linearly with the number of devices stacked. However, wiring parasitics are significant at RF and mmWave frequencies and there will be parasitic capacitance to ground from the intermediate drain/source and gate nodes which will prevent linear scaling of output capacitance with stacking. As a worst-case estimate, the overall output capacitance is taken to be the same as that of a single-device (=$C_{gd} + C_{ds}$, where $C_{ds} = \frac{C_{db} \times C_{sb}}{C_{db} + C_{sb}}$). This estimate assumes an SOI CMOS implementation, since the high body resistance in SOI technology causes C_{db} and C_{sb} to appear in series.

Based on this model, it is clear that stacking enables a Class-E-like PA with linearly larger voltage handling but with potentially linearly larger switch loss. The increased switch loss requires attention, particularly at mmWave frequencies. Several design methodologies have been proposed for the design of Class-E PAs with high levels of switch loss. Later in this chapter, we delve into the details of mmWave stacked Class-E PA design based on the loss-aware Class-E design methodology for mmWave PAs described in the previous chapter.

6.5 Application of stacking at microwave frequencies

The advantages of stacking transistors to increase voltage handling capability have been evident to numerous researchers over the years, and stacking enjoys a rich history. One of the earliest proposed techniques is the bean stalk amplifier proposed by K.J.Dean in 1964 [12]. As shown in Fig. 6.11, a ladder of resistors is used to bias and drive several BJTs connected in series. This technique, however, is directly suitable only for low frequency operation since the input signal has to propagate to the bases of all the devices through the resistor ladder, leading to poor synchronization of the base current inputs. For high frequency, high voltage applications, another technique was presented by Shifrin in 1992 [11], illustrated in Fig. 6.12. A resistive ladder is also used for DC biasing, and only the common source device at the input is driven. Interstage matching is carried out using series inductors, and capacitors at the gates of the stacked devices are included to allow RF swing. The reported power amplifier delivered 36dBm at 4.5GHz with 20% efficiency using 3 GaAs FETs. A related amplifier, with more general interstage matching networks, was analyzed and patented by Rodwell [13].

Numerous other variations of the stacking approach with different transistor drive approaches have been reported. On-chip transformers and couplers have been demonstrated to drive FETs and HBTs at different positions along the stack, such as shown in Figure 6.13 [14]. These techniques eliminate the synchronization problem encountered in the bean stalk amplifier, although the transformers can consume a considerable die area.

6.5 Application of stacking at microwave frequencies

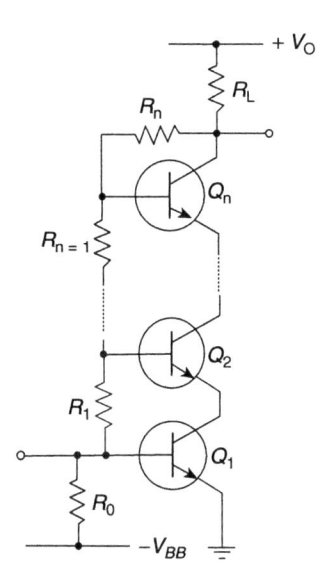

Fig. 6.11 Bean-stalk amplifier [12].

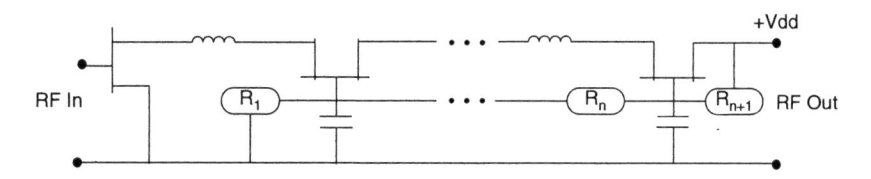

Fig. 6.12 Stacked FET PA of Shifrin [11].

A significant advance in transistor stacking was made in the work of Amin Ezzeddine, who demonstrated HiVP high voltage, high power amplifiers made initially with GaAs FETs, as shown in Fig. 6.14 [4]. These amplifiers utilize appropriately sized gate capacitors as described in Section 6.3, as well as feedback resistors to establish the input voltages.

The recent explosive development of the wireless communications industry and the associated need for low cost power amplifiers has placed an increased emphasis on stacking. Without stacking, CMOS power-amplifier efficiency in the 1-2-GHz frequency region at the level of 2-3W peak output power has not been sufficient to overcome the advantage of GaAs devices, which can inherently handle greater voltages due to higher electric breakdown field. Stacking CMOS FETs has led to demonstrations that may diminish the lead enjoyed by the III-V technology. Early work in use of CMOS stacking for RF includes a 2 stack amplifier (self-biased cascode) by T. Sowlati, in which feedback resistors and gate capacitors were used to induce an RF swing at the gate of the top device [15]. The amplifier, fabricated in $0.18 \mu m$ CMOS, provides 23 dBm with a PAE of 42% at 2.4GHz using a 2.4V supply. Additional circuits have been reported by Jeong, using Silicon-on-sapphire, by Pornpromlikit, using SOI [16], and by Leuschner, using

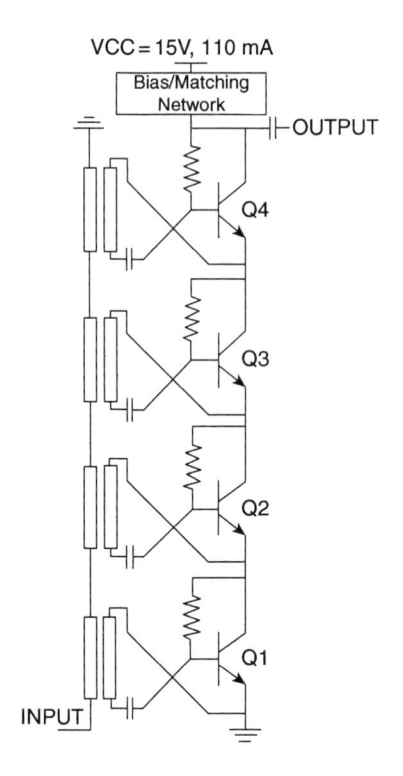

Fig. 6.13   Stacked bipolar amplifier with microwave couplers for base feed.

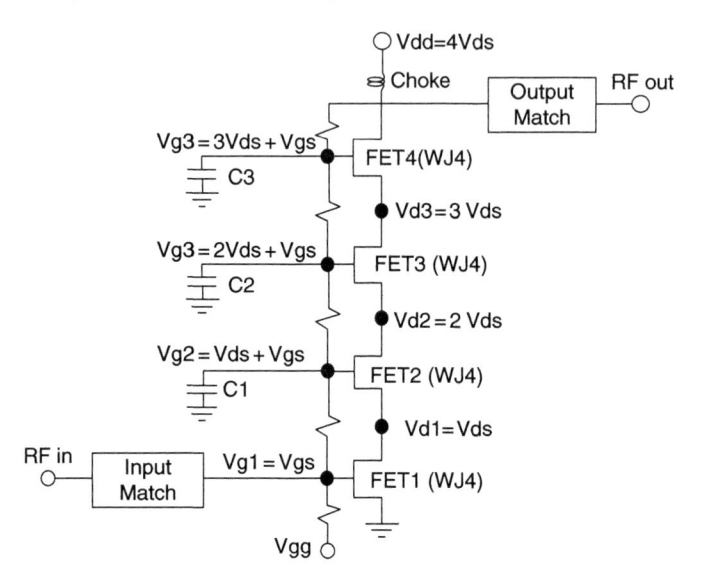

Fig. 6.14   HiV amplifier of Ezzedine.

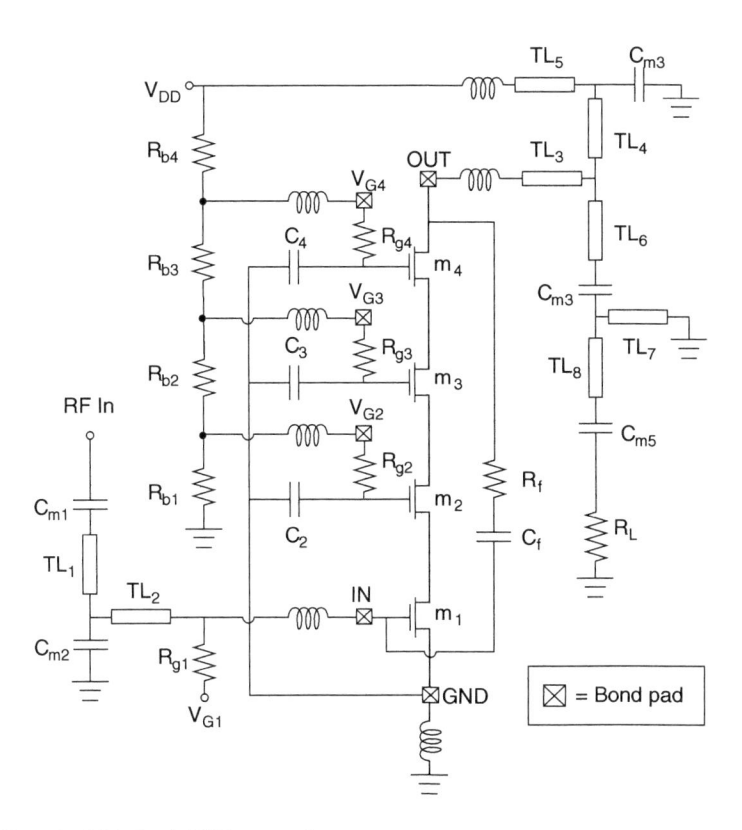

6.5 Application of stacking at microwave frequencies

Fig. 6.15 nMOS 1W stacked PA for 1.9GHz operation.

bulk CMOS [17], among others. An important objective is to attain watt-level output power at high efficiency in order to meet cell phone requirements. In the following, the design of reference [16] is described as a representative example.

A single-stage stacked-FET deep class-AB PA was designed to operate primarily at 1.9 GHz using 0.28-μm 2.5-V standard I/O FETs available in the STMicroelectronics 0.13-μm SOI CMOS process. The overall circuit diagram is shown in Figure 6.15. Four transistors are stacked in series, based on the topology presented in Section 6.3. Each transistor has a total gate width of 5 mm so the total device gate width in the amplifier is 20 mm. To avoid the device breakdown, the drain-to-source voltage of each transistor should be limited to 4.5 V, allowing a maximum voltage swing of 18 V at the top drain node with a 9-V drain bias. For a higher margin of safety, the PA was designed to achieve linear amplification up to the required >1W output power under a 6.5-V supply. The external gate capacitances are 9, 2.6, and 2 pF, respectively, setting the optimum load impedances seen by each transistor. For flexibility, the gate-bias voltages of each stacked device are implemented by an off-chip resistive voltage divider and applied through 1-k on-chip resistors. An *RC* feedback circuit is introduced to improve stability. The simulated Rollett stability factor is greater than 1 for all frequencies between dc and 50 GHz, except for the frequency band from 6.4 to 7.6 GHz,

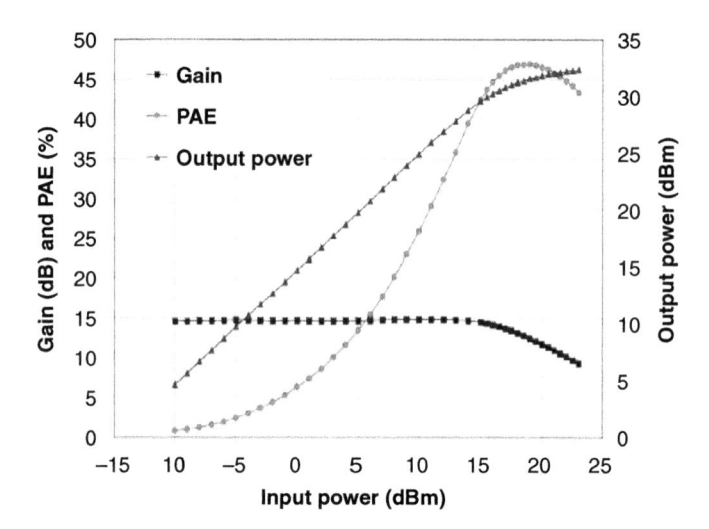

Fig. 6.16 Measured large-signal characteristics of 1W stacked CMOS PA at 1.9GHz.

where stability is ensured in the simulation for the designed input and output matching conditions. The optimum output load impedance is 11.5, which lies in a convenient range to match to 50 over broad bandwidth with high efficiency. The input impedance is 17.5, a comparatively large value corresponding to that of a single 5-mm transistor. The input and output matching circuits are implemented off-chip in this work to allow opportunities for varying the tuning. A high-pass L-match section, consisting of a series capacitor and a short stub, was used to transform the external 50 load to the optimum load. The on-board drain bias circuit included a quarter-wavelength microstrip line with a 39-pF shunt capacitor to ground, providing a short circuit at successive even harmonics.

The PA was tested under a CW input at 1.9 GHz and a supply voltage of 6.5 V. The measurement results as illustrated in Figure 6.16 are in line with the simulated results. The measured small-signal gain is 14.6 dB, with a fairly flat gain roll-off. At the 1-dB compression point, the output power is 30.8 dBm with a 46.1% power-added efficiency (PAE).A maximum PAE as high as 47% is achieved at a 31.6-dBm output power, whereas the saturated output power reaches 32.4 dBm. High power efficiency (PAE 40%) is also maintained over a wide range of supply voltages (4.59 V) with constant load impedance. To test its robustness under a high supply voltage as designed, the PA was continuously operated at the peak output power, under a 9-V supply. No performance degradation was observed after one week.

The broadband nature of the amplifier, a benefit of the high value of output load impedance achieved through stacking, is evident in Figure 6.17. The figure shows that, with a fixed matching circuitry, the saturated output power is maintained above 31 dBm and the drain efficiency is above 40% for the entire digital communication system (DCS), personal communication system (PCS), and Universal Mobile Telecommunications System (UMTS) frequency bands.

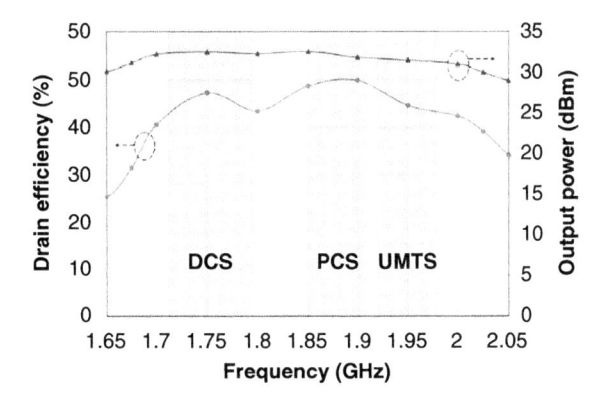

Fig. 6.17 Frequency response of output power and efficiency of 1W stacked CMOS PA.

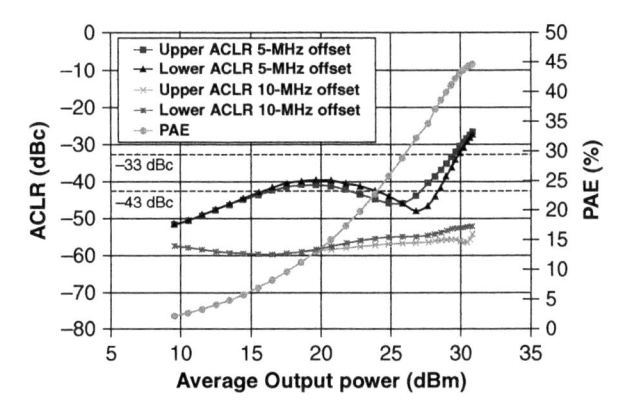

Fig. 6.18 Measured linearity characteristics of 1W stacked CMOS PA with WCDMA signals.

The stacked FET CMOS amplifier also exhibited a high degree of linearity. Fig. 6.18 shows the measured adjacent channel leakage ratios (ACLRs) and PAE performance of the PA as a function of average output power using an uplink WCDMA signal at 1.9 GHz. The ACLRs were measured at 5- and 10-MHz offsets from the center frequency. The modulated input signal has a chip rate of 3.84 Mc/s and a PAPR of 2.58 dB. The measured ACLR at 5-MHz offset is below the -dBc requirement up to an average output power of 29.4 dBm and a PAE of 41.4%. These results confirm that the implemented CMOS PA exhibits efficiency and linearity performance comparable to those of GaAs-based PAs.

6.6 Si device technology for stacked designs

Stacked transistor approaches can be applied in a variety of device technologies, including Si and compound semiconductor FETs and bipolar devices. In this chapter the focus is particularly on Si-based FETs, with frequency response adequate for mm-wave operation.

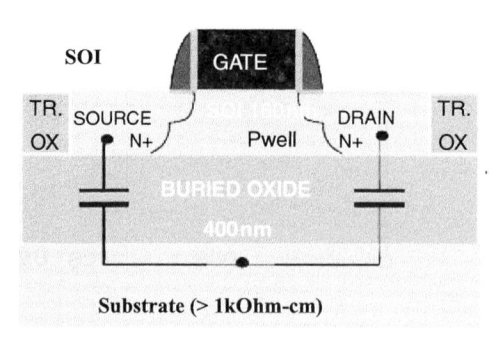

Fig. 6.19 Structure of representative SOI nMOS FET.

Silicon-on-insulator (SOI) CMOS technology (or the closely related silicon-on-sapphire technology) is particularly attractive for the stacked FET technique due to its lack of body effect and relatively small parasitic junction capacitances. Figure 6.19 illustrates the cross-section of a representative SOI FET. In bulk CMOS source-body capacitance and the body effect progressively reduce the gain of the transistors in the upper sections of the stack. Moreover, the maximum allowable supply voltage is limited by the breakdown voltage of the drain-bulk junction diode. An alternative solution is to use a triple-well CMOS process, where the p-well of each transistor can be tied to its source to avoid the body effect. The n-iso layer should also be tied to the source to prevent the p-well/n-iso junction diode from turning on. Nevertheless, the maximum allowable supply voltage is still limited by the breakdown voltage of the n-iso/p-sub junction diode, and the parasitic capacitances are greater than those incurred in the SOI technique.

Within SOI technology, various options are available regarding the conductivity of the Si substrate wafer below the buried oxide. High resistivity ($\rho_s > 1$ kOhm-cm) substrates are available in some process technologies, which minimize parasitic capacitances of transistors and interconnects to the substrate. At mm-wave frequencies, the tolerance to substrate conductivity increases, however, since the response of the substrate is resistive only at frequencies below $f_c = 1/(2\pi \times \rho_s \times \epsilon_s)$, where substrate resistivity ρ_s and dielectric constant ϵ_s define a characteristic time constant. For ρ_s of 13 Ohm cm, this time constant is approximately 14 psec and $f_c = 12$ GHz.

SOI transistors in frequent use are based on a partially depleted technology, so that there is a quasi-neutral body that can be left floating or tied to the source through a body contact. To secure the highest breakdown voltage and to avoid single transistor latch-up, body-contacted devices are preferred. However, layouts that succeed in connecting to the body incur substantial parasitic capacitance so that ft of the body-contacted transistors is smaller than that for floating-body transistors. As a result, many circuits have been reported in which the body is allowed to float.

With highly scaled CMOS devices in SOI technology, particularly with floating body, the bare device models indicate very high values of f_T and f_{max}. However, the back-end-of-the-line (BEOL) terminal connections introduce noticeable parasitic capacitance

6.6 Si device technology for stacked designs

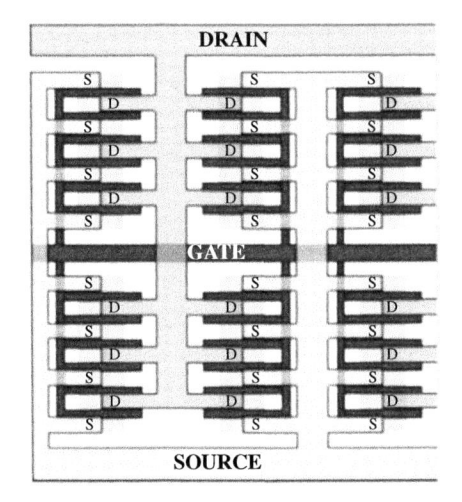

Fig. 6.20 Zipper layout of FET to minimize parasitics.

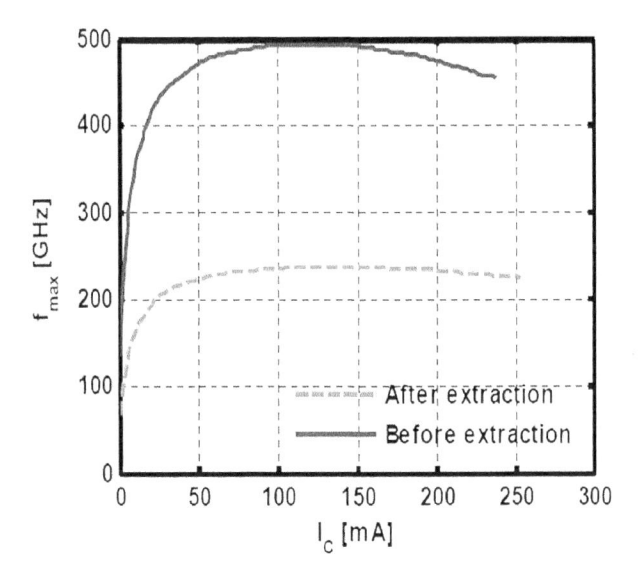

Fig. 6.21 Simulated f_{max} vs drain current for 45nm nMOS FET before and after parasitic extraction.

and resistance. By using a careful layout such as the zipper layout in Figure 6.20 the parasitics can be minimized. The f_{max}-I_d curves for an n-channel FET based on pre- and post-layout simulations are shown in Figure 6.21 for a 45nm floating-body SOI CMOS technology. The peak f_{max} drops from 500 GHz to 240GHz when the layout parasitics are accounted for. The reduction is likely associated with losses in the thin lower-level metal layers and via resistances. From the figure, one can determine that the current density for high gain is 0.4 to 1.7mA/μm.

An important issue underlying the stacked FET strategy is the fact that a composite FET consisting of a series connection of short gate length FETs can outperform a FET

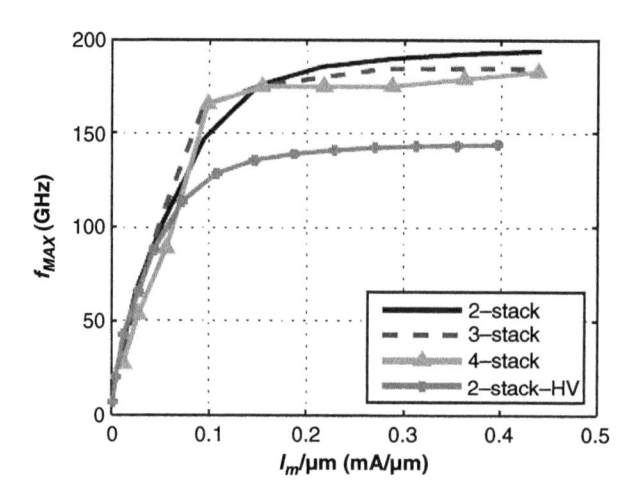

Fig. 6.22 Comparison of simulated f_{max} for stacked thin-oxide FETs with thick-oxide FET stack with the same voltage handling capability.

in the same technology that is not scaled as aggressively and thus has higher breakdown voltage. This fact is confirmed by simulations such as those illustrated in Figure 6.22, in which a comparison is made between simulated f_{max} of FET stacks containing 2, 3 and 4 scaled devices (with thin oxide) and a 2-stack implemented with thick oxide devices with greater gate length, capable of handling voltages comparable to those of the thin oxide 3 stack [5]. The advantage in gain of the scaled transistor stack is evident.

6.7 Stacked FET mm-wave design

The straightforward stacked FET amplifier design outlined in section 6.2 is generally not adequate to optimize circuits for mm-wave operation. A key issue is that the load impedances presented at intermediate nodes of the stack have a reactive part that is significant at mm-wave frequencies, caused by the transistor and interconnect capacitances. The reactive part of the node impedance reduces the efficiency for two reasons: (a) part of the transistor RF current flows out through C_{gs} and other capacitances at the drain of M1 and does not reach the load; (b) the voltage waveforms developed at the drains of the different transistors lose the phase alignment needed for the highest swing at the top drain. As a result it is appropriate to introduce impedance matching at intermediate nodes in the stack [5].

It can be shown that the optimal admittance to be presented at the drain of a device in the stack is given by

$$Y_{opt,k} \approx \frac{1}{kR_{opt}} - \frac{s}{k}(C_{ds,k} + kC_{dsub,k} + C_{gd,k}) \tag{6.10}$$

$$= \frac{1}{kR_{opt}} - \frac{s}{k}C_{eqv,k}, \; k = 1, 2, \ldots, K \tag{6.11}$$

6.7 Stacked FET mm-wave design

Here R_{opt} is the resistive contribution derived in section 6.3 for low frequencies (R_L/K), while $C_{eqv,k}$ is an equivalent capacitance that should be tuned out, given by an appropriately scaled composite of C_{ds}, C_{gd}, and C_{dsub} for the devices in the stack. This condition ensures that all drain-source voltages as well as drain currents are aligned, leading to highest output power and best efficiency. By contrast, if no extra impedance matching is done at the intermediate node, the admittance seen at the source of a transistor in the upper stack is in general different from this, and is given by

$$Y_{s,k+1} = \frac{1}{kR_{opt}} - \frac{sC_{ds,k+1}}{k} + \frac{sC_{gs,k+1}}{kg_{m,k+1}R_{opt}} \qquad (6.12)$$

Typically this has a positive imaginary part (capacitive loading) and it departs significantly from the desired value $Y_{opt,k}$. A consequence is that the phase angle between the drain current and the voltage is not optimal, and in fact it varies as one goes up the stack. The overall voltage at the top of the stack then falls short of its optimal value. It can be shown that if the phase angle difference between the actual admittance Y_k and the optimal $Y_{opt,k}$ is Φ_k, then there is an inefficiency in the power combining, given by the factor $\eta_{stacking}$, with

$$\eta_{stacking} \approx (\prod_{k=1}^{K} \cos \Phi_k)^2 \qquad (6.13)$$

The power combining efficiency can be dramatically reduced for the case of many stacked devices, and significant errors in phase angle, as detailed in Fig. 6.23.

In order to provide an appropriate amount of matching at intermediate nodes, various techniques using passive elements can be used, among which are (a) shunt inductance; (b) series inductance; and (c) shunt external drain-source capacitance C_{dsx}. These techniques are illustrated schematically in Figure 6.24.

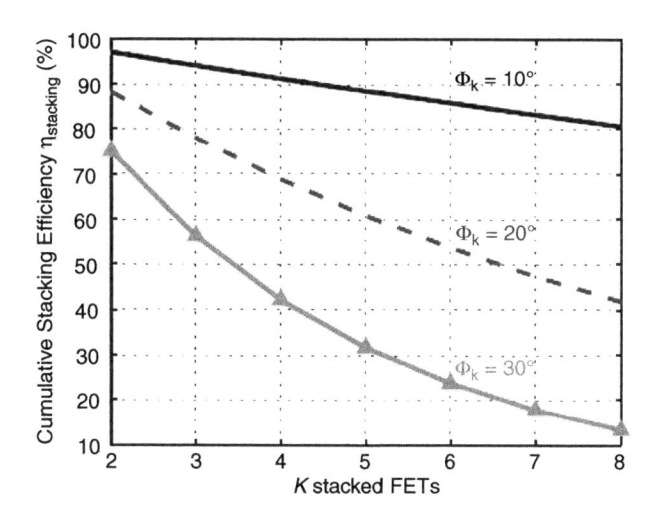

Fig. 6.23 Efficiency factor vs number of FETs in stack, for various phase angle differences per stage.

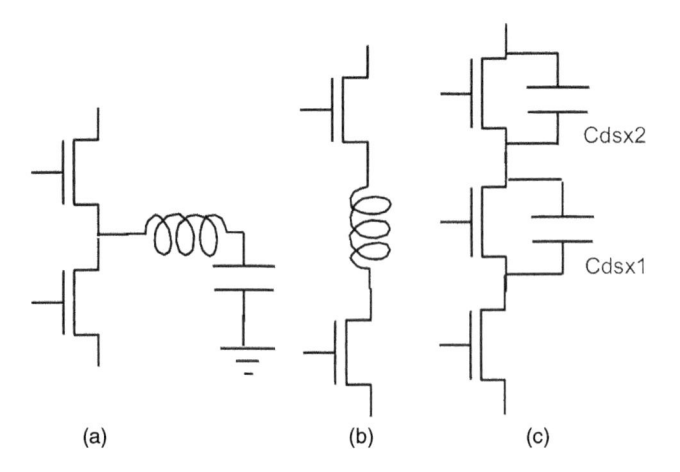

Fig. 6.24 Techniques for providing impedance matching at intermediate nodes: (a) shunt inductor; (b) series inductor; and (c) external shunt C_{ds}.

In the case of the shunt inductance (connected via a large dc blocking capacitor to ground), in order to achieve proper matching, the inductance L_k should be chosen according to

$$\frac{1}{L_k} = \frac{\omega^2(C_{ds,k} - C_{ds,k+1})}{k} + \frac{\omega^2 C_{gs,k+1}}{k g_{m,k+1} R_{opt}} + \frac{\omega^2(C_{gd,k} + k C_{dsub,k})}{k} \quad (6.14)$$

The first term shows that for equally-sized transistors, the drain-source capacitances cancel. The second term is the capacitive load of the $(k+1)$th transistor. Its effect is reduced by the voltage gain across the transistor. The third term relates to the gate-drain and drain-substrate capacitance of the kth transistor.

For the case of series inductance, L_k should be chosen according to

$$L_k \approx k R'_{opt,k} \frac{C_{gs,k+1}}{g_{m,k+1}} - k R'_{opt,k} R'_{opt,k} C_{ds,k+1} + k R'_{opt,k} R_{opt} C_{eqv,k} \quad (6.15)$$

The first two terms represent a series inductance tuning out the capacitive loading caused by the $(k+1)$th transistor. The third term tunes out the effective shunt-feedback C_{ds} of the kth transistor. Here $R'_{opt,k}$ is the modified value for R_{opt} which takes into account the parallel capacitances,

$$R'_{opt,k} = \frac{R_{opt}}{1 + (\omega C_{eqv,k} R_{opt})^2} \quad (6.16)$$

The case of tuning by addition of shunt drain-source capacitances is an interesting concept proposed by A. Ezzeddine [18]. It is based on the notion that the charge lost at a node by C_{ds} injection into a node lower in the stack, may be compensated by C_{ds} injection from a device higher in the stack. Thus (as observed above), if the stack consists of equally sized transistors and their voltage swings are equivalent, the C_{ds} contributions cancel out. If an appropriate additional C_{ds} is added externally at the higher nodes, then the additional injected charge can cancel also the charge lost by gate capacitances

6.7 Stacked FET mm-wave design

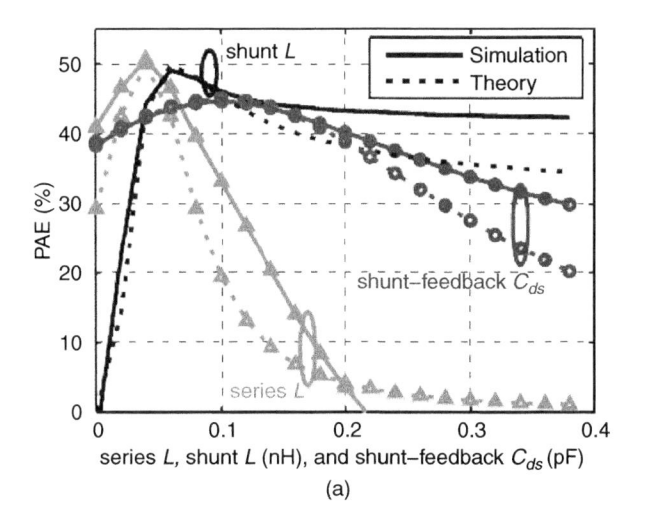

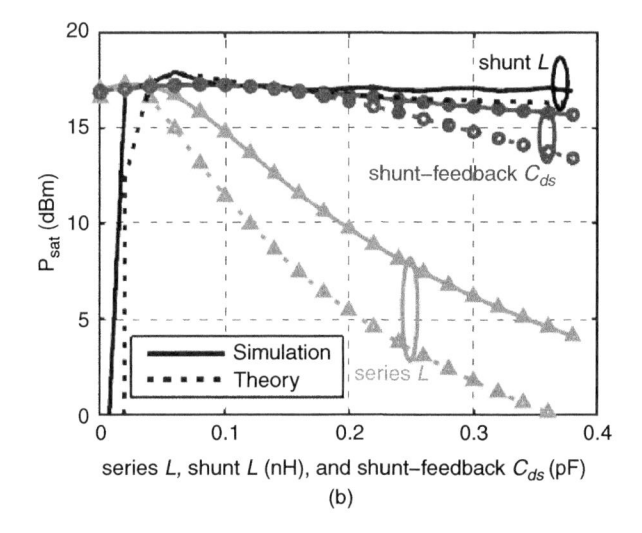

Fig. 6.25 Saturated output power and efficiency of 2-stack amplifiers at 45 GHz, with intermediate node matching carried out with the different techniques.

and capacitance to ground. An appropriate design value for the added capacitances is given by

$$C_{d,k+1} = C_{ds,k} - C_{ds,k+1} + k\frac{C_{gs,k+1}}{g_{m,k+1}R_{opt}} + kC_{gd,k} + k^2 C_{dsub,k} \quad (6.17)$$

It is of interest to compare the effectiveness of the different intermediate node matching techniques in optimizing amplifier performance. An analytical and numerical study has been carried out for 2-stack amplifiers at 45GHz with 45nm nMOS transistors. In Figure 6.25 are shown the output power and efficiency for the 3 techniques, computed according to the above analyses and by simulation, as a function of the matching element values. While the simulation indicates comparable performance when each technique is tuned appropriately, there are some practical differences between the tuning techniques. First, the required gate capacitance C_{xk} is larger when the series inductance is used, which reduces the gate swing and increases the gate-drain voltage. This would slightly reduce the saturated power P_{sat} and potentially the peak PAE in cases where the gate-drain breakdown voltage is limiting the reliable operation range. Second, the capacitive tuning technique requires a larger inductive impedance at the top drain to compensate for the additional capacitive loading. This might make the output matching more challenging. Furthermore, the efficiency benefits are sensitive to model accuracy, as one can see in Fig. 6.25. The shunt inductive tuning seems the least sensitive to mistuning. However, both the series L and shunt-feedback C_{ds} techniques have one advantage over the shunt L tuning technique. These tuning elements are frequency independent, making them suitable for broadband amplifiers.

A more in-depth investigation of the intermediate node matching can be done, which takes into account the gate resistance of the FETs in the stack. This gate resistance (associated with the FET as well as with the external capacitor and the interconnects between them) is one of the fundamental gain limiters for the transistors in the stack. The gate resistance decreases the magnitude of drain current flowing though the stack, and changes its phase angle at various stack levels, because it causes current to be diverted from the stack to the gate port. This impact is similar to the effect of capacitance loading the intermediate node, as expressed in the equations above, for example. For the purely reactive loading of the nodes, however, complete matching with exclusively reactive elements can in principle be done. The loading associated with the gate resistance cannot be compensated in this simple fashion. To identify the conductance loading the intermediate nodes, it is simplest to transform the series R_g, C_{gs} branch at the intermediate node to a parallel admittance form, as illustrated in Fig. 6.26. Here a simplified scenario is illustrated, which ignores C_{dg} effects for simplicity, and shunt inductance is used to compensate the load reactance. As described by Agah et al. [19], there is a shunt conductance stemming from the resistance component that leads to a reduction in drain current magnitude as one moves up the stack, given by the factor $\gamma = I_{d,n+1}/I_{d,n}$. The simplified case of Figure 6.26, γ is given by the simple expression

$$\gamma = \frac{1 + jQ_x}{1 + jQ_x + j\omega/\omega_T} \tag{6.18}$$

Where ω_T is $2\pi f_T$, the current gain cutoff frequency of the FET, and Q_x is the effective Q_x factor of the combined capacitance loading the node $(C_{gs} + C_x)/(\omega R_g C_{gs} C_x)$. For high mm-wave frequencies (f 100GHz), representative Si FETs lead to Q_x values of order 3 to 5, while $\omega/\omega_T \approx 0.5$. The R_g effect can thus significantly reduce the efficiency of current flow through the stack. This has motivated additional techniques to maintain the current flow, in particular the active drive discussed below.

The stability of stacked transistor amplifiers is an important design consideration, inasmuch as there are a variety of mechanisms that can lead to oscillatory behavior. One of the mechanisms, well-known for common-gate or cascode amplifiers, is related to capacitive loading of the source of a stacked transistor (as shown schematically in

6.7 Stacked FET mm-wave design

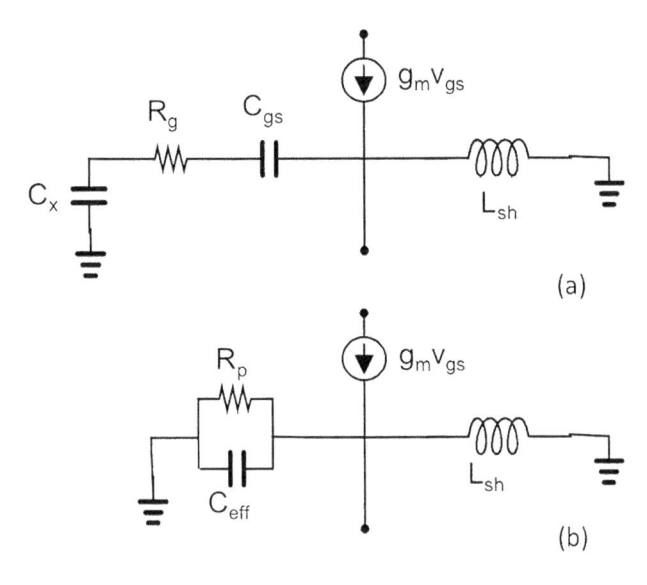

Fig. 6.26 Simplified circuit of transistor in stack, illustrating effect of Rg (a) in direct form; (b) expressed as parallel conductance for estimation of leakage current.

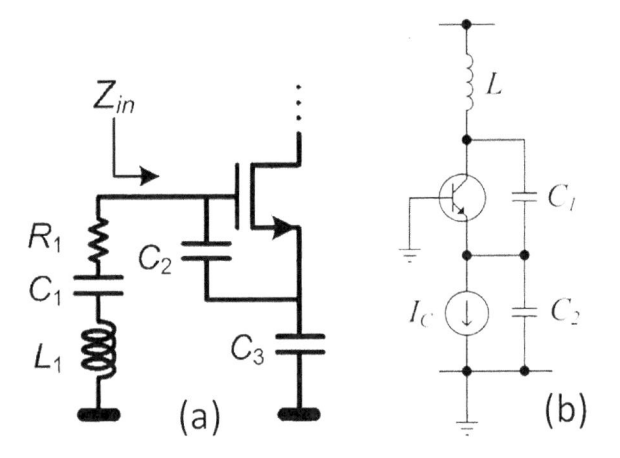

Fig. 6.27 (a) Possible configuration of upper transistor in stack exhibiting parasitic inductance; (b) circuit diagram of Colpitts oscillator implemented with common-base BJT.

Fig. 6.27). Under these circumstances the impedance Z_{in} seen from the gate can have a negative real part, and if there is parasitic inductance L_1 in series with the external gate capacitor, oscillations can take place at a frequency given by the resonance of L_1 and the overall capacitance. A related observation is the fact that the impedance seen looking into the source of the top transistor (the drain load impedance Z_{dk} described above) has the proper design value R_{opt} only for the design capacitance C_{xk} at the gate terminal; if there is parasitic inductance to ground, then at a high enough frequency the impedance from gate to ground will have positive reactance, and the drain load impedance Z_{dk} will have a negative real part. This mechanism leads to a potential for oscillations at frequencies typically higher than the design frequency. For mm-wave amplifiers, the transistors often do not have sufficient gain at the resonance frequency to lead to oscillations; for amplifiers at lower frequencies, the stability can be assured by including an extra resistance in series with the gate capacitor. A separate concern for instability is related to the feedback between transistors in the stack associated with C_{ds}. The potential for oscillation in the presence of excessive C_{ds} values can be appreciated from Fig. 6.27, a diagram of a Colpitts oscillator configured with a common-base circuit.

It is of interest to determine how many transistors can be usefully included in a stack, at various frequencies of operation. Practical limits to the degree of stacking are imposed by a number of factors, including the following:

(1) The external gate capacitors needed to provide an appropriate gate swing at the uppermost transistor nodes become very small (or, more generally, the external gate impedance levels become very high) to the point where uncertainties in parasitic capacitances and device C_{dg} values dominate the performance.
(2) The need to establish proper phasing of the voltages and currents at each level of a large stack becomes very challenging (particularly in the face of imperfect models).
(3) There are parasitic capacitances to ground throughout the stack. As the overall voltage relative to ground builds up along the stack, the losses associated with charging these capacitances (through unavoidable series resistance) become more severe. In a related fashion, the losses associated with the parasitic resistance (finite Q) of the added gate capacitors and of the intermediate node shunt inductors becomes a larger fraction of the power.
(4) Stacked transistors at mm-wave frequencies benefit from being close together, thereby reducing interconnect parasitics. With large stacks and large output powers, there can be thermal resistance and self-heating concerns.
(5) The added power per extra stacking stage becomes a relatively small fraction of the total power with large stacks. Power combining via passive interconnect structures appears to be a more effective way to significantly increase output power.

6.8 Design of mm-wave stacked-FET switching power amplifiers

The previous section dealt with the details of the design of linear stacked-FET power amplifiers at mmWave frequencies. In this section, we discuss the analysis and design of switching mmWave stacked-FET amplifiers, with a focus on Class-E-like operation.

6.8.1 Loss-aware mmWave stacked class-E-like PA analysis and design methodology

The model for a stacked class-E-like amplifier was depicted in Fig. 6.10(b). Based on this model, it is clear that stacking enables a class-E-like PA with linearly larger voltage handling but with linearly larger switch loss, which can degrade efficiency, particularly at mmWave frequencies.

6.8 Design of mm-wave stacked-FET switching power amplifiers

To facilitate a theoretical analysis at mmWave frequencies, the improved loss-aware class-E design methodology for mmWave PAs described in the previous chapter is employed. While we consider equal-sized devices here for simplicity, the devices can be of different sizes and there could potentially be some benefit in tapering device sizes as well ([20]), since the gate capacitors conduct a portion of the device current. Thus, progressive device size reduction up the stack would reduce parasitic capacitances and prevent capacitive discharge loss at intermediate nodes. Device ON-resistance, output capacitance and input-drive-power as functions of device size have been listed in the previous chapter for a 65nm low-power bulk CMOS technology as well as a 45nm SOI CMOS technology. The analysis presented here utilizes the 45nm SOI CMOS technology at 45GHz operating frequency. In order to incorporate the loss of the DC-feed inductance, an inductive quality factor of 15 is assumed at 45GHz [21]. For various levels of stacking (n), the design methodology is used to analytically vary device-size and DC feed inductance to find the design point(s) with optimal PAE under the constraint of a 50Ω load impedance to avoid impedance transformation losses. As an example, for a stack of four devices ($n=4$), we start with an initial device size of 100μm and set the tuning parameter $\omega_s = 0.8 \times \omega_0$. The design methodology then determines the optimal load impedance for highest PAE and the corresponding output power. The load impedance is then scaled (along with device size, and input and output powers) to have a real part of 50Ω. The procedure is repeated by changing the tuning parameter ω_s. Finally, amongst all these design points for a stack of four devices driving a 50Ω load, the one with the highest PAE is chosen. This yields a device size of 204μm for the 4-stack PA, with theoretical output power and PAE of 145mW and 48% respectively (as shown in Fig. 6.28). The procedure can similarly be used to determine the corresponding metrics for other levels of stacking.

Fig. 6.28(a) depicts the optimal output power and PAE for different levels of stacking in 45nm SOI CMOS at 45GHz. The optimal size of each stacked device and the associated device stress (defined as the ratio of the average current drawn from the power supply to the device width) are shown in Fig. 6.28(b). It is clear that due to the increasing achievable output voltage swing, stacking in class-E-like CMOS PAs enables dramatic increases in output power (near-quadratic due to the linear increase in output swing). The PAE reduces with increased stacking due to increasing total switch loss. However the methodology ensures that the PAE degradation is gradual. In order to do this, the design methodology requires the size of each stacked device to increase with n to reduce the individual (and hence overall) ON-resistance. Consequently, careful device layout is required for high levels of stacking as it is challenging to layout large devices while maintaining a high f_{max}. Another important consideration for device stacking is the current stress for the stacked devices, which increases with the level of stacking. The current stress (or large signal current-density) is the ratio of the average current drawn from the supply under large signal operation (I_{DC}) to the device width. Note that I_{DC} is different from the supply current I_{bias} drawn with no input power (i.e. under small signal operation), the latter being used to determine the current density for operating at highest f_{max} in linear PAs. The current drawn under large signal operation is typically 1.5-2 times higher than the small signal bias current in our

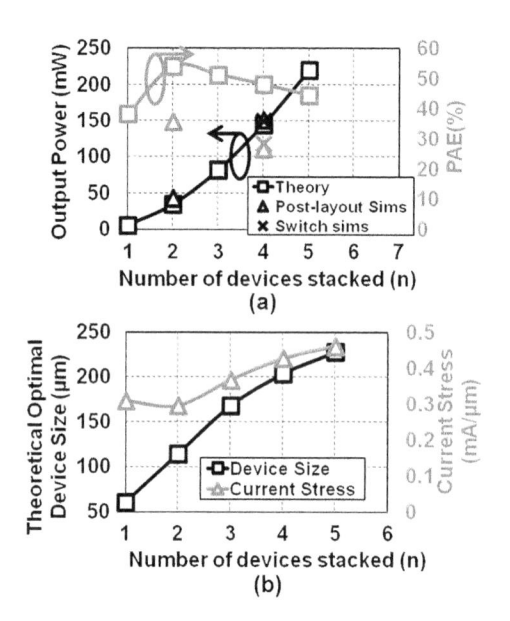

Fig. 6.28 (a) Theoretical and simulated (post-layout) output power and PAE and (b) device size and theoretical device stress for the optimal design as a function of number of devices stacked based on the loss-aware class-E design methodology at 45 GHz in 45nm SOI CMOS. Loss in DC-feed inductance is included for theoretical results. Output power and PAE for a switch+capacitor based model for the 4-stack configuration are also annotated.

implementations. This implies that, in a practical implementation, the metallization of the source and drain fingers of the MOS devices must be augmented with additional metal layers, if required, so that they can support the required currents while satisfying electromigration rules for the technology. While Fig. 6.28(a) shows an increasing trend for output power till 5 devices, at much higher levels of stacking the assumption in the theoretical analysis that the switch ON resistance is much smaller than the impedance of its output capacitance [22] would be violated. Furthermore, there would be diminishing returns in output power owing to increased losses with stacking. In practice, the maximum practical device size, the maximum current stress that can be tolerated as per electromigration requirements, and drain-bulk/buried-oxide breakdown mechanisms would determine the maximum number of devices that can be stacked. The post-layout simulated results for output power and PAE for a 2-stack and a 4-stack class-E-like PA that are discussed later in this chapter as case studies have been annotated on Fig. 6.28(a) as well and show excellent agreement with the theoretical output power. The post-layout simulated efficiency is lower by ≈20% owing to various implementation losses and soft-switching at mmWave as well as power-loss at intermediate nodes which are not accounted for in the theory. However, the theoretical and simulated trends in PAE are in agreement. Later in this chapter, a switch+capacitor based model for the device is constructed for simulation-based investigation of power loss at intermediate nodes for a 4-stack configuration. The resulting output power and PAE (Fig. 6.28(a)) show excellent agreement with post-layout device-based simulations and re-affirm the utility of a simplified theoretical analysis.

6.8.2 Interpretation using waveform figures of merit

An analysis using the unique properties of switching PAs facilitates a better understanding of the underlying phenomena associated with device stacking and an interpretation of the results of the loss-aware class-E design methodology. An excellent description of the characteristics of switching PAs can be found in [23]. We have

$$PAE = 1 - \frac{P_{loss}}{P_{DC}} - \frac{P_{in}}{P_{DC}} \qquad (6.19)$$

where P_{in} and P_{DC} are the input power to the PA and the DC power consumption respectively. The loss in the PA (P_{loss}) is given by

$$P_{loss} = P_{loss,switch} + P_{loss,cap} \qquad (6.20)$$
$$= I_{RMS,n}^2 \times n \times R_{ON,n} + P_{loss,cap} \qquad (6.21)$$
$$= I_{RMS,n}^2 \times n \times \frac{\overline{R_{ON,n}}}{W_n} + P_{loss,cap} \qquad (6.22)$$

where n is the number of devices stacked in series, $R_{ON,n}$ is the ON-resistance of each device in the n-stack PA, W_n is the width of each device/switch and $I_{RMS,n}$ is the RMS value of the current flowing through the stack of n switching devices (excluding the output capacitance). $P_{loss,cap}$ is the switching loss associated with the output capacitance of the PA and is dependent on the topmost drain voltage value at the switching instant. In general, at mmWave frequencies, the capacitive discharge loss is negligible compared with the conduction loss in the switching device(s). This is evident from Table 6.1, where the conduction loss in the switch and the capacitive discharge loss have been tabulated for the optimal designs at different levels of stacking described in Fig. 6.28. Indeed, this reinforces our assertion in the previous chapter that the conventional ZVS/ZdVS-based class-E design methodology is not applicable at mmWave frequencies. Therefore, for the purpose of simplifying our analysis, we shall ignore the contribution of the term $P_{loss,cap}$ to the overall loss. In a switching PA, the average current drawn from the supply is always proportional to $I_{RMS,n}$. The proportionality constant depends on the tuning of the load network [23]. The tuning of a class-E load network is determined by the DC-feed inductance (L_s) and the load impedance (Z_{load}) in relation to the device output capacitance. Since we are in a regime where conduction loss is significant, the proportionality constant will also depend on the value of the total switch ON-resistance ($n \times R_{ON,n}$) relative to the output capacitance $C_{out,n}$. Since $R_{ON,n} \times C_{out,n}$ is a technology constant, specifying n, $C_{out,n}$, L_s, and Z_{load} completely characterizes the tuning of the stacked class-E PA. It is also therefore clear that the optimal tuning is likely to vary for different levels of stacking due to the increasing total switch loss. Ignoring capacitive discharge loss, Eqn. (6.22) becomes

$$P_{loss} \approx I_{RMS,n}^2 \times n \times \frac{\overline{R_{ON,n}}}{W_n}$$
$$= \frac{I_{RMS,n}^2}{I_{DC,n}^2} \times I_{DC,n}^2 \times n \times \frac{\overline{R_{ON,n}}}{W_n}$$
$$= F_{I,n}^2 \times I_{DC,n}^2 \times n \times \frac{\overline{R_{ON,n}}}{W_n} \qquad (6.23)$$

Table 6.1 Conduction loss and capacitive discharge loss for the optimal designs at different levels of stacking described in Fig. 6.28. Values for the waveform figures of merit F_I^2 and F_C for loss-aware and ZVS-based designs are also tabulated.

n	W_n (μm)	$R_{ON,n}$ (Ω)	$C_{out,n}$ (fF)	$P_{loss,switch}$ (mW)	$P_{loss,cap}$ (mW)	F_I^2	F_C	F_I^2 (ZVS)	F_C (ZVS)
1	60	4.58	35.46	1.5	0	2.06	2.06	2.36	3.61
2	114	2.41	67.37	10.6	1.3	2.56	0.94	2.17	3.02
3	168	1.64	99.29	32.8	5.7	2.73	0.74	2.1	2.61
4	204	1.35	120.56	80.2	17.8	2.62	0.68	2.07	2.29
5	228	1.21	134.75	176.3	45.4	2.54	0.63	2.05	2.04

where $F_{I,n} = \frac{I_{RMS}}{I_{DC}}$ is a waveform figure of merit defined in [23] and $I_{DC,n}$ is the average supply current with n devices stacked.

For a stack of n devices, the supply voltage scales linearly with n. On the other hand, in a switching PA, the average supply current is proportional to the product of the output capacitance, the supply voltage, and the operating frequency (ω_0), the constant of proportionality being dependent on the tuning of the circuit. The linear dependence on output capacitance is simply an artifact of circuit scaling properties, while the linear scaling with supply voltage arises from the fact that switching PAs are linear with respect to excitations at the drain node (e.g. supply voltage) [23]. Denoting the impedance of the device output capacitance at the fundamental frequency by

$$Z_C = \frac{1}{\omega_0 \times (\overline{C_{out}} \times W_n)}$$

the waveform figure of merit F_C is defined in [23] as

$$F_C = \frac{P_{DC}}{\frac{V_{DC}^2}{Z_C}} \tag{6.24}$$

For an n-stack PA, $V_{DC} = (n \times V_{DD})$ where V_{DD} is the supply voltage for a single device PA. Substituting

$$P_{DC} = V_{DC} \times I_{DC,n} = (n \times V_{DD}) \times I_{DC,n}$$

in Eqn. 6.24, we get

$$I_{DC,n} = F_{C,n} \times \omega_0 \times (\overline{C_{out}} \times W_n) \times (n \times V_{DD}). \tag{6.25}$$

Consequently,

$$PAE = 1 - \frac{F_{I,n}^2 \times I_{DC,n}^2 \times n \times \frac{\overline{R_{ON,n}}}{W_n}}{(n \times V_{DD}) \times I_{DC,n}}$$
$$- \frac{W_n \times \overline{P_{in}}}{(n \times V_{DD}) \times I_{DC,n}} \tag{6.26}$$
$$= 1 - n \times F_{I,n}^2 \times F_{C,n} \times \omega_0 \times \overline{C_{out}} \times \overline{R_{ON}}$$
$$- \frac{k \times \overline{C_{in}} \times (V_{high} - V_{low})^2}{n^2 \times V_{DD}^2 \times F_{C,n} \times \overline{C_{out}}} \tag{6.27}$$

6.8 Design of mm-wave stacked-FET switching power amplifiers

where k is a technology- and frequency-dependent constant of proportionality that results from the input power functions discussed in the previous chapter.

Table 6.1 lists the values of the waveform figures of merit for designs based on the loss-aware and ZVS methodologies for different levels of stacking. While the waveform metric F_I for these two methodologies is comparable, the loss-aware methodology shapes the waveforms to minimize F_C, thereby yielding optimal designs with highest possible PAE.

The foregoing expression captures the variation in PAE in terms of technology constants and number of devices stacked. The only design-related variables in this expression are the waveform figures of merit. The third term captures the PAE benefit of stacking, since output power is increased quadratically but input power is only provided to the bottom device in the stack. This explains why PAE improves when one goes from a single-device class-E-like PA to a 2 stack class-E-like PA in Fig. 6.28(a). However, as stacking is increased beyond 2 devices, the benefits from the third term wear off and the second term causes a reduction in PAE due to a reduction in drain efficiency. It is well known that $\overline{R_{ON}} \times \overline{C_{out}}$ is the technology constant (which we shall refer to as the switch time constant) that determines the drain efficiency of a class-E PA [23] for a given operating frequency and this constant degrades linearly for an n-stack device, since the ON-resistances add in series while the output capacitance remains that of a single device. However, the loss-aware class-E design methodology optimizes the output network tuning to ensure that the PAE degradation is gradual as stacking is increased by minimizing the $F_{I,n}^2 \times F_{C,n}$ product. The benefits of the loss-aware class-E tuning methodology over the ZVS/ZdVS-based tuning methodology can be appreciated in Fig. 6.29, where the $F_{I,n}^2 \times F_{C,n}$ product for the loss-aware design technique can be observed to be lower than that corresponding to the ZVS design methodology by a factor of 2-3, depending on the number of devices stacked.

The preceding analysis also highlights the importance of the switch time constant as a technology metric that determines the efficiency of switching PAs. For linear-type PAs, f_{max} is a sufficient metric to gauge the PA efficiency. For switching-type PAs, f_{max} determines the input power requirements (via the technology- and frequency-dependent constant k) while the Switch Time Constant determines the drain efficiency. As the levels

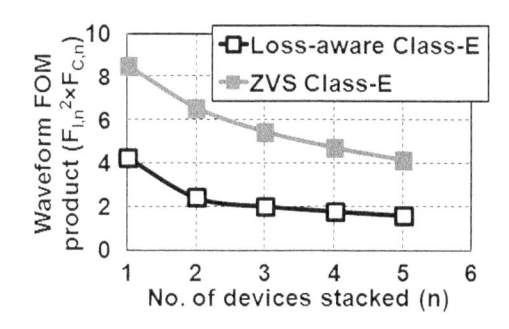

Fig. 6.29 Product of waveform figures of merit $F_{I,n}^2$ and $F_{C,n}$ for stacked Class-E-like PAs in 45nm SOI CMOS at 45GHz based on the loss-aware and ZVS based design methodologies.

of stacking are increased, the Switch Time Constant becomes more significant than f_{max}. As can be seen in the previous chapter and in the literature [24], 65nm low-power bulk CMOS exhibits the same f_{max} as 45nm SOI CMOS but has a significantly lower Switch Time Constant. Consequently, it can be expected that switching-type PAs in 45nm SOI CMOS will achieve higher efficiencies than those in 65nm low-power bulk CMOS. This is validated by case studies presented later in this chapter.

6.9 Stacking versus passive power-enhancement techniques

In order to appreciate the benefits of device stacking, it is imperative to contrast this approach to conventional impedance transformation and power combining techniques. In this section, we perform a comparison for stacked mmWave Class-E-like PAs, but the conclusions hold true for stacked linear PAs as well.

6.9.1 Stacking vs. power combining

To evaluate the performance of power combining, a 45nm SOI 2-stacked class-E-like PA (resulting from the loss-aware design methodology) with a theoretical output power of 34mW and a corresponding theoretical PAE of 54% at 45GHz is chosen, since it has reasonable output power as well as the highest efficiency (Fig. 6.28(a)). Since the 2-stack PA is designed for an optimal load impedance of 50Ω, a cascaded tree of 2-way 50Ω Wilkinson power-combiners is chosen. A 2-way 50Ω Wilkinson power-combiner with 70.7Ω $\frac{\lambda}{4}$ transmission lines in 45nm SOI CMOS technology has an EM-simulated efficiency $\eta=0.87$ at 45GHz. An N-way cascaded-Wilkinson-tree power-combiner (where N is an even multiple of two) will therefore have an overall efficiency of $\eta^{\log_2 N}$. Fig. 6.30(a) compares the theoretical PAE, as a function of output power, for different levels of device stacking with that of 2, 4 and 8-way Wilkinson-tree power-combining. For a given output power, stacked class-E-like PAs implemented using the loss-aware class E design methodology offer ≈10-20% higher efficiency than Wilkinson power-combining (using 2-stack PAs). Power-combining using transformers is a better alternative at mm-wave frequencies, since ideally transformer-based series power-combining has a constant efficiency with number of elements combined. However, interwinding and self-resonant capacitances introduce asymmetry in transformer power-combiners, degrade efficiency and cause stability problems, usually permitting a maximum of two transformer sections to be combined in series [25]. Ignoring the effect of parasitic capacitances, a 2-section series transformer-combiner is used to power-combine two 2-stack PAs. The secondary inductance is chosen for maximum efficiency subject to a 50Ω load, and the PAs are appropriately scaled to drive the load impedance presented by the primary of the transformer. As shown in Fig. 6.30(a), transformer power-combining utilizing 2-stack PAs can yield results similar to stacking only under ideal conditions and is fundamentally limited to 2-way combining. The corresponding results for Wilkinson and transformer-based power-combining using 1-stack (single-device) class-E-like PAs obtained from the loss-aware design methodology are

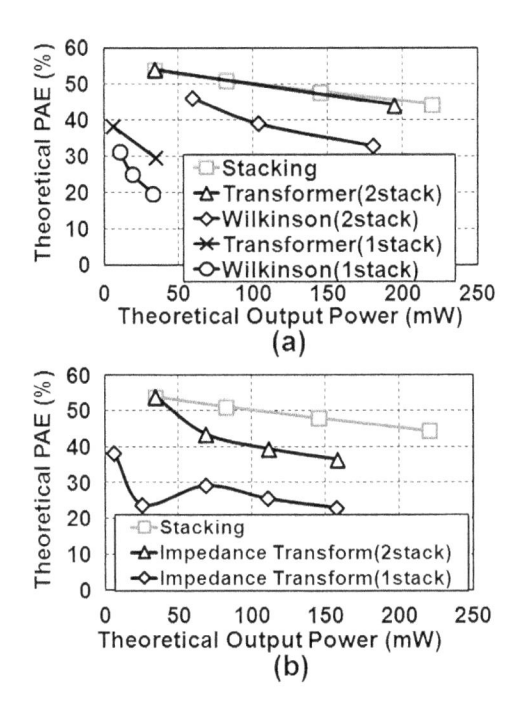

6.9 Stacking versus passive power-enhancement techniques   239

Fig. 6.30   Comparison of device stacking in class-E-like PAs (based on loss-aware class-E design methodology) at 45GHz in 45nm SOI CMOS with (a) Two, 4 and 8-way Wilkinson-tree-based power-combining and transformer-based series power-combining (the 2-stack and 1-stack class-E-like PAs obtained from the loss-aware class-E design methodology are used with both the N-way Wilkinson-tree-based and transformer power-combiners) and (b) impedance transformation at 45GHz in 45nm SOI CMOS (the 2-stack and 1-stack class-E-like PAs obtained from the loss-aware class-E design methodology are scaled to increase output power and a 2-element L–C network is used to transform the 50Ω load to the optimal load impedance for the scaled PAs. The quality factors for the inductor and capacitor are assumed to be 15 and 10, respectively, at 45GHz).

also included to emphasize the inefficacy of the traditional design technique of using single-device PAs for high-power amplification.

6.9.2   Stacking vs. impedance transformation

The efficiency of the alternative technique of impedance transformation is dependent on the steepness of transformation as well as the topology of the impedance transformation network. The 2-stack and 1-stack Class-E-like PAs in 45nm SOI obtained from the loss-aware Class-E design methodology at 45GHz are again employed for the purpose of comparison. In order to achieve output power comparable to those obtained from device stacking, the Class-E-like PAs are scaled appropriately while an impedance transformation network is used to transform the 50Ω load to the corresponding lower load impedance for the scaled PAs. A 2-element L–C impedance transformation network is designed and used in each case. The quality factors of the inductor and capacitor are assumed to be 15 and 10 respectively at 45GHz, based on measured characterizations of inductors, capacitors, and transmission lines in the 45nm SOI CMOS technology [21]. A comparison of the PAEs of impedance transformation and device stacking is summarized in Fig. 6.30(b). Device stacking results in designs with ≈10-30% higher efficiency for the same output power compared with using impedance transformation.

Once device stacking is exploited to the limit as dictated by secondary breakdown mechanisms (e.g. that of the buried oxide in SOI), it is interesting to consider the combination of device stacking with impedance transformation and/or power combining to achieve watt-class output power levels at mmWave frequencies.

6.10 Harmonic matching in stacked structures

To obtain the highest output power and efficiency in conventional power amplifiers, careful attention is given to waveform engineering so that the output voltage and current waveforms for the transistors have minimal overlap, and the harmonic content is appropriately recycled in the amplifier (instead of being attenuated subsequently in lossy filters). Control over the waveforms is established by a combination of input waveform selection and gate bias (leading to transistor cutoff) to govern the current waveform, and harmonic impedance design (and transistor saturation) in order to govern the voltage waveform. For mm-wave amplifiers, it is generally difficult to achieve desired impedance levels at harmonic frequencies because of the low Q and resistive losses of capacitors, inductors and transmission lines at multiples of the fundamental frequency. However, the transistors can have sufficiently fast response to the input signal that they provide waveforms with a significant harmonic content. For example, biasing in Class B reduces gain, but leads to current waveforms that cut off for a significant part of the waveform period (together with capacitive contributions that in general do not cut off). The voltage waveform typically contains only the fundamental because of capacitance loading at the second and higher harmonics. The drain efficiency is then improved over class A as a result of the smaller overlap of voltage and current waveforms. The PAE may or may not improve, because of the corresponding reduction in gain. Accordingly, as described below, stacked mm-wave amplifiers can provide their highest PAE in either class A or class AB operation.

6.11 Active drive for stacked structures

In most of the stacked transistor designs, the upper levels of the stack have purely passive connections to their gates, and the gate voltages are induced by the signals introduced at their corresponding source terminals. There are additional possibilities in which an active circuit is connected to the gates of the higher-level transistors, providing additional opportunities for gain. One advantage of these arrangements is that they can overcome the losses of current as one goes up the stack brought about by the loading of the intermediate nodes. As discussed in Section 6.6, the gate resistance provides

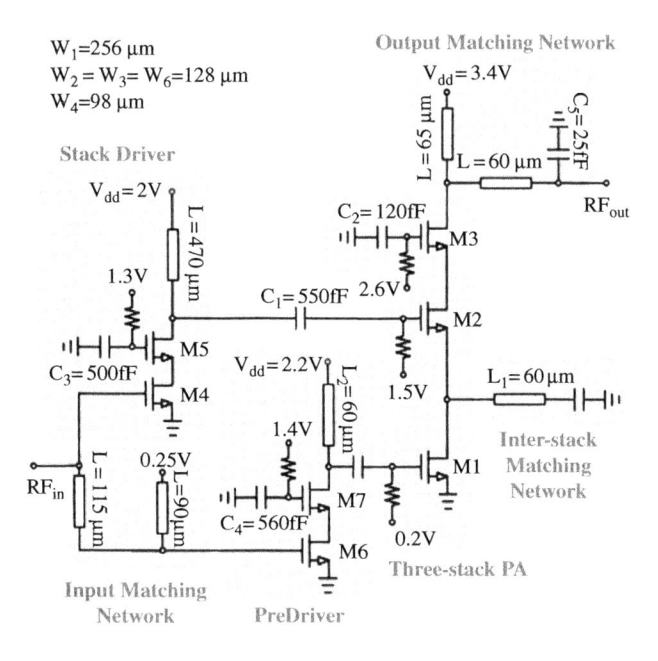

Fig. 6.31 Circuit diagram of 94GHz stacked FET PA using active drive for the second level transistor in the stack.

a component of loss that cannot be compensated by appropriate passive terminations at the gate or source of the transistors, but the losses can be mitigated by active drive. An example of such a circuit is given in Fig. 6.31, which illustrates a 94GHz amplifier consisting of a 3-stack final stage, and driver stages attached both to the lowest level transistor and the second level transistor of the final stack [19]. With appropriate choice of gain and phase of the stack driverĺ the output power and efficiency of the circuit were found to increase considerably, up to 19dBm and 14%, respectively, as described in detail below (case study 3).

6.12 Case studies and experimental demonstrations

6.12.1 Case study 1: Stacked FET PAs at 45GHz

Multiple power amplifiers have been designed for operation at 45 GHz, with the goal of maximizing the output power and power-added efficiency. In this case study, we describe a series of experimental amplifiers implemented in the same technology, IBM 45nm SOI CMOS, with various degrees of stacking (2-, 3- and 4-transistor stacks) [5]. The results allow a comparison of the advantages and disadvantages of stacking. The circuit designs employed are pictured in Fig. 6.32. Matching at intermediate nodes was carried out by a combination of series and shunt inductors (or shorted stub transmission lines). Transmission lines were implemeneted on the chips as grounded coplanar lines. Transistor sizes (noted in Fig. 6.32) were chosen such that the admittance for optimal matching

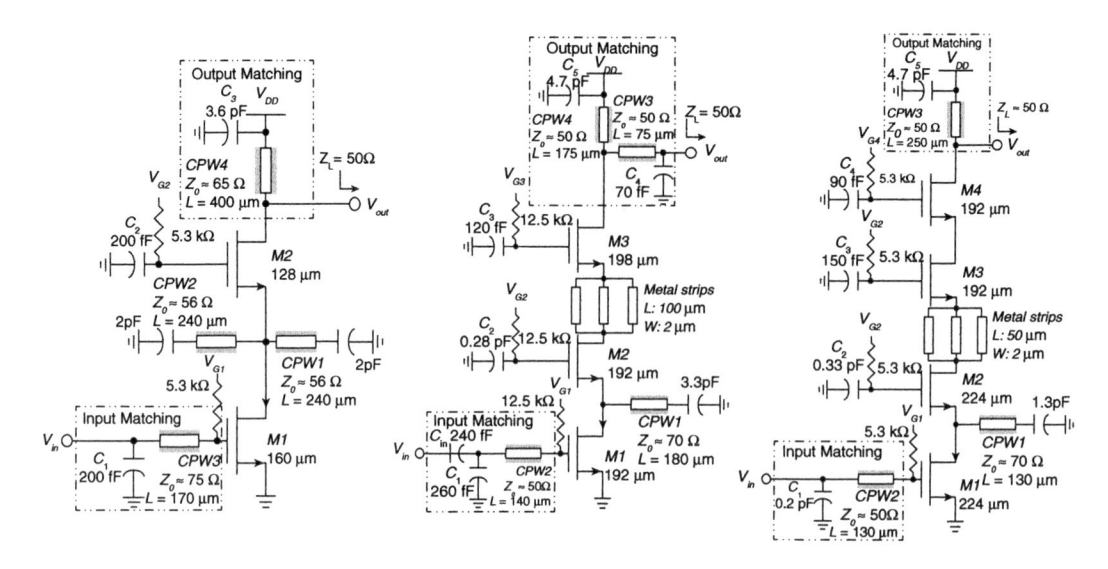

Fig. 6.32 Circuit diagrams for 2-, 3- and 4-stack 45 GHz PAs.

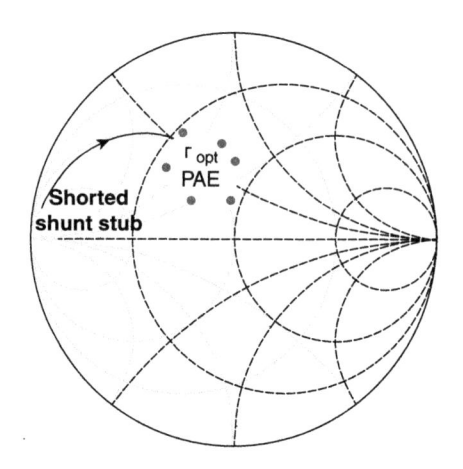

Fig. 6.33 Smith chart showing output matching strategy for 45 GHz PAs.

had a real part of 1/50 ohms. The overall matching of the top drain to 50 ohm could then be accomplished with a shunt inductor (shorted transmission line) which also served as dc feed; Figure 6.33 shows a Smith chart describing the output matching strategy. A microphotograph of a representative fabricated IC is shown in Fig. 6.34.

Small-signal measurements of the amplifiers showed that, when biased such that the quiescent currents are equal for the three PAs, the gains are approximately equal (although there was some mistuning, dependent on the particular amplifier). The 2-, 3- and 4-stack amplifiers respectively have gains of 9.6 dB at 47 GHz, 8.6 dB at 53 GHz and 10.6 dB at 45 GHz. Large signal measurement results versus output power are shown in Fig. 6.35. The best performance for the 2-stack and 4-stack PAs was measured in class-A / AB regime. The peak gain is approximately 9.4 dB for both amplifiers. The

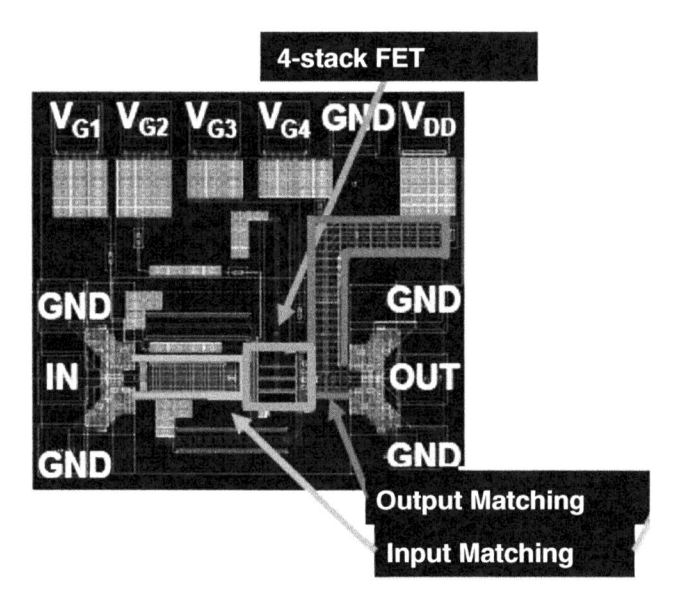

Fig. 6.34 Layout of 4-stack 45 GHz amplifier.

best PAE for the 3-stack amplifier was observed when operating closer to the class B regime. Its peak gain is 8.9 dB. The 2-stack has a peak PAE of 32.7 % at 14.6 dBm, the 3-stack has a PAE of 26.3 % at 18.5 dBm and the 4-stack achieves a peak PAE of 25.1 % at 20.5 dBm. The 2- and 3-stack were measured at 46 GHz and the 4-stack at 41 GHz. The measurement results are in good agreement with simulations. Accurate modeling of the source and drain interconnects and associated inductance was found to be critical for this agreement, as well as EM simulation of interconnects between stages. The saturated output powers of the 2-, 3- and 4-stack were 15.9 dBm, 19.8 dBm and 21.6 dBm. Figure 6.36 compares the output powers of the different amplifiers. The saturated output power of the amplifiers increases with each added transistor as expected by analysis, with a 5-6 dB increase in output power from the 2-stack relative to the 4-stack. The theoretical prediction assumes a peak current of 1.05 mA per m gate width, a knee voltage of 0.15 V, and a drain voltage swing of 2.45 V per device. The stacking concept enables these high output powers without requiring low load impedances. The 2-, 3-, and 4-stack cases have loadline impedances of approximately 15Ω, 18.5Ω and 21Ω. This allows a relatively low impedance transformation with quality factors of 1.1 to 1.5 and enables a wideband on-chip matching network. When increasing the number of stacked transistors, one can trade off the bandwidth and saturated power by keeping the current constant and not increasing the device size (or only moderately increasing the device size) versus aggressively increasing the device size at constant load resistance. The constant current approach would provide the widest bandwidth, but low output power. The constant R_L case would provide the highest output power, but a relatively small bandwidth. Furthermore, a small reduction in efficiency is expected due to the additional losses in the device parasitics. In this work a moderate increase in device

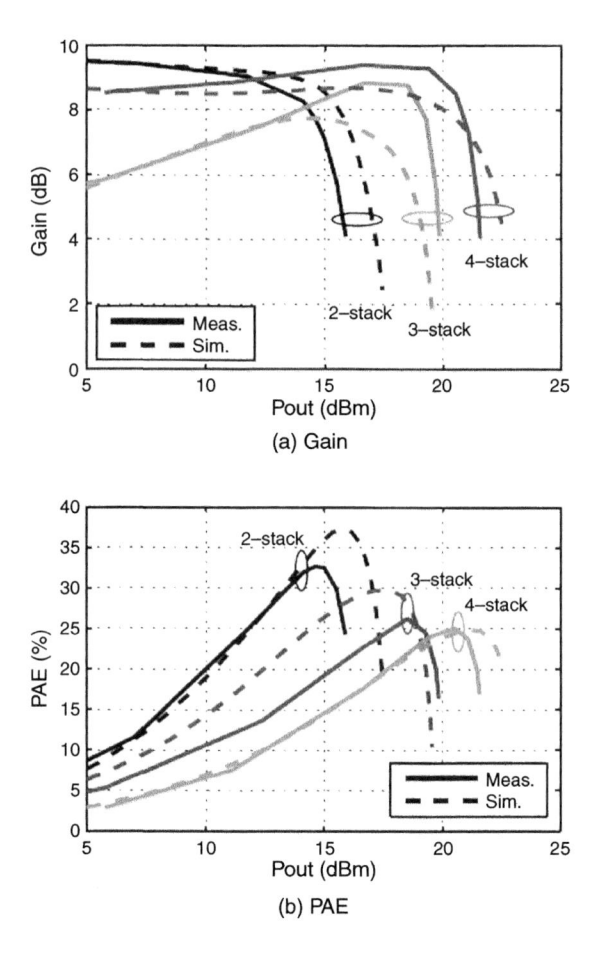

Fig. 6.35 Measured and simulated gain and PAE for 45 GHz stacked FET amplifiers.

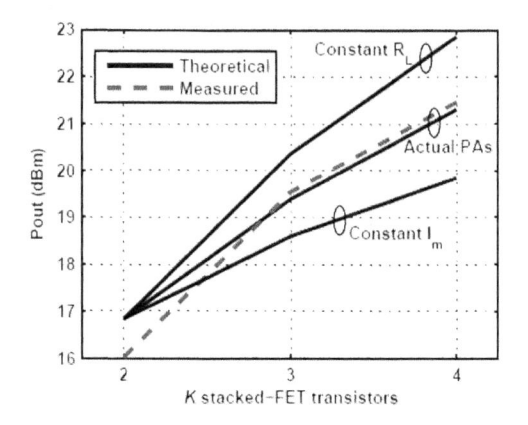

Fig. 6.36 Measured saturated output power vs number of transistors in stack, and for comparison, expected result for constant R_L and constant transistor width (and I_d).

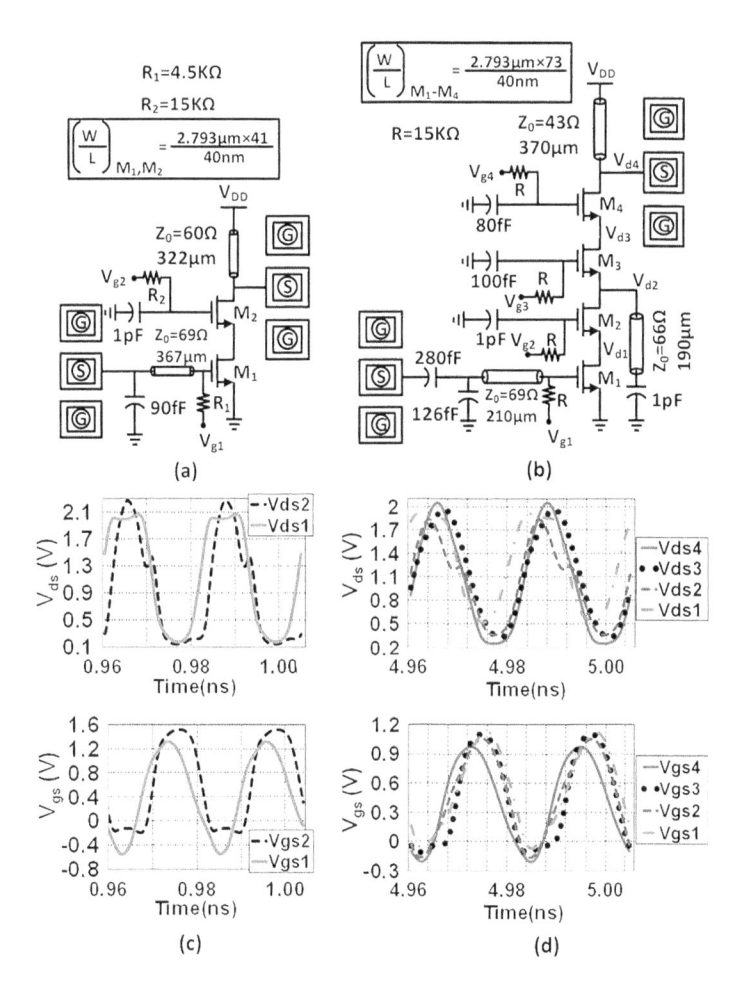

Fig. 6.37 Schematics of 45nm SOI CMOS Q-band Class-E-like PAs with (a) 2 devices stacked and (b) 4 devices stacked. Simulated drain-source and gate-source voltage waveforms of the Q-band (c) 2-stack Class-E-like PA (V_{g1}=0.4V, V_{g2}=1.7V, V_{DD}=2.4V) and (d) 4-stack Class-E-like PA in 45nm SOI CMOS (V_{g1}=0.4V, V_{g2}=1.8V, V_{g3}=2.8V, V_{g4}=4V, V_{DD}=4.8V).

size was chosen for the 3- and 4-stack amplifiers, trading off saturated output power and bandwidth. While reduction of efficiency is expected, the 3 and 4-stack amplifiers nevertheless still achieve PAE around to 22-25 %. The 3- and 4-stack amplifiers maintain good performance over a wide frequency range from approximately 40 GHz to 48 GHz.

6.12.2 Case study 2: Class-E-like stacked FET PAs at 45 GHz in 45-nm SOI CMOS and 65-nm low-power bulk CMOS

The schematics in Figs. 6.37(a) and (b) depict class-E-like PAs implemented by stacking 2 and 4 floating-body devices in 45nm SOI CMOS technology. Device sizes and DC feed inductance values are chosen based on the theoretical analysis, while supply

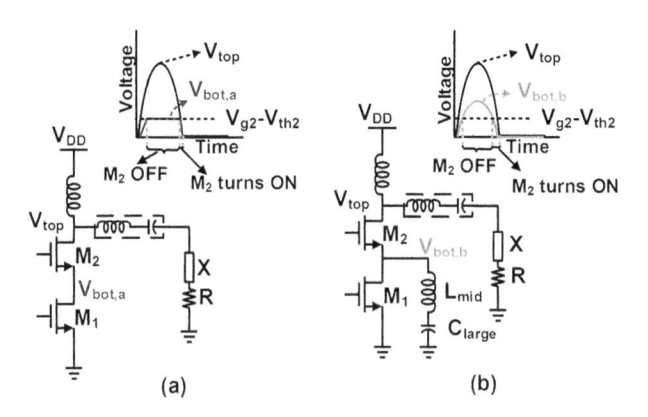

Fig. 6.38 Simulated voltage profiles for 2-stack class-E-like PA (a) without tuning inductor, and (b) with tuning inductor. (c) Close-up of voltage profiles with (bottom) and without (top) tuning inductor.

and gate bias voltages and gate capacitor values are selected based on the considerations described earlier. For the first stacked device (M_2 in both designs), the gate voltage must be held to a constant bias. This can be accomplished through a large bypass capacitor placed as close as possible to the gate to mitigate stray inductance that can result in oscillations. DGNCAPs (which are device capacitors) are suitable for this purpose since their wiring is in the lowest metal layer and they provide higher capacitance density than VNCAPs (interdigitated finger capacitors). All other capacitors, including gate capacitors for the higher stacked devices, which are not large in value, are implemented using VNCAPs. For both the designs, the output harmonic filter is eliminated to avoid passive loss, with minimal impact on performance.

As was mentioned earlier, a tuning inductor may be placed at intermediate nodes to improve their voltage swing and make them more Class-E-like. Simulation results indicate that the improvement in swing for the 2-stack PA is offset by an increase in the conduction loss of the top device. This can be explained as follows. The voltage swing at the intermediate node controls the turn-on and turn-off of the top device. As shown in Fig. 6.38(a), in the absence of the tuning inductor, the intermediate node voltage gets clipped to $V_{g2} - V_{th2}$ once the top device turns off during the OFF half-cycle [8]. The voltage remains unchanged at $V_{g2} - V_{th2}$ till the end of the OFF half-cycle, when the drain voltage of the top device reduces to $V_{g2} - V_{th2}$ and the top and bottom node voltages roll off in tandem thereafter. Introducing an inductor at the intermediate node results in a Class-E-like voltage profile (Fig. 6.38(b)), which causes the top device to turn back on earlier during the latter part of the OFF half-cycle, as shown in Fig. 6.38(c). This leads to additional power loss in the top device. Consequently, no tuning inductor is used in designing the 2-stack PA. For the 4-stack PA, a tuning inductor at V_{d2} is seen to provide benefit. Intuitively, a 4-stack configuration can be viewed as a stack of two 2-stack PAs with the inductor serving as an inter-stage tuning element. Fig. 6.37(c) shows the drain waveforms for the 4-stack PA. As is evident, drain–source voltage swings are almost equally shared across all four devices. The lack of a tuning inductor at V_{d1} results in a relatively flat-topped waveform. This is to be expected, in

6.12 Case studies and experimental demonstrations

view of the foregoing discussion for the 2-stack PA. The situation is somewhat different for node V_{d3}. Despite the absence of a tuning inductor, we can observe a Class-E-like waveform even when device M_4 is off. This is a consequence of capacitive coupling through C_{gs} and C_{gd} of M_4 (in conjunction with capacitive voltage division due to the presence of the 80-fF gate capacitor), which induces voltage swing at V_{d3} when M_4 is not conducting. This eliminates the need for a tuning inductor at V_{d3}. A similar voltage coupling does occur for V_{d2} as well. However, in that case the coupling is through two levels of devices and the resulting series connection of intrinsic capacitances reduces the strength of the voltage coupled to V_{d2}. Since M_1 and M_2 can be viewed as a 2-stack PA with the tuning inductor serving as the choke inductance in large signal, a tuning inductor is not required at V_{d1} (as discussed before).

An important characteristic of switching PAs, which sets them apart from the linear classes, is the non-overlapping nature of switch voltage and switch current waveforms and the high harmonic content of these waveforms compared with linear PAs. In a device-based implementation, it is difficult to isolate the current flowing through the device capacitances from that flowing through the "switch." As a first order approximation, the currents through the external wiring parasitic capacitances C_{gs}, C_{gd}, C_{ds} and C_{d0} are scaled in proportion to the ratio of the intrinsic to external wiring parasitic capacitance, and their sum is subtracted from the total device current to arrive at the switch current in simulation. Fig. 6.39 shows the V_{DS} and the corresponding I_{switch} for the

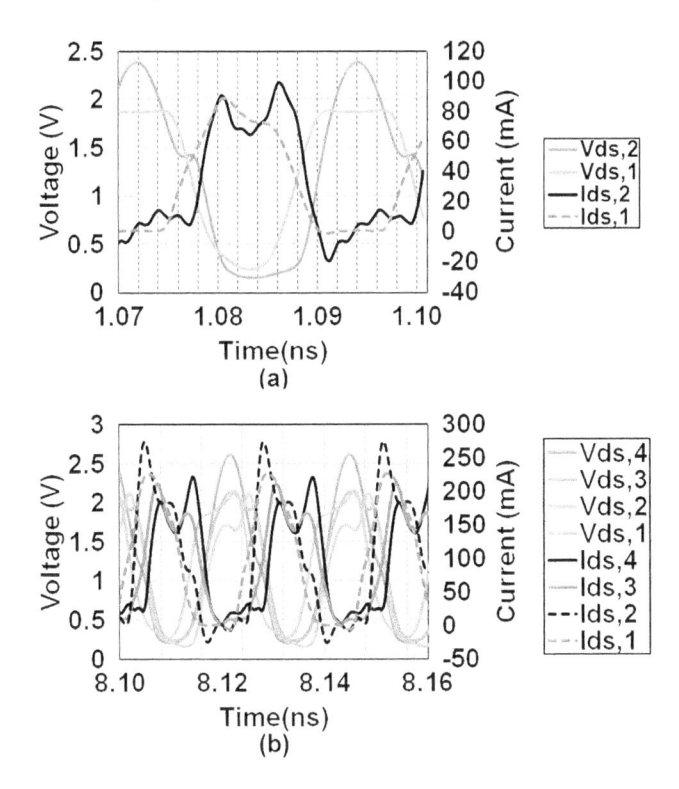

Fig. 6.39 Post-layout simulated drain-source voltages and corresponding switch currents for (a) 2-stack PA and (b) 4-stack PA in 45-nm SOI CMOS.

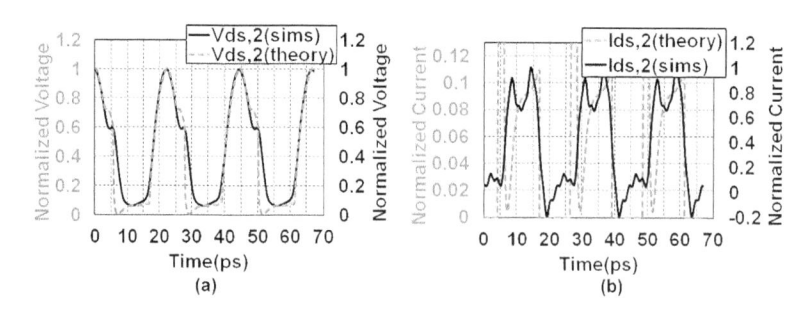

Fig. 6.40 Comparison of post-layout simulated waveforms for device M_2 of the 2-stack PA in 45-nm SOI CMOS with theory.

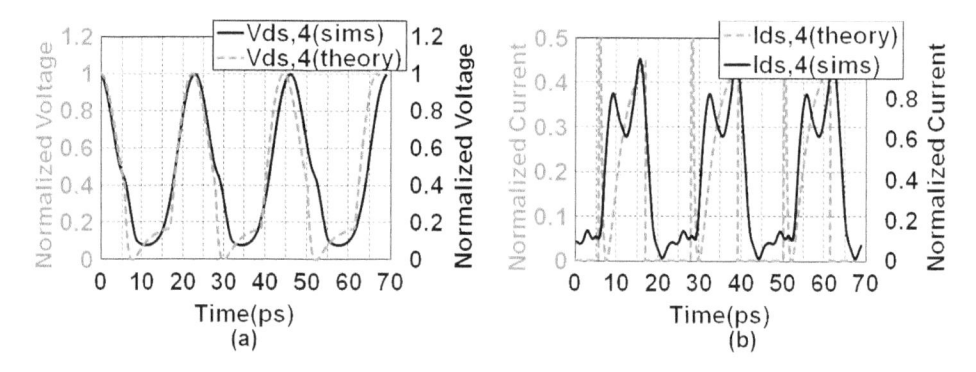

Fig. 6.41 Comparison of post-layout simulated waveforms for device M_4 of the 4-stack PA in 45-nm SOI CMOS with theory.

various devices in the 2-stack and 4-stack PAs implemented in 45nm SOI CMOS, from which the non-overlapping characteristic of voltage and current waveforms is clearly evident. Figs. 6.40 and 6.41 compare the switch-voltage and switch-current waveforms for devices M_2 and M_4 of the 2-stack and 4-stack PAs respectively with theory. Aside from the sharp current spikes in the theoretical waveforms at switch turn-on, there is excellent correspondence between theory and simulation. The current spikes arise from the assumption of hard-switching, which is not possible at mm-wave. However, the soft-switching in simulation does not compromise the shaping of voltages and currents and their harmonic content for the rest of the switching cycle. These results clearly indicate the feasibility of switching operation at mmWave frequencies.

The measured peak small-signal gain of the 2-stack PA is 13.5dB at 46GHz, with a -3dB bandwidth extending from 32GHz to 59GHz. The -1dB bandwidth extends from 42GHz to 52GHz, making it suitable for wideband applications. The measured small-signal peak gain of the 4-stack PA is 12.3dB at 48.5GHz, with a -3dB bandwidth extending from 37GHz to 56GHz. The measured -1dB bandwidth spans a wide frequency range from 43.5GHz to 52.5GHz. The large-signal characteristics of the 45-nm SOI PAs are shown in Fig. 6.42 and Fig. 6.43. Large-signal measurements yield a peak PAE of 34.6% for the 2-stack PA with a saturated output power of 17.6dBm at 47GHz.

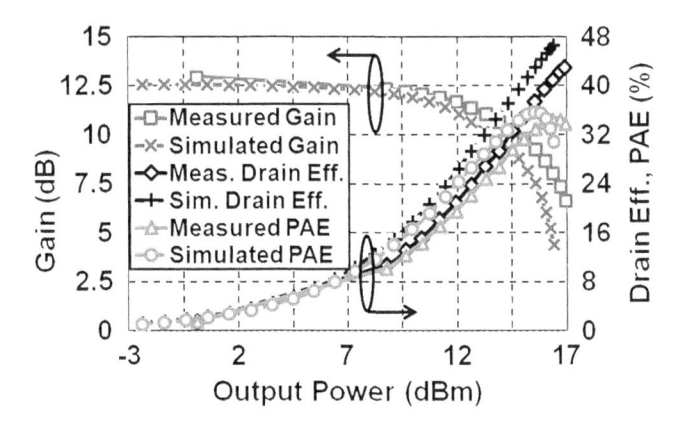

Fig. 6.42 Measured gain, drain efficiency and PAE as a function of output power for the 45-nm SOI 2-stack class-E-like PA at 47GHz ($V_{g1} = 0.4$V, $V_{g2} = 1.7$V, $V_{DD} = 2.4$V).

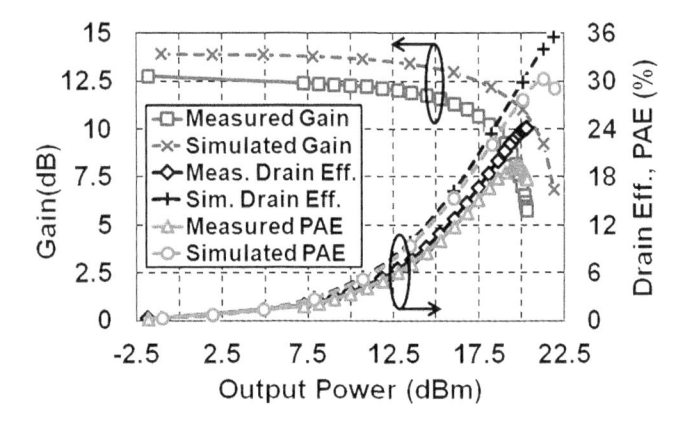

Fig. 6.43 Measured gain, drain efficiency and PAE as a function of output power for the 45-nm SOI 4-stack class-E-like PA at 47.5GHz ($V_{g1} = 0.4$V, $V_{g2} = 1.8$V, $V_{g3} = 2.8$V, $V_{g4} = 4$V, $V_{DD} = 4.8$V).

The 4-stack PA has measured saturated output power of 20.3dBm at 47.5GHz at a peak PAE of 19.4%.

Unlike the 2-stack PA, the measured performance metrics of the 4-stack PA (particularly efficiency) are somewhat lower than those predicted by simulations. This has been correlated to unmodeled losses in the active device layout. As was mentioned earlier, the optimal device size increases with an increase in the number of devices in the stack, rendering low-loss power device layout challenging.

6.12.3 Case study 3: 45nm SOI CMOS stacked class-E-like FET PAs at 45GHz with additional power combining

As was mentioned earlier, once stacking is exploited to the limits dictated by secondary breakdown mechanisms as well as by practical implementation details, it is interesting to consider the possibility of using additional passive power combining to achieve even

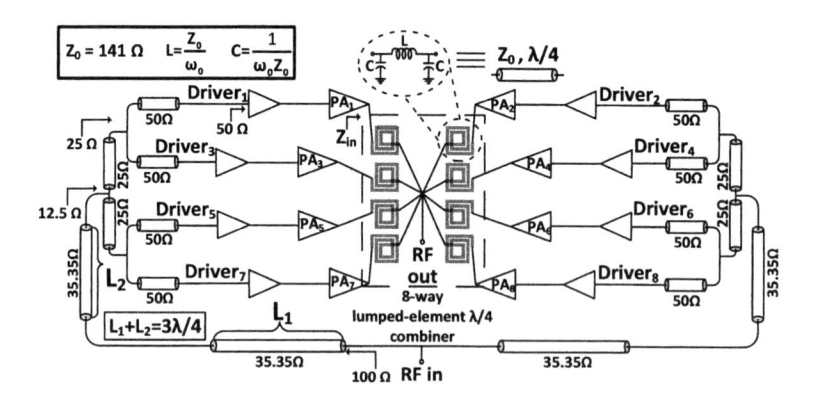

Fig. 6.44 Schematic of the 33-46 GHz 45-nm SOI CMOS watt-class PA.

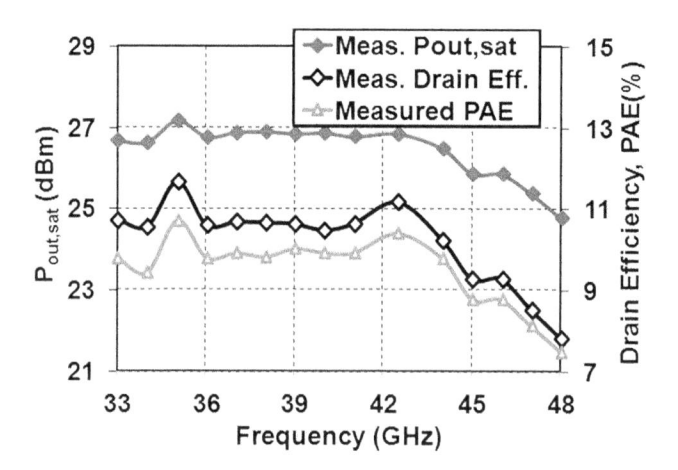

Fig. 6.45 Large-signal saturated output power, peak PAE and drain efficiency at peak PAE across frequency.

higher output power levels. Fig. 6.44 depicts the schematic of a 33-46 GHz 45 nm SOI CMOS watt-class PA array. Eight class-E-like PA unit-cells are combined using an 8-way lumped quarter-wave combiner – essentially an 8-way Wilkinson-like combiner sans the isolation resistors where spiral inductors are used to mimic high-Z_0 quarter-wave transmission lines. Each unit-cell PA used in the power-combined array is based on a 2-stage design, where the driver is the 2-stack class-E-like PA described earlier and the output stage is the 4-stack class-E-like PA.

A peak S_{21} of 19 dB was measured in small-signal operation at 50GHz. The efficiency and saturated output power of the PA array across frequency are shown in Fig. 6.45. The PA maintains 1 dB-flatness in saturated output power (26-27 dBm) from 33-46 GHz while the measured PAE varies between 8.8% to 10.7% in this range. The peak saturated output power of 27.2 dBm is the highest reported for any CMOS mmWave PA by a factor of approximately 3 [26].

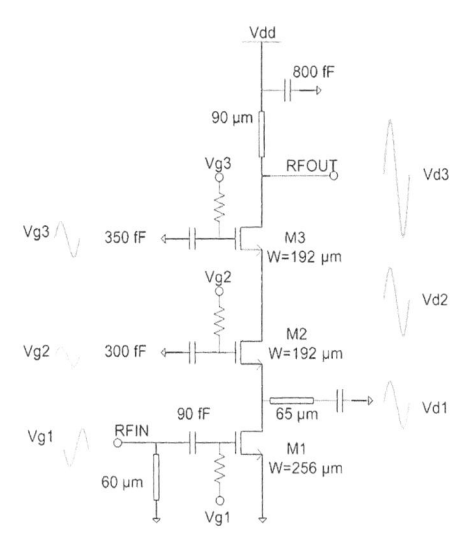

Fig. 6.46 Circuit schematic of 90 GHz stacked FET PA.

6.12.4 Case Study 4: stacked FET PAs at 90GHz

Stacked FET amplifiers have been implemented at frequencies above 90 GHz, with good results. The output power reaches record levels for CMOS ICs (at the time of writing), even though no passive power-combining structures are used. The circuit schematic for one such amplifier is shown in Figure 6.46 [27]. The technology chosen is 45-nm SOI CMOS, which reduces the source/drain to body parasitic capacitance. A single side contacted gate layout for the FET with the gate contact on the source side is used to reduce the gate to drain capacitance (C_{gd}). The simulated f_T and f_{max} of the extracted FET with the contacts to the top metal are near 240 GHz. The design uses smaller-size FETs for upper level transistors in the stack, since this arrangement leads to higher transconductance from the common source-like device while reducing the parasitics associated with the FETs stacked on top of it. The PA design uses a 256-μm-wide thin oxide NFET as the common source device, and 196-μm-wide stacked FETs as common gate devices. The real part of the output impedance of the stack is chosen to be close to 50Ω, so the output matching is achieved using a single shunt inductance to reduce output matching loss.

As is typical for lower frequency designs, gate capacitances are chosen so that the impedance seen at the drains has a real part that is approximately $R_L/3$ and $2R_L/3$ for the bottom two FETs in the stack, where R_L is the load resistance seen by the top device. Shunt elements at the drain of the bottom FET allow further optimization to the load impedance, allowing phase alignment of the drain-source voltages of the FETs. In this work, a shunt inductor at the drain node of the common-source device is used to improve the efficiency of the PA. The input match to the PA is achieved with a series capacitor and shunt inductor. Shunt inductances are implemented using coplanar waveguides on the top-most metal layer. The capacitors are implemented using metal finger structures.

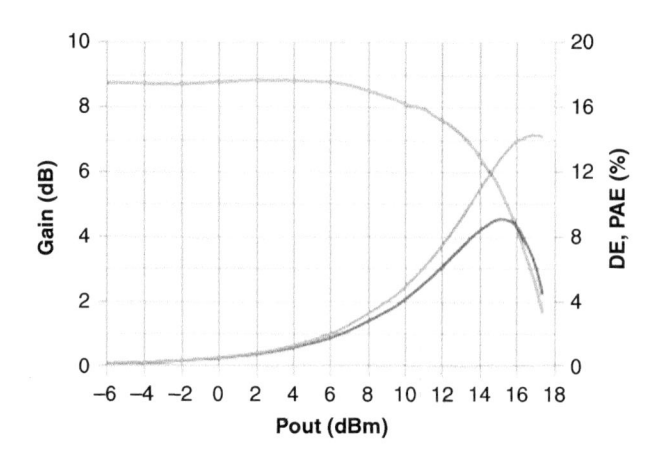

Fig. 6.47 Measured gain, drain efficiency and PAE of 90GHz stacked FET PA.

The chip size is 0.61 mm × 0.42 mm including the pads. The active circuit occupies only 0.06 mm² excluding the pads. The W-band PA was measured by on-wafer probing. The measured small-signal performances agree very well with the simulations (which included extensive electromagnetic simulations). The PA demonstrates maximum small signal gain of 8 dB at 91 GHz and is well matched at the input. The 3 dB gain-bandwidth of the PA is about 18 GHz (81–99 GHz). The PA is biased with a supply voltage of 4.2 V and gate voltages of $V_{g1} = 0.3$ V, $V_{g2} = 2$ V, and $V_{g3} = 3.25$ V.

For the large signal measurements, the signal source comprises a signal generator at 30GHz, an amplifier, a frequency tripler and a commercial W-band amplifier. The measured gain and PAE versus output power at 89 GHz are shown in Fig. 6.47. The PA has a linear gain of more than 8.5 dB and saturated output power of more than 17 dBm. The peak PAE is 9 % at output power of 15 dBm. The P1dB is higher than 11.5 dBm. At maximum power the PA consumes 91 mA of current from 4.2 V supply, resulting in a DC power consumption of 382 mW. The large signal frequency response shows a 1.5 dB bandwidth of 5 GHz. The output power peaks at 89 GHz (17.3 dBm) and efficiency peaks at 88 GHz (9.6%).

A modification of the circuit was carried out using an active drive approach to maximize the current that could be driven through the stack, using the circuit configuration shown in Fig. 6.31 [19]. The transistor widths within the final stack were chosen to be 256μm for the bottom device, and 128μm each for the upper transistors. The driver transistor widths were 128μm for the predriver and 98μm for the stack driver. The amplifier layout is shown in Fig. 6.48, which illustrates the compact nature of the overall structure (active area of order 200μm × 200μm). Figure 6.49 shows the measured output power and efficiency for the PA, illustrating that more than 19 dBm can be obtained, with a PAE of 14%. At the time of writing, this chip demonstrates the highest output power reported from a CMOS power amplifier unit cell at its frequency of operation.

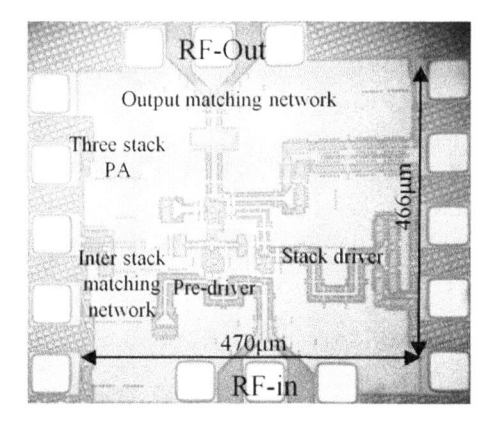

Fig. 6.48 Chip microphotograph of 94GHz stacked FET PA with active drive.

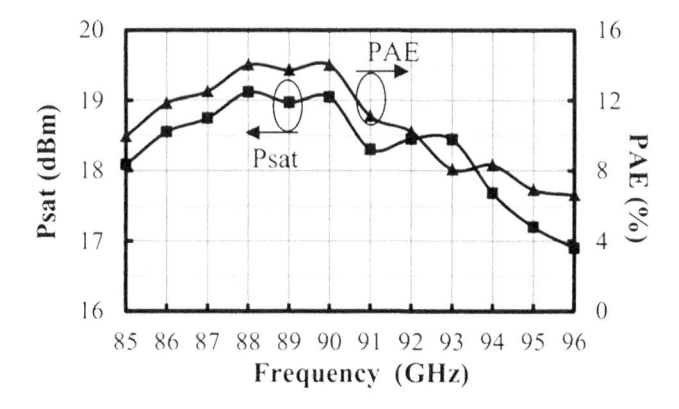

Fig. 6.49 Measured characteristics of 90 GHz stacked FET PA with active drive.

6.13 Summary and conclusions

The stacked FET technique is a useful approach to enable power amplifiers with high output power and efficiency to be implemented with scaled silicon technologies, despite the relatively low voltage handling capability of individual transistors. The usefulness of this technique has been demonstrated at low microwave frequencies over the past several years. The corresponding application at frequencies up to 100GHz has now been verified as a result of the work reported in this chapter. Design considerations for the mm-wave regime differ somewhat from those at the lower frequencies, such as the need to provide reactive matching at intermediate nodes in the stack, as discussed in detail here. Examples have been described that reach efficiencies (PAE) up to 34% at 45GHz, and up to 14% at 90 GHz. Output power up to 600mW has been demonstrated at 45GHz, and up to 19dBm at 90 GHz (in a single unit cell). This work demonstrates the considerable promise of using silicon transistor technologies for implementation of transmitter circuits at mm-wave frequencies. The stacked transistor technique can provide the required power amplification for many systems directly, or it can be used to drive external PAs based on III-V semiconductors. Using the high integration level of the silicon technologies, the PAs can be readily combined with other system blocks, such as modulators, pre-distortion components, frequency converters, built-in test circuits, and others, as described in other chapters of this book, leading to compact, highly functional and potentially low-cost mm-wave systems.

6.14 Acknowledgments

Much of the material contained in this chapter was developed during the course of the ELASTx program, sponsored by DARPA. The authors are grateful to Dr. Sanjay Raman, Dr. Dev Palmer and Dr. James Harvey for sponsorship and guidance throughout this program. They are also grateful to the graduate students and post-doctoral fellows who carried out much of the work, in particular, Joo-Hwa Kim, Sataporn Pornpromlikit, Hayg Dabag, Bassel Hanafi, Jefy Jayamon, Amir Agah, Fatih Golcuk, and Ritesh Bhat and to their faculty colleagues within the ELASTx program, many of whom made vital contributions to the work described here, in particular, Prof. James Buckwalter and Gabriel Rebeiz (UCSD); Lawrence Larson (Brown University); Sorin Voinigescu (University of Toronto); and Hossein Hashemi (USC).

References

[1] C.-H. Jan, M. Agostinelli, H. Deshpande, M. El-Tanani, W. Hafez, U. Jalan, L. Janbay, M. Kang, H. Lakdawala, J. Lin, Y.-L. Lu, S. Mudanai, J. Park, A. Rahman, J. Rizk, W.-K. Shin, K. Soumyanath, H. Tashiro, C. Tsai, P. Vandervoorn, J. Y. Yeh, and P. Bai, "RF CMOS technology scaling in High-k/metal gate era for RF SoC (system-on-chip) applications," in *Electron Devices Meeting (IEDM), 2010 IEEE International*, Dec 2010, pp. 27.2.1–27.2.4.
[2] "2011 International Technology Roadmap for Semiconductors," www.itrs.net.
[3] M. Apostolidou, M. van der Heijden, D. Leenaerts, J. Sonsky, A. Heringa, and I. Volokhine, "A 65 nm CMOS 30 dBm Class-E RF Power Amplifier With 60% PAE and 40% PAE at 16 dB Back-Off," *Solid-State Circuits, IEEE Journal of*, vol. 44, no. 5, pp. 1372–1379, May 2009.
[4] A. Ezzeddine and H. Huang, "The high voltage/high power FET (HiVP)," in *Radio Frequency Integrated Circuits (RFIC) Symposium, 2003 IEEE*, June 2003, pp. 215–218.
[5] H. Dabag, B. Hanafi, F. Golcuk, A. Agah, J. Buckwalter, and P. Asbeck, "Analysis and Design of Stacked-FET Millimeter-Wave Power Amplifiers," *Microwave Theory and Techniques, IEEE Transactions on*, vol. 61, no. 4, pp. 1543–1556, April 2013.
[6] N. Sokal and A. Sokal, "Class E – A new class of high-efficiency tuned single-ended switching power amplifiers," *Solid-State Circuits, IEEE Journal of*, vol. 10, no. 3, pp. 168–176, Jun 1975.
[7] O. Ogunnika and A. Valdes-Garcia, "A 60GHz Class-E Tuned Power Amplifier with PAE >25% in 32nm SOI CMOS," in *Radio Frequency Integrated Circuits Symposium (RFIC), 2012 IEEE*, June 2012, pp. 65–68.

[8] A. Mazzanti, L. Larcher, R. Brama, and F. Svelto, "Analysis of reliability and power efficiency in cascode class-E PAs," *Solid-State Circuits, IEEE Journal of*, vol. 41, no. 5, pp. 1222–1229, May 2006.

[9] O. Lee, J. Han, K. H. An, D. H. Lee, K.-S. Lee, S. Hong, and C.-H. Lee, "A Charging Acceleration Technique for Highly Efficient Cascode Class-E CMOS Power Amplifiers," *Solid-State Circuits, IEEE Journal of*, vol. 45, no. 10, pp. 2184–2197, Oct 2010.

[10] A. Chakrabarti, J. Sharma, and H. Krishnaswamy, "Dual-Output Stacked Class-EE Power Amplifiers in 45nm SOI CMOS for Q-Band Applications," in *Compound Semiconductor Integrated Circuit Symposium (CSICS), 2012 IEEE*, Oct 2012, pp. 1–4.

[11] M. Shifrin, Y. Ayasli, and P. Katzin, "A new power amplifier topology with series biasing and power combining of transistors," in *Microwave and Millimeter-Wave Monolithic Circuits Symposium, 1992. Digest of Papers, IEEE 1992*, June 1992, pp. 39–41.

[12] K. J. Dean, *Transistors, Theory and Circuitry*. McGraw-Hill, New York, 1964.

[13] M. Rodwell, S. Jaganathan, and S. T. Allen, "Series-connected microwave power amplifiers with voltage feedback and method of operation for the same," U.S. Patent 5 945 879.

[14] J. McRory, G. Rabjohn, and R. Johnston, "Transformer coupled stacked FET power amplifiers," *Solid-State Circuits, IEEE Journal of*, vol. 34, no. 2, pp. 157–161, Feb 1999.

[15] T. Sowlati and D. Leenaerts, "A 2.4-GHz 0.18- um CMOS self-biased cascode power amplifier," *Solid-State Circuits, IEEE Journal of*, vol. 38, no. 8, pp. 1318–1324, Aug 2003.

[16] S. Pornpromlikit, J. Jeong, C. Presti, A. Scuderi, and P. Asbeck, "A Watt-Level Stacked-FET Linear Power Amplifier in Silicon-on-Insulator CMOS," *Microwave Theory and Techniques, IEEE Transactions on*, vol. 58, no. 1, pp. 57–64, Jan 2010.

[17] S. Leuschner, J.-E. Mueller, and H. Klar, "A 1.8GHz wide-band stacked-cascode CMOS power amplifier for WCDMA applications in 65nm standard CMOS," in *Radio Frequency Integrated Circuits Symposium (RFIC), 2011 IEEE*, June 2011, pp. 1–4.

[18] A. Ezzeddine, H.-C. Huang, and J. Singer, "UHiFET - A new high-frequency High-Voltage device," in *Microwave Symposium Digest (MTT), 2011 IEEE MTT-S International*, June 2011, pp. 1–4.

[19] A. Agah, J. Jayamon, P. Asbeck, L. Larson, and J. Buckwalter, "Multi-Drive Stacked-FET Power Amplifiers at 90 GHz in 45 nm SOI CMOS," *Solid-State Circuits, IEEE Journal of*, vol. 49, no. 5, pp. 1148–1157, May 2014.

[20] A. Balteanu, I. Sarkas, E. Dacquay, A. Tomkins, and S. Voinigescu, "A 45-GHz, 2-bit Power DAC with 24.3 dBm Output Power, >14 V_{pp} Differential Swing, and 22% Peak PAE in 45-nm SOI CMOS," in *Radio Frequency Integrated Circuits Symposium (RFIC), 2012 IEEE*, June 2012, pp. 319–322.

[21] A. Chakrabarti and H. Krishnaswamy, "High power, high efficiency stacked mm-Wave Class-E-like power amplifiers in 45nm SOI CMOS," in *Custom Integrated Circuits Conference (CICC), 2012 IEEE*, Sept. 2012, pp. 1–4.

[22] ——, "An Improved Analysis and Design Methodology for RF Class-E Power Amplifiers with Finite DC-feed Inductance and Switch On-Resistance," in *Circuits and Systems (ISCAS), 2012 IEEE International Symposium on*, May 2012, pp. 1763–1766.

[23] S. Kee, "The Class E/F Family of Harmonic-Tuned Switching Power Amplifiers," Ph.D. dissertation, California Institute of Technology, Pasadena, California, 2001. [Online]. Available: http://resolver.caltech.edu/CaltechETD:etd-04262005-152703

[24] A. Chakrabarti and H. Krishnaswamy, "High-Power, High-Efficiency, Class-E-like, Stacked mmWave PAs in SOI and bulk CMOS: Theory and Implementation," *IEEE Transactions on Microwave Theory and Techniques*, (accepted) to appear.

[25] J. wei Lai and A. Valdes-Garcia, "A 1V 17.9dBm 60GHz Power Amplifier In Standard 65nm CMOS," in *Solid-State Circuits Conference Digest of Technical Papers (ISSCC), 2010 IEEE International*, Feb 2010, pp. 424–425.

[26] R. Bhat, A. Chakrabarti, and H. Krishnaswamy, "Large-scale power-combining and linearization in watt-class mmWave CMOS power amplifiers," in *Radio Frequency Integrated Circuits Symposium (RFIC), 2013 IEEE*, June 2013, pp. 283–286.

[27] J. Jayamon, A. Agah, B. Hanafi, H. Dabag, J. Buckwalter, and P. Asbeck, "A W-band stacked FET power amplifier with 17 dBm Psat in 45-nm SOI MOS," in *Radio and Wireless Symposium (RWS), 2013 IEEE*, Jan 2013, pp. 256–258.

7 On-chip power-combining techniques for mm-wave silicon power amplifiers

Tian-Wei Huang, Jeng-Han Tsai, and Jin-Fu Yeh

7.1 On-chip power-combining techniques

7.1.1 Introduction

The millimeter-wave (mm-wave) silicon power amplifier (PA) can provide a cost-effective solution for many potential commercial applications, like 60-GHz Wi-Fi, 77-GHz car radar, or the future B4G/5G 10-Gbps cellular backhaul links. The output power requirements are distance dependent from short-range Wi-Fi using 10 dBm to long-range backhaul links with more than 20-dBm output power. Design and implementation of mm-wave PAs in advanced CMOS suffer from various challenges. Most of all, the capability of a mm-wave PA to deliver watt-level output power is restricted by the low supply voltage and the limited gain performance for large-size transistors at the mm-wave frequency band. Moreover, the thick gate devices cannot be employed in mm-wave PAs due to poor gain performance. Figure 7.1 shows the decreasing trend of supply voltage of the advanced CMOS processes.

To achieve effective mm-wave power combining, the advanced CMOS processes with multiple metal layers can provide design freedom for the layout symmetry and increased metal thickness by stacking multiple metals for loss reduction. As a result, power-combining techniques play an important role in mm-wave PA design.

Over the past decade, although the industry and academia have been devoted to developing monolithic mm-wave power-combining techniques to increase the output power, the mm-wave power-combining techniques are still challenging. The objective of this chapter is to provide the necessary background, design principles of the mm-wave power-combining techniques, and various case studies for the readers.

7.1.2 Performance indicators

In this section, we will introduce two performance indicators and their usages in the mm-wave power-combining techniques.

Area per way (mm^2)
Generally, monolithic power-combining techniques in the literature are focused on how to generate maximum output power in a minimum chip area. As shown in Fig. 7.2, the chip area of multi-way combined PAs can be increased exponentially due to complicated

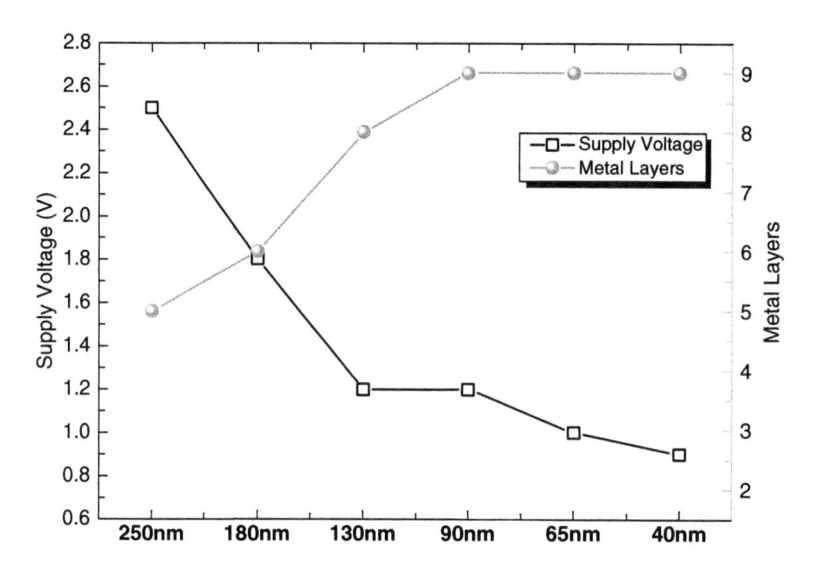

Fig. 7.1 Evolution of supply voltages and metal layers for different CMOS technologies.

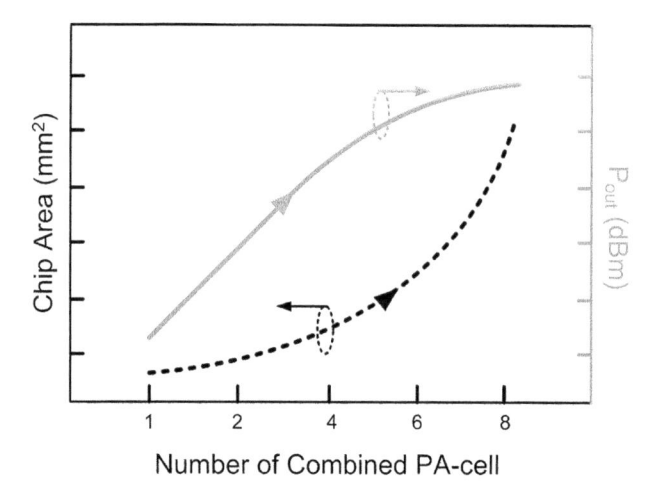

Fig. 7.2 The trend of output power and chip area for the increasing number of combined PA-cells.

multi-way power splitters and combiners. On the other hand, the output power has a saturation trend of multi-way combined PAs due to the increased loss in a multi-way power combiner.

As mentioned in the foregoing section, mm-wave PAs rely on the M-way power-combining techniques to deliver the high output power. Figure 7.3 plots the trends of the chip area of the CMOS mm-wave PAs in the literature [1–23]. As we observe from Fig. 7.3, the chip area grows significantly as the number of the combiner PA-cells increases. We can observe that there is conflict in maintaining the active area per way from 1-way to 4-way. As a result, the area efficiency can be defined as in Eqn (7.1) to be one of the performance indicators for mm-wave power-combining technique:

7.1 On-chip power-combining techniques   259

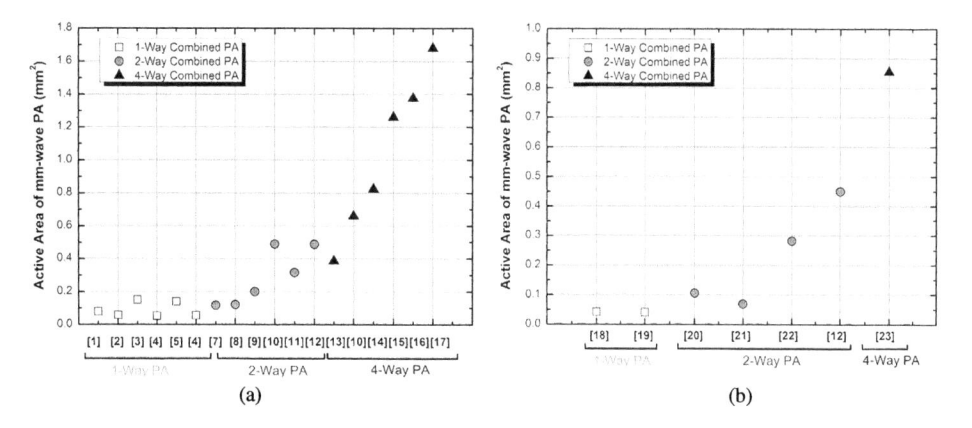

Fig. 7.3 Trend of active area of CMOS mm-wave power amplifiers from 1-way to 4-way power combining: (a) 30–70 GHz; (b) >70 GHz (1-way stands for a differential amplifier).

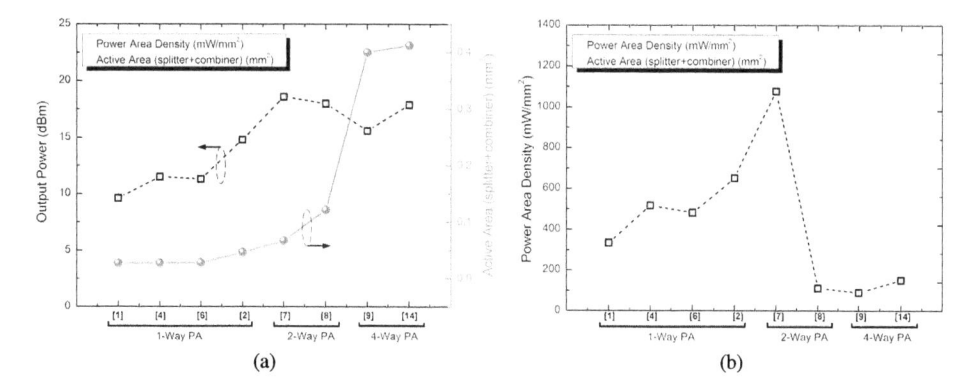

Fig. 7.4 Trend of (a) P_{out}/active area and (b) power-area densities of a 60-GHz CMOS power amplifier in a 65-nm CMOS from 1-way to 4-way power combining.

$$\text{Area per way } (mm^2) = \frac{\text{Active area of (power splitter + power combiner)}}{\text{Number of combined PA-cells}}. \quad (7.1)$$

Power area density (mW/mm²)

Figure 7.4 demonstrates the trend of output-power, active-area, and power-area densities of 60-GHz PAs in 65-nm CMOS [1–14]. As depicted in Fig. 7.3, one has to use more chip area to increase mm-wave output power. Therefore the power-area density (mW/mm²) can be the other important performance indicator to evaluate the mm-wave power-combining techniques. Moreover, the chip area has to include not only the power-combining network but also the power-splitting network. Then the power-area density here can be defined as Eqn (7.2):

$$\text{Power-area density } (mW/mm^2) = \frac{P_{out}}{\text{Active area of (power splitter + power combiner)}}. \quad (7.2)$$

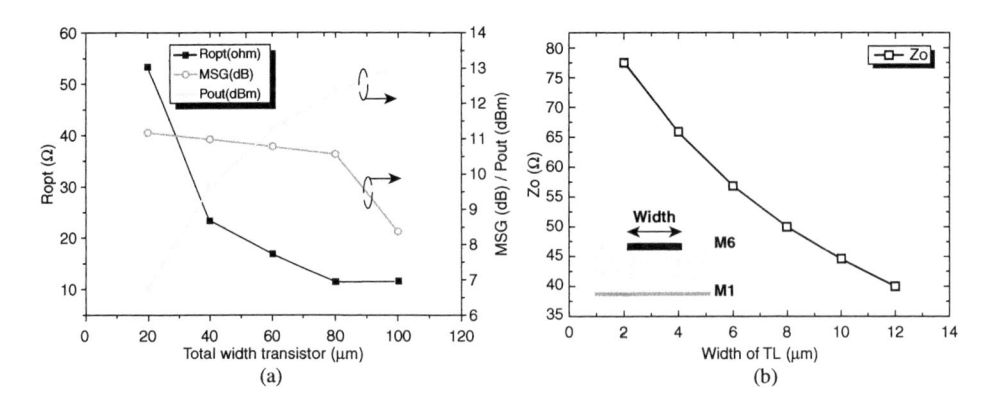

Fig. 7.5 (a) Optimal loads/MSG/P_{out} of power transistors with various size; (b) characteristic impedances of TFML (thin-film microstrip line) with various metal widths in 65-nm CMOS.

7.1.3 Challenges

The monolithic power combiners can be categorized as 50-Ω matched combiners (e.g. Wilkinson and 90° coupler) and non-50-Ω matched ones (e.g. transformers). The 50-Ω matched combiner can easily be implemented, but the matching networks for 50-Ω consume a larger chip area. On the other hand, the non-50-Ω matched power combiners have the advantage of achieving the power combining and impedance transformation simultaneously in a compact size. Nevertheless, the design of those combiners suffers from more challenges, such as impedance restriction and complicated power-splitting networks.

Impedance freedom
The monolithic transformer has been popularly employed as a power combiner since Dr. Akoi made a significant contribution for it in 2002 [24]. The output voltage and impedance can be transferred to the other side of the transformer and stacked to the output load (i.e. 50 Ω). In terms of the mm-wave PA, the size of the transistor and the maximum available gain cannot be satisfied simultaneously. In other words, a large transistor raises the output power but its maximum available gain decreases, as shown in Fig. 7.5. For a design of 100-GHz PA or beyond, the output power will be restricted because the transistor sizes are small to limit the number of combined PA-cells. So as to employ the transformer-based power-combining technique, the impedance freedom of transistor size is seriously restricted in PA design. Figure 7.5b shows the limited range of line impedance in a 65-nm CMOS process. Besides the line impedance constraint, the maximum current density of a CMOS process also imposes another constraint on the DC-feed-line design.

Layout symmetry
Layout-symmetry issues always play an important role in M-way combined PAs. It is worth restating it for mm-wave PAs. Figure 7.6 depicts conventional Wilkinson and transformer power-combined architectures. Two reasons account for the layout symmetry. The first one is that the mm-wave frequency leads to a much shorter

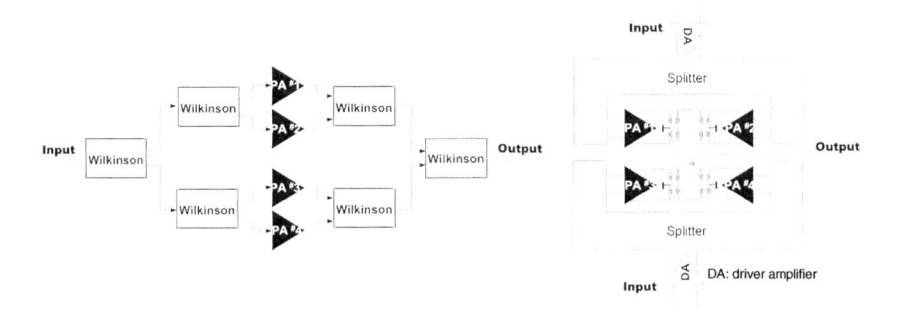

Fig. 7.6 Conventional Wilkinson and transformer power-combined architectures [25].

wavelength, so the performance of mm-wave PAs becomes more sensitive to an asymmetric layout. The second one is that the locations of PA-cells in the power combiner in two dimensions need complicated routing to maintain signals in phase. As mentioned in the previous paragraph, an mm-wave power combiner is required to assemble more PA-cells (i.e. 4-way) to deliver high output power due to the small transistor size of PA-cells being employed for mm-wave bands. However, the layout of a 4-way (or more) combined PA would encounter the problem of layout symmetry and suffers from costs such as the trace loss and the occupied chip area.

Miniaturization

As mentioned in Section 7.1.2, mm-wave amplifiers require full integration, including the input/output power-splitting/combining networks. Miniaturization of mm-wave PAs is demanded for cost reduction. Moreover, miniaturization of PA architecture eases the difficulty of mm-wave system integration such as phase-array systems. As demonstrated in Fig. 7.6, the reader may observe that the occupied active area of a power-splitting network becomes larger than the power combiner in transformer power-combining design. Moreover, the placement of PA-cells in a 2D (two-dimensional) transformer PA makes the miniaturization challenging. In the last section of this chapter, we will introduce a 3D (three-dimensional) PA architecture for miniaturizing the mm-wave PA architecture.

7.1.4 Future mm-wave high-power PA

In general, one can obtain more mm-wave output power by increasing the chip area. However, as mentioned in Section 7.1.2, on-chip power combining has potential limitations of output power and power-area efficiency for a large number of combined PA-cells (e.g. 8-way). In other words, one will seek another power-combining technique to produce the mm-wave high-power PA (e.g. watt-level mm-wave PA). For example, the loss of microstrip-line on 65-nm CMOS is about 1.0 dB/mm [26]. If a high-Q LTCC is used as an off-chip combiner, then the loss can be ten times smaller, like 1.2 dB/cm [27]. Compared with the on-chip power-combining techniques, the flip-chip high-Q combiner provides a low-loss high-current combining capability, as shown in Fig. 7.7.

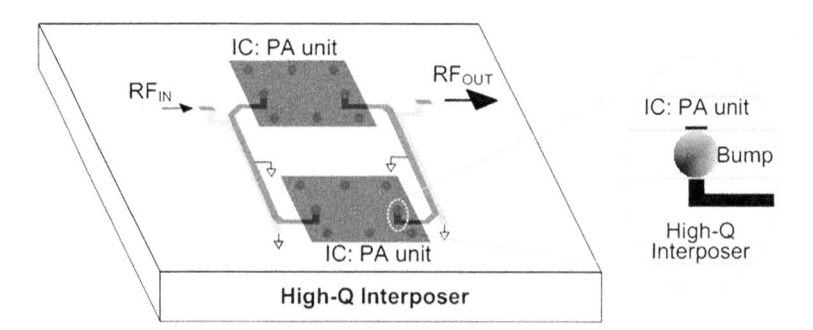

Fig. 7.7 Flip-chip PAs employing a high-quality package for power combining.

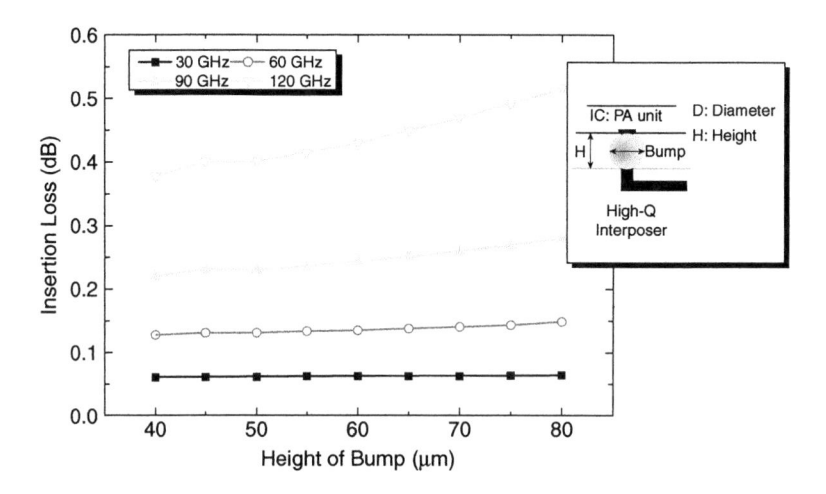

Fig. 7.8 Effects of the bump between IC and high-Q interposer with various heights.

The critical determining factor of choosing an on-chip or flip-chip power combiner is the insertion loss due to the flip-chip bump. Figure 7.8 shows the insertion loss for various flip-chip bump heights, with a fixed bump diameter of 40 μm. The flip-chip mm-wave PA can lower the chip area cost through an off-chip high-Q combiner, and provide high DC current for a high-power cell. Therefore, a flip-chip PA with low-loss bump process can be a future candidate for mm-wave high-power amplifiers.

7.2 Direct-shunt power combining

To satisfy the system specification of the output power, the power amplifiers require large device size, especially on advanced CMOS technology with low-voltage characteristic. Owing to the bandwidth, maximum available gain, power-added efficiency, and impedance-matching considerations, several smaller-power devices are adopted first and combined on-chip. Direct-shunt combining is the simplest combining technique for

7.2 Direct-shunt power combining

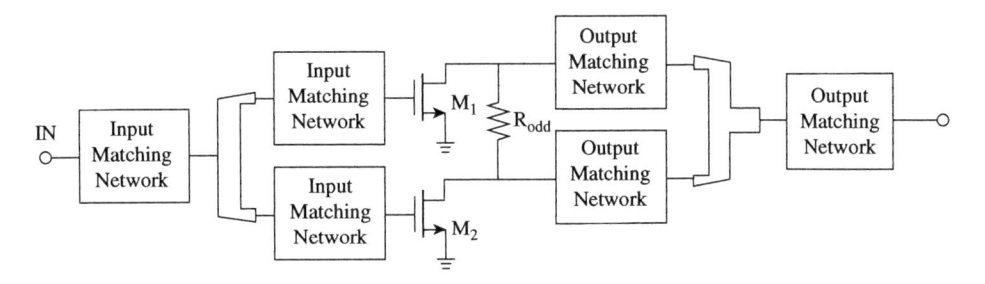

Fig. 7.9 Schematic of a 2-way direct-shunt combining power amplifier.

PA design [28–30]. The power devices are parallel, and the output current and output power of each power device are combined directly. Three important design issues will be discussed.

7.2.1 Impedance-matching networks design

Figure 7.9 is the schematic diagram of a 2-way direct-shunt combining PA. The transistors M_1 and M_2 are directly shunted by 3-port junction and output-matching networks. The output matching networks are used to transfer the transistor optimal output power impedance R_{opt} to the 50-Ω system load impedance. The input-matching network is used to transfer the transistor-input impedance to the system-source impedance. Several series and shunt transmission lines are adopted for input- and output-matching networks in mm-wave frequency. The transmission lines and direct-shunt combiner junction of an output- and input-matching network can be realized using thin-film microstrip (TFMS) line structure on CMOS technology. The TFMS consists of a bottom metal layer in a multi-layer CMOS process as ground plane and a top metal layer as the microstrip signal line with a SiO_2 layer as substrate. Owing to the thin oxide layer as the substrate, the TFMS lines can be meandered in a very compact manner without suffering the coupling effect.

The fundamental limitation of the direct-shunt combining technique is the decreasing output impedance of the number of shunted transistors. The impedance transfer ratio of R_{opt} to the system impedance increases. Therefore, it is more difficult to realize the output matching with an increasing number of combined devices.

7.2.2 Current density

The current density is an issue in the PA design. In particular, the advanced CMOS technology with continuously device gate length down-scaling has a low-breakdown-voltage characteristic. The low-voltage supply of the power device requires a specific amount of current to maintain output power. Therefore, the current density of a PA should be considered when designing a matching network with DC current feed. After calculating the current density of the power device, the width of the metal lines of the DC feed networks of the PA, should afford maximum current density.

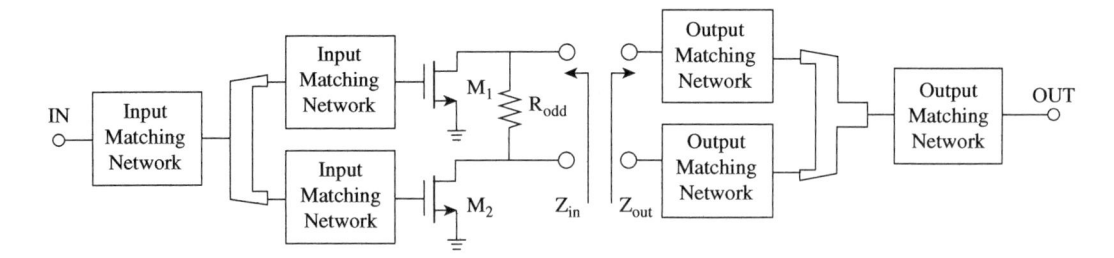

Fig. 7.10 Schematic of a 2-way direct-shunt combining power amplifier.

7.2.3 Odd-mode suppression

Ideally, the N direct-shunt transistors are identical and input/output-matching networks are perfectly symmetric. Nevertheless due to the process variation and potential layout asymmetry, the direct-shunt combining technique will potentially introduce additional modes of oscillation [31, 32]. Generally, when N devices are combined in parallel, one even mode and $N - 1$ odd modes may co-exist.

Using the K-factor to determine the amplifier stability is only suitable for even modes. However, the PA with N parallel combined devices could still oscillate in odd modes.

For analysis of the odd-mode oscillation, 2-way direct-shunt combined PA for the $n = 2$ case is discussed here. The analysis is applicable to n-way direct-shunt combining. Figure 7.10 is the schematic diagram of 2-way direct-shunt combined devices. Using 2-port symmetrical z-matrix analysis, the even- and odd-mode impedances seen at each port are

$$Z_e = Z_{11} + Z_{12} \quad \text{even mode} \quad (7.3)$$
$$Z_o = Z_{11} - Z_{12} \quad \text{odd mode.} \quad (7.4)$$

Each half of the amplifier has even and odd modes. The odd-mode impedance seen at the power device is denoted by $Z_{in,o}$, and the odd-mode impedance seen at the out matching network is denoted by $Z_{out,o}$. Then, the well-known oscillation conditions for VCO design are

$$\text{Re}\{Z_{in,o} + Z_{out,o}\} < 0 \quad (7.5)$$
$$\text{Im}\{Z_{in,o} + Z_{out,o}\} = 0. \quad (7.6)$$

Therefore, if the above oscillation conditions are satisfied at some frequency, the odd-mode oscillation will happen. To avoid odd-mode oscillation between the two parallel combined power devices in direct-shunt combining PA design, an odd-mode suppression resistor, R_{odd}, is required as shown in Fig. 7.10. In odd-mode operation, the virtual ground at the center of the resistor is achieved. If the half of R_{odd} is equal to the output impedance of a single transistor, M_1 or M_2, then the odd-mode oscillation can be terminated. After fine-tuning the value of the odd-mode oscillation resistor, the sum of the real parts of the odd-mode impedance at each port is greater than 0 so that the odd-mode oscillation conditions will not happen.

7.3 2D power combining

Since the single-transistor PA units in the various semiconductor technologies have limited output power with reasonable power gain and efficiency, to further achieve high-output power at mm-wave frequency, power-combining techniques using 2D power combiners are adopted (as will be mentioned in this section). The power-combining techniques combine the output power of several single-transistor PA units to achieve larger output power. Three important power-combination schemes, Wilkinson combiner, balanced amplifier, and transformer combiner, will be discussed.

7.3.1 Wilkinson combiner

The in-phase power-combination scheme utilizing Wilkinson power combiners is illustrated in Fig. 7.11. The Wilkinson power combiner has several advantages, such as easy implementation, good port-to-port isolation, and low insertion loss. It can also be a power divider for in-phase power-combining techniques. As shown in Fig. 7.11, a Wilkinson power divider splits the input power of the PA equally in-phase into two single-transistor PA units for amplification. Two amplified signals at the output ports of two single-transistor PA units are combined by a Wilkinson power combiner, and double output power can be achieved. The isolation resistor is added to terminate odd-mode signals for port-to-port isolation. Good isolation is required to avoid odd-mode oscillation between the two parallel combined PA units.

Figure 7.12 shows a conventional Wilkinson power combiner, which is composed of two quarter-wave-length transmission lines and an isolation resistor [33]. All impedances are normalized to the characteristic impedance Z_0 for simplicity. For an in-phase combination power combiner, the characteristic impedance of these quarter-wave-length transmission lines should be $\sqrt{2}Z_0$ and the value of the isolation resistor should be $2Z_0$ as given in Fig. 7.12. The detailed analysis of the Wilkinson power combiner has been described in [33]. The isolation resistor helps in isolating the combining ports from each other at in-band and prevents generation of odd-mode oscillations in the combined circuits. The quarter-wave-length transmission line is easy to implement on

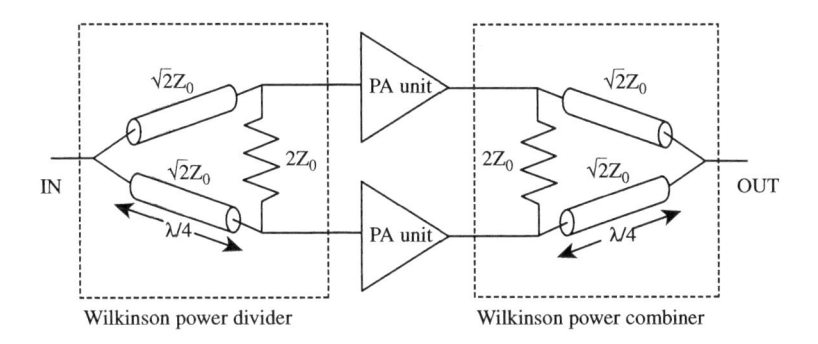

Fig. 7.11 In-phase power-combination scheme using Wilkinson power combiner/divider.

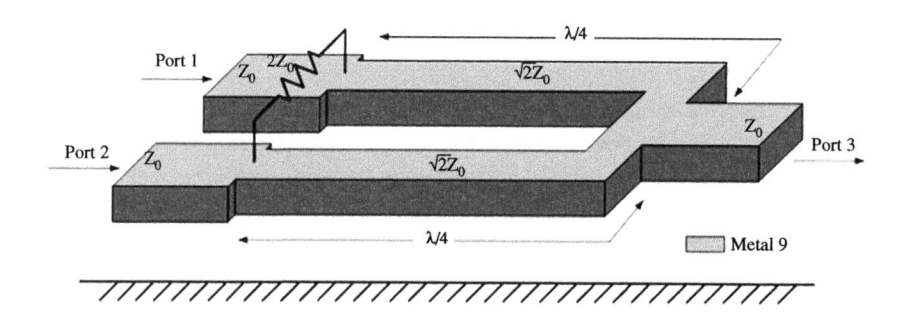

Fig. 7.12 Physical layout of a Wilkinson power combiner/divider [33].

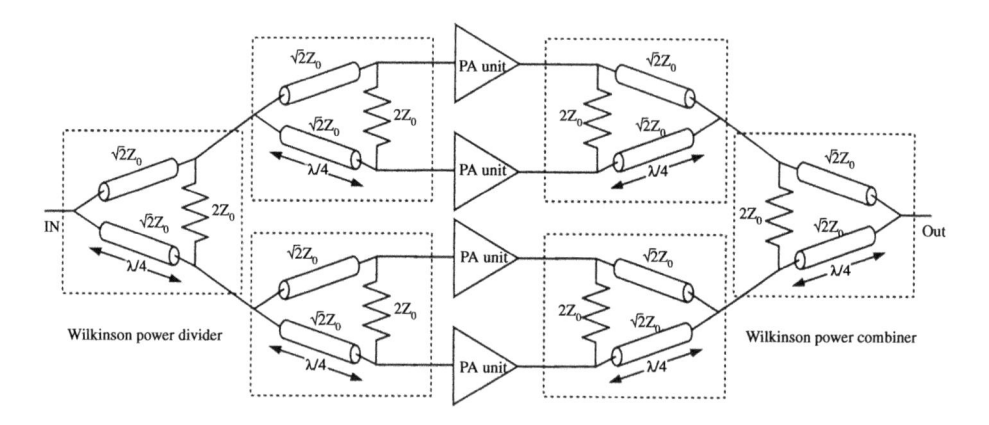

Fig. 7.13 Four single-transistor PA units in-phase combination PA using cascaded 2-way Wilkinson power combiners/divders [16].

CMOS process for mm-wave PA design due to the scale-down wavelength in mm-wave frequency bands. In addition, the thin-film microstrip (TFMS) line structure can be selected for the quarter-wave-length transmission lines on CMOS technology. The 2-way Wilkinson power combiner combines power from input port 1 and port 2 to output port 3. On the other hand, the Wilkinson power combiner is also usually adopted for an in-phase power splitter.

For higher-output power requirements, four or more single-transistor PA units can be in-phase combined by cascading six or more 2-way Wilkinson power combiners [16]. Figure 7.13 is an example of a four-single-transistor PA units in-phase combination PA using cascaded 2-way Wilkinson power combiners/dividers for quadruple output power. The input signal of the PA is split equally in-phase into four identical single-transistor PA units by cascaded 2-way Wilkinson power dividers. Finally, the four amplified signals are combined by a set of cascaded Wilkinson power combiners (three 2-way Wilkinson power combiners) and quadruple output power can be obtained. In addition, the insertion loss of the Wilkinson power combiner may decrease the final combined output power of the whole PA.

7.3.2 Balanced amplifier

The quadrature-phase power-combining scheme using 90° 3-dB hybrid couplers is also known as the balanced amplifier [22, 33–35]. Figure 7.14 illustrates the balanced amplifier configuration which adopts two identical PA units and two 90° 3-dB hybrid couplers. A 90° 3-dB hybrid coupler produces two equal-amplitude and quadrature-phase signals from the input power of the PA. The two signals are fed into two identical PA units for amplification. Two amplified signals at each of the output ports of two PA units are recombined by means of the same 90° 3-dB hybrid coupler. It should be noted that the connected ports of couplers in the input and output terminals, respectively, need to be exchanged for in-phase combining.

The balanced amplifier configuration with 90° couplers can absorb reflections and improve input/output matching, so the individual PA units can be optimized for output power and gain flatness. In addition, the balanced configuration provides high stability over a wide bandwidth, and it is easy to cascade with another driver stage because of isolation from couplers.

There are many ways to accomplish a quadrature 3-dB hybrid coupler. Several frames, as shown in Fig. 7.15, are well described in the microwave literature [33]. The quarter-wave-length branch-line coupler composed of four quarter-wave-length transmission lines with characteristic impedances of Z_0 and $Z_0/\sqrt{2}$ is illustrated in Fig. 7.15. The input signal at port 1 is equally divided into port 2 and port 3 with quadrature-phase. However, the disadvantages of large size and narrow bandwidth limit its practical implementation for MMIC integration. Figures 7.16a, b show the edge and broadside quadrature couplers, respectively. The design and analysis of a pair of coupled quarter-wave-length transmission lines have been established in the literature [33]. When the

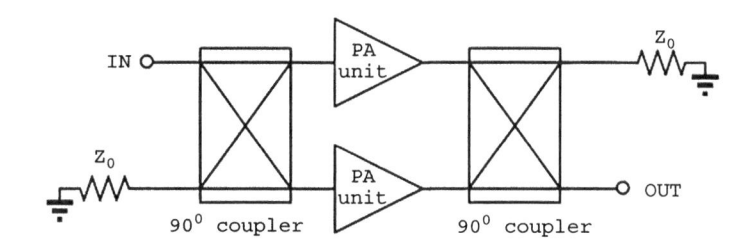

Fig. 7.14 Balanced amplifier configuration [34].

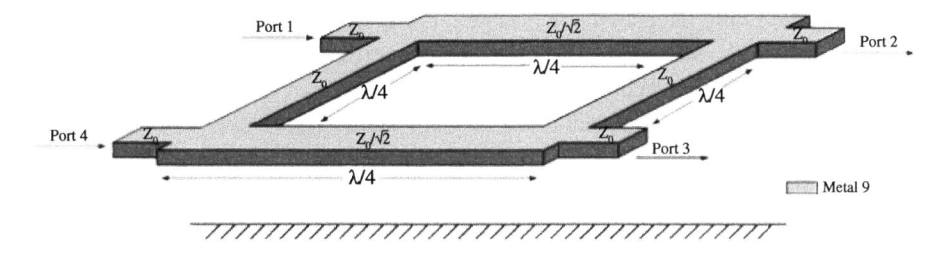

Fig. 7.15 Branch-line coupler [33].

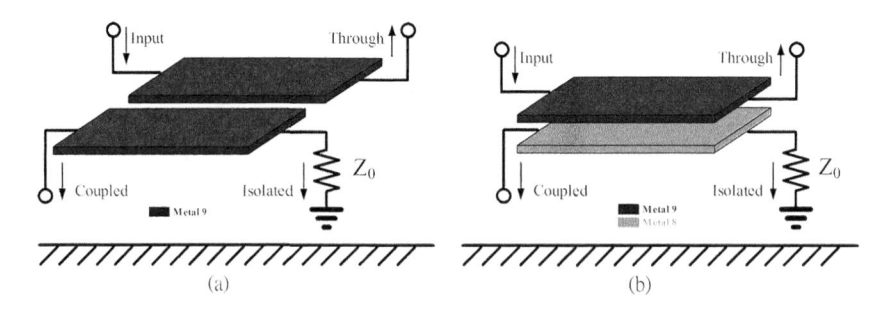

Fig. 7.16 (a) Edge coupler and (b) broadside coupler [33].

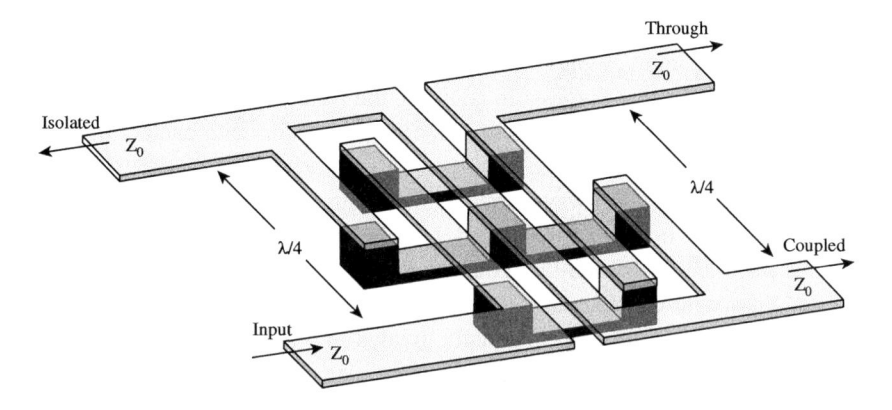

Fig. 7.17 A 3-dB Lange coupler using multi-layer interconnects [33].

input signal is fed into the input port of the quadrature coupler, there is an equal-amplitude and 90° phase shift between the direct and coupled port signals. Sometimes, the coupling between two coupled quarter-wave-length transmission lines is too small to realize a 3-dB coupling factor. To increase the coupling of the coupled lines, a Lange coupler using several coupled lines in parallel is presented. Figure 7.17 illustrates a Lange coupler with four coupled lines to provide tight coupling.

7.3.3 Transformer combiner

A transformer-based power combiner has recently become an attractive solution for mm-wave fully integrated silicon-based PA implementations. Although it is an old technique for push–pull amplifier applications from the early years, many benefits have let the transformer combiner attain more attention in recent years. One of the key benefits is that the transformers can perform single-ended-to-differential and differential-to-single-ended conversion for splitting and combining RF power while transforming the impedance simultaneously for input-impedance match and output-power match. Differential-signal techniques are advantageous for electronic circuits, including power amplifiers. The second benefit is that the inherent virtual ground of the differential signal operation can achieve RF grounding easily for practical implementation and thus

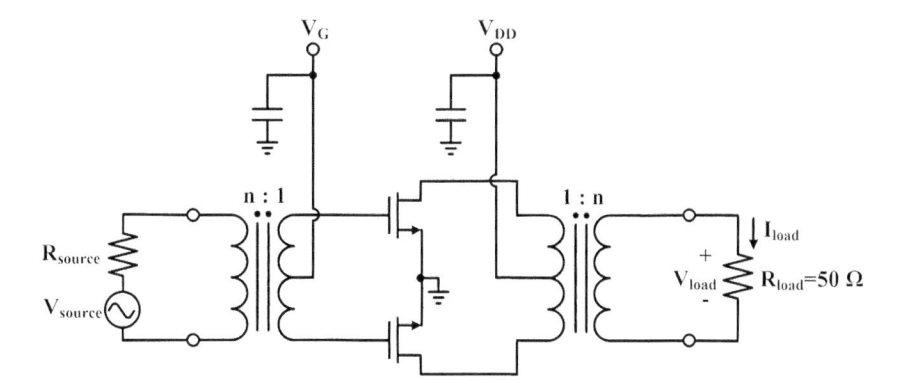

Fig. 7.18 Schematic of the transformer-combining power amplifier.

reduces the use of bypass capacitance. Particularly in mm-wave PA design, imperfect RF grounding results in parasitic inductance, which seriously degrades the gain performance of the power device. The third benefit is that the DC bias can be fed to the power device via the center tap of the transformer directly, which mitigates the DC bias network design issue of the PA. Therefore, using the transformer-combining technique avoids the use of RF chokes and DC blocks. Finally, the transformer can be wound into two coils with a very compact circuit size which is suitable for silicon-based integration.

Figure 7.18 illustrates the basic differential PA with transformer combining [4, 5]. It consists of two power devices with differential drive and two transformers for differential-to-single-ended conversion. Via the optimized transformers, the input and output matching of the PA can be achieved simultaneously. The drain and gate bias of the devices can be fed through the center tap of the transformer directly.

Figure 7.19a is the schematic diagram of a $1 : n$ coupled-inductor transformer for a PA system. The turn ratio between primary and secondary inductor coils of the transformer is n. Via magnetically coupling two inductors, the magnetic field caused by the power device current I_{in} through the primary inductor L_p produces a voltage in the secondary inductor L_s. Assuming an infinite coupling coefficient, the magnetically coupled voltage and current at the secondary inductor are

$$V_{load} = n \cdot 2 \cdot V_o \tag{7.7}$$

and

$$I_{load} = \frac{1}{n} \cdot I_o, \tag{7.8}$$

where V_o and I_o are the output voltage and current of the power devices, respectively. V_{load} and I_{load} are the magnetically coupled voltage and current, respectively. The optimal impedance for maximum output power seen by the power device is

$$R_{opt} = \frac{V_o}{I_o} = \frac{1}{2} \cdot \frac{1}{n^2} \cdot R_{load}. \tag{7.9}$$

The power from each power device is

$$P_{opt} = \frac{1}{2} \cdot \frac{V_o^2}{R_{opt}}. \tag{7.10}$$

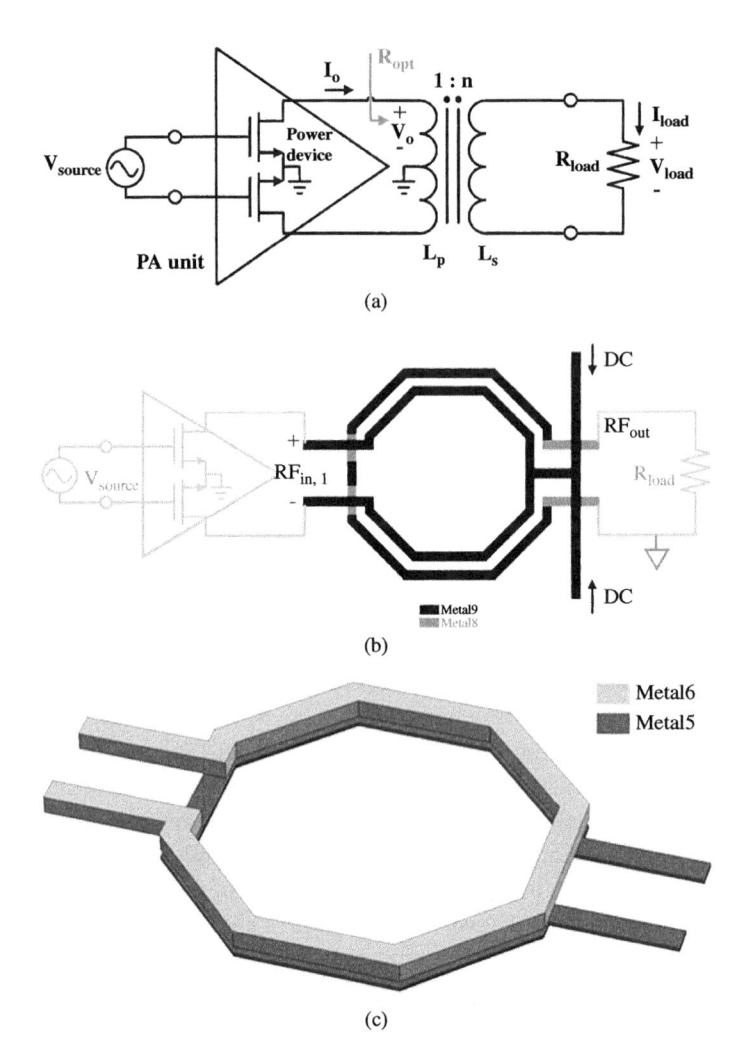

Fig. 7.19 (a) Schematic diagram of a 1 : n coupled-inductor transformer for PA system. (b) Physical layout of a 1:1 edge-coupled transformer. (c) Physical layout of a 1:1 broadside-coupled transformer.

The power delivered to the load is

$$P_{load} = \frac{1}{2} \cdot \frac{V_{load}^2}{R_{load}} = \frac{1}{2} \cdot \frac{n^2 \cdot 2^2 \cdot V_o^2}{2 \cdot n^2 \cdot R_{opt}} = 2 \cdot \frac{1}{2} \cdot \frac{V_o^2}{R_{opt}} = 2 \cdot P_{opt}. \quad (7.11)$$

As can be observed, the output powers of the two power devices are combined by the transformer and the output power can be doubled. Figure 7.19b shows a practical transformer layout with a turn ratio of 1:1. A nine-metal-layer process, like 90-nm or 65-nm CMOS, is used as the layout example. Typically, only one thick top metal layer is available in a mixed-signal RF CMOS process. Therefore, the primary and secondary inductor both use thick top metal for high-Q winding inductors with edge-coupled structure. In addition, the primary and secondary inductors realized the top two metal layers with broadside-coupled structure as an alternative for higher coupling coefficient at mm-wave frequency as shown in Fig. 7.19c.

To obtain higher output power, more power devices can be combined by using voltage-combining transformers (VCTs) or current-combining transformers (CCTs). Several hybrid-type transformer-combining techniques that overcome the design limitations of the traditional transformers are also being developed, such as current–voltage-combining transformers (CVCTs).

Voltage-combining transformer

The voltage-combining transformer technique, also called the series-combining transformer technique, accumulates the voltage of each power device using the stacked transformer to boost the output voltage and power of low-voltage CMOS devices [7, 10, 17, 36–39]. The schematic diagram of the voltage-combining transformer, which stacks several $1:n$ coupled-inductor transformers, is illustrated in Fig. 7.20. The voltage-combining transformer PA structure is composed of M PA units. Each PA unit consisting of two power devices has a differential impedance of $2R_{opt}$, where R_{opt} is the optimal impedance of each power device for output-power matching. Assuming an infinite coupling coefficient, the magnetically coupled voltage and current at the output load are

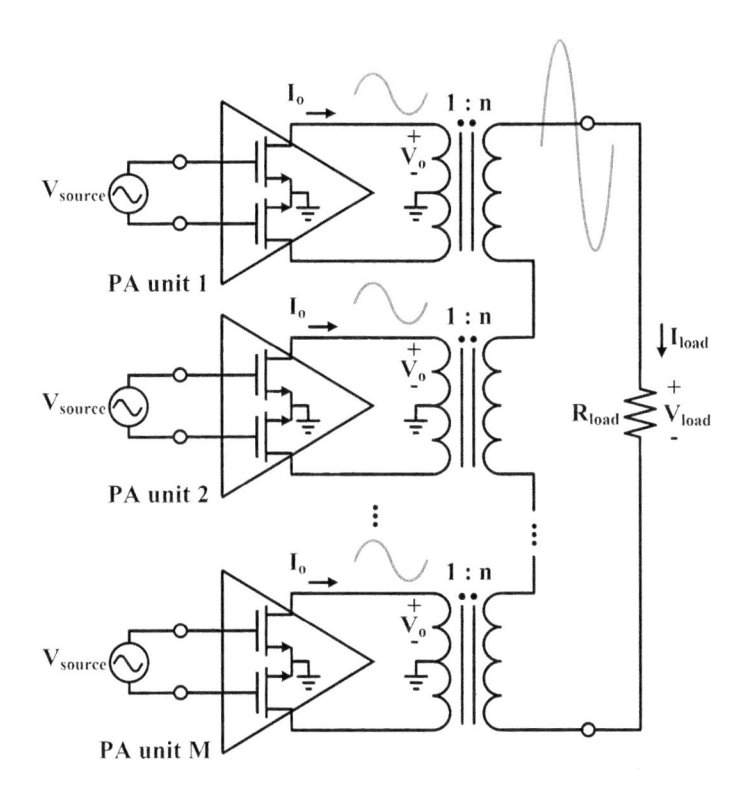

Fig. 7.20 Schematic diagram of the voltage-combining transformer.

$$V_{load} = M \cdot n \cdot 2 \cdot V_o \tag{7.12}$$

and

$$I_{load} = \frac{1}{n} \cdot I_o, \tag{7.13}$$

where V_o and I_o are the output voltage and current of the power devices, respectively. V_{load} and I_{load} are magnetically coupled voltage and current, respectively. M is the number of combined PA units. The optimal impedance for maximum output power seen by the power device is

$$R_{opt} = \frac{V_o}{I_o} = \frac{1}{2} \cdot \frac{1}{M} \cdot \frac{1}{n^2} \cdot R_{load}. \tag{7.14}$$

The power from each power device is

$$P_{opt} = \frac{1}{2} \cdot \frac{V_o^2}{R_{opt}}. \tag{7.15}$$

The total output power delivered to the load is

$$P_{load} = \frac{1}{2} \cdot \frac{V_{load}^2}{R_{load}} = \frac{1}{2} \cdot \frac{M^2 \cdot n^2 \cdot 2^2 \cdot V_o^2}{2 \cdot M \cdot n^2 \cdot R_{opt}} = \frac{1}{2} \cdot M \cdot 2 \cdot \frac{V_o^2}{R_{opt}} = M \cdot 2 \cdot P_{opt}. \tag{7.16}$$

As can be observed, the output powers of the M PA units ($2 \cdot M$ power devices) are combined by the voltage-combining transformer, and the output power can be increased M times. Figure 7.21 shows two types of the practical voltage-combining-transformer layout that is usually adopted in mm-wave frequency.

However, the voltage-combining-transformer technique reveals a practical design limitation on the maximum number of combined PA units, especially at mm-wave frequencies. From Eqn (7.14), for a given $R_{load} = 50\ \Omega$, the optimal load of each power device R_{opt} is $R_{load}/(2 \cdot M \cdot n^2)$. For example, four PA units (eight power devices) are series combined by four 1 : 1 transformers, and the calculated optimal load R_{opt} of each power device is 6.25 Ω. Such a low impedance of the optimal load implies that a large power-device size is required in modern low-voltage CMOS technology. However, because parasitics do not scale with the size of a multi-finger CMOS device, the large device size results in low maximum available gain at high frequency. If more PA units (more than four) have to combine using the voltage-combining-transformer technique,

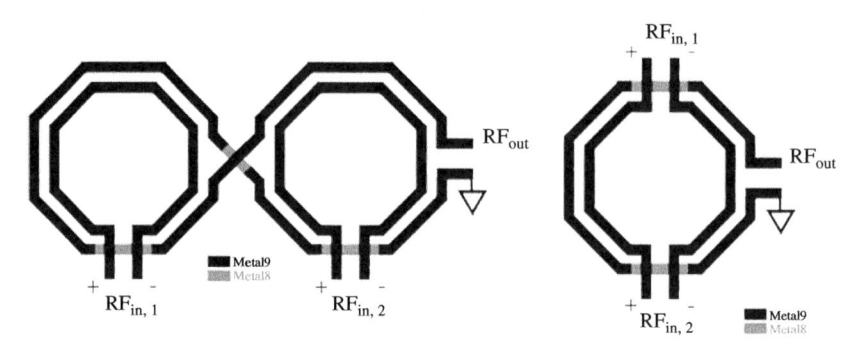

Fig. 7.21 Physical layout of a voltage-combining transformer with transformer turn ratio of 1 : 1.

the optimal load R_{opt} of each power device will be too small to be designed practically. Therefore, the voltage-combining transformer is not preferred for combining with a large number of PA units.

Current-combining transformer

The current-combining-transformer technique, also called parallel-combining transformers, is an alternative for efficient mm-wave power combining. The current-combining transformer collects the current of each power device using the parallel coupled transformer to increase the output current and boost the output power. The schematic diagram of the current-combining transformer, which employs in parallel several $1 : n$ coupled-inductor transformers is illustrated in Fig. 7.22a. The current-combining-transformer PA structure consists of N PA units. Assuming an infinite coupling coefficient, the magnetically coupled voltage and current at output load are

$$V_{load} = n \cdot 2 \cdot V_o \tag{7.17}$$

and

$$I_{load} = N \cdot \frac{1}{n} \cdot I_o, \tag{7.18}$$

where V_o and I_o are the output voltage and current of each power device in the PA unit, respectively. V_{load} and I_{load} are magnetically coupled voltage and current, respectively. N is the number of the combined power devices. The optimal impedance for maximum output power seen by the power device is

$$R_{opt} = \frac{1}{2} \cdot N \cdot \frac{1}{n^2} \cdot R_{load}. \tag{7.19}$$

The power from each power device is

$$P_{opt} = \frac{1}{2} \cdot \frac{V_o^2}{R_{opt}}. \tag{7.20}$$

The power delivered to the load is

$$P_{load} = \frac{1}{2} \cdot \frac{V_{load}^2}{R_{load}} = \frac{1}{2} \cdot \frac{n^2 \cdot 2^2 \cdot V_o^2}{2 \cdot (\frac{1}{N}) \cdot n^2 \cdot R_{opt}} = \frac{1}{2} \cdot N \cdot 2 \cdot \frac{V_o^2}{R_{load}} = N \cdot 2 \cdot P_{opt}. \tag{7.21}$$

As can be observed, the output powers of the N PA units ($2 \cdot N$ power devices) are combined by the current-combining transformer and the output power can be increased N times. Figure 7.22b shows a practical current-combining-transformer layout with a transformer turn ratio of $1 : 1$ for mm-wave PA applications.

Another more-common type for current-combining-transformer implementation in mm-wave PA design is illustrated in Fig. 7.23a [20]. The $1 : n$ transformers are directly combined at the output of the PA and the output currents of all of the transformers are collected to gain higher output power. The analysis of the impedance seen by each power device and the power delivered to the load is similar to that for the current-combining-transformer structure in Fig. 7.23. The practical layout of this current-combining-transformer type is plotted in Fig. 7.23(b).

Compare the two types of current-combining topologies: the current-combining topology in Fig. 7.22 has a smaller circuit size than that of the topology in Fig. 7.23.

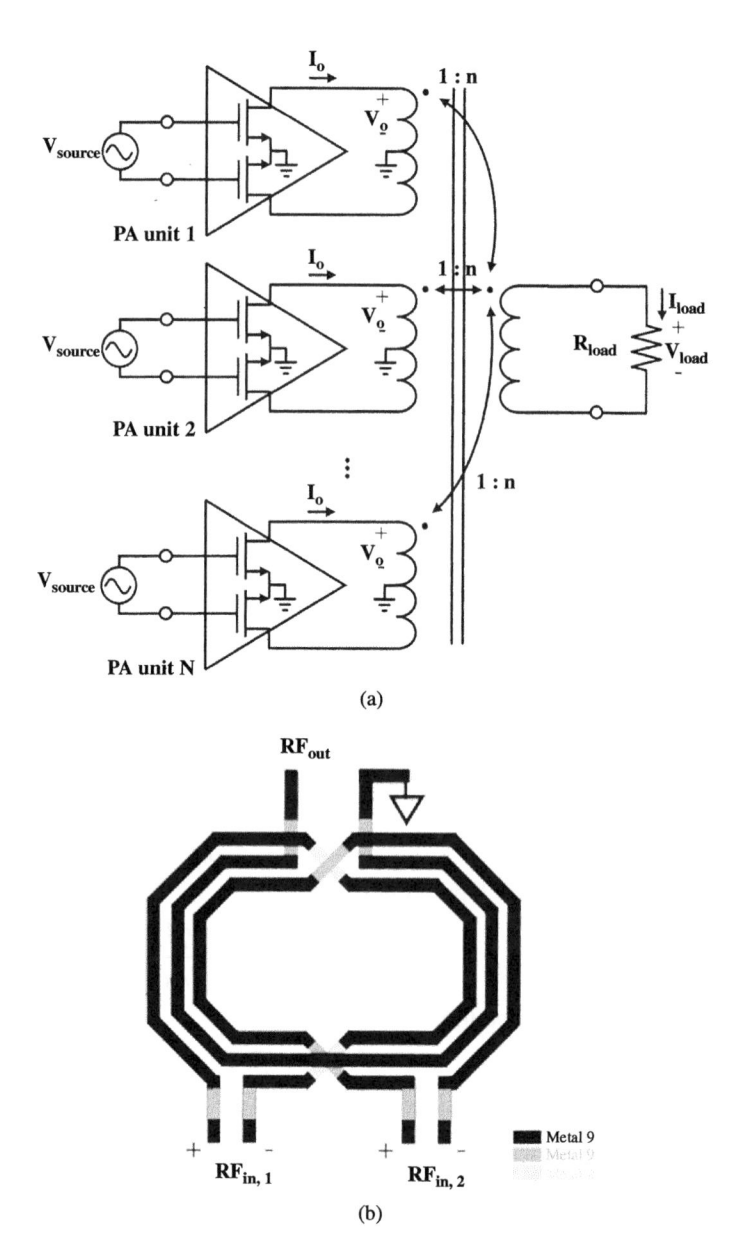

Fig. 7.22 (a) Schematic diagram of the current-combining transformer. (b) Physical layout of a current-combining transformer with transformer turn ratio of 1 : 1.

As shown in Fig. 7.22, the two secondary inductors are wound into one coil with a primary inductor to achieve a compact circuit size. However, the layout structure in Figure 7.22b requires a cross-over line at DC current path. Since most of the standard CMOS processes provide only one thick top metal for the interconnect, the metal current density of the DC current cross-over line is a serious design problem. In addition,

7.3 2D power combining

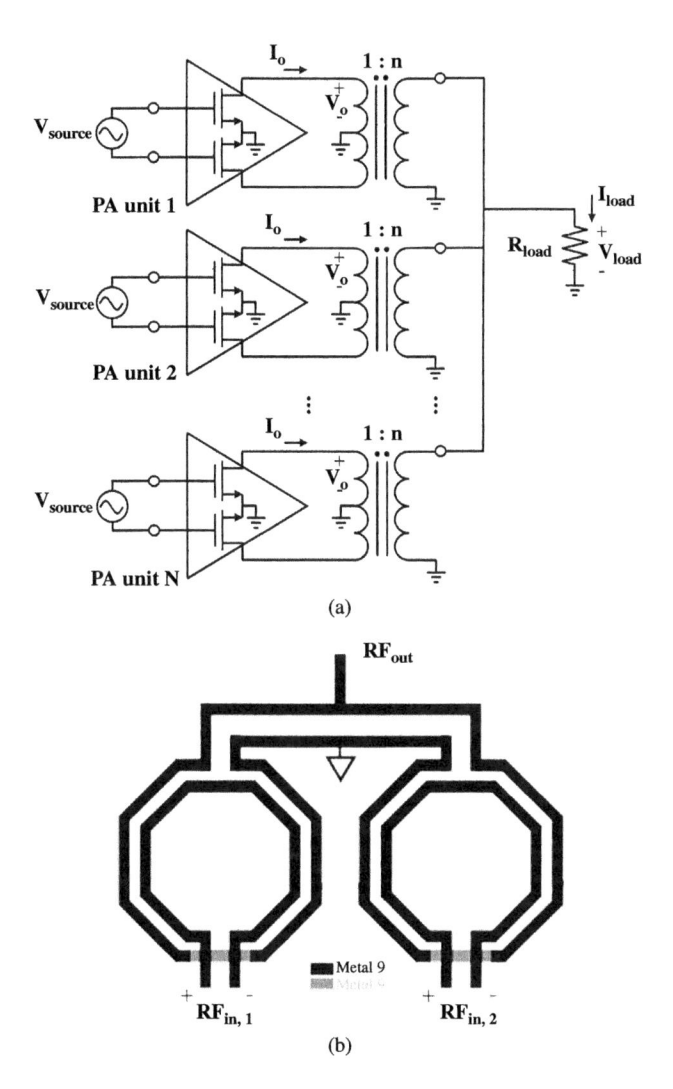

Fig. 7.23 (a) Schematic diagram of the current-combining transformer. (b) Physical layout of a current-combining transformer with transformer ratio of 1 : 1.

a non-symmetric layout occurs when the topology in Fig. 7.22b is extended to combine more than two PA units. As can be observed in Fig. 7.23, the direct-current-combining topology, which combines a number of the 1 : n transformers directly at the output of the PA, can be extended to combine more PA units with more symmetric structure than the topology in Fig. 7.22. Moreover, the layout structure of the direct-current-combining topology in Figure 7.23b avoids the cross-over line at the DC current path. Nevertheless, the required circuit size is larger than the topology in Fig. 7.22.

From Eqn (7.9), the relation between R_{opt} and R_{load} can be expressed as $R_{opt} = N/(2 \cdot n^2) \cdot R_{load}$. If the turn ratio of the transformer is 1 : 1, the current-combining-transformer technique requires $N/2$ times the output impedance R_{opt} of each power device for an output power match. Therefore a transformer with 1 : n turn ratio is used to raise the output impedance of each power device to $n^2 R_o$. Compared with the voltage-combining-transformer technique, the current one has more design freedom in impedance selection of power devices. A larger turn ratio of the transformer can be utilized for more PA unit combinations. However, a large turn ratio of the transformer will lead to other problems, such as low self-resonance frequency and low passive power transferring efficiency of the transformer, especially for mm-wave PA design. In addition, the appropriate turn ratio of the transformer for N-parallel combining transformers may not be an integer number, which is also a design issue for practical implementation.

Current–voltage-combining transformer

The voltage-combining transformer can boost the voltage at the output. However, the maximum number of combined PA units is limited by the optimal impedance R_{opt} of each power device. On the other hand, the current-combining transformer can collect the current at the output. However, the required large turn ratio of the transformer for the number of combined PA cells leads to low passive power transferring efficiency and low self-resonance frequency of the transformer. A hybrid-type current–voltage-combining transformer (PSCT) that takes advantage of the voltage-combining transformer and current-combining transformer has been developed [14, 36–42]. The schematic diagram of the two types of current–voltage-combining transformers using several 1 : n coupled-inductor transformers are illustrated in Figs. 7.24a,b, respectively. In Fig. 7.24a, there are N PA units that are parallel coupled to one secondary. In addition, M sets of N-parallel combining transformers are stacked in series. In Fig. 7.24b, there are M PA units that are series coupled to one secondary. In addition, N sets of M-series combining transformers are combined directly and each PA unit consists of two power devices with differential drive. Therefore, the total combining power device number in the current–voltage-combining transformer is $2 \cdot M \cdot N$. Assuming an infinite coupling coefficient, the magnetically coupled voltage and current at output load are

$$V_{load} = M \cdot n \cdot 2 \cdot V_o \tag{7.22}$$

and

$$I_{load} = N \cdot \frac{1}{n} \cdot I_o, \tag{7.23}$$

where V_o and I_o are the output voltage and current of the power devices, respectively. V_{load} and I_{load} are magnetically coupled voltage and current, respectively. The impedance seen by the power device is

$$R_{opt} = \frac{V_o}{I_o} = \frac{1}{2} \cdot \frac{N}{M} \cdot \frac{1}{n^2} \cdot R_{load}. \tag{7.24}$$

The power from each power device is

$$P_{opt} = \frac{1}{2} \cdot \frac{V_o^2}{R_{opt}}. \tag{7.25}$$

The power delivered to the load is

$$P_{load} = \frac{1}{2} \cdot \frac{V_{load}^2}{R_{load}} = \frac{1}{2} \cdot \frac{M^2 \cdot n^2 \cdot 2^2 \cdot V_o^2}{2 \cdot (\frac{M}{N}) \cdot n^2 \cdot R_{opt}} = \frac{1}{2} \cdot M \cdot N \cdot 2 \cdot \frac{V_o^2}{R_{opt}} = M \cdot N \cdot 2 \cdot P_{opt}. \tag{7.26}$$

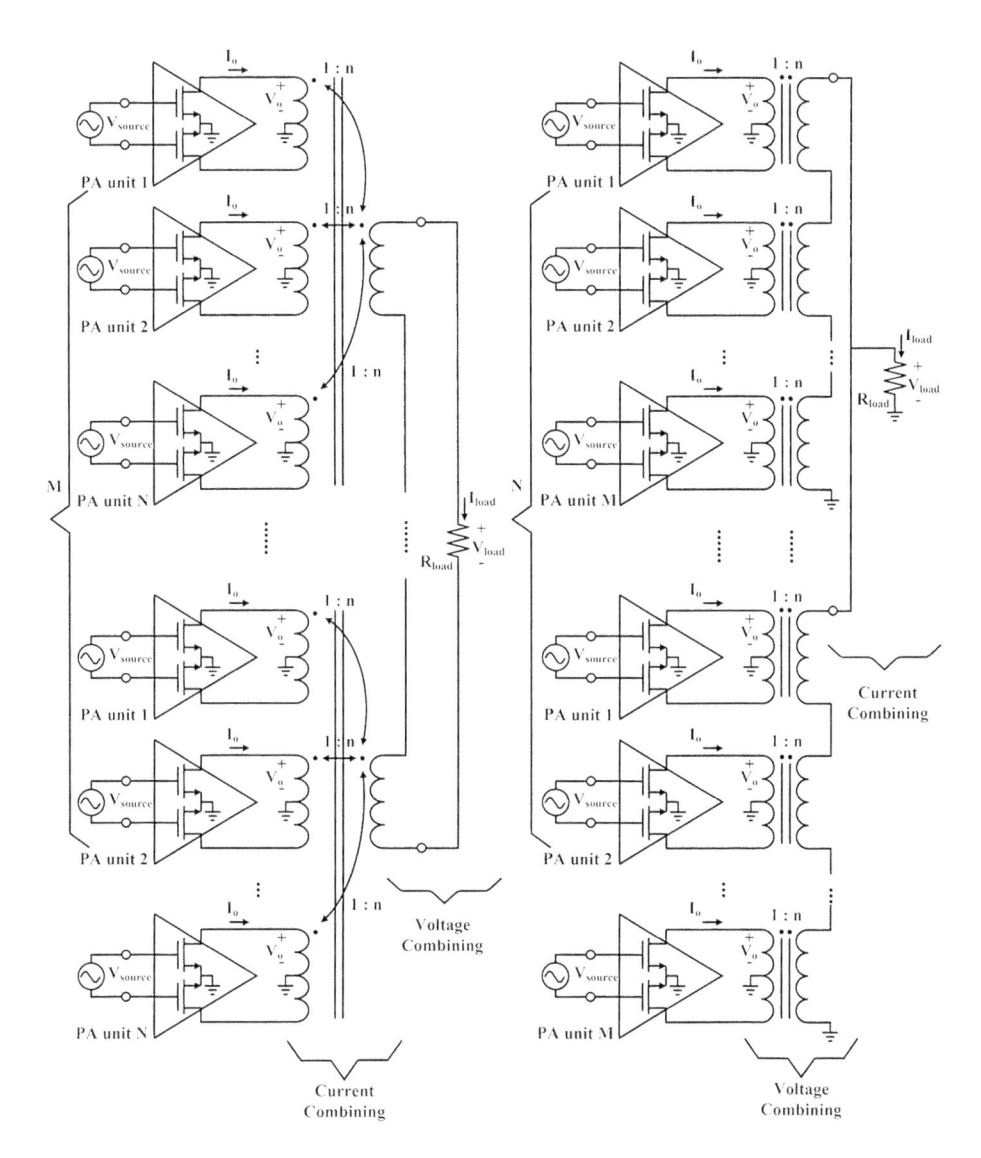

Fig. 7.24 Schematic diagram of two types of current–voltage-combining transformer.

As can be observed, the output powers of the $M \cdot N$ PA units ($2 \cdot M \cdot N$ power devices) are combined by the current–voltage-combining transformer and the output power can be increased $M \cdot N$ times. The current–voltage-combining-transformer technique has more design freedom in impedance selection of power devices than voltage or current techniques, which is favorable for mm-wave PA design. Figure 7.25 is a practical current–voltage-combining transformer layout for mm-wave PA applications.

Finally, a comparison of characteristics of several transformer power-combining techniques is summarized in Table 7.1. M is the number of PA units combined in a series combining transformer, N is the number of PA units combined in a parallel combining

Table 7.1 Comparison of characteristics of several transformer power-combining techniques.

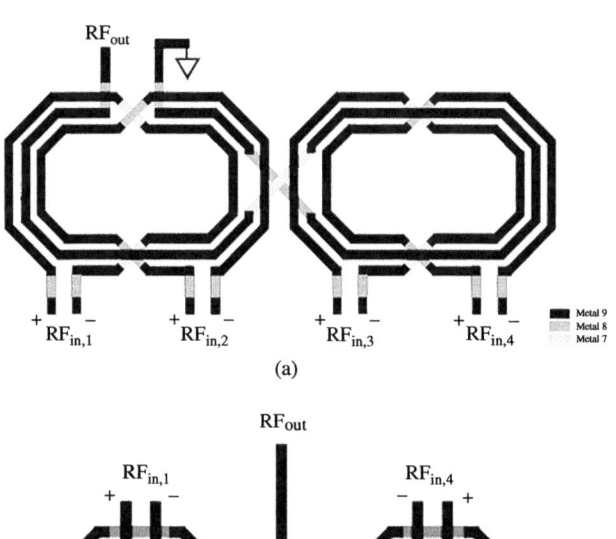

	V_{load}	I_{load}	R_{load}	P_{load}
Two-way transformer	$n \cdot 2 \cdot V_o$	$\frac{1}{n} \cdot I_o$	$2 \cdot n^2 \cdot R_{opt}$	$2 \cdot \frac{1}{2} \cdot \frac{V_o^2}{R_{opt}} (2 \cdot P_{opt})$
Voltage-combining transformer	$M \cdot n \cdot 2 \cdot V_o$	$\frac{1}{n} \cdot I_o$	$2 \cdot M \cdot n^2 \cdot R_{opt}$	$M \cdot 2 \cdot \frac{1}{2} \cdot \frac{V_o^2}{R_{opt}} (M \cdot 2 \cdot P_{opt})$
Current-combining transformer	$n \cdot 2 \cdot V_o$	$N \cdot \frac{1}{n} \cdot I_o$	$2 \cdot \frac{n^2}{N} \cdot R_{opt}$	$N \cdot 2 \cdot \frac{1}{2} \cdot \frac{V_o^2}{R_{opt}} (N \cdot 2 \cdot P_{opt})$
Current–voltage-combining transformer	$M \cdot n \cdot 2 \cdot V_o$	$N \cdot \frac{1}{n} \cdot I_o$	$2 \cdot \frac{M}{N} \cdot n^2 \cdot R_{opt}$	$M \cdot N \cdot 2 \cdot \frac{1}{2} \cdot \frac{V_o^2}{R_{opt}}$ $\cdot (M \cdot N \cdot 2 \cdot P_{opt})$

Fig. 7.25 Physical layout of the two types of current–voltage-combining transformers with transformer turn ratio of 1 : 1.

transformer, each PA unit consists of two power devices with differential drive, n is the turn ratio between the primary and secondary coils of the transformer, V_o and I_o are the output voltage and current of each power device, respectively, and R_{opt} is the optimal impedance of each power device for output power match.

Case study: A 60-GHz power amplifier using a series combining transformer

In this section, an example of a 60-GHz power amplifier using a series combining transformer is presented [10]. The 60-GHz power amplifier is designed and fabricated in a 90-nm CMOS process. Using optimized CMOS topology and a deep N-well (DNW), this topology provides an f_{max} of 142 GHz at maximum-transconductance bias. The process provides a single poly layer for the gates of the MOS and nine metal layers for inter-connection. An ultra-thick metal layer as top metal (metal 9) is also available, which is favorable for PA implementation. The transmission lines were implemented using thin-film microstrip (TFMS) lines. The TFMS consists of metal 1 (bottom layer) in the 1P9M CMOS process as the ground plane and metal 9 (top layer) as the microstrip signal line with a thick SiO_2 layer as substrate. A broadside-coupled transformer with greater coupling coefficient than an edge-coupled one at mm-wave frequency is selected for the 60-GHz PA design. The primary and secondary inductors are realized using metal 9 and metal 8, respectively, with a broadside-coupled structure. In addition, metal 7 is used for cross-over of the interconnection.

Figure 7.26 shows the schematic of the 60-GHz power amplifier. The size ratio of the first driver amplifier stage (DA1), the second driver amplifier stage (DA2), and the final PA stage is 1 : 2 : 4. Two 1 : 1 transformer baluns drive four power devices in a differential manner. Finally, the output powers of these four power devices are combined using a series combining transformer. The schematic of the driver amplifiers (DAs) and PA units is plotted in Fig. 7.27. The cascode configuration combining a common source and a common gate transistor is adopted in the PA design. The cascode device has better high-frequency performance than the common source one, which is favorable for mm-wave amplifier design.

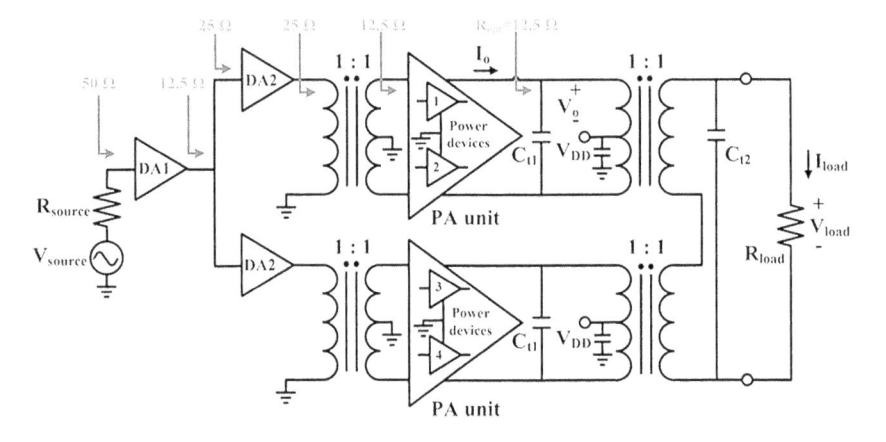

Fig. 7.26 Schematic diagram of the three-stage PA.

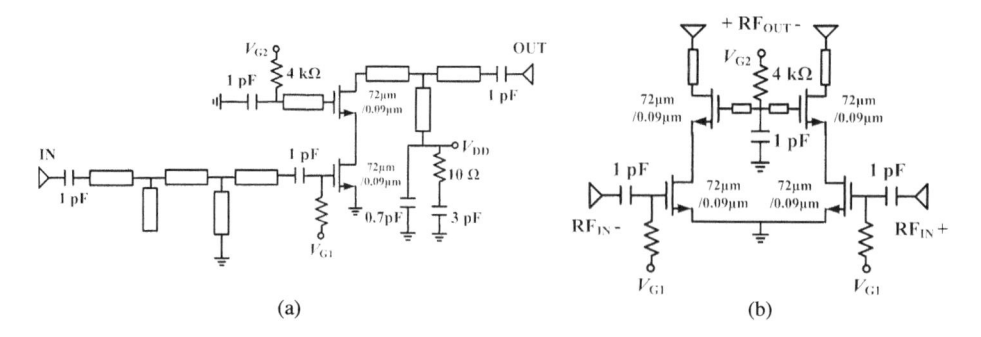

Fig. 7.27 Schematic of (a) the driver amplifier (DA) units and (b) the PA unit.

The impedance allocation plan of the PA is illustrated in Fig. 7.26. At the power stage, the voltage-combining transformer provides two-way combination capability. The turn ratio of the transformer is 1 : 1. Therefore, from Eqn (7.14) for a given R_{load} = 50 Ω, the optimal load R_{opt} of each power device is 12.5 Ω. The on-chip balun between the second stage and the power stage provides 2 : 1 impedance transformation, so input-termination impedance of the PA unit is transferred to 25 Ω. Then, the input/output-termination impedance of DA2 is selected as 25 Ω accordingly. At the first stage, the output is terminated at 12.5 Ω, and the input is terminated at 50 Ω load. The impedance allocation plan helps us to get rid of the use of an area-consuming and high-loss power splitter. The circuit schematic of DA1 and DA2 is plotted in Figure 7.27(a). The cascode configuration combining a common source and a common gate transistor is adopted in the design. The cascode device has better high-frequency performance than the common source one, which is favorable for mm-wave amplifier design. According to the impedance allocation plan, the input port of DA1 is terminated to 50 Ω and the output port is terminated to 12.5 Ω. In DA2, both input and output ports are terminated to 25 Ω. The circuit schematic of the PA unit is plotted in Figure 7.27(b). Each PA unit consists of two power devices with the differential optimal load impedance of $2R_{opt}$.

The 60-GHz power amplifier is implemented on a 90-nm CMOS process. The entire chip size including pads and chip-matching components is 0.78 mm × 0.82 mm as shown in Fig. 7.28. The quiescent DC current is 159 mA from 1.8-V supply voltage. The chip has on-wafer probing. The simulated and measured S-parameters are shown in Fig. 7.29. As can be observed, the PA demonstrates flat gain performance of 26 dB (±1 dB) from 57 to 69 GHz. The input and output return loss are under 7 dB from 54 to 65 GHz. The measured and simulated power performances are shown in Fig. 7.30. Under 1.8-V supply voltage, the PA delivers saturation power of 14.5 dBm to the output 50-Ω load at 60 GHz. The highest PAE at P_{sat} is 10.2% and the OP1dB is 10.5 dBm. To further investigate the power performance at high supply voltage of the mm-wave CMOS PA, the supply voltage is then raised to 3 V. After increasing the supply voltage of the power device, the measured output saturation power is 18 dBm, and OP1dB is 14.5 dBm with the maximum PAE of 12.2% at 60 GHz.

7.3 2D power combining

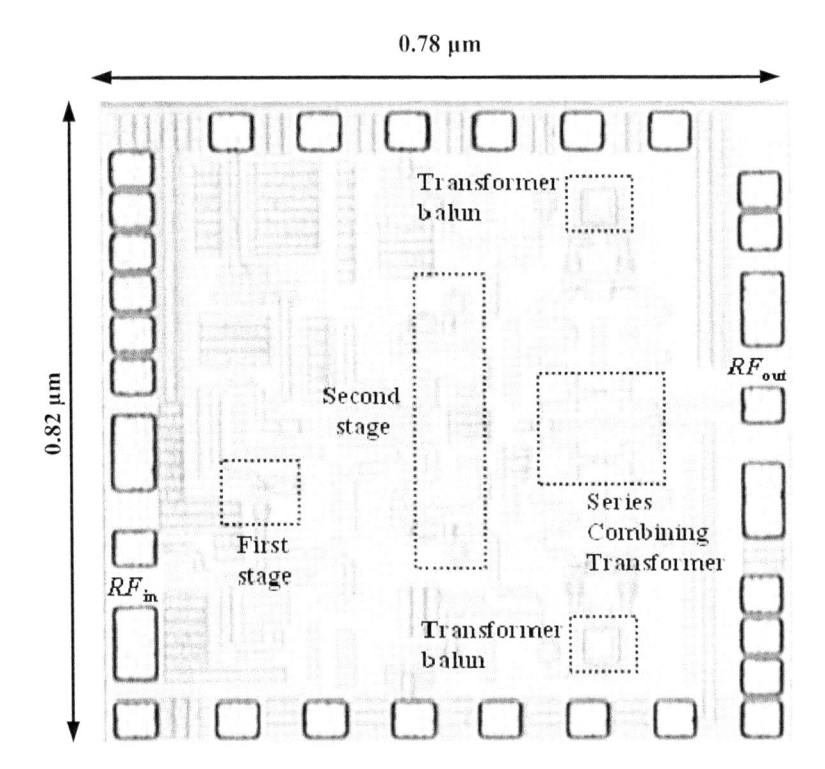

Fig. 7.28 Chip photograph of the 60-GHz PA.

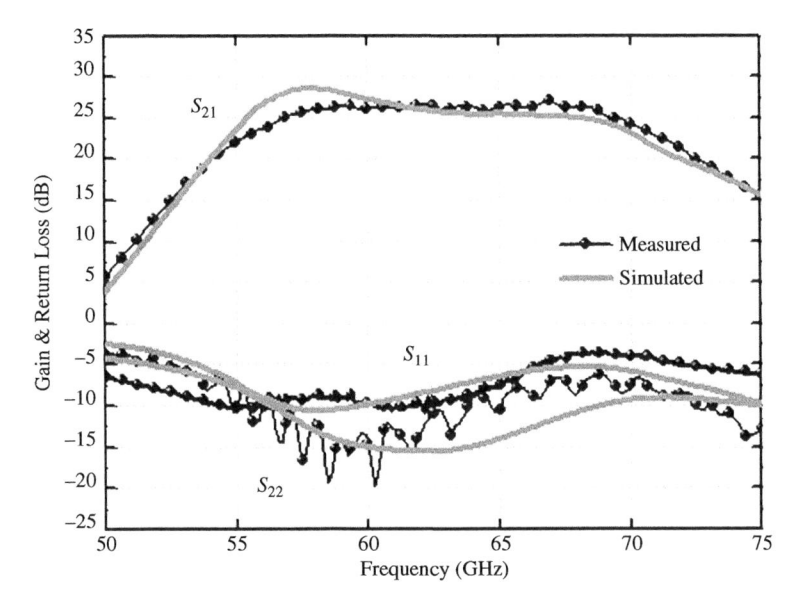

Fig. 7.29 Measured and simulated S-parameters of the PA under 1.8-V drain bias [10]. Reprinted with permission from the IEEE.

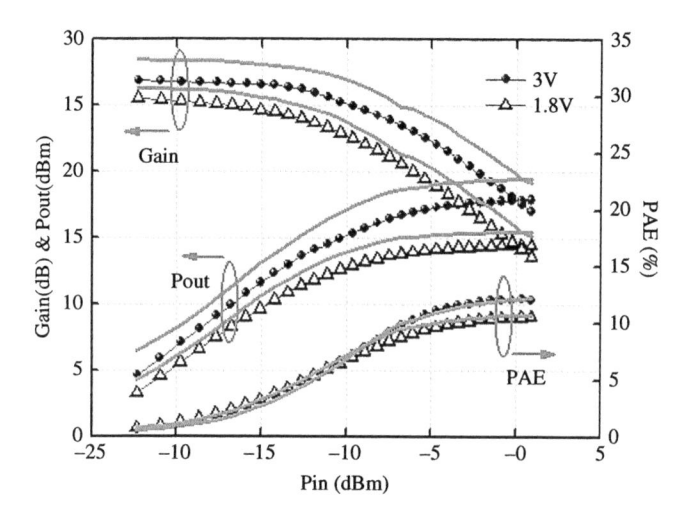

Fig. 7.30 Measured and simulated power performance of the PA at 60 GHz under 1.8-V and 3-V supply. The thick lines are simulated data and the symbols with the thin line are the measured data [10]. Reprinted with permission from the IEEE.

7.4 3D power-combining technique

The mm-wave transformer-based combined PAs have demonstrated great advantage in the active area of the combiner over the conventional binary combined PAs [1, 4–10, 12, 13, 15, 20, 23–25]. However, as mentioned in Section 7.1, one evaluation of the combined PA can be the area efficiency per 1-way. An attribute of the area efficiency of the combined PA architecture is that it becomes dramatically lower as the number of combined PA-cells increases. Moreover, this evaluation factor is more important for mm-wave PA design because mm-wave PAs are required to assemble many PA-cells to deliver the high output power. This section describes new on-chip PA architecture to provide high area efficiency. Furthermore it is capable of multi-mode operation for different usage scenarios.

7.4.1 From 2D to 3D PA architecture

In general, the most commonly used layout profiles for on-chip PAs can be categorized as 2D architecture. The power-combining techniques demand the layout symmetry for in-phase combination, especially for mm-wave PA; however, the 2D architecture suffers from the drawback of less design freedom for symmetric layout. Figure 7.31a illustrates the most commonly utilized architecture for the transformer-based M-way combined mm-wave PA. Although the mm-wave M-way transformer-based power combiners are compact, the active area of the power splitter is much larger than that of power combiners. We should make some remarks about the challenge: in order to mitigate the amplitude and phase difference in 2D power-splitting networks, the

7.4 3D power-combining technique

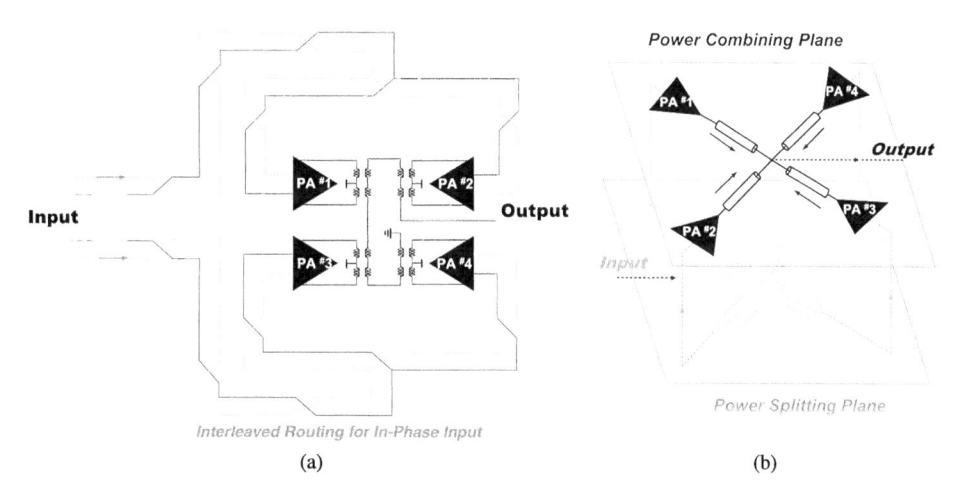

Fig. 7.31 Comparison of PA architectures. (a) Traditional 2D PA architecture. (b) Proposed 3D PA architecture [25].

differential feeding lines need to be interleaved with complicated routing to maintain symmetry.

In other words, the power-distribution networks are much larger and more complex than power-combining ones. In order to miniaturize an M-way combined PA without compromising symmetry and the compact size of the layout, an on-chip 3D PA architecture as shown in Fig. 7.31b is proposed [13]. As shown in Fig. 7.31b, it is obvious that the 3D PA architecture employs the additional dimension, such as a lower plane under the power-combining plane, then a built-in symmetry on the radial power-distribution layout. Furthermore, the concept of a folded-transformer is also proposed to enable 3D connection between different planes [13]. The folded-transformer with a radial combining network can mitigate the impedance restriction of mm-wave power-device selection.

In terms of a CMOS process with nine metal layers, the folded-transformers can be realized in M9 and M8 for high-Q consideration. Moreover, the dual-radial symmetric networks in Fig. 7.31b can be also implemented in M9 and M7. Then the input signal can be fed in to power the center distribution network of the 3D PA architecture along the axis of symmetry. Then the 3D PA architecture can complete the power distribution, input/output-impedance matching, signal amplification, and power combining in the same active area.

7.4.2 Dual-radial symmetric networks for power-splitting/combining

Following the above discussion about the PA architectures in implementation, it is not enough for monolithic mm-wave PAs to focus merely on power-combining techniques. One of the most challenging tasks is that mm-wave PAs are required to fully integrate

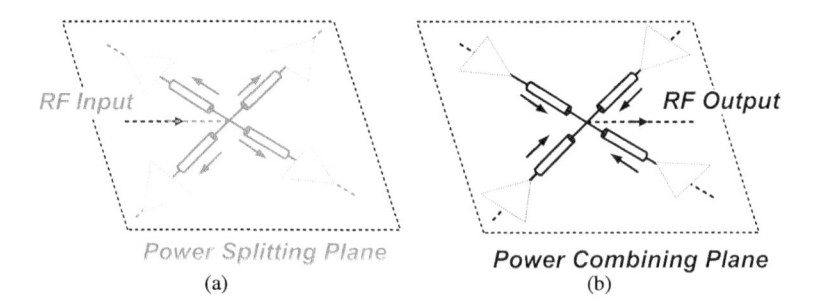

Fig. 7.32 Radial symmetric layouts for (a) power-splitting plane and (b) power-combining plane.

not only the power combiners but also the power splitters, to validate the devices in advanced CMOS technologies.

Figure 7.32 illustrates the commonly used on-chip layout topologies of transformer-based mm-wave power-combining PAs [1, 6–10, 12, 15, 20, 23, 25]. One of the major functions of power-distribution networks is accurate phase control, which is area consuming, especially as the number of PA-cells is greater (i.e. 4-way). As a result, the power-distribution network in 2D architecture possesses limited layout options to route the signal to achieve phase accuracy. In order to mitigate the amplitude and phase differences, the differential feeding lines need to be interleaved with complicated routing. Furthermore, the occupied active area of the power-splitting network of M-way transformer-based PA architecture is much larger than that of the power-combiner, and dominates the active area. In addition, two drawbacks would also come with complex power-splitting networks. The first one is that the complex splitting networks must introduce more loss to degrade the total PAE performance of a PA. The second one is making the occupied active area significantly grow as the number of combined PA-cells increases. In other words, it limits the number of combined PA-cells.

To enhance the area efficiency of a combined power amplifier, the dual-radial symmetric architecture is proposed [13]. Figure 7.33 depicts the conceptual diagram. This idea of a radial-prototype is adopted as the highly symmetric power-distribution/combining geometry by taking both amplitude and phase accuracy into account. Moreover, the 3D architecture is designed to achieve the maximum output power in the minimum active area. This idea can have three distinct features. The first one is that the architecture can be single-ended as input and output ports but not differential paths, and it can ease the control of amplitude and phase, especially for M-way combined PAs ($M > 2$). The second one is that the radial lines can be sorted in the design of transformers. The third one is that the radial lines can increase the impedance for the current-combining technique, which can provide more impedance freedom than the conventional transformer-based PAs and without the problems of self-resonance frequency of transformers.

7.4.3 Conceptualization of folded-transformers for 3D PA architecture

In the study of dual-radial symmetric architecture as shown in Fig. 7.33, we can perceive the need for the structure of the PA-cell to be integrated with the above dual-radial

7.4 3D power-combining technique

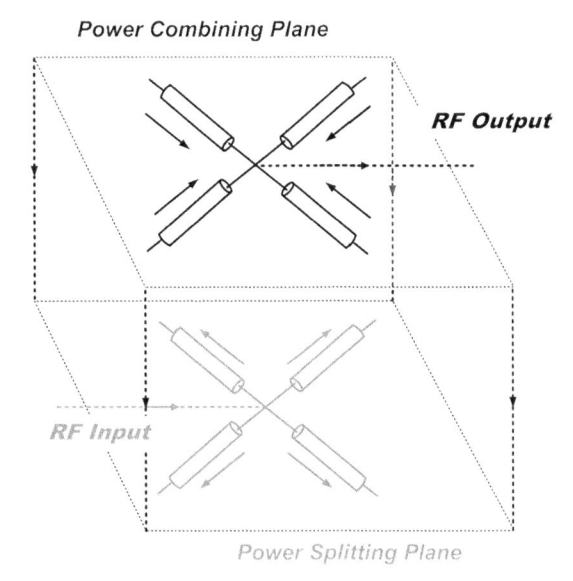

Fig. 7.33 Dual-radial symmetric layout incorporating power-splitting and power-combining planes.

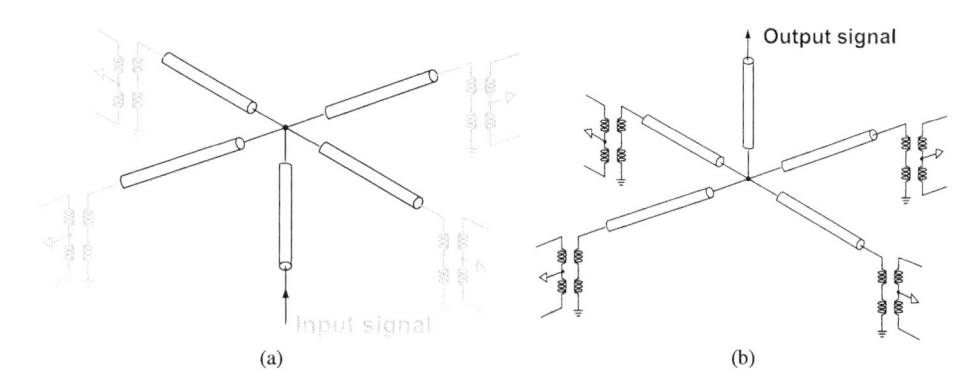

Fig. 7.34 Symmetric radial distribution networks incorporated with the transformers to serve as (a) input power-distribution network and (b) output power-combining network.

symmetric architecture. A conceptualization of a folded-transformer is proposed in [13]. In general, there are two candidates for the PA-cell, which are single-ended and differential PAs, respectively. Differential PAs are popular as PA-cells due to the advantages of full voltage swing and the attribute of virtual ground to ease the biasing circuit. Figure 7.34(a) illustrates the input power-distribution planes adopting the symmetric radial distribution network incorporated with transformers for feeding the input signals into each differential PA-cell. Furthermore, Fig. 7.34(b) depicts the output power-combining plane by means of the radial distribution network incorporating transformers for assembling the output signals from each differential PA-cell.

Nevertheless, in general, input signals are fed into the the mm-wave transformer-based PA-cells from one side and deliver the amplified output power to the other side via

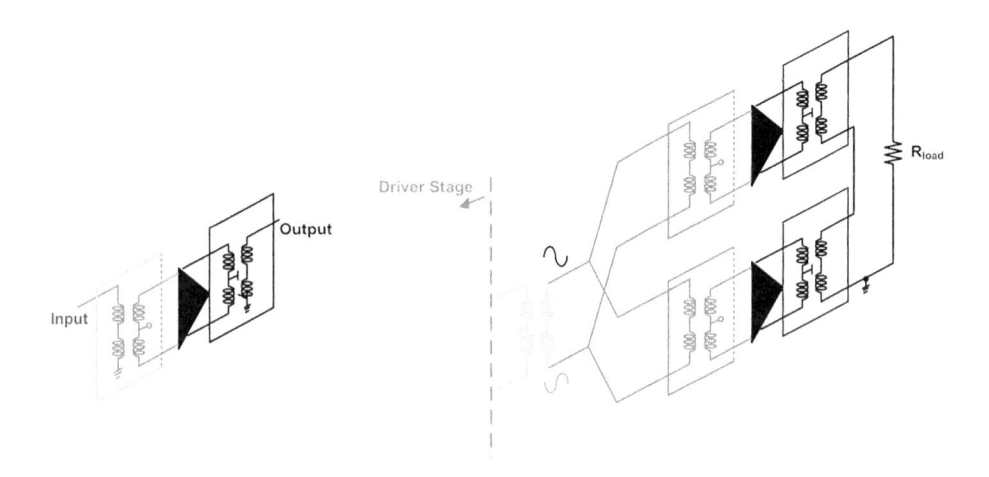

Fig. 7.35 Typical transformer-based mm-wave PAs: (a) 1-way combined PA; (b) 2-way combined PA.

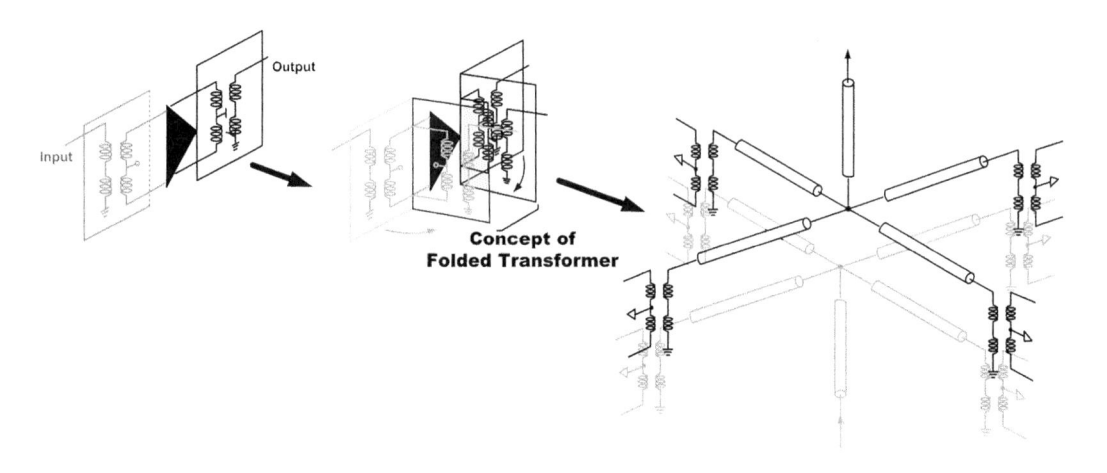

Fig. 7.36 Conceptualization of a folded-transformer.

the transformers. Two typical topologies of 1-way and 2-way transformer-based combined PAs are depicted in Fig. 7.35. By contrast, the dual-radial symmetric architecture is a 3D one. As shown in Fig. 7.36, we can observe that the PA-cell with transformers has to make a breakthrough so as to realize the 3D PA architecture.

In order to take advantage of the dual-radial symmetric architecture as illustrated in Fig. 7.36, here a little bit of imagination is needed to introduce the idea of the folded-transformer. If we can fold the input/output transformers of the PA-cell in Fig. 7.36 to make a folded-transformer-based PA-cell as shown, the PA-cell can be integrated with the 3D dual-radial symmetric architecture.

After introducing the dual-radial symmetric architecture and folded-transformer, now we summarize the 3D PA architecture. As shown in Fig. 7.34, the radial power distribution and combining networks are proposed for layout symmetry and compact size.

7.4 3D power-combining technique

3D PA architecture

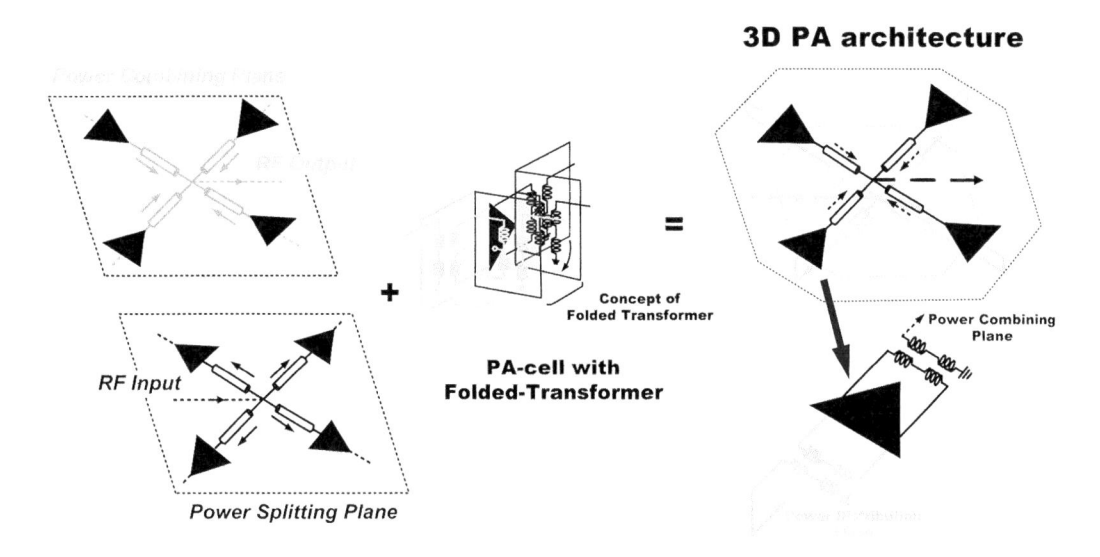

Fig. 7.37 Evolution of 3D PA architecture

Moreover, the dual-radial symmetric architecture is used to miniaturize the active area of the M-way combined PA. Owing to the locations of PA-cells in two different planes, it is difficult to integrate the traditional 2D transformer-based PA-cell into the dual-symmetric architecture. The conceptualization of the folded-transformer in PA-cell design is proposed to realize the 3D PA architecture.

Figure 7.37 demonstrates the evolution of the 3D PA architecture incorporating the dual-radial symmetric architecture and folded-transformers. As demonstrated in Fig. 7.37, the input signal can be fed into the lower radial power-splitting plane and the split signal is converted into differential signals fed into each PA-cell. The amplified differential signals are converted into single-ended ones and assembled in the upper radial combining network. Finally, high output can be delivered to the output load from the center of the PA architecture.

7.4.4 Impedance design: optimization of 3D power-combined PA

The foregoing studies show that M-way 3D PA architecture possesses higher area efficiency than that of the ones using conventional 2D PA architecture as $M > 2$. Nevertheless, as mentioned in Section 7.3, the reported studies reveal a design limitation on impedance freedom of each PA-cell for transformer-based voltage-combining or current-combining techniques, especially for mm-wave PA designs. Figure 7.38 illustrates the simplified schematic of a 3D PA for explaining the impedance transformation. As depicted in Fig. 7.38, the 3D PA adopts the current-combining technique and impedance-transformation lines to mitigate the impedance restriction of PA-cells. The turn ratio of the transformer used in this topology is fixed to $n = 1$ to avoid a low self-resonance frequency and to lower insertion loss. As depicted in Fig. 7.38, TL1 can be designed to accomplish the transformation from a low-impedance power device to

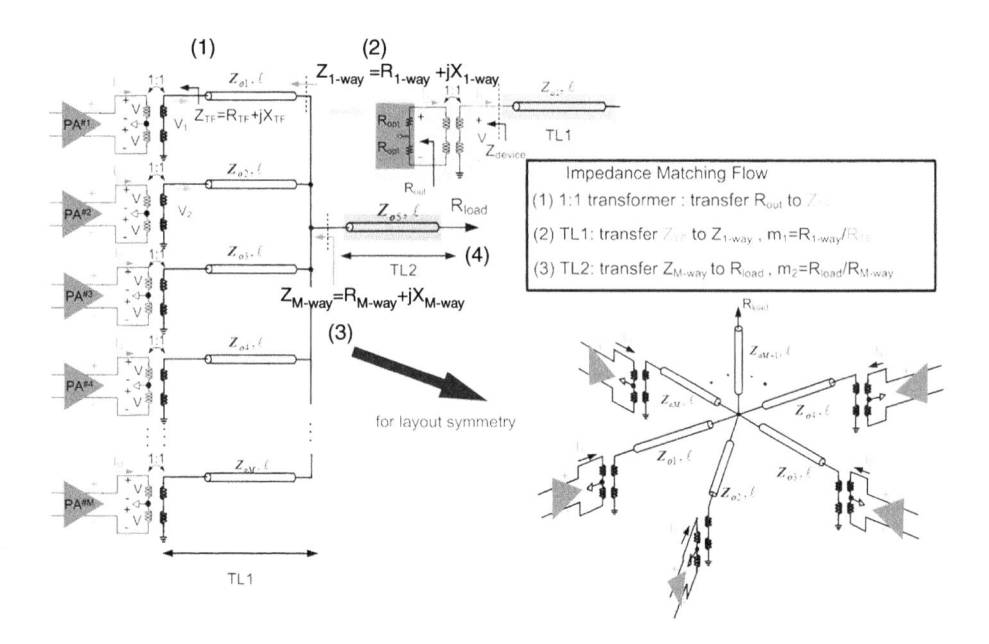

Fig. 7.38 Schematic of 3D PA architecture for impedance transformation.

high impedance for the current-combining matching. After the current-combining node, the other (TL2) can raise the current-combining impedance to the output 50-Ω load. Therefore design flexibility in impedance-transformation ratios can be provided by the transmission-line theory.

Furthermore, in order to optimize the combining efficiency, we have to formulate the impedance-transformation ratios and optimize the impedance matching of the 3D PA architecture. Since the efficiency of a matching network is known [24, 43], as depicted in Fig. 7.38, the passive power-transfer efficiency of the combining technique can be estimated as the cascaded efficiencies of the transformer, η_{TF}, and two TL-based matching networks, η_{TL1} and η_{TL2}, as shown below:

$$\eta_{Total} \equiv \eta_{TF} \times \eta_{TL1} \times \eta_{TL2}, \quad (7.27)$$

where η_{TF} can be estimated as in [24] with $n = 1$, and the efficiencies of the TL-based matching networks can be modeled as [24, 43]

$$\eta_{TL\text{-}network} = \frac{1}{1 + \frac{\sqrt{m-1}}{Q_{TL}}}, \quad (7.28)$$

where m represents the impedance-transformation ratio of the matching network, and Q_{TL} stands for the quality factor of the passive component (i.e. transmission line). Because the efficiency of the transformer with a fixed 1 : 1 turn ratio plays no role in efficiency optimization, the total power-transferring efficiency can be maximized by optimizing transmission-line efficiencies. The passive power-transferring efficiency of cascaded matching networks of TL1 and TL2 can be estimated as

7.4 3D power-combining technique

$$\eta_{TL,cascaded} = \eta_{TL1} \times \eta_{TL2} = \frac{1}{1 + \frac{\sqrt{m-1}}{Q_{TL1}}} \cdot \frac{1}{1 + \frac{\sqrt{m-1}}{Q_{TL2}}}. \quad (7.29)$$

The impedance-transformation ratios of TL1 and TL2, m_1 and m_2, are defined in Fig. 7.38. Here we can make reasonable assumptions, $\sqrt{m_1 - 1}/Q_{TL1} \ll 1$ and $\sqrt{m_2 - 1}/Q_{TL2} \ll 1$, then (7.29) can be approximated as

$$\eta_{TL,cascaded} \approx \left(1 - \frac{\sqrt{m-1}}{Q_{TL1}}\right) \cdot \left(1 - \frac{\sqrt{m-1}}{Q_{TL2}}\right)$$

$$= 1 - \left(\frac{\sqrt{m-1}}{Q_{TL1}} + \frac{\sqrt{m-1}}{Q_{TL2}}\right) + \frac{\sqrt{(m_1-1)\cdot(m_2-1)}}{Q_{TL1} \cdot Q_{TL2}}. \quad (7.30)$$

Here $\frac{\sqrt{(m_1-1)\cdot(m_2-1)}}{Q_{TL1} \cdot Q_{TL2}}$ can be neglected. And then (7.29) could be simplified as

$$\eta_{TL,cascaded} \approx \left(1 - \frac{\sqrt{m-1}}{Q_{TL1}}\right) \cdot \left(1 - \frac{\sqrt{m-1}}{Q_{TL2}}\right). \quad (7.31)$$

Since $R_{1-way} = MR_{M-way}$ the relation between m_1 and m_2 can be shown as

$$m_1 \times m_2 = \frac{M \times R_{load}}{R_{IF}} = X, \quad (7.32)$$

where X is a constant for a fixed M, R_{IF}, and 50-Ω load. The value of m_2 can be further substituted by m_1 and X. To determine the optimal value of m_1 for the maximum passive power-transfer efficiency, we must solve the criteria of optimal m_1. As indicated in [13], m_1 has three non-trivial solutions for optimized $\eta_{TL,cascade}$ and they are $\sqrt{X}$ and $(X \pm \sqrt{X^2 - 4X})/2$, assuming $X \geq 4$. Nevertheless, for practical on-chip realization, a large ratio m_1 or m_2 means that TL1 or TL2 requires a transmission line with low characteristic impedance for implementation. Transmission lines with low characteristic impedance need wide line width, and they are not suitable to be realized in CMOS technology. Therefore we need to select $\sqrt{X}$ such that the impedance-transformation ratios m_1 and m_2 are not too large, and to mitigate CMOS design challenges. If a specific R_{TF} of a power device is selected, the impedance transformation ratios m_1 or m_2 can be determined for various numbers of combined PA-cells. Figure 7.39 provides the design guidelines to determine the required impedance transformation ratios for different M-way combined PAs. For instance, for an 8-way combined PA with the R_{TF} of 15 Ω, the required impedance-transformation ratios, m_1 and m_2, can be determined as 5.48 by the PA designer.

7.4.5 Multiple-power-mode operation of 3D PA architecture

In terms of multiple-power-mode operation, this technique is proposed to extend the service time of mobile devices. Many RF transmitters are equipped with the ability to support efficient transmissions at various power levels. For example, for short-distance communication, the output power required is less than that for long-distance communication. In other words, the power consumption can be saved if the PA can dynamically

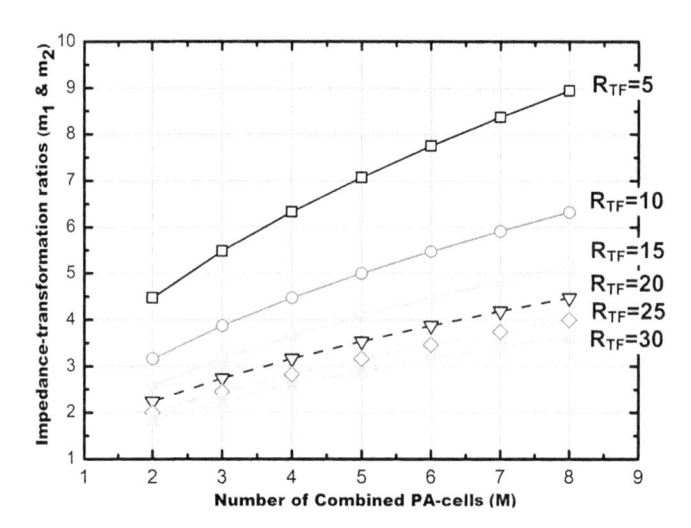

Fig. 7.39 Optimal impedance ratios (m_1 and m_2) vs number of combined PA-cells (M), ($R_{TF} = 2R_{opt}$ for $n = 1$).

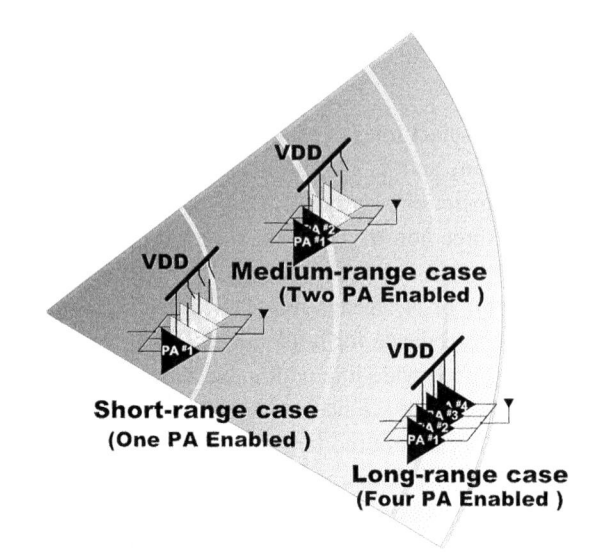

Fig. 7.40 PA with multiple-power-mode operation for different distances.

operate with different power modes for long/medium/short ranges. Figure 7.40 illustrates the scenario of multiple-power-mode operation for a power amplifier. Although the multiple-power-mode operation is a popular function in low-frequency bands [44–52], it is still rare for it to be employed in mm-wave PAs. Moreover, in the near future, the mobile devices will require a low-cost multi-Gb/s link, which demands a highly integrated CMOS SOC solution. In this section, we will introduce another significant feature of multiple-power-mode operation of 3D PAs. The most challenging tasks to implement PAs with multiple-power-mode operation must be building the tunable matching

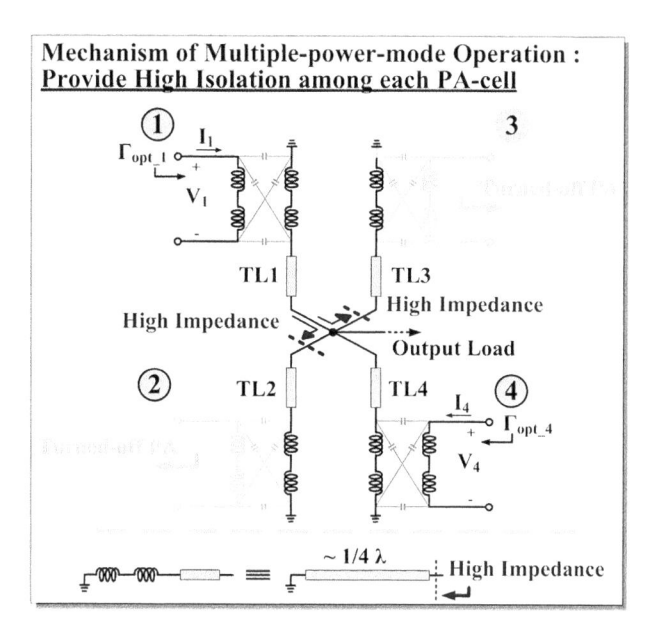

Fig. 7.41 Mechanism of blocking power leakage in power back-off mode by means of the 3D PA architecture.

network and blocking power leakage to the PA-cell in power-off mode. Figure 7.41 shows the mechanism for how the 3D PA architecture blocks the power leakage to the other PA-cells in power-off mode. As illustrated in Fig. 7.41, the power combiner consists of the radial network and transformers, and adopts the idea of quarter-wave-length to deal with the problem of power leakage. As we observe, the secondary-side inductor and TL1 form a quasi-quarter-wave-length transmission line to transform the short at the secondary side to an open at the current-combining node. The combined power can be prevented from leaking into the PA-cell(s) in power-off mode by providing high impedance at the current-combining node.

Figure 7.42 depicts the trend of optimal impedances at the current-combining nodes from 1-way to 4-way combining by utilizing the 3D PA architecture. As we observe from Fig. 7.42, intuitively the PA has distinct optimal loads with respect to the PA in different operation modes. However, the matching network as shown in Fig. 7.42 is designed to deliver maximum power (e.g. 4-way power-on mode). It is also obvious that the required optimal loads of the PA in the other operation modes depart from the $Z_{4\text{-}way}$ and the output power is definitely reduced due to the impedance mismatch as well. In other words, the advance of the multiple-power-mode operation in power efficiency will be also reduced. Traditionally, varactors are employed for impedance tuning for different modes in low-frequency PAs. Nevertheless varactor-based impedance tuning is difficult to implement in mm-wave PAs due to the capacitance variation of femtofarad capacitors.

Interestingly, Fig. 7.43 demonstrates the significant feature of inductive compensation in the 3D PA architecture. As shown in Fig. 7.43, the secondary-side inductor

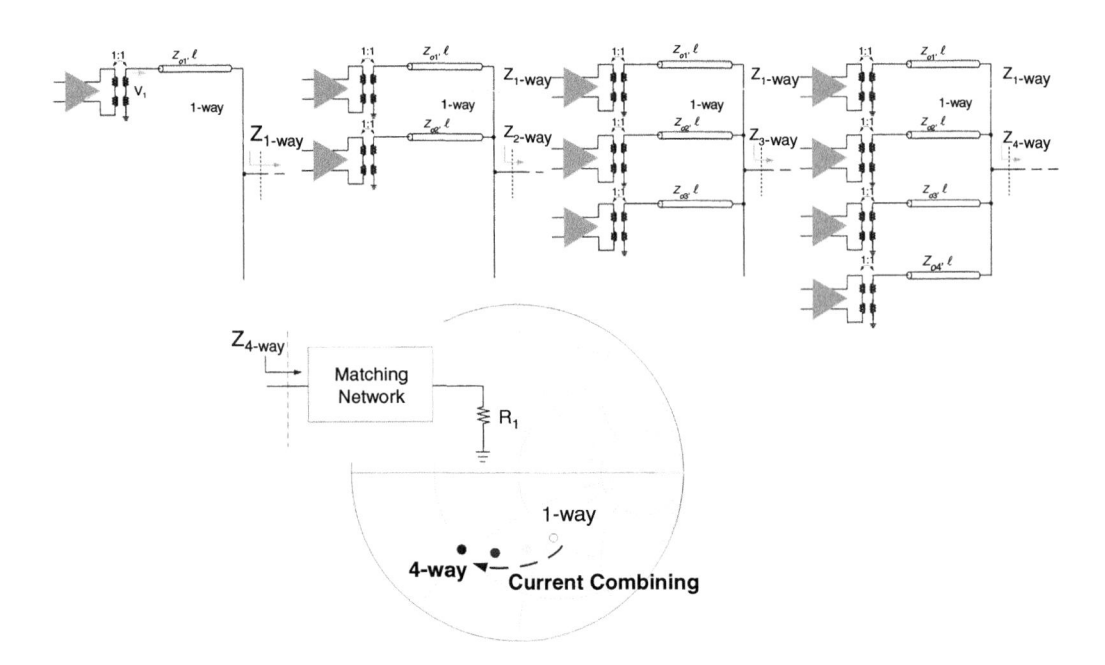

Fig. 7.42 Trends of the optimal loads of 3D PA architecture from 1-way to 4-way combined topologies.

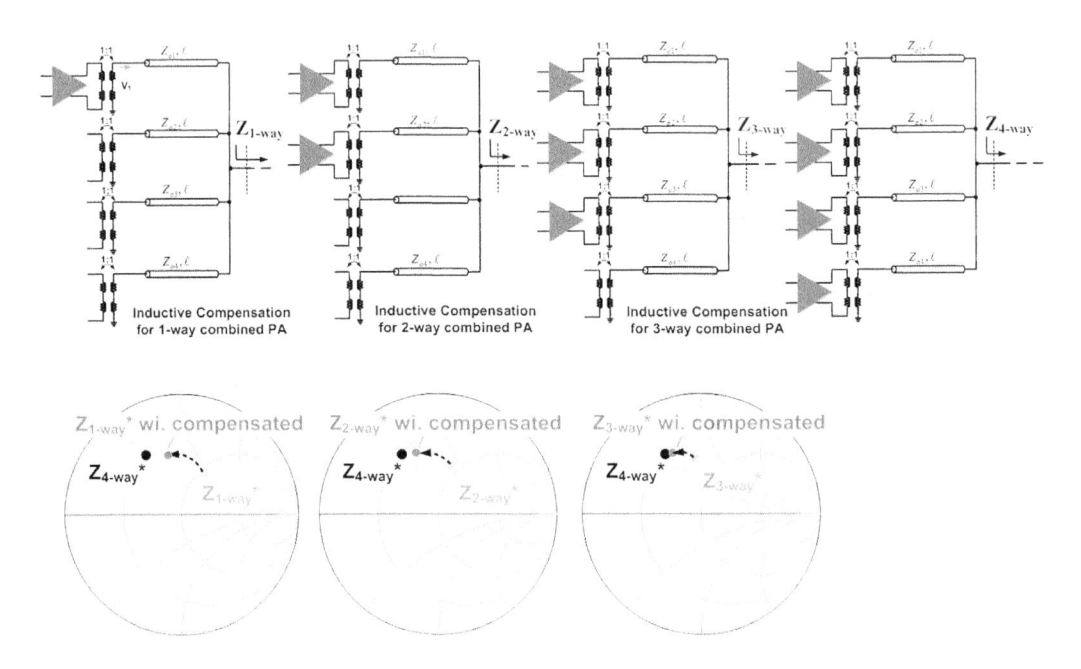

Fig. 7.43 Inductive compensation of 3D PA architecture for 3-/2-/1-way power back-off operations.

and TL1 form an inductive component to move the complex conjugate impedance of $Z_{M\text{-}way}$ (i.e. $Z_{1\text{-}way}*$, $Z_{2\text{-}way}*$, $Z_{3\text{-}way}*$) near the impedance of $Z_{4\text{-}way}*$. Therefore the mismatch can be effectively reduced. The reader may suspect that required inductive values need to be smaller for turning off more PA-cells. As we observe (Fig. 7.43), the

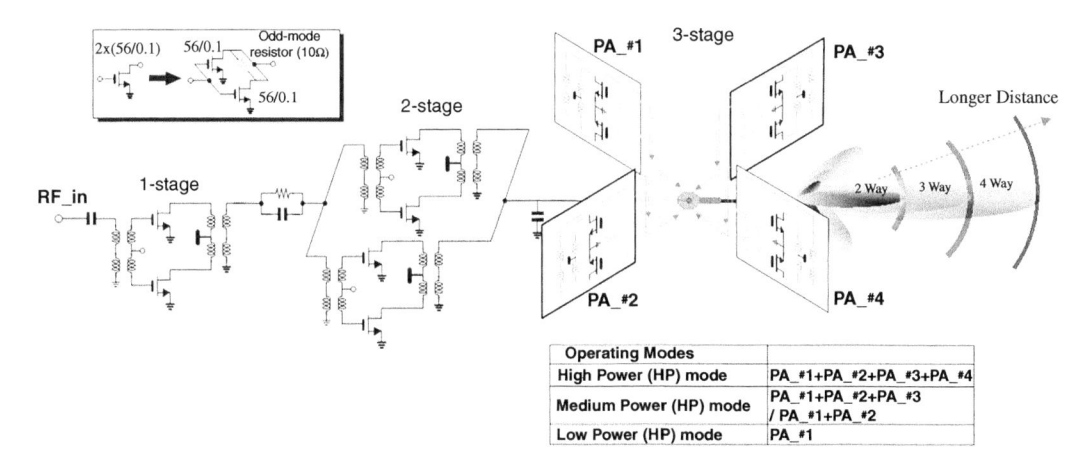

Fig. 7.44 Completed schematic of the multiple-power-mode 60-GHz PA employing 3D PA architecture.

inductive values do indeed decrease with more equivalent compensative components in parallel.

7.4.6 Case study: 60-GHz 3D PA

In this section, an example of a 60-GHz PA employing 3D PA architecture will be introduced. Figure 7.44 illustrates the proposed 3-stage 60-GHz PA. The 3D PA architecture was employed in the power stage to serve as a 4-way combined PA. For linearity considerations, the size ratio of the first gain stage, second driver stage, and final PA stage is 1 : 2 : 4. As shown in Fig. 7.44, the gain stage is composed of a PA-cell that is a pseudo-differential amplifier with a transformer to perform the single-ended-to-differential and differential-to-single-ended transformations for power splitting and combining. The driver stage adopts 2-way TF-based current-splitting/combining techniques for the consideration of symmetry layout and feasibility to feed the signal into the next 4-way combined PA.

Figure 7.45a depicts the simplified circuit model of single PA-unit in the 4-way combined PA, which includes a PA-cell, input/output transformers, and TL matching networks. Figure 7.45b illustrates the detailed components utilized in this work. The readers may realize the corresponding relationship between Figs. 7.45a,b according to the denotations of P_{in} and P_{out}. As shown in Fig. 7.45b, the slab inductors can be modeled as the input/output inductances ($L_{m,in}$, $L_{m,out}$, $L_{s,in}$, $L_{s,out}$) and parasitic resistances ($R_{m,in}$, $R_{m,out}$, $R_{s,in}$, $R_{s,out}$). The output impedance of the power device, Z_{device}, is transferred to Z_{TF} as $13.6 + j23.33$ Ω by the output transformer. Moreover the TL1 as depicted in Fig. 7.38 can be optimized by the aforementioned analysis in Section 7.4.3 to provide the required impedance transformation ratio of $R_{1\text{-}way}/R_{TF}$ as 3.65. For optimizing the whole PA, $Z_{1\text{-}way}$ is eventually determined as $46.23 + j77.56$ Ω for power matching. In other words, the impedance ratio $R_{1\text{-}way}/R_{TF}$ is 3.39. After the EM layout optimization, the impedance of $Z_{4\text{-}way}$ at the current-combining node will

294 **On-chip power-combining techniques for mm-wave silicon power amplifiers**

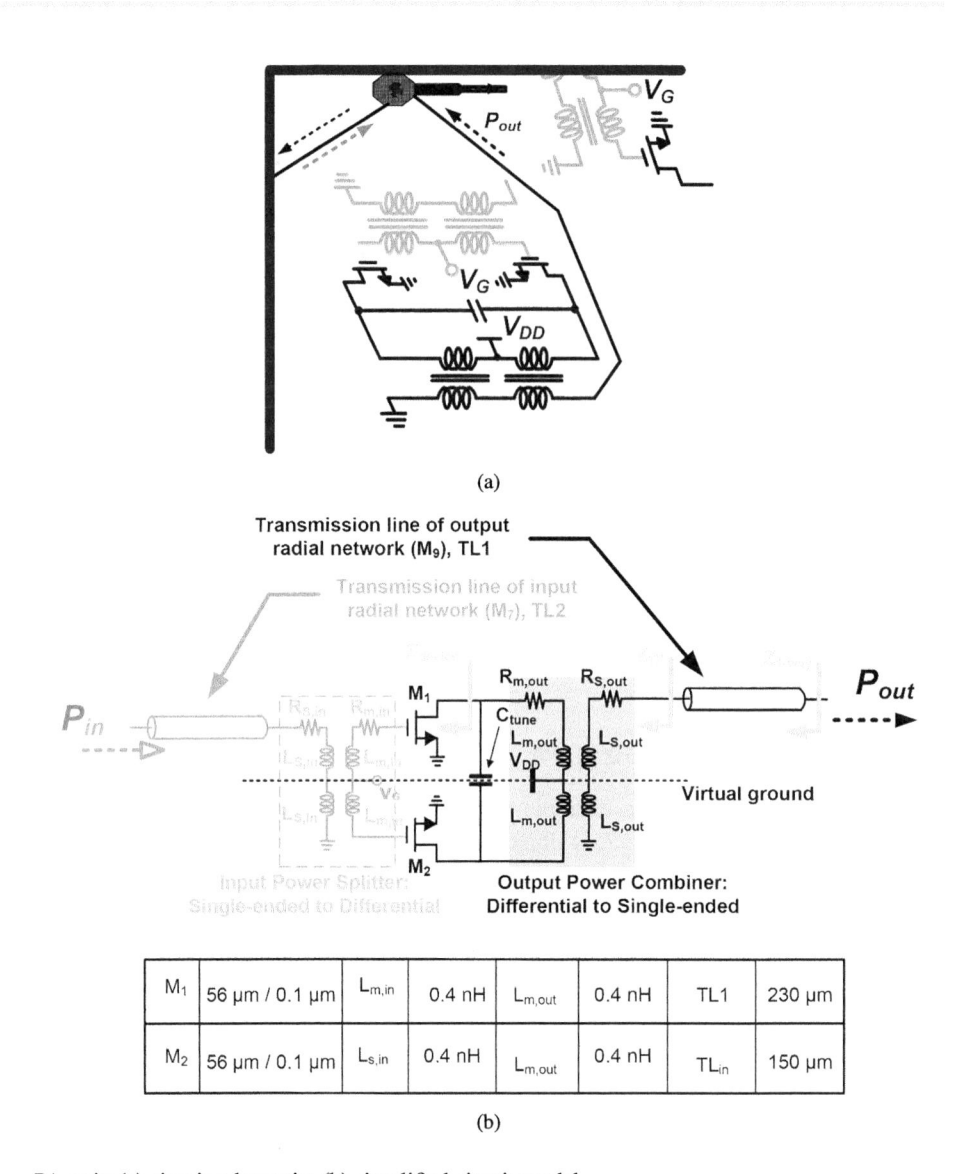

(a)

(b)

Fig. 7.45 PA-unit: (a) circuit schematic; (b) simplified circuit model.

be $14.59 + j19.47\ \Omega$, which is close to the ideal value of $R_{1\text{-}way}/4$. The TL2 matching network as shown in Fig. 7.38 is composed of a small shunted capacitor with R_{load} of 50 Ω and a transmission line. Similarly the impedance ratio of $R_{4\text{-}way}/50$ is around 3.42, which is close to the analysis data of 3.65 in Section 7.3.

Figure 7.46 demonstrates the multi-layer view and output-stage circuit schematic of the 4-way combined 60-GHz PA. The RF signal can be fed into the center of the highly symmetric architecture and then distributed into four differential pairs, and the amplified signals can be combined with the radial distributive networks. Primary and secondary windings utilizing the slab inductors and vertical topology consist of the input/output

7.4 3D power-combining technique　　　295

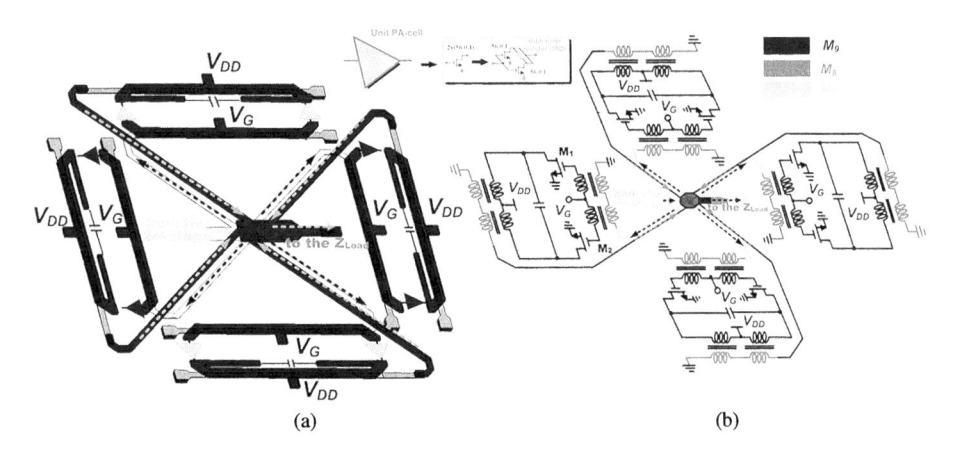

Fig. 7.46　(a) Multi-layer view of the implemented folded-transformer based combiner. (b) The simplified schematic of the output stage.

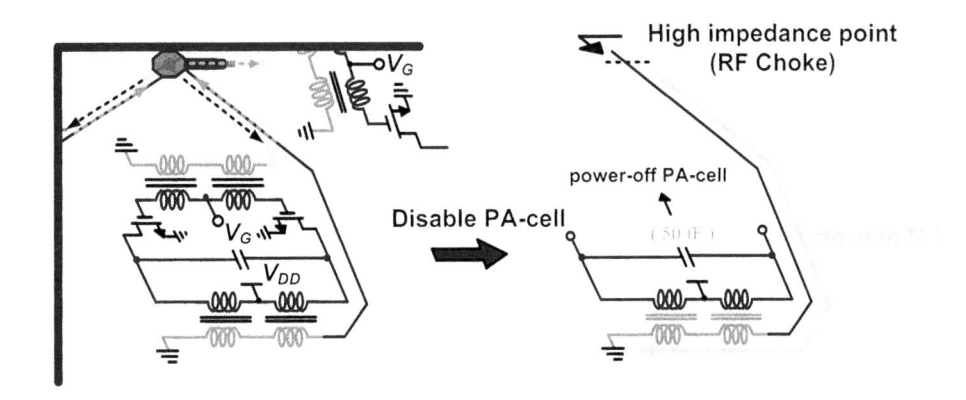

Fig. 7.47　Simplified equivalent circuit model for PA-cell in power-off mode.

transformers for high quality factors and coupling factors. The proposed PA architecture incorporating both the combiner and splitter networks has a compact total area of 0.35×0.35 mm^2.

As depicted in Fig. 7.47, the secondary-side inductor and TL1 will form a quasi-quarter-wave-length transmission line to transform the short at the secondary side to an open at the current-combining node. The combined power can be prevented from leaking into the PA-cell(s) in power-off mode by providing high impedance at the current-combining node. Furthermore, Fig. 7.48 explains the inductive compensation mentioned in the previous section. As depicted in Fig. 7.48, the compensated impedance of the 2-way power-on mode can be moved near that of the 4-way power-on mode.

The prototype PA is fabricated in the TSMC 1.2-V 90-nm low-power (LP) 1P9M CMOS process. Figure 7.49 illustrates the chip micrograph. The 3-stage PA consumes an active area of 0.94×0.41 mm^2. As shown in Fig. 7.50, with 1.2-V supply, the PA has a peak S_{21} of 15.9 dB at 59 GHz. The measured saturation power can achieve

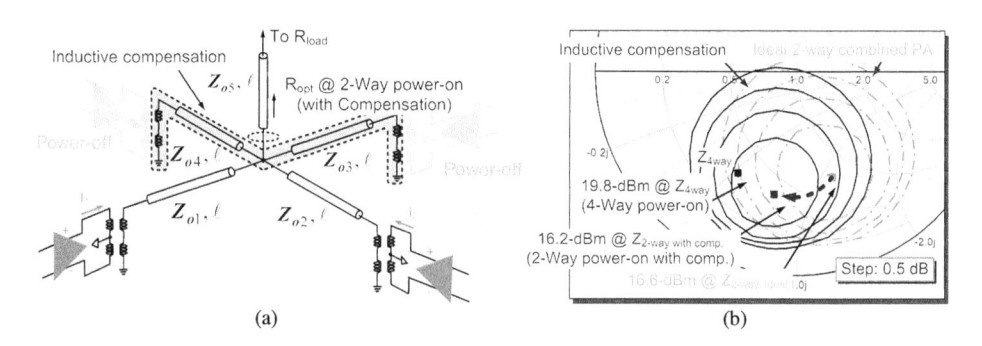

Fig. 7.48 The 2-way operation mode: (a) equivalent circuit topology of the proposed inductive compensation in the 2-way power-on mode; (b) simulated optimal loads at 60-GHz of the 4-way power-on mode and power contours of the 2-way power-on mode.

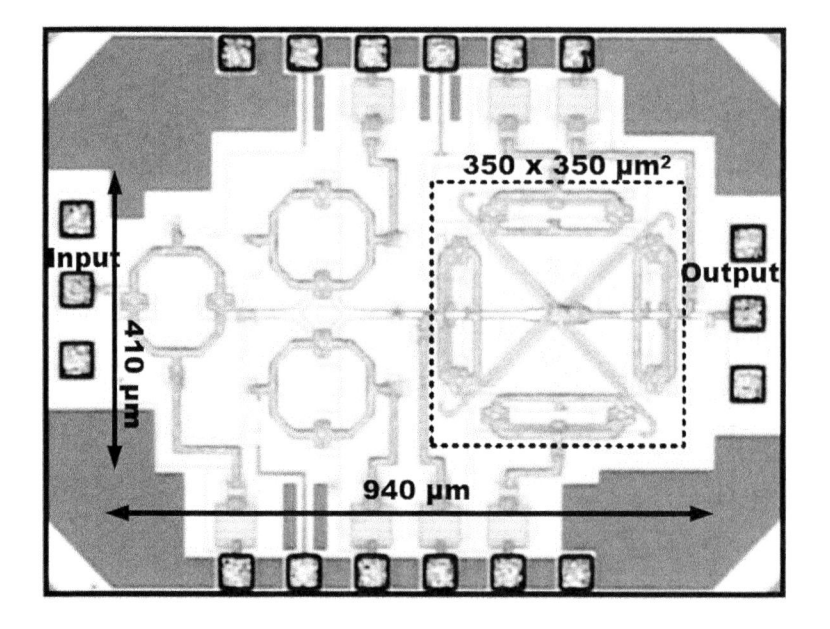

Fig. 7.49 Chip micrograph of 3-stage 60-GHz PA employing 3D PA architecture.

18.5 dBm with the PAE and drain efficiency of the power stage reaching 10.2% and 24.18%, respectively. Figure 7.51a exhibits measured power gain of 15.7/14.7/12.1/9.3 dB at four power-level operation modes, respectively, by turning off the 1-/2-/3-way PA in the power stage. Moreover, the measured drain efficiencies of the power stage are apparently improved at power back-off to verify the multiple-power-mode operation mechanism of the 3D PA architecture. Figure 7.51b demonstrates measured *IM3* results and current consumption of the power stage in four power-level modes. Furthermore, the current consumption can be effectively reduced at the power back-off modes. It should be noted that the *IM3* performances also decrease at the power back-off modes. Nevertheless, this work can dynamically adjust the power consumption for limited battery

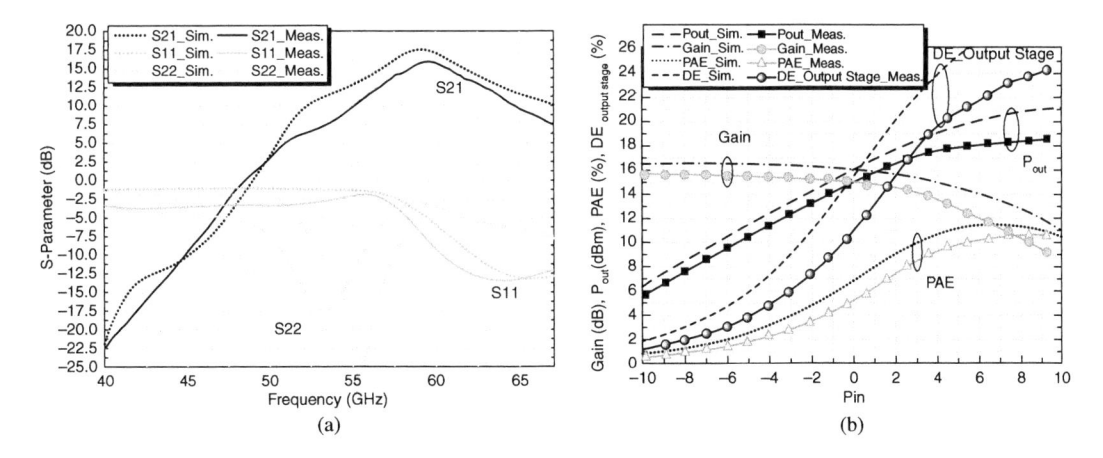

Fig. 7.50 Measured performances of (a) small-signal and (b) large-signal.

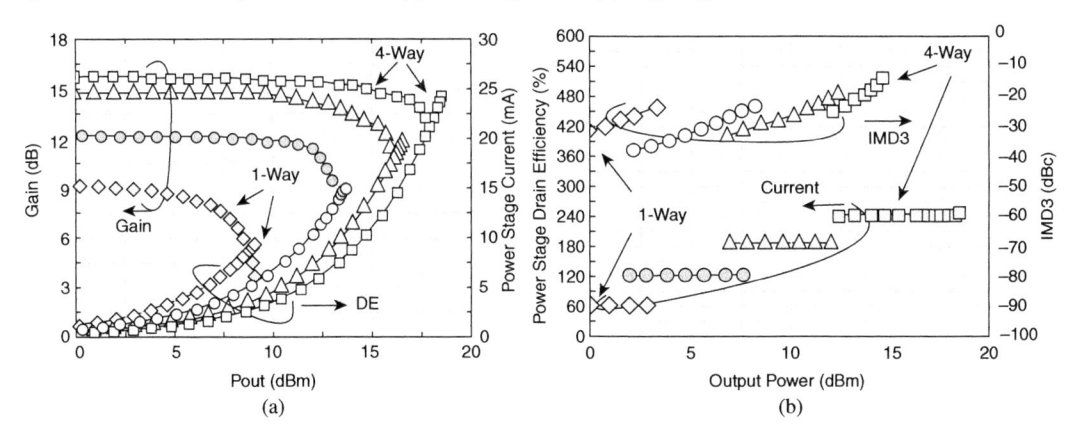

Fig. 7.51 The measured (a) power gain/drain efficiencies and (b) IMD3/current consumption of four operation modes.

lifetime under a specific linearity requirement (e.g. $IM3 = -35$ dBc). Finally, Fig. 7.52 reveals the important factor that the traditional 2D TF-based power-combined PAs have significant growth in the active area per unit PA-cell from 1-way, 2-way, up to 4-way, combined topology. Such an area growth is mainly caused by the 2D power-distribution networks. The 3D PA architecture can achieve area efficiency (area per way) as good as that of 2-way. Moreover, the power area density ($P_{sat}/Area_{4-wayPA}$) of this 3D 60-GHz PA can reach 577.9 mW/mm^2.

7.5 Conclusion

To overcome the output power limitation of mm-wave low-voltage CMOS circuits, many 2D and 3D on-chip power-combining techniques have been developed. Traditionally, direct shunt combining (Section 7.2) is the simplest combining technique for PA design, but the impedance-matching constraint and poor isolation limit its

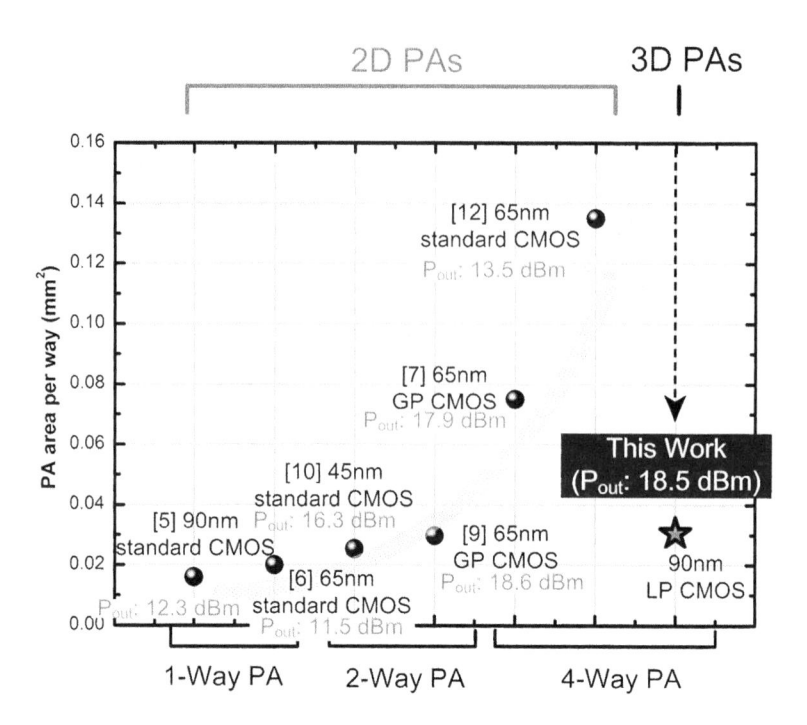

Fig. 7.52 PA area per way vs the number of TF-based combined PAs.

applications. In Sections 7.3.1 and 7.3.2, 2D 50-Ω combiners, like the Wilkinson combiner or 90° hybrid coupler, provide good port-to-port isolation and 50-Ω impedance-matching. Nevertheless, 2D 50-Ω combiners still require impedance-matching circuits between transistor output and 50-Ω combiner input. To reduce the size of power combiners, non-50-Ω transformers are introduced in Section 7.3.3 to finish impedance matching and power combining in a single combiner design. Recently, many miniature transformer combiners have been developed, but the power splitters still require large 2D chip area to provide in-phase signal input routing, especially for the multi-way power-combining circuits. In Section 7.4, the 3D power splitter and combiner are introduced to significantly reduce the chip size. A dual-radial symmetry layout for power combiner and splitter is carefully designed on the top and bottom metal planes to enforce the in-phase input and output for all power cells. From 2D to 3D power-combining designs, an additional design dimension is introduced to provide a multi-way symmetry layout for equal power splitting and combining through a center point. A 90-nm 4-way 3D power amplifier is successfully implemented to show the feasibility of 3D mm-wave on-chip power combining.

References

[1] J. Y.-C. Liu, R. Berenguer, and M.-C. F. Chang, "Millimeter-wave self-healing power amplifier with adaptive amplitude and phase linearization in 65-nm CMOS," *IEEE Trans. Microw. Theory Tech.*, vol. **60**, no. 5, pp. 1342–1352, May 2012.

[2] T. Wang, T. Mitomo, N. Ono, and O. Watanabe, "A 55–67 GHz power amplifier with 13.6% PAE in 65 nm standard CMOS," in *RFIC Dig.*, June 2012.

[3] T. LaRocca and M. C. Frank Chang, "60 GHz CMOS differential and transformer-coupled power amplifier for compact design," in *RFIC Dig.*, June 2008.

[4] W. L. Chan, J. R. Long, M. Spirito, and J. J. Pekarik, "A 60 GHz-band 1 V 11.5 dBm power amplifier with 11% PAE in 65 nm CMOS," in *ISSCC Dig. Tech. Papers*, Feb. 2009, pp. 380–381.

[5] D. Chowdhury, P. Reynaert, and A. M. Niknejad, "A 60 GHz 1 V + 12.3 dBm transformer-coupled wideband PA in 90 nm CMOS," in *ISSCC Dig. Tech. Papers*, Feb. 2008, pp. 560–635.

[6] J. Y.-C. Liu, Q. J. Gu, A. Tang, N.-Y. Wang, and M.-C. Frank Chang, "A 60 GHz tunable output profile power amplifier in 65 nm CMOS," *IEEE Microw. Wireless Compon. Lett.*, vol. 21, no. 7, pp. 377–379, July 2011.

[7] J. Chen and A. M. Niknejad, "A compact 1 V 18.6 dBm 60 GHz power amplifier in 65 nm CMOS," in *ISSCC Dig. Tech. Papers*, Feb. 2011, pp. 432–433.

[8] L. Chen, L. Li, and T. J. Cui, "A 1 V 18 dBm 60 GHz power amplifier with 24 dB gain in 65 nm LP CMOS," *Asia-Pacific Microwave Conference (APMC)*, Dec. 2012.

[9] D. Zhao, S. Kulkarni, and P. Reynaert, "A 60 GHz outphasing transmitter in 40 nm CMOS with 15.6 dBm output power," in *ISSCC Dig. Tech. Papers*, Feb. 2012, pp. 170–172.

[10] Y.-N. Jen, J.-H. Tsai, T.-W. Huang, and H. Wang, "Design and analysis of a 55–71-GHz compact and broadband distributed active transformer power amplifier in 90-nm CMOS process," *IEEE Trans. Microw. Theory Techn.*, vol. 57, no. 7, pp. 1637–1646, July 2009.

[11] M. Bohsali and A. M. Niknejad, "Current combining 60 GHz CMOS power amplifier," in *RFIC Dig.*, June 2009.

[12] Y. Zhao, and J. R. Long, "A wideband, dual-path millimeter-wave power amplifier with 20 dBm output power and PAE above 15% in 130 nm SiGe-BiCMOS," *IEEE J. Solid-State Circuits*, vol. 47, no. 9, pp. 1981–1997, Dec. 2012.

[13] J.-F. Yeh, J.-H. Tsai, and T.-W. Huang, "A 60-GHz power amplifier design using dual-radial symmetric architecture," *IEEE Trans. Microw. Theory Tech.*, vol. 61, no. 3, pp. 1280–1290, Mar. 2013.

[14] J.-W. Lai and A. Valdes-Garcia, "A 1 V 17.9 dBm 60 GHz power amplifier in standard 65 nm CMOS," in *ISSCC Dig. Tech. Papers*, Feb. 2010, pp. 424–425.

[15] S. Aloui, B. Leite, N. Demirel, *et al.*, "High-gain and linear 60-GHz power amplifier with a thin digital 65-nm CMOS technology," *IEEE Trans. Microw. Theory Tech.*, vol. 61, no. 6, pp. 2425–2437, June 2013.

[16] C. Y Law and A.-V. Pham, "A high gain 60 GHz power amplifier with 20 dBm output power in 90 nm CMOS," in *IEEE ISSCC Dig. Tech. Papers*, Feb. 2010, pp. 426–427.

[17] U. R. Pfeiffer and D. Goren, "A 23-dBm 60-GHz distributed active transformer in a silicon process technology," *IEEE Trans. Micro-wave Theory Tech.*, vol. 55, no. 5, pp. 857–865, May 2007.

[18] V. Giammello, E. Ragonese, and G. Palmisano, "A 15-dBm SiGe BiCMOS PA for 77-GHz automotive radar," *IEEE Trans. Microw. Theory Tech.*, vol. 59, no. 11, pp. 2910–2918, Nov. 2011.

[19] Z. Xu, Q. J. Gu, and M.-C. F. Chang, "A 100–117 GHz W-band CMOS power amplifier with on-chip adaptive biasing," *IEEE Microw. Wireless Compon. Lett.*, vol. 21, no. 10, pp. 547–649, July 2011.

[20] Q. J. Gu, Z. Xu, and M.-C. F. Chang, "Two-way current-combining W-band power amplifier in 65-nm CMOS," *IEEE Trans. Microw. Theory Techn.*, vol. **60**, no. 5, pp. 1365–1374, May 2012.

[21] N. Deferm, J. F. Osorio, A. de Graauw, and P. Reynaert, "A 94 GHz differential power amplifier in 45 nm LP CMOS," in *RFIC Dig.*, June 2011.

[22] Y.-S. Jiang, J.-H. Tsai, and H. Wang, "A W-band medium power amplifier in 90 nm CMOS," *IEEE Microwave Wireless Comp. Lett.*, vol. **18**, no. 12, pp. 818–820, Dec. 2008.

[23] K.-Y. Wang, T.-Y. Chang, and C.-K. Wang, "A 1 V 19.3 dBm 79 GHz power amplifier in 65 nm CMOS," in *ISSCC Dig. Tech. Papers*, Feb. 2012, pp. 260–262.

[24] I. Aoki, S. D. Kee, D. B. Rutledge, and A. Hajimiri, "Fully integrated CMOS power amplifier design using the distributed active-transformer architecture," *IEEE J. Solid-State Circuits*, vol. **37**, no. 3, pp. 371–383, Mar. 2002.

[25] J.-F. Yeh, J.-H. Tsai, and T.-W. Huang, "A multi-mode 60-GHz power amplifier with a novel power combination technique," in *RFIC Dig.*, June 2012, pp. 61–64.

[26] Jing-Lin Kuo, Yi-Fong Lu, Ting-Yi Huang, et al., "60-GHz four-element phased-array transmit/receive system-in-package using phase compensation techniques in 65-nm flip-chip CMOS process," *IEEE Trans. Microw. Theory Techn.*, vol. **60**, no. 3, pp. 743–756, Mar. 2012.

[27] A. E. I. Lamminen, J. Saily, and A. R. Vimpar, "60-GHz patch antennas and arrays on LTCC with embedded-cavity substrates," *IEEE Trans. Antennas and Propagation*, vol. **56**, no. 9, pp. 2865–2874, Sep. 2008.

[28] Y.-N. Jen, J.-H. Tsai, C.-T. Peng, and T.-W. Huang, "A 20 to 24 GHz +16.8-dBm fully integrated power amplifier using 0.18-μm CMOS process," *IEEE Microwave and Wireless Components Letters*, vol. **19**, no. 1, pp. 42–44, Jan. 2009.

[29] T. Suzuki, Y. Kawano, M. Sato, T. Hirose, and K. Joshin, "60 and 77 GHz power amplifiers in standard 90 nm CMOS," in *IEEE ISSCC Dig. Tech. Papers*, Feb. 2008, pp. 562–573.

[30] C.-C. Hung, J.-L. Kuo, K.-Y. Lin, and H. Wang, "A 22.5-dB gain, 20.1-dBm output power K-band power amplifier in 0.18-μm CMOS," *IEEE RFIC Symp. Dig.*, pp. 557–560, May 2010.

[31] R. G. Freitag, "A modal analysis of MMIC power amplifier stability," *Proceedings of the 35th Midwest Symposium on Circuits and Systems*, vol. **2**, pp. 1012–1015, 1992.

[32] R. G. Freitag, "A unified analysis of MMIC power amplifier stability," *1992 MTT-S International Microwave Symposium Digest*, vol. **1**, pp. 297–300, 1992.

[33] D. M. Pozar, *Microwave Engineering*, 3rd edn. Hoboken, NJ: Wiley, 2004.

[34] G. Gonzalez, *Microwave Transistor Amplifiers Analysis and Design*, 2nd edn. Upper Saddle River, NJ: Prentice-Hall, 1997.

[35] J. Lee, C.-C. Chen, J.-H. Tsai, K.-Y. Lin, and H. Wang, "A 68–83 GHz power amplifier in 90 nm CMOS," in *IEEE MTT-S Int. Microwave Symp. Dig.*, June 2009, pp. 437–440.

[36] A. Pallotta, W. Eyssa, L. Larcher, and R. Brama, "Millimeter-wave 14 dBm CMOS power amplifier with input–output distributed transformers," in *Proc. CICC*, Sep. 2010, pp. 1–4.

[37] J. Essing, R. Mahmoudi, Pei Yu, and A. van Roermund, "A fully integrated 60 GHz distributed transformer power amplifier in bulky CMOS 45 nm," in *RFIC Dig.*, June 2011, pp. 1–4, 5–7.

[38] Z. Yi, J. R. Long, and M. Spirito, "A 60 GHz-band 20 dBm power amplifier with 20% peak PAE," in *RFIC Dig.*, June 2011, pp. 1–4, 5–7.

[39] J. P. Comeau, E. W. Thoenes, A. Imhoff, and M. A. Morton, "X-Band +24 dBm CMOS power amplifier with transformer power combining," in *SiRF Dig.*, Sep. 2010, pp. 49–52.

[40] B. Martineau, V. Knopik, A. Siligaris, F. Gianesello, and D. Belot, "A 53-to-68 GHz 18 dBm power amplifier with an 8-way combiner in standard 65 nm CMOS," in *ISSCC Dig. Tech. Papers*, Feb. 2010, pp. 428–429.

[41] D. Sandstrom, B. Martineau, M. Varonen, *et al.*, "94 GHz power-combining power amplifier with +13 dBm saturated output power in 65 nm CMOS," in *RFIC Dig.*, June 2011, pp. 1–4.

[42] M. Thian, M. Tiebout, N. B. Buchanan, V. F. Fusco, and F. Dielacher, "A 76–84 GHz SiGe power amplifier array employing low-loss four-way differential combining transformer," *IEEE Trans. Microw. Theory Techn.*, vol. **61**, no. 2, pp. 931–938, Feb. 2013.

[43] A. M. Niknejad, D. Chowdhury, and J. Chen, "Design of CMOS power amplifiers," *IEEE Trans. Microw. Theory Tech.*, vol. **60**, no. 6, pp. 1784–1796, June 2012.

[44] D. Chowdhury, C. D. Hull, O. B. Degani, Y. Wang, and A. M. Niknejad, "A fully integrated dual-mode highly linear 2.4 GHz CMOS power amplifier for 4G WiMax applications," *IEEE J. Solid-State Circuits*, vol. **44**, no. 12, pp. 3393–3402, Dec. 2009.

[45] G. Liu, P. Haldi, T.-J. King Liu, and A. M. Niknejad, "Fully integrated CMOS power amplifier with efficiency enhancement at power back-off," *IEEE J. Solid-State Circuits*, vol. **43**, no. 3, pp. 600–609, Mar. 2008.

[46] J. Kim, Y. Yoon, H. Kim, *et al.*, "A linear multi-mode CMOS power amplifier with discrete resizing and concurrent power combining structure," *IEEE J. Solid-State Circuits*, vol. **46**, no. 5, pp. 1034–1048, May 2011.

[47] J. Kim, W. Kim, H. Jeon, *et al.*, "A fully-integrated high-power linear CMOS power amplifier with a parallel-series combining transformer," *IEEE J. Solid-State Circuits*, vol. **47**, no. 3, pp. 599–614, Mar. 2012.

[48] A. Scuderi, C. Santagati, M. Vaiana, F. Pidala, and M. Paparo, "Balanced SiGe PA module for multi-band and multi-mode cellular-phone applications," in *ISSCC Dig. Tech. Papers*, Feb. 2008, pp. 572–637.

[49] G. Hau and M. Singh, "Multi-mode WCDMA power amplifier module with improved low-power efficiency using stage-bypass," in *RFIC Dig.*, May 2010, pp. 163–166.

[50] H. Jeon, Y. Park, Y.-Y. Huang, *et al.*, "A triple-mode balanced linear CMOS power amplifier using a switched-quadrature coupler," *IEEE J. Solid-State Circuits*, vol. **47**, no. 9, pp. 2019–2032, Sep. 2012.

[51] Y. Yoon, J. Kim, H. Kim, *et al.*, "A dual-mode CMOS RF power amplifier with integrated tunable matching network," *IEEE Microw. Wireless Compon. Lett.*, vol. **60**, no. 1, pp. 77–88, Jan. 2012.

[52] B. Koo, T. Joo, Y. Na, and S. Hong, "A fully integrated dual-mode CMOS power amplifier for WCDMA applications," in *ISSCC Dig. Tech. Papers*, Feb. 2012, pp. 82–84.

8 Outphasing mm-wave silicon transmitters

Patrick Reynaert and Dixian Zhao

8.1 Introduction

Efficiency enhancement and linearization techniques are always the focus of research in power amplifier (PA) design. As the last building block of the transmitter (TX) and the most power-hungry one, the PA has to maintain sufficient linearity to ensure the signal integrity and achieves high efficiency for low power. The conventional TX based on I/Q modulation has variable envelopes in both I and Q paths. It needs to operate at 3–6 dB back-off from the 1-dB compression point (P_{1dB}) depending on the applied modulation for linearity, which inevitably results in low average efficiency. Digital pre-distortion (DPD) has been proposed to correct the AM–AM and AM–PM distortions of the PA by a closed feedback loop [1]. However, the loop stability is usually difficult to ensure, especially for wideband signals. The reported improvement of DPD is limited to approximately 1–2 dB in P_{1dB} and 3%–5% in PAE [1]. The polar modulation (or envelope elimination and restoration) [2] separates the envelope and phase components so that the phase information can be amplified by the highly efficient switching amplifier. The envelope modulator operates at relatively low speed (on the order of the modulation bandwidth) and thus consumes low power. It will modulate the supply of the switching amplifier and restore the amplitude modulation at PA output. The potential challenges of a polar transmitter for high-speed communication are to design an efficient wide-band envelope modulator and balance the delay between the envelope and phase paths. The Doherty PA [3] is another efficiency-enhancement technique. It can maintain high efficiency over a broad power range by combining the outputs of a class-AB PA (main amplifier) and a class-C PA (auxiliary amplifier). The issue here is that the main amplifier stays in deep saturation at high power level. Although the gain saturation of the main amplifier can be compensated by the gain expansion of the auxiliary amplifier. The compensation in phase distortion is rather difficult to predict and sensitive to the bias current. Besides, the main and auxiliary paths are very unbalanced, potentially causing problems for high-speed (i.e. Gb/s) communication.

Outphasing has become one of the most popular techniques in the past five years [4–6]. In outphasing, the linear amplification is achieved by vector summing the outputs of two identical branch PAs. Both PAs only deal with constant-envelope phase-modulated signals, so a highly efficient switching PA or saturated linear PA can be used. As a result, the whole outphasing system has the potential to maximize the linear output power ($P_{LIN} = P_{SAT}$) and associated efficiency ($PAE_{LIN} = PAE_{MAX}$) simultaneously.

In addition, the two signal paths in outphasing are identical and thus well balanced. Therefore, relatively high-speed communication can be expected compared with the polar transmitter.

Advances in CMOS technology have permitted the integration of mm-wave systems with a data rate of gigabits per second (Gb/s). On top of the transistor scaling, the design and modeling methodologies developed have steadily improved the performance of the mm-wave building blocks over the past few years. For instance, the saturated output power (P_{SAT}) and peak PAE of 60-GHz power amplifiers (PAs) have reached 18 dBm and 20%, respectively [7–9]. One of the remaining technical problems related to mm-wave CMOS transmitters is the poor average efficiency when transmitting complex amplitude- and phase-modulated signals (e.g. 16QAM). The cause of this low efficiency is the required back-off from the P_{1dB} to meet EVM and transmit spectral mask specifications. A conventional PA only provides maximum efficiency near P_{SAT}. For a 6-dB back-off from P_{1dB}, the output power and PAE of the state-of-the-art PAs remain below 9 dBm and 5%, respectively. Although this issue is well known by mm-wave designers, the optimization of mm-wave TXs is still limited to the circuit level (i.e. optimization of the PA and the up-conversion chain). Low-GHz linearization or efficiency enhancement techniques are usually not applied, mainly due to the inferior performance of active and passive devices at mm-wave frequencies and the wideband processing bandwidth required by the mm-wave system.

Outphasing a TX allows the branch PAs to operate at peak output power and peak efficiency, which significantly enhances the average efficiency for complex modulated signals. Such attributes of outphasing can also be adopted in the mm-wave design. In 2011, Liang [10] first combined the outphasing technique with beamforming at 60 GHz. Each PA in the beamforming system shows 9.7 dBm saturated output power with 11% PAE_{MAX} and 200 Mb/s 16QAM is achieved by spatially combining the two outphased TX outputs. Li [11] demonstrated an outphasing signal generator with a throughput of 3.4 GS/s and 12-b phase accuracy, which is suitable for mm-wave outphasing TXs. The authors of [12] reported a 60-GHz outphasing transmitter with integrated power combiner in 40-nm CMOS. It achieves a 500 Mb/s 16QAM modulation with 12.5 dBm average output power and 15% average efficiency (PA) at an EVM of −22 dB without any calibration applied. This design performs two times better than the state-of-the-art 60-GHz transmitters. Note that such an efficiency enhancement is not attainable by other techniques at mm-wave frequencies where only linear class-A or -AB PAs are available due to the limited transistor power gain and lossy passives. The outphasing TX does need a combiner at the output to reconstruct the amplitude modulation, but it does not require extra design effort as the power-combining technique is normally utilized to achieve high output power and efficiency at mm-wave frequencies. All the aforementioned reasons make the outphasing technique a promising solution for an efficient transmitter at mm-wave frequencies.

Considering the fact that very limited mm-wave outphasing TXs have been reported, we will not restrict our discussion only to the mm-wave designs. Some examples of low-GHz outphasing systems will also be included. The feasibility of re-using these techniques at mm-wave will be discussed. Different aspects of outphasing will be covered in the chapter, including the outphasing signal generation and combining techniques, along with some state-of-the-art design examples.

8.2 Outphasing basics

The outphasing concept is based on the fact that any amplitude- and phase-modulated signal $S(t)$ can be decomposed into two outphased constant-envelope signals $S_1(t)$ and $S_2(t)$, as shown in Fig. 8.1. The module that performs signal decomposition is called the outphasing signal generator (OPG). The term signal component separator (SCS) is also used in some literature. We use the term OPG in the text as some generation mechanisms are not just based on signal separation. Figure 8.2 shows the phasor diagram of the original signal $S(t)$ and the two outphasing vectors $S_1(t)$ and $S_2(t)$. If we represent $S(t)$ as

$$S(t) = A(t) \cdot e^{j(\omega_C t + \theta(t))}, \tag{8.1}$$

then the two outphasing vectors are given by

$$S_1(t) = \frac{A_M}{2} \cdot e^{j(\omega_C t + \theta(t) + \varphi(t))}, \tag{8.2}$$

$$S_2(t) = \frac{A_M}{2} \cdot e^{j(\omega_C t + \theta(t) - \varphi(t))}, \tag{8.3}$$

where ω_C is the carrier frequency, A_M is the peak value of $A(t)$, $\theta(t)$ indicates the modulated phase, and the outphasing angle $\varphi(t)$ equals

$$\varphi(t) = \cos^{-1}[A(t)/A_M]. \tag{8.4}$$

These two signals will be amplified separately and then combined at the transmitter output. After vector-summing the two amplified signals, the amplitude- and phase-modulation is restored. As mentioned, the unique feature of outphasing is that both $S_1(t)$ and $S_2(t)$ have constant envelope. Therefore, highly efficient switching or saturated PAs can be used, which gives outphasing a huge efficiency benefit over other approaches for the same linearity.

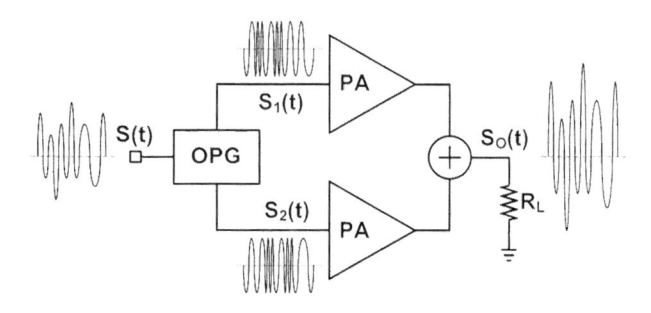

Fig. 8.1 Simplified diagram of outphasing amplifier.

Table 8.1 Comparison of class-A PA and outphasing PA at mm-wave.

	PA$_1$ (Outphasing)	PA$_1$ (I/Q)	PA$_2$ (I/Q)
P_{SAT} [dBm]	15	15	18
η at P_{SAT} [%]	20	20	20
P_{LIN} [dBm]	15	12 (P_{1dB})	15 (P_{1dB})
η at P_{LIN} [%]	20	10	10
Gain [dB]	15	20	20
P_{DC} [mW]	158	158	316

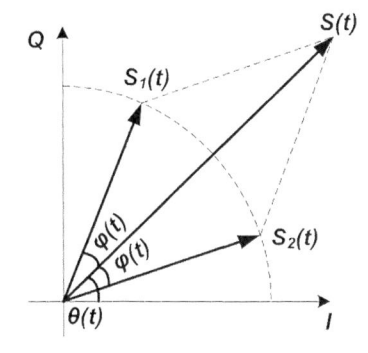

Fig. 8.2 Outphasing phasor diagram.

8.2.1 Outphasing for mm-wave applications

The outphasing concept can also be used for mm-wave application, which proves to achieve better efficiency and linearity when transmitting a modulated signal [12]. We will first illustrate the benefit of outphasing by a mm-wave PA model. We noticed that the right part of Fig. 8.1, which is composed of two branch PAs and a lossless power combiner, can also function as a conventional power-combining PA if we apply

$$S_1(t) = S_2(t) = \frac{1}{2} A(t) \cdot e^{j(\omega_c t + \theta(t))}. \tag{8.5}$$

Therefore, the performance of the outphasing PA and conventional power-combining PA (based on I/Q modulation) can be directly compared. Let us assume we have two such power-combining PAs, namely PA$_1$ and PA$_2$. PA$_1$ has 3-dB lower output power than PA$_2$ and it will work as both outphasing and conventional PAs. Based on prior power amplifiers in the 60-GHz band [7–9], Table 8.1 summarizes the performance of the two PAs. Figure 8.3a is extrapolated according to Raap's nonlinear PA model (the load impedance of 50 Ω is assumed). The output voltage V_{OUT} is expressed by

$$V_{OUT} = \frac{G \cdot V_{IN}}{\left[1 + \left(\frac{G \cdot V_{IN}}{V_{SAT}}\right)^{2m}\right]^{1/2m}}, \tag{8.6}$$

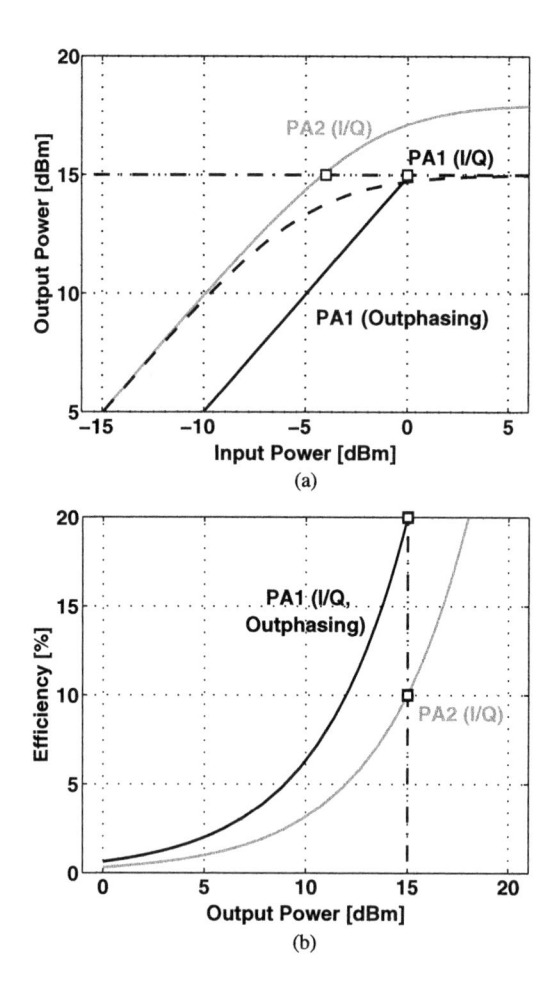

Fig. 8.3 Comparison of PA$_1$ (outphasing and I/Q) and PA$_2$ (I/Q): (a) output power; (b) efficiency.

where G and V_{SAT} denote the small-signal gain and saturated output voltage. The fitting parameter m is chosen to be 1.8, which makes P_{1dB} approximately 3 dB lower than P_{SAT} in Fig. 8.3a. Figure 8.3b is plotted according to class-A PA behavior, where the DC power dissipation is constant and thereby the efficiency is proportional to the output power. PA$_1$ has 3-dB lower P_{SAT} and P_{1dB} than PA$_2$ and is employed in both conventional I/Q, where $S_1(t) = S_2(t)$, and outphasing PAs. The outphasing PA has 5-dB lower gain as it operates in saturation. Although the achievable peak PAE relates to the output power and CMOS technology used, the same value is assumed here for PA$_1$ and PA$_2$, without losing the generality of the conclusion. PA$_1$ in the outphasing mode operates linearly up to the maximum output power (Fig. 8.3a) and it has the same efficiency as in the I/Q mode (Fig. 8.3b) if no interaction between two PAs occurs (the back-off efficiency enhancement provided by the outphasing will be discussed later). Compared with PA$_2$, the outphasing PA achieves the same P_{LIN} with only half the power consumption. In addition, it has 10% higher efficiency at the output power of 15 dBm, which leads to higher average efficiency for modulated signals. The above benefits of the outphasing PA are verified by a 60-GHz outphasing TX that will be discussed in Section 8.6.

8.3 Outphasing signal generation

Generation of the outphasing vectors involves nonlinear operations (e.g. to generate the outphasing angle (8.4)) and thus requires substantial analog or digital signal processing. It may limit the linearity and signal bandwidth of the whole system. By investigating the phasor diagram of the original signal and the outphasing vectors (Fig. 8.2), we see that the outphasing vectors can be derived either by vector rotation or vector addition. In this section, both methods will be discussed.

8.3.1 Vector-rotation method

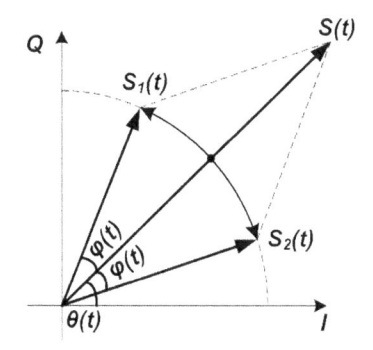

Figure 8.4 shows the phasor diagram of the outphasing signal generation based on the vector-rotation method. The two outphasing vectors $S_1(t)$ and $S_2(t)$ have the mathematical form of (8.2) and (8.3). From Fig. 8.4, we can identify the two steps that are required to generate the outphasing signal. The first step is to obtain the desired phases (i.e. $\pm\varphi(t)$ or $\theta(t) \pm \varphi(t)$) based on the original modulated signal. Owing to the nonlinear trigonometric functions involved, it is usually processed in digital baseband, which can be realized by the CORDIC[1] processor [13], look-up table (LUT) mapping [7], polynomial approximation [11], or a combination of the above techniques. The second step is to rotate the carrier or amplitude-normalized $S(t)$ (i.e. $S'(t) = e^{j(\omega_c t + \theta(t))}$) by the angle obtained in the previous step and generate the outphasing signal. In other words, a phase modulator is required to apply the modulated phase onto the carrier or $S'(t)$. A PLL-based modulator [14] is extensively used in low-GHz designs for frequency and phase modulation. However, due to its narrowband nature, the signal bandwidth is usually limited to 10 MHz.

Fig. 8.4 Phasor diagram of the outphasing signal generation based on the vector-rotation method.

[1] CORDIC stands for COordinate Rotation DIgital Computer, which is an efficient algorithm to calculate hyperbolic and trigonometric functions.

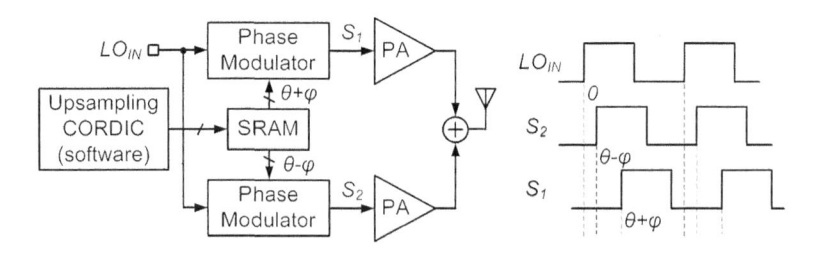

Fig. 8.5 Delay-based digital SCS architecture [4].

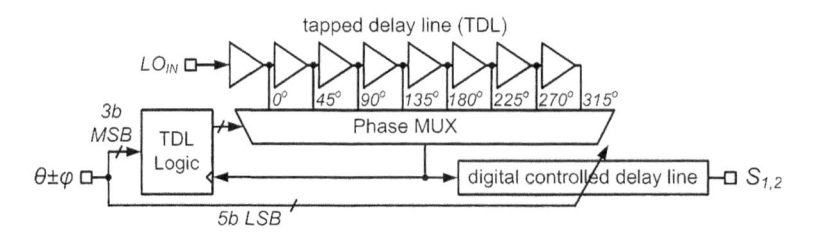

Fig. 8.6 Open-loop delay-based phase modulator [4].

Figure 8.5 shows the architecture of an open-loop delay-based outphasing transmitter [4], which can afford relatively wide-signal bandwidth. The phase modulation data are computed digitally from the original I/Q signals through the up-sampling stages and CORDIC processor. The results are fed to the phase modulator through an on-chip SRAM and then applied to each of the LO edges by the phase modulator. The sampling rate of the phase modulator is as high as the LO frequency (i.e. 2.4-GHz carrier) to minimize the quantization noise. Figure 8.6 shows the block diagram of the delay-based phase modulator. It consists of a coarse 3-b tapped delay line (TDL) and a fine 5-b digitally controlled delay line (DCDL), achieving a total resolution of 8 bits. The phase MUX is used to select the proper phase from TDL, as illustrated in Fig. 8.6. The DCDL is composed of three identical cascaded delay stages. Each of them consists of an inverter and 4-b binary weighted bank of switched MOS capacitors. It achieves a delay resolution of about 1.6 ps (i.e. 1.4° at 2.4 GHz). One of the advantages of the delay-based modulator is its straightforward calibration scheme of correcting the mismatch between delay elements. The output of DCDL is looped back to the input (i.e. LO_{IN}) and the phase modulator is now configured as a ring oscillator. By stepping the TDL tap and counting the pulses over a fixed time window, the delay of each tap is measured and the delay mismatch of each tap between the two paths can then be corrected. The calibration data that are measured on startup are also stored in the SRAM. Designed in 32-nm CMOS, the outphasing transmitter achieves 20-dBm average output power with a system efficiency of 18.6% while transmitting 54-Mb/s 64QAM. It is capable of delivering wideband OFDM 64QAM up to 40 MHz.

With high-speed logic circuits and bandwidth enhancement technique [15], the delay-based phase modulator also has the potential to work in the mm-wave range. The limiting factor here is the phase resolution that can be achieved. At 60 GHz, to achieve

8.3 Outphasing signal generation

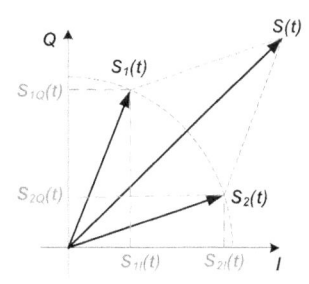

Fig. 8.7 Phasor diagram of outphasing vectors and their I/Q components.

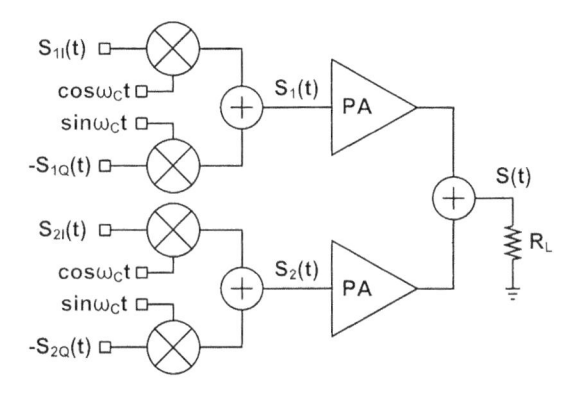

Fig. 8.8 I/Q modulator-based outphasing signal generation.

a phase resolution of 2°, the incremental delay should be as small as 0.1 ps. In addition, it inevitably consumes relatively high power [16].

Instead of generating outphasing signals directly at carrier frequency, the constant-envelope outphasing vectors $S_1(t)$ and $S_2(t)$ (see Fig. 8.7) can be further decomposed into four components at baseband (see Fig. 8.8) given by

$$S_{1I}(t) = \frac{1}{2} A_M \cos\left[\theta(t) + \varphi(t)\right], \tag{8.7}$$

$$S_{1Q}(t) = \frac{1}{2} A_M \sin\left[\theta(t) + \varphi(t)\right], \tag{8.8}$$

$$S_{2I}(t) = \frac{1}{2} A_M \cos\left[\theta(t) - \varphi(t)\right], \tag{8.9}$$

$$S_{2Q}(t) = \frac{1}{2} A_M \sin\left[\theta(t) - \varphi(t)\right]. \tag{8.10}$$

As shown in Fig. 8.8, a conventional quadrature modulator will be used in each branch for up-conversion and forming the outphasing signals $S_1(t)$ and $S_2(t)$ represented by

$$S_1(t) = S_{1I}(t)\cos(\omega_C t) - S_{1Q}(t)\sin(\omega_C t), \tag{8.11}$$

$$S_2(t) = S_{2I}(t)\cos(\omega_C t) - S_{2Q}(t)\sin(\omega_C t). \tag{8.12}$$

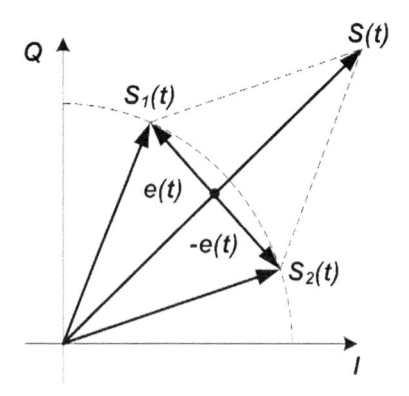

Fig. 8.9 Phasor diagram of the outphasing signal generation based on the vector-addition method.

It is also important to realize that the outphasing baseband signals $S_{1I}(t), S_{1Q}(t)$ and $S_{2I}(t), S_{2Q}(t)$ that constitute the outphasing vectors do not have a constant envelope. Its properties will be analyzed in more detail in Section 8.5.1.

8.3.2 Vector-addition method

Alternatively, the two outphasing signals can also be constructed by vector addition as shown in Fig. 8.9. We have

$$S_1(t) = \frac{1}{2} S(t) + e(t), \tag{8.13}$$

$$S_2(t) = \frac{1}{2} S(t) - e(t), \tag{8.14}$$

where $e(t)$ is in quadrature to the input signal $S(t)$, given by

$$e(t) = j \cdot \frac{1}{2} S(t) \cdot \sqrt{\frac{A_M^2}{|S(t)|^2} - 1}. \tag{8.15}$$

We can see that $e(t)$ needs to be shifted in phase by 90° and scaled properly according to the amplitude of the signal. The quadrature phase error between $S_{1,2}(t)$ and $e(t)$ will severely degrade the matching between two outphasing paths and thus limit the system linearity [17]. The nonlinear square-root operation involved in (8.15) can be realized by feedback technique [18], translinear circuit [17], or simply a look-up table (LUT).

With the phase-addition method, $S_1(t)$ and $S_2(t)$ can also be generated based on quadrature up-conversion with the four baseband signals represented by

$$S_{1I}(t) = \frac{1}{2} S_I(t) + e_I(t), \tag{8.16}$$

$$S_{1Q}(t) = \frac{1}{2} S_Q(t) + e_Q(t), \tag{8.17}$$

$$S_{2I}(t) = \frac{1}{2} S_I(t) - e_I(t), \tag{8.18}$$

8.3 Outphasing signal generation

$$S_{2Q}(t) = \frac{1}{2}S_Q(t) - e_Q(t). \tag{8.19}$$

Figure 8.10 shows the $S_1(t)$ and $e(t)$ vectors and their I/Q components. The decomposed signal $e(t)$ can be written as

$$e(t) = \left[-\frac{1}{2}S_Q(t) + j \cdot \frac{1}{2}S_I(t)\right]\sqrt{\frac{A_M^2}{|S(t)|^2} - 1}. \tag{8.20}$$

Interestingly, after decomposition the phase difference between $S_{I,Q}(t)$ and $e_{Q,I}(t)$ becomes 180° (or 0°), which can be simply generated by an inverter (or a delay module). The following example shows a way to generate the four baseband outphasing signals based on the analog feedback technique.

Figure 8.11 shows the block diagram of the analog baseband outphasing signal generator [18]. By following the signal flow in the diagram, it is not difficult to recognize the corresponding parts that realize Eqns (8.16)–(8.20). When the loop is closed, the difference between $|S_1(t)|^2 + |S_2(t)|^2$ and $0.5A_M^2$ will be amplified by the gain of the

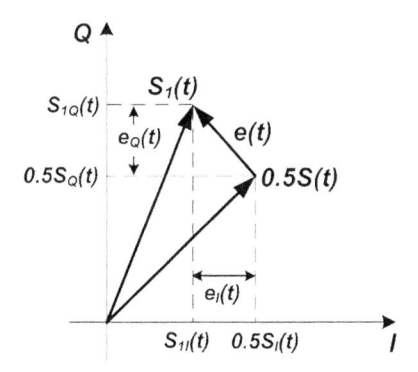

Fig. 8.10 The phasor diagram of $S_1(t)$, $e(t)$, and their I/Q components.

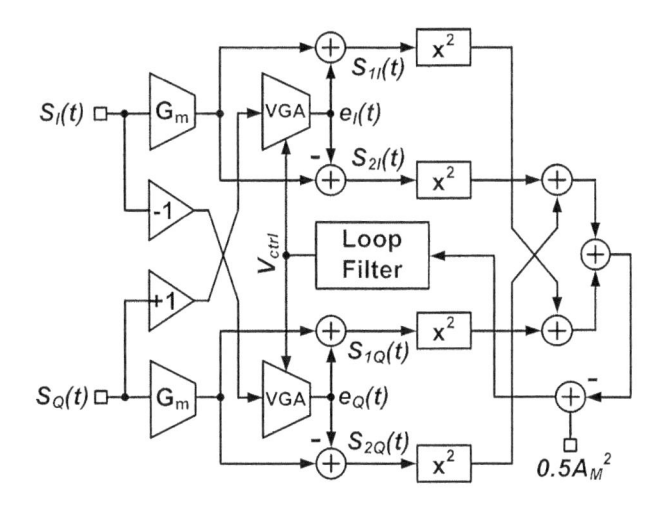

Fig. 8.11 Baseband outphasing signal generation based on the analog feedback technique [18].

312 Outphasing mm-wave silicon transmitters

VGA. The signals $e_{I,Q}(t)$ will then be adjusted to force $|S_1(t)|^2 + |S_2(t)|^2$ to equal $0.5A_M^2$. Note that the loop has to respond fast enough to cope with the fast variation of the input signal and avoid glitches. To fulfill the 802.11a/g WLAN specification, the loop bandwidth is designed to be 200 MHz ($>10\times$ signal bandwidth) with a maximum loop gain of 40 dB.[2] In the measurement, the generated outphasing baseband signals will then be up-converted by an off-chip I/Q modulator and combined by a Wilkinson power combiner. Designed in 0.25 μm CMOS (a relatively old CMOS technology node), it achieves 54 Mbps OFDM 64QAM with measured EVM of -34.1 dB and ACPR of -33.9 dB when no calibration is applied. The measured results are well within the 802.11a/g specification, which leaves some margin for potential performance degradation with the RF front-end integrated.

8.3.3 Asymmetric multi-level outphasing

The efficiency of outphasing can be further improved by allowing the PA to have dynamic supply voltage rather than a single supply level. Figure 8.12 shows the phasor diagram of the asymmetric multi-level outphasing (AMO) technique [6], which demonstrates such a concept.[3] The corresponding equations of AMO are given below:

$$S(t) = S_1(t) + S_2(t) = A(t) \cdot e^{j(\omega_c t + \theta(t))}, \tag{8.21}$$

$$S_1(t) = A_1(t) \cdot e^{j(\omega_c t + \theta(t) + \varphi_1(t))}, \tag{8.22}$$

$$S_2(t) = A_2(t) \cdot e^{j(\omega_c t + \theta(t) - \varphi_2(t))}, \tag{8.23}$$

$$\varphi_1(t) = \cos^{-1}\left[\frac{A_1(t)^2 + A(t)^2 - A_2(t)^2}{2A_1(t)A(t)}\right], \tag{8.24}$$

$$\varphi_2(t) = \cos^{-1}\left[\frac{A_2(t)^2 + A(t)^2 - A_1(t)^2}{2A_2(t)A(t)}\right]. \tag{8.25}$$

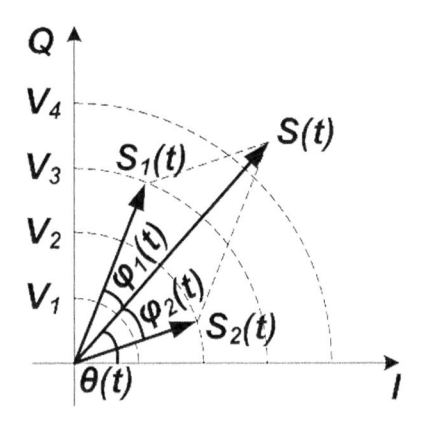

Fig. 8.12 Phasor diagram of asymmetric multi-level outphasing technique.

[2] The loop gain is nonlinear to the input signal and reaches its maximum value when $|S(t)| = A_M/2$.
[3] Such a concept can be considered as a combination of outphasing and polar techniques.

In the AMO system, the supply voltages of the two PAs can be different and will be selected dynamically from a discrete set of voltages (four levels of voltages V_1–V_4 shown in Fig. 8.12). The switching PAs (e.g. class-E) are used so that the signal amplitudes $A_1(t)$ and $A_2(t)$ are in proportion to the supply voltage applied. We can also see the outphasing angles of AMO do not equal (i.e. $\varphi_1(t) \neq \varphi_2(t)$), but depend on the values of $A_1(t)$ and $A_2(t)$. For a given $S(t)$, multiple sets of $A_1(t)$ and $A_2(t)$ can be chosen. The values of $A_1(t)$ and $A_2(t)$ can then be optimized to minimize the total power consumption. The work in [11] proposed a fixed-point piece-wise linear approximation algorithm to compute the nonlinear functions and generate AMO signals. It achieved a throughput of 3.4 GS/s with 12-b phase accuracy but consumes relatively high DC power (323 mW). This technique may also be adopted in a symmetric outphasing system. Lower power consumption can be expected due to reduced system complexity. We would like to mention that as the AMO technique relies on the use of the switching PAs, its benefits may diminish with increase of the operating frequency. It is because the switching PA has low power gain and does not outperform the saturated class-A/AB PA at mm-wave frequencies.

8.4 Outphasing signal combining

8.4.1 Isolating and non-isolating combiners

In the outphasing system, a three-port power combiner is needed to combine the two outphasing signals and reconstruct the original amplitude- and phase-modulated signal at the transmitter output. As a lossless isolated three-port combiner does not exist [19], the combiner of the outphasing system falls into two categories: the lossy isolating combiner and the lossless non-isolating one. Considering that the PA in outphasing operates with peak efficiency, the combining structure employed will then determine the efficiency of the whole system. Figure 8.13 shows the Wilkinson combiner that provides sufficient isolation between ports 1 and 2. The outphasing PA with an isolated combiner is usually referred to as the LINC (linear amplification with nonlinear component) system [20, 21].

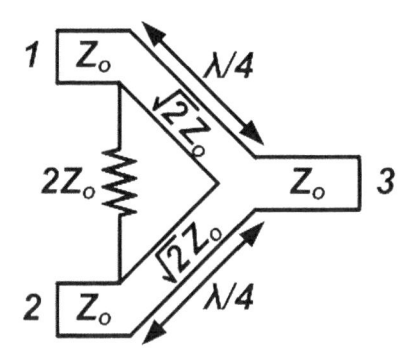

Fig. 8.13 Wilkinson power combiner for LINC system (ports 1 and 2 are isolated).

With an isolating combiner, both branch PAs see a constant load. However, such a combiner has low power-transfer efficiency due to the dissipation in the isolation resistor of the Wilkinson combiner. The average efficiency of the combiner depends on the PDF of the signal and is usually much less than 50% for the complex modulated signal (i.e. 16QAM or 64QAM). For instance, a 5.8-GHz fully integrated outphasing power amplifier with two class-E PAs and Wilkinson combiner [22] achieved 47% peak efficiency while its back-off efficiency is even lower than a conventional class-A PA due to the dissipated power in the isolation resistor. The work in [23] proposed a method to recycle the dissipated power. However, it requires a complex passive network and resonant rectifiers, which are not favored for IC implementation.

The non-isolating combiner can be a lossless Wilkinson-type combiner (i.e. without the isolation resistor) or two quarter-wave-length ($\lambda/4$) transmission lines (T-lines) [24]. The $\lambda/4$ T-lines transform the output of each unit PA (assumed to be a voltage source in outphasing) to a current source which can be summed directly at the load (see Fig. 8.14). The $\lambda/4$ T-lines can also be realized by a lumped LC network to facilitate the on-chip integration [25]. With a non-isolating combiner, the impedance seen by the PA varies due to the interaction between the two ports. As a result, the power delivered by each PA is no longer constant but depends on the outphasing angle (i.e. signal amplitude), which significantly improves the back-off efficiency compared with the LINC system. The load modulation of outphasing will be further discussed in Section 8.4.2. Chireix [26] introduced compensation reactances to further improve the back-off efficiency. As shown in Fig. 8.14, additional reactances $+jB$ and $-jB$ are introduced at the PA outputs to cancel the parasitic reactances that appear at back-off due to the interaction between two PAs. The value of B can be chosen depending on the PDF of the modulated signal for a high average efficiency. In [27], a direct comparison between Chireix (non-isolating) and Wilkinson (isolating) combiners shows that the Chireix combiner has considerable improvement in back-off efficiency while still achieving good linearity. Note that the additional efficiency improvement from the compensation reactances may be compromised at RF or mm-wave frequencies due to the loss of the on-chip passives, complexity of the layout floor plan, and degradation of the symmetry between two

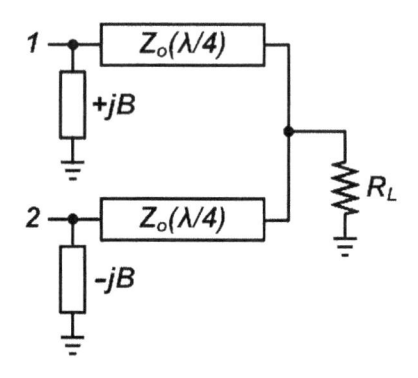

Fig. 8.14   Chireix outphasing power combiner.

outphasing signal paths, which limits the Chireix combiner being fully integrated. The work in [5] proposed realizing the compensation reactances by two capacitor banks to facilitate the integration and on-chip tunability. The effective capacitance at the output of each PA can be adjusted separately and dynamically. The outphasing PA [5] achieves 13 (13.6) dBm average output power and 30% (44%) average efficiency while meeting the WCDMA (EDGE) specification at 900 MHz. However, such a technique is also not preferred at mm-wave since it considerably increases the complexity at mm-wave PA outputs, reduces the peak output power that can be potentially achieved, and limits the signal bandwidth far below GHz.

8.4.2 Transformer-based combiner and load-modulation effect

The transformer-based power combiner is extensively used in RF and mm-wave PA designs [7, 9] as it ensures a compact layout and thus reduces the losses associated with the long interconnects. The series transformer-based combiner can also be employed in an outphasing system [12, 28] which readily sums the outputs from two PAs at its secondary winding. Figure 8.15 shows the simplified schematic of the outphasing PA. The saturated PA output stage is modeled by a voltage source with an output impedance of R_O [27, 29]. The parasitic capacitance C_P of the output stage is resonated out by the magnetizing inductance L_M of the ideal coupled transformer, which does not affect the analysis. From Fig. 8.15, the combined signal at the output equals

$$V_o = \frac{R_L}{R_L + 2R_O} \cdot A(t) \cdot e^{j(\omega_c t + \theta(t))}. \qquad (8.26)$$

We can see the modulated signal at the transformer output is correctly reconstructed and the amplitude of the signal is only linearly scaled with $R_L/(R_L + 2R_O)$. For outphasing operation, a PA with low R_O is preferred, which delivers more output power. Besides, the output impedance of unit PA is nonlinear. The time-varying load impedances will generate distortion at the output [28], while the distortion can be kept to a minimum if $R_O \ll R_L$. In this sense, a class-D PA is considered to be best for outphasing as its output impedance remains low and relatively constant during operation. However, with current advanced CMOS technology, the switching class-D PA still underperforms the conventional class-A or -AB PAs [30]. The distortion of the mm-wave outphasing PA will be demonstrated with a 60-GHz design example in Section 8.6.

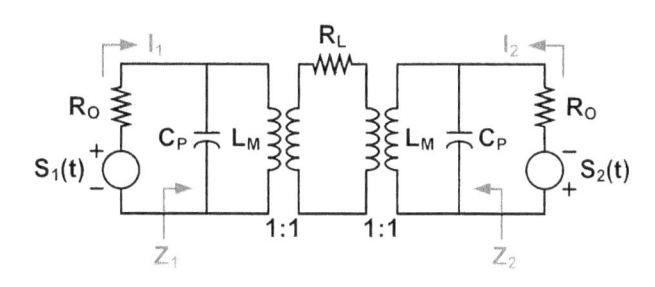

Fig. 8.15 Simplified schematic of the outphasing PA output stage.

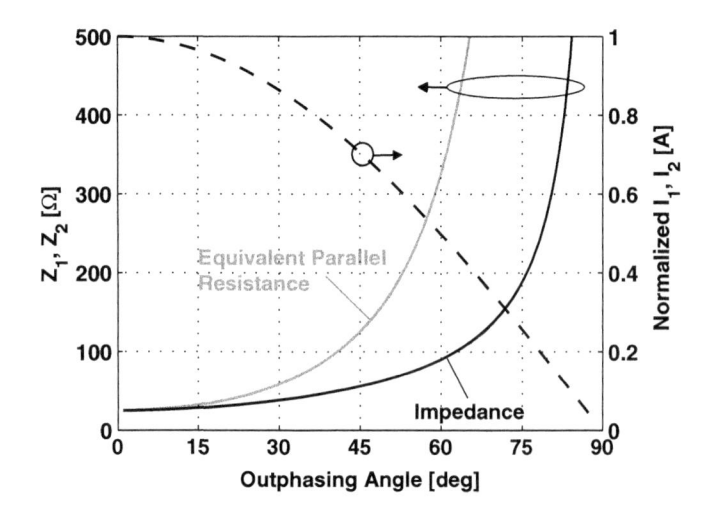

Fig. 8.16 PA load impedance and supply current vs outphasing angle.

In an outphasing PA, although the amplifiers are driven into saturation and operate with constant voltage amplitude, the supply current heavily depends on the outphasing angle due to the limited isolation provided by the combiner. Owing to the interaction between two PAs, the load impedances seen by each PA vary with the outphasing angle, given by

$$Z_1 = \frac{R_L}{2} \cdot \left[1 - j \left(1 + \frac{2R_O}{R_L} \right) \tan \varphi \right], \quad (8.27)$$

$$Z_2 = \frac{R_L}{2} \cdot \left[1 + j \left(1 + \frac{2R_O}{R_L} \right) \tan \varphi \right]. \quad (8.28)$$

As shown in Fig. 8.16, the impedances and equivalent parallel resistances of Z_1 and Z_2 increase at larger outphasing angle (i.e. lower output power), which results in a drop of the supply currents (I_1 and I_2) for the same voltage amplitude.[4] This load modulation effect reduces power consumption at back-off, providing the outphasing PA efficiency benefit over the LINC system where a lossy isolating combiner is used.

8.4.3 Signal combining by beamforming

Beamforming is commonly used at mm-wave frequencies, which provides spatial directivity and array gain to increase the spectral efficiency and channel capacity. It can also be adopted in the outphasing system. Instead of combining the outphasing signals on-chip, the outputs of the two transmitters can be combined in space. Figure 8.17 shows the architecture of a 60-GHz dual-transmitter outphasing system. The outphasing baseband signals S_{1I} and S_{1Q} are applied to one transmitter while S_{2I} and S_{2Q} are applied to the other. They are up-converted separately by the quadrature modulator in each transmitter.

---

[4] In Fig. 8.16, $R_L = 2R_O = 50 \, \Omega$ is assumed, which gives more realistic numbers for mm-wave PAs and also achieves the impedance-matching condition.

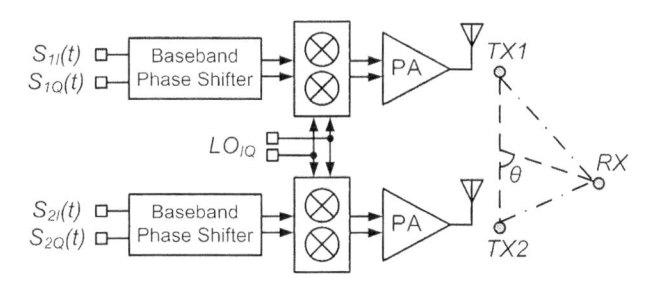

Fig. 8.17 Dual-transmitter outphasing architecture that realizes outphasing signal combining by beamforming [10].

Both transmitters operate at peak output power with constant envelope to achieve high efficiency. We noticed that, by applying conventional outphasing signals, the original amplitude- and phase-modulated signals can only be correctly reconstructed if $\theta = 90°$ (see Fig. 8.17) where two transmitted beams reach the receiver with equal delay. For reception at other spatial angles, the corresponding phases must be adjusted in the two transmitters to compensate the unequal propagation delays of the the beams, which was realized by the phase shifters in Fig. 8.17. Thanks to beamforming, the mm-wave on-chip combiner can be avoided and the total transmitted power is increased due to the additional array gain. The disadvantage of such architecture is that the spectrum transmitted by each transmitter is increased due to the property of the outphasing signal (see Section 8.5.1). This may reduce the spectral efficiency and interfere with other users.

8.5 Outphasing non-idealities

8.5.1 Outphasing signal bandwidth

As shown in Fig. 8.18, the spectrum of the baseband and up-converted outphasing signals (i.e. before the combiner) is increased. This is mainly due to the nonlinear inverse cosine operation involved when generating the baseband outphasing signals. Although the larger bandwidth is relatively easy to achieve for the 60-GHz PAs (due to the high carrier frequency), it will have an influence on the analog baseband (or digital baseband, depending on how the outphasing signal is generated). The wideband baseband circuits inevitably dissipate a large portion of DC power in a 60-GHz system. For instance, the work in [31] presents that a 6-b 3456-MS/s DAC consumes 21 mW. It is therefore instructive to identify the required bandwidth in the baseband signal paths.

System models of the outphasing TX (Fig. 8.8) with a switching PA and I/Q TX with a linear PA are developed in MATLAB and tested with a pulse-shaped 16QAM signal with a peak-to-average power ratio (PAPR) of 7.4 dB (raised-cosine filter with a roll-off factor of 0.35). The simulations are performed by sweeping the baseband bandwidth, leading to the following observations (see Fig. 8.19 and Fig. 8.20).

(1) When narrowing the system bandwidth, the PAPR of the outphasing baseband signal increases. (2) The PAPR of the outphasing baseband signal flattens out to 3 dB

318   **Outphasing mm-wave silicon transmitters**

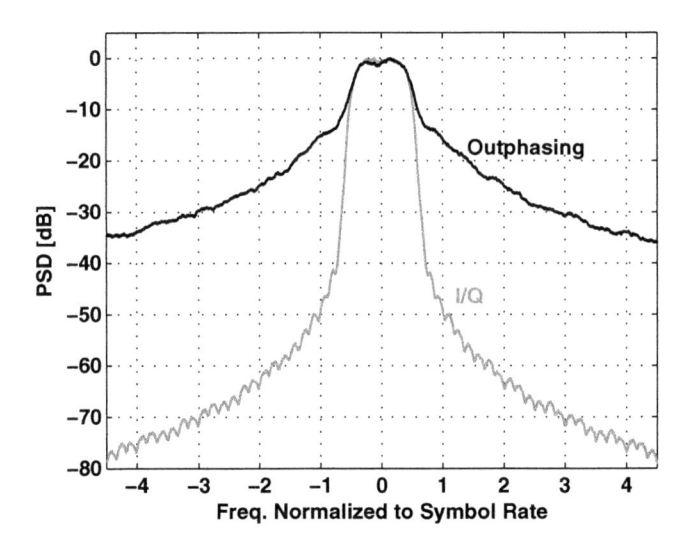

Fig. 8.18   Spectrum of the 16QAM outphasing and I/Q signals.

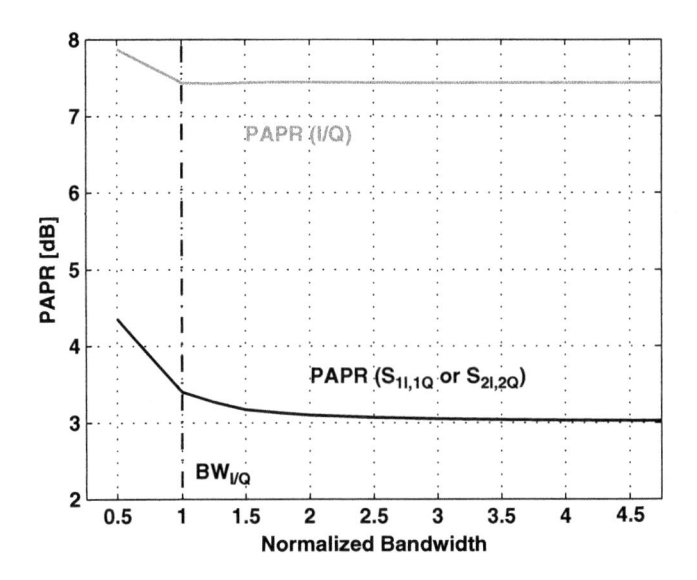

Fig. 8.19   Simulated PAPR of baseband outphasing and I/Q signals vs system bandwidth.

at the double of the required bandwidth of a conventional I/Q signal ($BW_{I/Q}$). (3) To reach an EVM of −40 dB, the outphasing TX requires twice as much bandwidth as the conventional I/Q TX.

From the above considerations, it can be concluded that, for an outphasing TX, the analog baseband circuits, including DAC and low-pass filter, need a bandwidth of $2BW_{I/Q}$, of which the power consumption is in proportion to the signal bandwidth. However, the linearity requirement of the DAC is mitigated since the outphasing baseband signals have lower PAPR than the original I/Q signals, indicating the DAC resolution

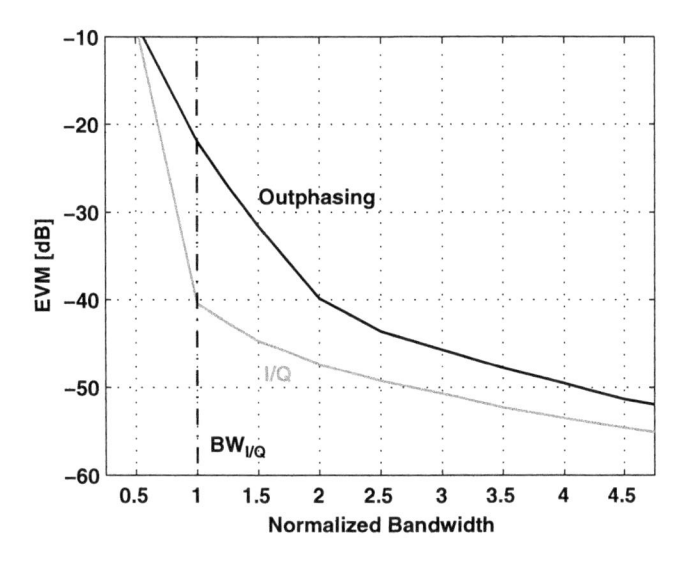

Fig. 8.20 Simulated EVM of outphasing and I/Q TXs vs system bandwidth.

can be reduced by about 1 bit [32]. Therefore, although the outphasing TX requires four DACs [18], its analog baseband consumes comparable power to the conventional I/Q TX, considering that the DAC power consumption is in exponential proportion to its resolution.

As indicated before, the bandwidth of the mm-wave front-end is sufficiently wide to accommodate outphasing while wide bandwidth is easier to achieve than high PAPR in nanometer CMOS. In this regard, the outphasing architecture trades amplitude linearity for phase (time) resolution and bandwidth. This approach matches very well with CMOS technology scaling.

8.5.2 Mismatches between signal paths

As the outphasing technique relies on the combination of the two signals, the amplitude and phase mismatches between two signal paths will make the signal quality deteriorate [33]. Compared with the phase mismatch, gain mismatch can be tolerated for the integrated outphasing TX because both unit PAs are placed in close proximity and driven into saturation. If we consider the phase mismatch between two paths is $\Delta\varphi$, the combined signal is given by

$$S(t)' = S_1(t)' + S_2(t) = (S_1(t) + S_2(t)) + \left(S_1(t)' - S_1(t)\right) \tag{8.29}$$

$$= S(t) + \frac{A_M}{2} \cdot e^{j(\omega_c t + \theta(t) + \varphi(t))}(e^{j\Delta\varphi} - 1) \tag{8.30}$$

$$= S(t) + S_1(t) \cdot (e^{j\Delta\varphi} - 1) \tag{8.31}$$

$$\approx S(t) + j\Delta\varphi \cdot S_1(t). \tag{8.32}$$

Since $S_1(t)$ is a constant-envelope signal, the minimum output signal is now different from zero and in proportion to the phase mismatch between the two outphasing signal

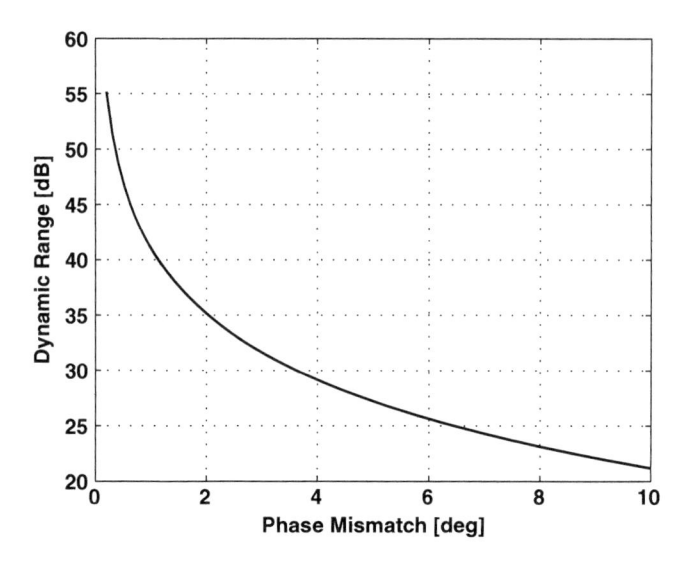

Fig. 8.21 The decrease of dynamic range due to the phase mismatch.

paths. If we take the ratio of the maximum signal power to the minimum signal power, the dynamic range of the outphasing system is given by

$$DR = \left|\frac{S_c(t)'_{MAX}}{S_c(t)'_{MIN}}\right|^2 \approx \left|\frac{2}{e^{j\Delta\varphi}-1}\right|^2 \approx \frac{4}{\Delta\varphi^2}. \tag{8.33}$$

Therefore, a phase mismatch of 2° will limit the dynamic range to about 35 dB. Besides, the additional term in (8.32) will increase the EVM and cause spectral regrowth. Figure 8.22 shows that the EVM of the 16QAM degrades to about −20 dB and the adjacent channel power ratio (ACPR) is worsened by 20 dB with a phase mismatch of 6°. Note that mismatches may come from the baseband, LO, and RF signal paths. At baseband, sufficient accuracy must be guaranteed when generating the outphasing signals from conventional I/Q signals. The design and layout techniques, such as proper transistor sizing and symmetrical layout floor plan, can be employed to minimize the mismatch in the LO and RF signal paths [12]. The work in [33] proposed introducing an additional feedback loop to correct the mismatch and the measured results showed the ACPR was further improved by about 7–10 dB. It is worth mentioning that a certain phase mismatch will translate to different delay mismatch at baseband and RF. A phase mismatch of 6° corresponds to 16.6 ps at 1 GHz (baseband signal path) and 0.3 ps at 60 GHz (RF signal path). Such small delay offset at 60 GHz can only be calibrated in the phase domain at the digital baseband, which is not good for wideband operation where true-time-delay calibration is required [34]. Therefore, the wideband performance of the mm-wave outphasing TX is limited by the layout mismatch due to the process variation, especially in the mm-wave part.

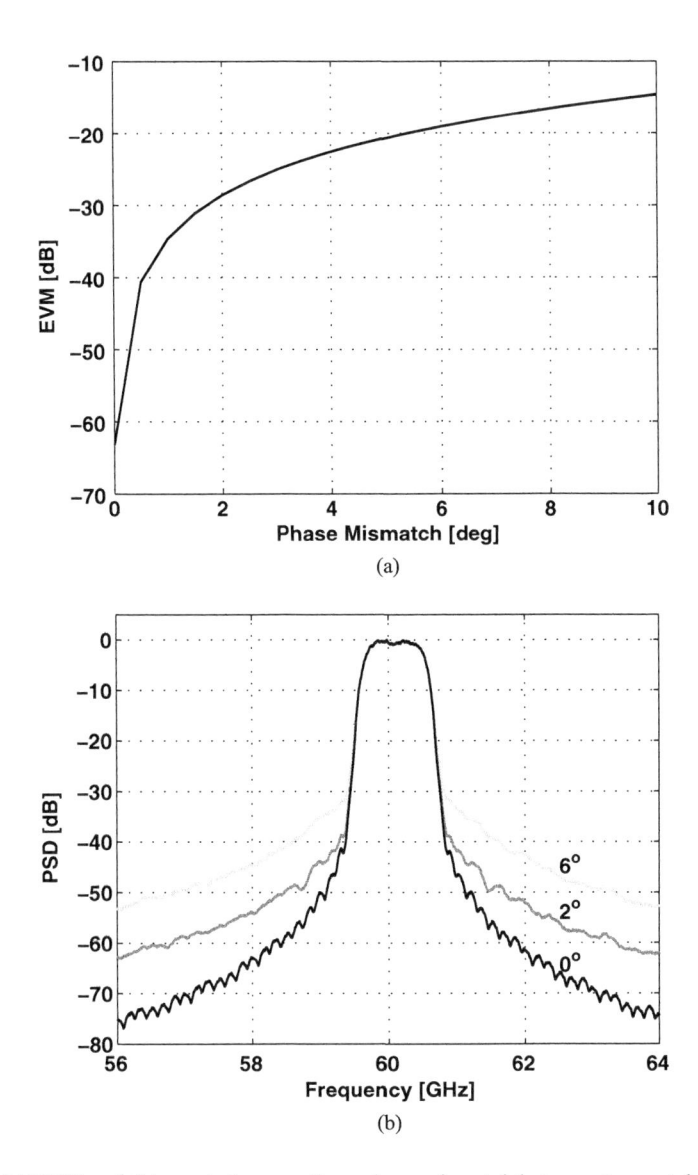

Fig. 8.22 Simulated (a) EVM and (b) spectral regrowth vs phase mismatch between two outphasing signal paths.

8.6 Case study: 60-GHz outphasing transmitter

From the above discussions, we have understood that an outphasing TX is able to achieve higher average efficiency for the same linear output power when a complex modulated signal is applied. It is suitable to be fully integrated in most advanced CMOS technologies, which makes it a good candidate for mm-wave applications. There have been a few implementations reported that adopt the outphasing concept at mm-wave frequencies [10, 12]. In this section, the design and implementation of a 60-GHz outphasing TX in 40-nm bulk CMOS technology will be discussed [12].

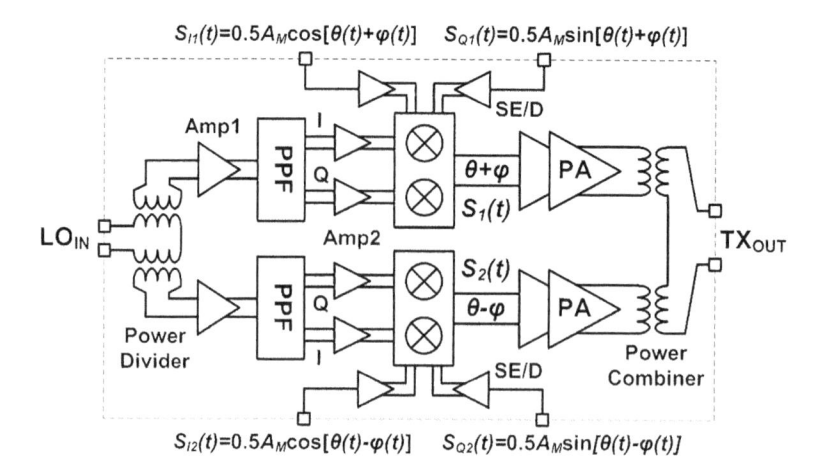

Fig. 8.23 System diagram of the 60-GHz outphasing transmitter [12].

8.6.1 Transmitter architecture

Figure 8.23 shows the system diagram of the outphasing transmitter. The constant-envelope outphasing signals at PA inputs are generated based on I/Q modulator architecture (see Fig. 8.7) where four baseband signals $S_{1I}(t)$, $S_{1Q}(t)$ and $S_{2I}(t)$, $S_{2Q}(t)$ are provided by the external arbitrary waveform generator (AWG).

The TX consists of single-ended to differential (SE/D) amplifiers, poly-phase filters (PPFs), LO buffer amplifiers, an I/Q modulator, two PAs, and a power combiner. The SE/D amplifier simplifies the interfacing and is terminated with 50-Ω resistor at the input to facilitate the measurement. The differential baseband signals are then fed to the four I/Q modulators for up-conversion and constructing the outphasing signals. The I/Q modulator consists of two double-balanced Gilbert mixers. It is worth mentioning that although the outphasing TX needs one additional I/Q modulator compared with the conventional I/Q TX with power combining PA, the required output power of the modulator in outphasing is also reduced by a factor of two. Therefore, the outphasing TX dissipates comparable power in the up-conversion modulator to the I/Q TX, assuming the same driving power required by the PA. The potential disadvantage of this IQ modulator-based architecture is that the TX linearity is sensitive to the quadrature errors of the modulator [33]. To minimize the quadrature errors, relatively large transistor size is used in the transconductance stage of the modulator to mitigate the random mismatch and DC offset. Besides, a two-stage parasitic-compensated PPF is used to generate the I/Q LOs. The measured image-rejection of the PPF is about 35 dB around 60 GHz [35]. The 60-GHz outphasing signals at PA inputs (i.e. S_1 and S_2) are amplified by two highly efficient saturated PAs and then summed at the output by a power combiner to reconstruct the amplitude modulation. The non-isolating transformer-based power combiner is used in this design which offers efficiency benefit at back-off.

Note that this TX can also operate in conventional I/Q direct-conversion mode when $S_{1I}(t) = S_{2I}(t) = I(t)$ and $S_{1Q}(t) = S_{2Q}(t) = Q(t)$. To show the benefits of the outphasing

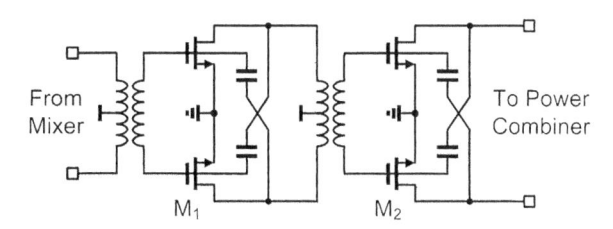

Fig. 8.24 Simplified schematic of 60-GHz neutralized power amplifier.

technique, the measurement results of both operating modes, namely the outphasing and conventional I/Q modes, will be compared.

8.6.2 Power amplifier and combiner

The power amplifier in the outphasing TX incorporates two branch PAs with a power combiner at the output. Figure 8.24 shows the schematic of the branch PA. In the design, the neutralization technique is adopted by cross-connecting the interdigitated MOM capacitors between the drain and gate terminals of the differential stage. The capacitor value is chosen to maximize the reverse isolation, which helps to reduce the design iterations as the optimum load impedance of the output stage is less affected by the driver stage and the matching network. Each neutralized amplifier stage has a simulated power gain of 14.5 dB. The driver stage is sized down by a factor of two, providing sufficient power for the output stage. The outphasing topology leads the branch PA to be optimized for saturated output power and peak efficiency. Owing to the limited reverse isolation, design iterations, including load-pull simulation and matching network design, are performed to assess the device parameters progressively. An optimum load impedance (i.e. $Z_{L,opt} = 13.9 + j20.8$) is chosen to provide an output power of 15 dBm with 39% PAE.

Transformer-based passives are extensively employed in the design, which prove to have better power transfer efficiency than the resonant LC impedance transformation network [36]. In addition, the use of transformers simplifies routing and ensures compact layout. In the design, transformers and the power combiner are implemented in an overlay structure with the top two metals to improve the coupling factor (simulated $k_m = 0.7$). Based on the load-pull simulations, the input and interstage transformers are designed for optimum power matching. The desired inductance values in different transformers can be obtained by changing the diameters and widths of the primary and secondary coils. The input and interstage matching networks, including the interconnects, have insertion losses of 3.5 dB and 2.5 dB, respectively. Figure 8.25 shows a micrograph of the transformer-based power combiner with its cross section. The combiner occupying 0.2×0.1 mm^2 sums the outphasing vectors at the TX output. Since the outphasing angle between these two vectors varies over time, a combiner as in [7] cannot be used. Instead, a combiner with two separate transformers is developed for outphasing operation. The primary side of the combiner has two pairs of differential ports excited by outphasing signals while the center taps provide DC access (see Fig. 8.25). The unwanted magnetic coupling of adjacent windings is minimized by the orthogonal

324   **Outphasing mm-wave silicon transmitters**

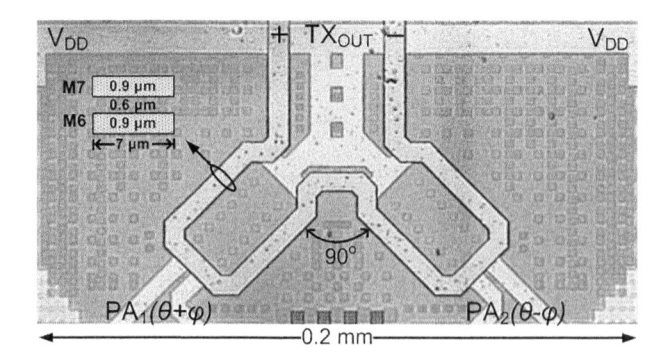

Fig. 8.25   Micrograph of the output power combiner.

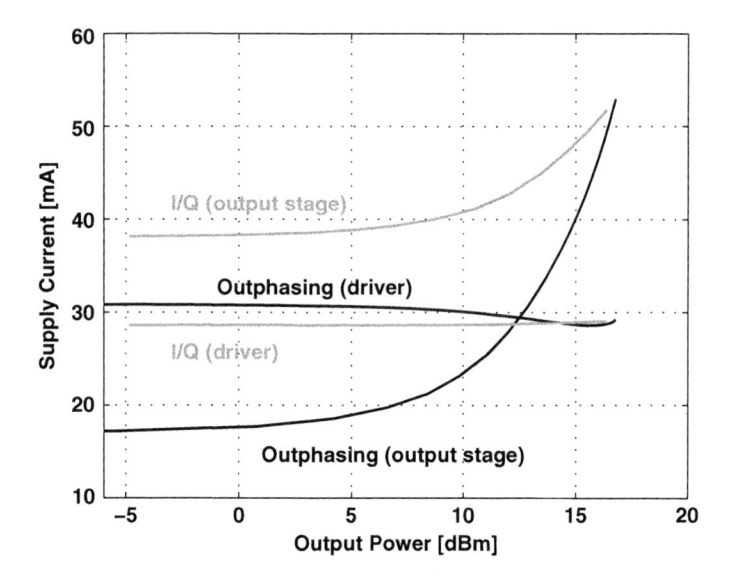

Fig. 8.26   Supply current of the output and driver stages of the unit PA in outphasing and I/Q modes as a function of TX output power.

placement of the two coils. Pad parasitics are used to tune the secondary winding of the combiner. The combiner achieves an insertion loss of 1.2 dB, which reduces the simulated peak *PAE* of the complete PA to 29%. The complete PA with the combiner shows a simulated power gain of 22 dB and saturated output power of 16.5 dBm.

Figure 8.26 compares the current consumption of the output and driver stages of the unit PA when the transmitter operates in the outphasing and conventional I/Q modes. While consuming a similar amount of current in the driver stage, the output stage in outphasing mode saves about 50% current consumption due to the load modulation effect. Note that the consumed current cannot be further reduced at large outphasing angle for two reasons: (1) the increase in the PA load impedance is limited by the loss of the transformer, which presents an equivalent loss resistance in parallel with the PA

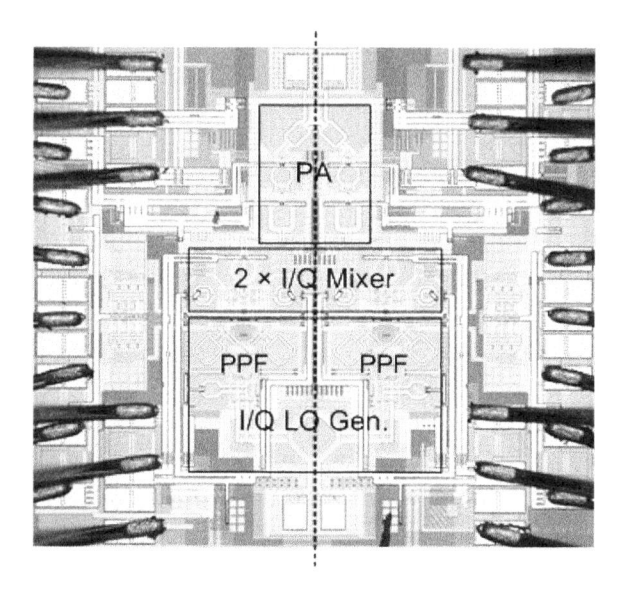

Fig. 8.27 Micrograph of the 60-GHz outphasing transmitter [12].

output stage; (2) the output resistance of the transistor inevitably shunts some supply currents to ground and dissipates DC power.

8.6.3 Floor plan

Mismatch between the two outphasing signals $S_1(t)$ and $S_2(t)$ (see Fig. 8.23) may make signal constellations deteriorate and cause spectral regrowth. Therefore, special attention is paid to the TX floor plan and the physical layout, to minimize the asymmetries in baseband, in I/Q LO distribution, and between the $S_1(t)$ and $S_2(t)$ signal paths. Figure 8.27 shows the die micrograph. It is seen that the two outphasing signal paths are perfectly symmetrical against the central dashed line. The floor plan is designed in a compact manner so that the building blocks in the up-conversion chain are connected naturally, which minimizes the length of 60-GHz interconnects, avoids potentially asymmetrical interconnects, and hence alleviates mismatches. Signal losses in the 60-GHz band can also be reduced by the compact floor plan. Measurement confirms that the impact of mismatches between two outphasing signal paths is small.

8.6.4 Measurement results

The outphasing transmitter is fabricated in an ST 40-nm bulk CMOS technology. The chip occupies 0.96 mm^2 with an active area of only 0.33 mm^2 (see Fig. 8.27). The measurements of the 60-GHz outphasing TX are performed on a high-frequency probe station. The DC and baseband signal pads are wire-bonded to a PCB, while LO and RF pads are directly accessed by 50-Ω GS probes which facilitate the differential design approach [8]. As mentioned in Section 8.6.1, the TX can operate in two modes, the

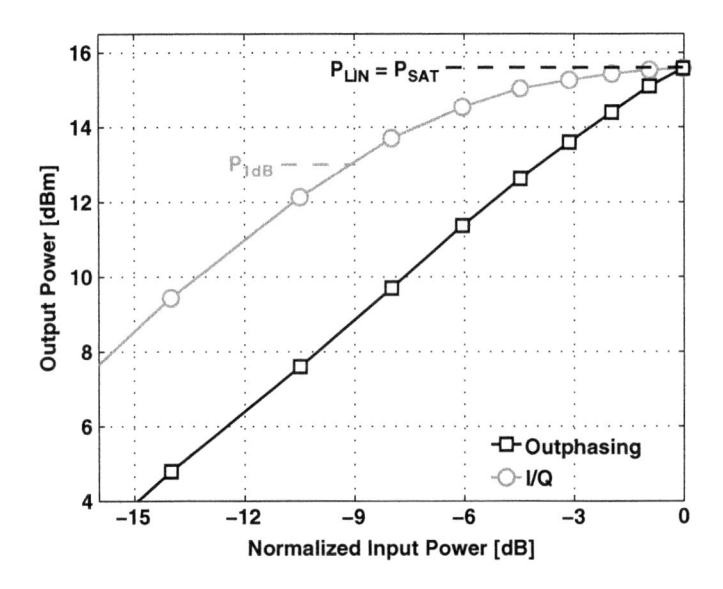

Fig. 8.28 Measured output power in outphasing and I/Q modes.

outphasing mode and the conventional I/Q mode. In this section, we will show the CW and modulated signal measurement results, demonstrating the benefits of outphasing. Readers can refer to [12] for the complete characterization of the transmitter.

CW measurements

The single-tone measurement was performed with a V-band power sensor and a high-precision power meter. The two major benefits of the TX in outphasing mode are demonstrated in Fig. 8.28 and Fig. 8.29, which show the measured output power and the PA power consumption for outphasing and I/Q modes. In Fig. 8.29, the normalized input power in outphasing mode is calculated from the applied outphasing angle according to (8.4). The TX in I/Q mode starts to saturate at a P_{1dB} of 13 dBm and shows a P_{SAT} of 15.6 dBm while the outphasing TX operates linearly up to the maximum output power of 15.6 dBm ($P_{LIN} = P_{SAT}$). Consequently, the outphasing TX does not need to operate at back-off when processing variable-envelope signals. Besides, the complete outphasing PA consumes about 13% less DC power at back-off because of the load-modulation effect. The *PAE*s of both modes are shown in Fig. 8.30. The outphasing PA shows only 2%–3% improvement in the back-off efficiency due to the parasitic effects discussed in Section 8.6.2. However, the key point here is that the outphasing PA can operate with P_{SAT} and peak *PAE*, leading to higher average output power (P_{AVG}) and better average efficiency with complex modulation.

Linearity and mismatch measurements

Distortions in PAs and TXs can be characterized by the AM–AM and AM–PM effects. For an outphasing TX this translates into φ–AM and φ–PM distortions. The distortion curves were measured by slowly changing the outphasing angle and capturing the I/Q

8.6 Case study: 60-GHz outphasing transmitter

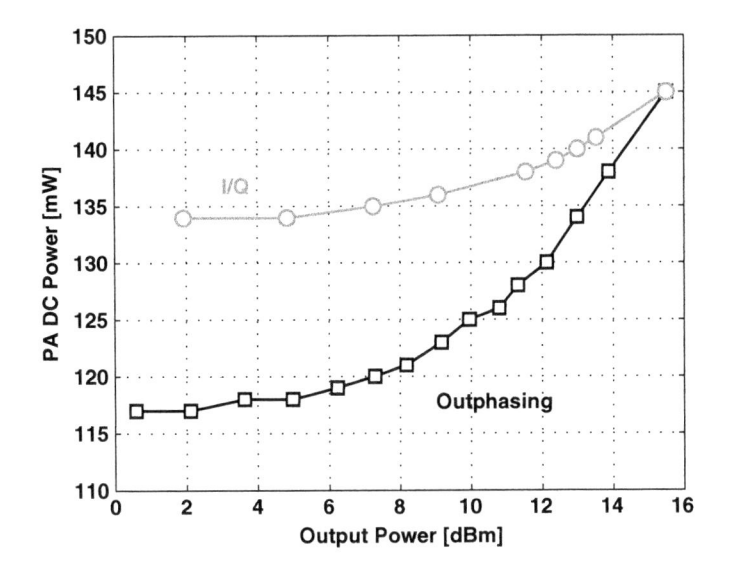

Fig. 8.29 Measured DC power consumption of the PA in outphasing and I/Q modes.

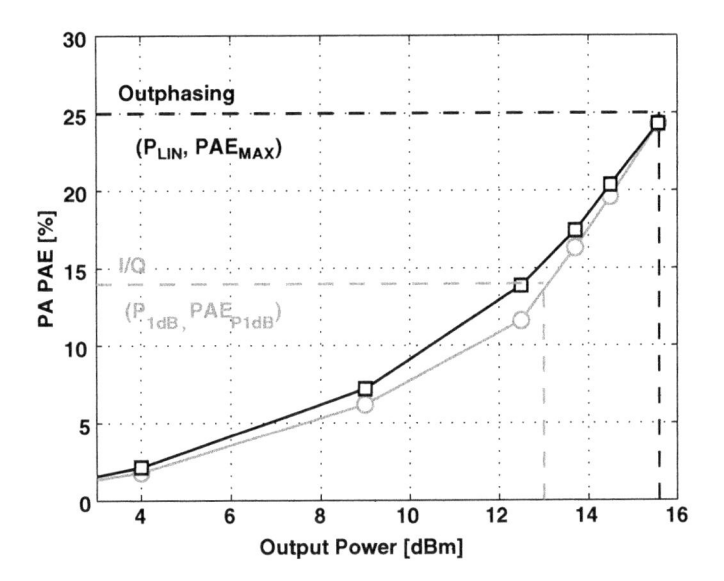

Fig. 8.30 Measured *PAE* of the PA in outphasing and I/Q modes.

down-converted data by a vector signal analyzer. From this data, the φ–AM and φ–PM curves were generated.

Figure 8.31 shows the measured results and compares them with the simulated ones when PA operates at P_{SAT}. The measured φ–AM distortion is less than 0.6 dB across the whole signal range. It is much smaller than the AM–AM distortion in conventional PAs, which is normally larger than 5 dB at P_{SAT} [7, 8, 36]. The measured φ–PM distortion is about 15°. There is good correspondence between the measurement and simulation

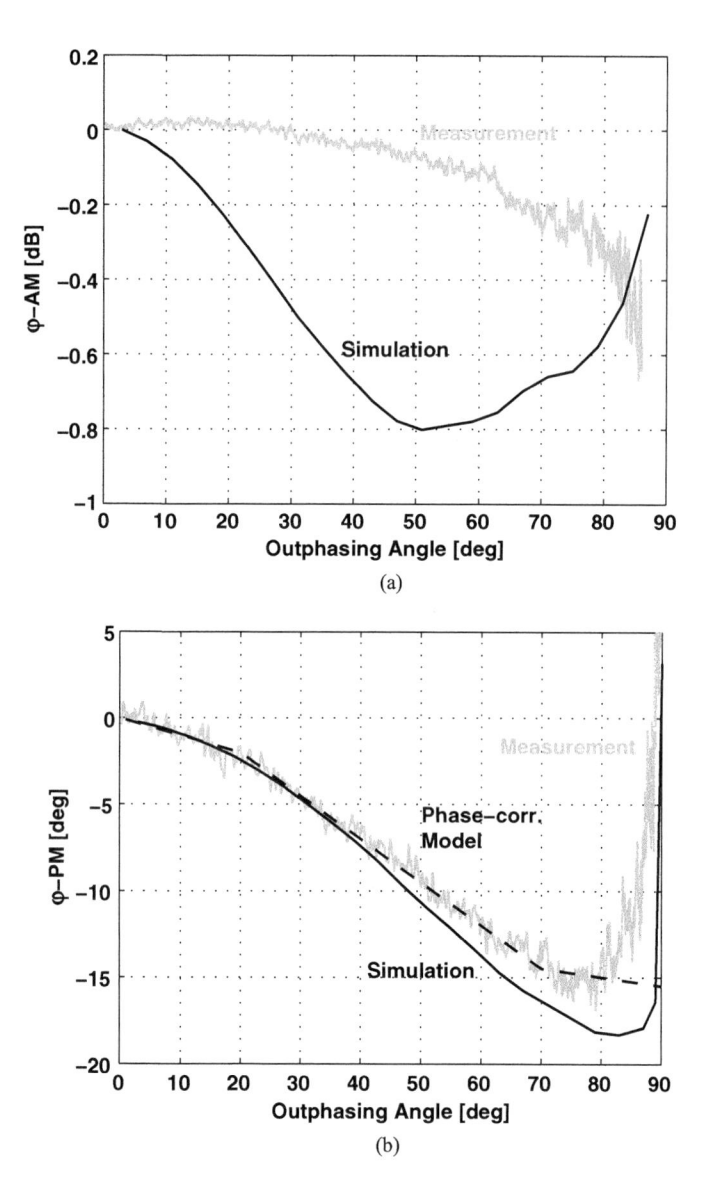

Fig. 8.31 Measured φ–AM and φ–PM distortions.

results. Note that for 16QAM modulation (raised-cosine filter with a roll-off factor of 0.35), 97.5% signal power is contained in the range with outphasing angles between 45° and 90° [12], the effective φ–PM distortion is less than 7°. Based on the measured φ–PM behavior, a simple first-order piecewise linear phase correction model is constructed and shown in Fig. 8.31. This phase correction (PC) can easily be applied on the modulated signals at baseband inputs.

In addition, the method in [37] is used to detect the phase mismatch between the two outphasing paths, giving a phase error of only 1.5°, including the mismatch errors in the measurement setup. This very low phase mismatch clearly indicates the excellent matching between the two signal paths. It will be shown that mismatch compensation (MC) indeed has very low impact on the modulated signal measurement results.

Modulated signal measurements

The benefits of outphasing for higher average output power and efficiency are confirmed by 500-Mb/s 16QAM EVM measurements. The outphasing signals are generated by two synchronized Agilent N6030A AWGs (500-MHz signal bandwidth). To simplify the measurement, the baseband signals are generated in a low-IF manner with 250-MHz IF carrier to avoid the influence of LO leakage in both outphasing and I/Q modes. The maximum data rate is limited by the AWG's bandwidth and synchronization error (100–150 ps delay difference between two AWGs). The transmitted output is down-converted by a fundamental V-band mixer. An oscilloscope is used to evaluate the constellation and EVM with build-in software. For the raised-cosine filter with a roll-off factor of 0.35, the PAPR of the 16QAM is 7.4 dB. The high PAPR limits the maximum measured average output power of the outphasing TX to about 10 dBm. The output power can be further improved by clipping the outphasing angle to reduce the effective PAPR. It is shown in [12] that EVM has negligible degradation when the effective PAPR is reduced by about 3 dB by clipping the input signal.

Figure 8.32 shows the measured 16QAM constellation, transmitter average output power, and PA average efficiency (η_{AVG}) as a function of EVM. The TX in outphasing mode (OP) achieves 12.5 dBm P_{AVG} and 15% η_{AVG} at the EVM of -22 dB. It performs two times better than the conventional I/Q mode (IQ) for both P_{AVG} and η_{AVG} with the same EVM (Fig. 8.32). By applying both PC and MC, the P_{AVG} and η_{AVG} are improved by 1.6 dB and 4% at EVM of -22 dB, shown in Fig. 8.32. Notice that MC has less influence on performance improvement.

Figure 8.33 shows the measured 500 Mb/s 16QAM spectrum at the transmitter output with an output power of 12 dBm. The spectrum measurements are performed with a harmonic mixer and a spectrum analyzer. The TX in I/Q mode underperforms as it requires about 7-dB back-off for 16QAM. The outphasing TX presents a tight spectrum and suppresses the out-of-band emissions by 7 dB (Fig. 8.33), indicating excellent matching between two outphasing paths. The PC and MC techniques further linearize the transmitter and reduce the effect of spectral regrowth.

8.7 Conclusions

This chapter has focused on the design and implementation of the RF and mm-wave outphasing TXs. Outphasing relaxes the linearity–efficiency trade-offs and has become a popular technique at low-GHz frequencies (e.g. 2.4 GHz for WLAN applications) to achieve high efficiency with no compromise in linearity. Recent published studies on mm-wave outphasing transmitters as well as high-speed outphasing signal generators show that outphasing is a promising technique to achieve highly efficient and linear mm-wave transmitters for future wireless communications. Several techniques that realize

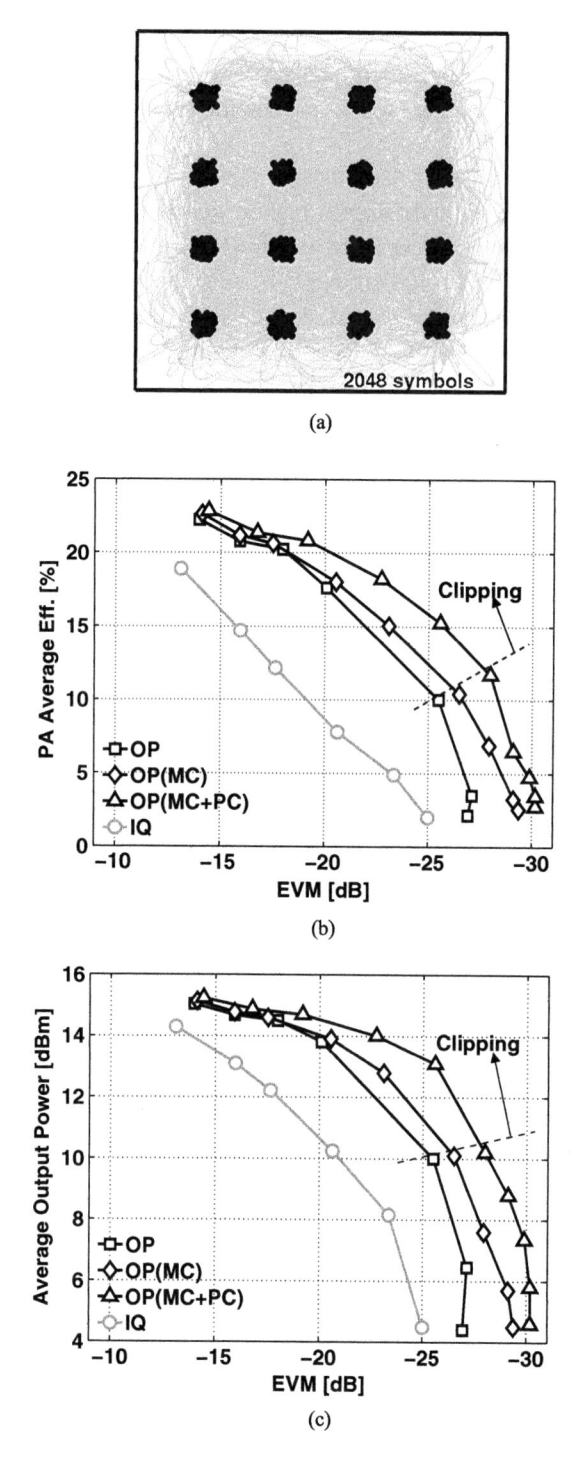

Fig. 8.32 Measured 16QAM constellations, average P_{OUT} and PAE of outphasing TX (with/without mismatch compensation or phase correction), and comparison with I/Q TX.

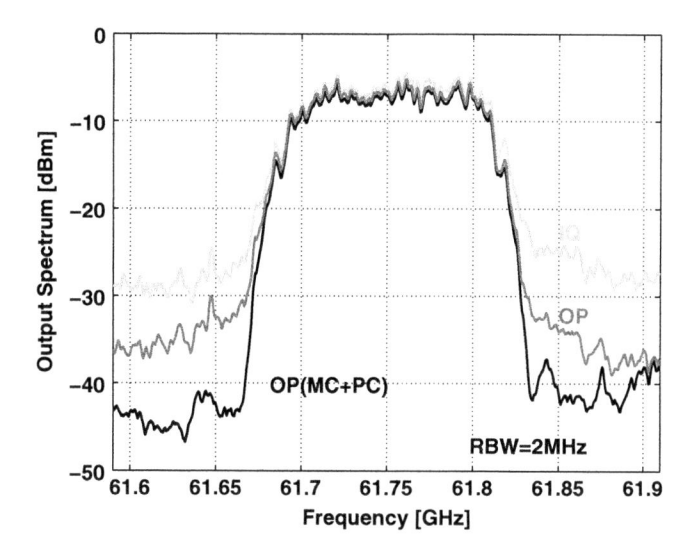

Fig. 8.33 Measured 16QAM spectrum with 12-dBm average output power.

the outphasing at low-GHz frequencies are discussed and the feasibility of implementing them at mm-wave is analyzed. It reveals that two challenges will be encountered for a fully integrated mm-wave outphasing transmitter with multi-GHz signal bandwidth: (1) sub-ps delay resolution is required in the baseband processor, and (2) excellent matching between the two outphasing paths must be ensured. Thankfully, both requirements will benefit from the scaling trend of CMOS. Efforts at extending the outphasing concept into the mm-wave range are well under way, with some good attempts discussed in this chapter.

References

[1] S. Kousai, K. Onizuka, T. Yamaguchi, Y. Kuriyama, and M. Nagaoka, "A 28.3 MW PA-closed loop for linearity and efficiency improvement integrated in a +27.1 dBm WCDMA CMOS power amplifier," *Solid-State Circuits, IEEE Journal of*, vol. 47, no. 12, pp. 2964–2973, 2012.

[2] P. Reynaert and M. S. J. Steyaert, "A 1.75-GHz polar modulated CMOS RF power amplifier for GSM-EDGE," *Solid-State Circuits, IEEE Journal of*, vol. 40, no. 12, pp. 2598–2608, 2005.

[3] E. Kaymaksut and P. Reynaert, "Transformer-based uneven Doherty power amplifier in 90 nm CMOS for WLAN applications," *Solid-State Circuits, IEEE Journal of*, vol. 47, no. 7, pp. 1659–1671, 2012.

[4] A. Ravi, P. Madoglio, H. Xu, et al., "A 2.4-GHz 20–40-MHz channel WLAN digital outphasing transmitter utilizing a delay-based wideband phase modulator in 32-nm CMOS," *Solid-State Circuits, IEEE Journal of*, vol. 47, no. 12, pp. 3184–3196, 2012.

[5] S. Moloudi, K. Takinami, M. Youssef, M. Mikhemar, and A. Abidi, "An outphasing power amplifier for a software-defined radio transmitter," in *Solid-State Circuits Conference, 2008. ISSCC 2008. Digest of Technical Papers. IEEE International*, 2008, pp. 568–636.

[6] P. Godoy, S. Chung, T. Barton, D. Perreault, and J. Dawson, "A 2.4-GHz, 27-dBm asymmetric multilevel outphasing power amplifier in 65-nm CMOS," *Solid-State Circuits, IEEE Journal of*, vol. **47**, no. 10, pp. 2372–2384, 2012.

[7] T.-W. Chen, P.-Y. Tsai, J.-Y. Yu, and C.-Y. Lee, "A sub-MW all-digital signal component separator with branch mismatch compensation for OFDM LINC transmitters," *Solid-State Circuits, IEEE Journal of*, vol. **46**, no. 11, pp. 2514–2523, 2011.

[8] T. LaRocca, J.-C. Liu, and M.-C. Chang, "60 GHz CMOS amplifiers using transformer-coupling and artificial dielectric differential transmission lines for compact design," *Solid-State Circuits, IEEE Journal of*, vol. **44**, no. 5, pp. 1425–1435, 2009.

[9] D. Zhao, S. Kulkarni, and P. Reynaert, "A 60 GHz dual-mode power amplifier with 17.4 dBm output power and 29.3% PAE in 40-nm CMOS," in *ESSCIRC, 2012 Proceedings of the*, 2012, pp. 337–340.

[10] C. Liang and B. Razavi, "Transmitter linearization by beamforming," *Solid-State Circuits, IEEE Journal of*, vol. **46**, no. 9, pp. 1956–1969, 2011.

[11] Y. Li, Z. Li, O. Uyar, *et al.* "High-throughput signal component separator for asymmetric multi-level outphasing power amplifiers," *Solid-State Circuits, IEEE Journal of*, vol. **48**, no. 2, pp. 369–380, 2013.

[12] D. Zhao, S. Kulkarni, and P. Reynaert, "A 60-GHz outphasing transmitter in 40-nm CMOS," *Solid-State Circuits, IEEE Journal of*, vol. **47**, no. 12, pp. 3172–3183, 2012.

[13] W. Gerhard and R. Knoechel, "LINC digital component separator for single and multicarrier W-CDMA signals," *Microwave Theory and Techniques, IEEE Transactions on*, vol. **53**, no. 1, pp. 274–282, 2005.

[14] R. Staszewski, J. Wallberg, S. Rezeq, *et al.*, "All-digital PLL and transmitter for mobile phones," *Solid-State Circuits, IEEE Journal of*, vol. **40**, no. 12, pp. 2469–2482, 2005.

[15] S. Voinigescu, *High-Frequency Integrated Circuits*, The Cambridge RF and Microwave Engineering Series. Cambridge University Press, 2013. [Online]. Available: http://books.google.be/books?id=71dHe1yb9jgC

[16] H. Veenstra, M. Notten, D. Zhao, and J. Long, "A 3-channel true-time delay transmitter for 6 GHz radar-beamforming applications," in *ESSCIRC, 2011 Proceedings of the*, 2011, pp. 143–146.

[17] B. Shi and L. Sundstrom, "A 200-MHz IF BiCMOS signal component separator for linear LINC transmitters," *Solid-State Circuits, IEEE Journal of*, vol. **35**, no. 7, pp. 987–993, 2000.

[18] L. Panseri, L. Romano, S. Levantino, C. Samori, and A. Lacaita, "Low-power signal component separator for a 64-QAM 802.11 LINC transmitter," *Solid-State Circuits, IEEE Journal of*, vol. **43**, no. 5, pp. 1274–1286, 2008.

[19] D. Pozar, *Microwave Engineering*, 3rd Edn. John Wiley and Sons, Inc., 2009.

[20] D. Cox, "Linear amplification with nonlinear components," *Communications, IEEE Transactions on*, vol. **22**, no. 12, pp. 1942–1945, 1974.

[21] A. Birafane, M. El-Asmar, A. Kouki, M. Helaoui, and F. Ghannouchi, "Analyzing LINC systems," *Microwave Magazine, IEEE*, vol. **11**, no. 5, pp. 59–71, 2010.

[22] A. Pham and C. Sodini, "A 5.8 GHz, 47% efficiency, linear outphase power amplifier with fully integrated power combiner," in *Radio Frequency Integrated Circuits (RFIC) Symposium, 2006 IEEE*, 2006, pp. 4–160.

[23] P. Godoy, D. Perreault, and J. Dawson, "Outphasing energy recovery amplifier with resistance compression for improved efficiency," *Microwave Theory and Techniques, IEEE Transactions on*, vol. **57**, no. 12, pp. 2895–2906, 2009.

[24] F. Raab, "Efficiency of outphasing RF power-amplifier systems," *Communications, IEEE Transactions on*, vol. **33**, no. 10, pp. 1094–1099, 1985.

[25] S. Hamedi-Hagh and C. Salama, "CMOS wireless phase-shifted transmitter," *Solid-State Circuits, IEEE Journal of*, vol. **39**, no. 8, pp. 1241–1252, 2004.

[26] H. Chireix, "High power outphasing modulation," *Radio Engineers, Proceedings of the Institute of*, vol. **23**, no. 11, pp. 1370–1392, 1935.

[27] I. Hakala, D. Choi, L. Gharavi, et al., "A 2.14-GHz Chireix outphasing transmitter," *Microwave Theory and Techniques, IEEE Transactions on*, vol. **53**, no. 6, pp. 2129–2138, 2005.

[28] H. Xu, Y. Palaskas, A. Ravi, et al., "A flip-chip-packaged 25.3 dBm class-D outphasing power amplifier in 32 nm CMOS for WLAN application," *Solid-State Circuits, IEEE Journal of*, vol. **46**, no. 7, pp. 1596–1605, 2011.

[29] S. C. Cripps, *RF Power Amplifiers for Wireless Communications*, 2nd Edn. Artech House Microwave Library. Norwood, MA, USA: Artech House, Inc., 2006.

[30] I. Sarkas, A. Balteanu, E. Dacquay, A. Tomkins, and S. Voinigescu, "A 45 nm SOI CMOS class-D mm-wave PA with >10 VPP differential swing," in *Solid-State Circuits Conference Digest of Technical Papers (ISSCC), 2012 IEEE International*, 2012, pp. 88–90.

[31] K. Okada, K. Kondou, M. Miyahara, et al., "Full four-channel 6.3-Gb/s 60-GHz CMOS transceiver with low-power analog and digital baseband circuitry," *Solid-State Circuits, IEEE Journal of*, vol. **48**, no. 1, pp. 46–65, 2013.

[32] M. Gustavsson, J. J. Wikner, and N. N. Tan, *CMOS Data Converters for Communications*. Norwell, MA, USA: Kluwer Academic Publishers, 2000.

[33] X. Zhang, L. Larson, P. Asbeck, and P. Nanawa, "Gain/phase imbalance-minimization techniques for LINC transmitters," *Microwave Theory and Techniques, IEEE Transactions on*, vol. **49**, no. 12, pp. 2507–2516, 2001.

[34] T.-S. Chu, J. Roderick, and H. Hashemi, "An integrated ultra-wideband timed array receiver in 0.13 μm CMOS using a path-sharing true time delay architecture," *Solid-State Circuits, IEEE Journal of*, vol. **42**, no. 12, pp. 2834–2850, 2007.

[35] S. Kulkarni, D. Zhao, and P. Reynaert, "Design of an optimal layout polyphase filter for millimeter-wave quadrature LO generation," *Circuits and Systems II: Express Briefs, IEEE Transactions on*, vol. **60**, no. 4, pp. 202–206, 2013.

[36] W. Chan and J. Long, "A 58–65 GHz neutralized CMOS power amplifier with PAE above 10% at 1-v supply," *Solid-State Circuits, IEEE Journal of*, vol. **45**, no. 3, pp. 554–564, 2010.

[37] S.-S. Myoung, I.-K. Lee, J.-G. Yook, K. Lim, and J. Laskar, "Mismatch detection and compensation method for the LINC system using a closed-form expression," *Microwave Theory and Techniques, IEEE Transactions on*, vol. **56**, no. 12, pp. 3050–3057, 2008.

9 Digital mm-wave silicon transmitters

Ali M. Niknejad and Sorin P. Voinigescu

9.1 Motivation

As tablets and smartphones have become the main drivers for high-data-rate wireless communications, bandwidth demand and energy efficiency have also emerged as major concerns in wired and fiber-optic data-center and last-mile links. Energy consumption is particularly poor in wireless links and typically exceeds 500 pJ/bit. Since the transmit power amplifier in most wireless devices radiates 500 mW or higher, its energy efficiency can be improved significantly by switching it off and on in deep saturation at tens of Gb/s. Additionally, the output power and power consumption of the entire transmitter must be scalable with the data rate and with the link range.

In a conventional transmitter, shown in Fig. 9.1, the in-phase (I) and quadrature-phase (Q) baseband data bits are converted to an analog signal at baseband, filtered, and then up-converted to the RF carrier. This signal is amplified by the PA, which either directly drives the antenna or drives another filter to limit emissions at harmonics of the carrier. The baseband clock is either Nyquist rate or oversampled by a modest factor, depending on how much signal processing (filtering) is performed in the digital domain. The role of the baseband filter is critical in this architecture as it smooths the data transitions, attenuating the spectral images of the signal. The I and Q signals traverse the constellation points of the modulation scheme, and need to be processed appropriately to minimize distortion (which leads to EVM degradation). This single-sideband radio transmitter architecture is versatile, transparent to the modulation scheme, and makes the most efficient use of the available bandwidth. However, for modulation formats that involve different amplitude levels, such as 16QAM and 64QAM, this architecture imposes very stringent linearity constraints on the up-converter and on the entire mm-wave path which are increasingly difficult to satisfy at multi-gigabit per second data rates. Furthermore, once the I and Q signals are combined, most spectrally efficient modulation schemes generate a signal with envelope variation, which requires the power amplifier to be linear. For example, to avoid degradation of the 16QAM signal (e.g. EVM increase and spectral regrowth) the power amplifier is typically backed-off by up to 6 dB from its 1-dB output compression point. Because it exhibits a quadratic efficiency loss with back-off, if a class-A PA is employed the efficiency drops by a factor of 4. Considering that millimeter-wave (mm-wave) power amplifiers have lower efficiency than their microwave counterparts, mostly arising from the limited transistor power gain, this efficiency loss can result in very low back-off efficiency, on the order

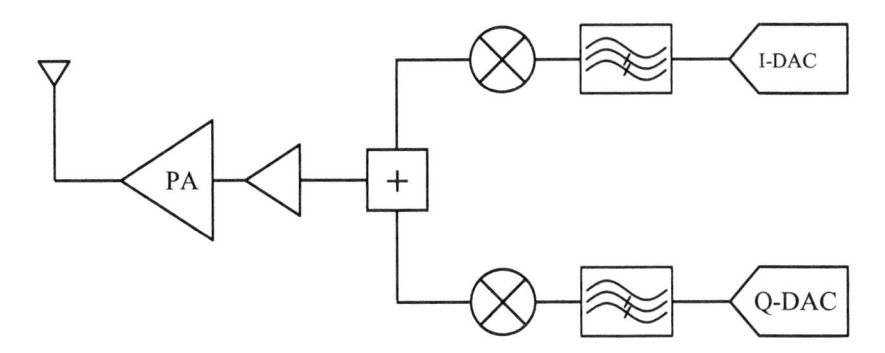

Fig. 9.1 A conventional transmitter uses a baseband DAC, filtering, and linear up-conversion and amplification.

of 5%. The penalty on the PA efficiency at this back-off power cannot be tolerated in mobile devices.

Historically, transmitters developed for optical-fiber communication links have relied on on–off (OOK) direct digital modulation of the optical carrier and are truly digital in nature, including the large-swing, broadband switching optical modulator driver [1]. Currently installed fiber-optic links use OOK and QPSK modulation formats at 40 Gb/s and 25–28 Gbaud/s, respectively, to reach aggregate data rates of 110 Gb/s per polarization-multiplexed optical carrier. The entire C-Band (4.5 THz of bandwidth centered on 193.5 THz) is now occupied, saturating at a maximum of 9 Tb/s per optical fiber with 50-GHz channel spacing. Further capacity increase must come from the deployment of higher-order modulation formats. As a result, fiber-optic companies are scrambling to develop 4PAM, 8PAM, 16PAM and synchronous 16QAM and OFDM, radio-like modulation schemes to improve bandwidth efficiency. As illustrated in Fig. 9.2, advanced long-haul fiber-optic transmitters now employ baseband 56 GS/s DACs (digital-to-analog converters) followed by linear power amplifiers to modulate the optical carrier by linear up-conversion [2, 3]. As in the linear up-converter radio architecture, they require a linear, inefficient, and extremely broadband power amplifier following the DAC to boost its signal to a level suitable for the optical modulator. These wireless and fiber-optic transmitters have practically identical architectures, with the carrier frequencies and data rates being typically three orders of magnitude larger in the case of the optical transmitters. As we shall see in this chapter, truly digital transmitter architectures are possible for both applications.

In this chapter we define digital transmitters as large-power, high-efficiency, multi-bit DACs, which directly modulate a mm-wave carrier and/or employ a mm-wave sampling clock to create an arbitrary mm-wave or optical signal with complex amplitude and phase modulation. They represent a perfect marriage between high-efficiency mm-wave power amplifiers and digital techniques. Such digital transmitters find application in wireless, fiber-optics, and wireline communications, as well as in instrumentation.

An attractive solution is an adaptable M-ary QAM power modulator (Fig. 9.3) which directly modulates the carrier signal and whose modulation format, output power, data

336   **Digital mm-wave silicon transmitters**

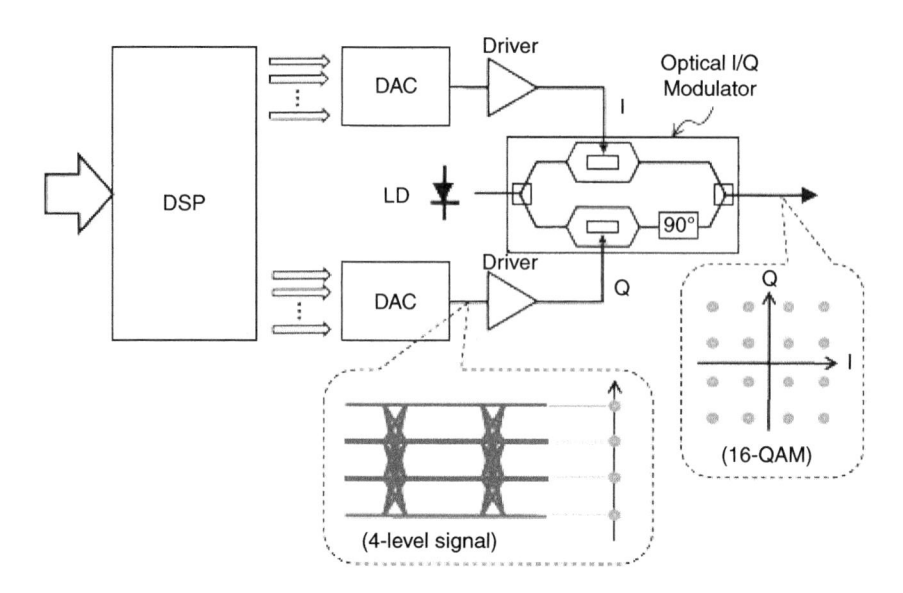

Fig. 9.2   A 16QAM fiber-optic transmitter with linear driver and optical linear up-conversion [3].

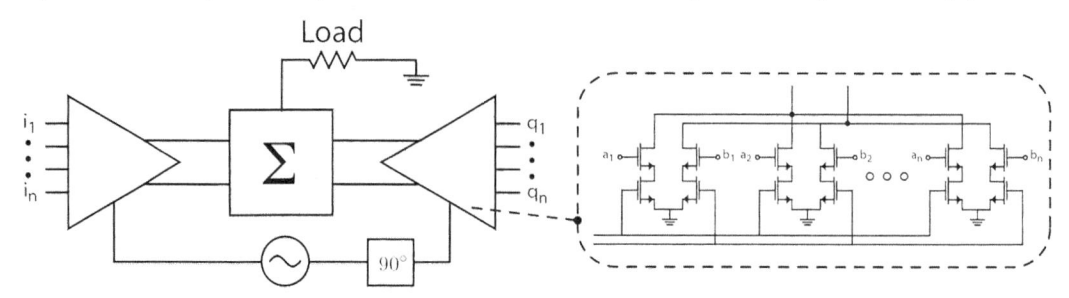

Fig. 9.3   A Cartesian digital transmitter is an adaptable M-ary QAM power modulator.

rate, and power consumption are adjustable to maintain the highest possible transmitter PAE. Since the modulator is the last block in the transmitter, and if certain design conditions are met, it can be operated in saturation mode, with an output power larger than +20 dBm, without requiring back-off. In fact, such a modulator can be regarded as a switched power amplifier or as a multi-bit transmitter power-DAC. If the efficiency, scalability, and silicon integrability are to be improved, the DAC–PA combination should be replaced by a power-DAC. Ideally, the power-DAC should have the following features.

- It should operate in saturated-power mode to maximize its efficiency.
- The output stage should be an efficient class-D, -E, or -F (including inverse variants and combinations such as class-EF [4]) PA.
- The power consumption should scale with the data rate, the modulation format, and the constellation point.

This can be accomplished with a fully segmented power-DAC architecture where only the DAC segments needed to generate a particular constellation point are on, thereby saving a considerable amount of power at back-off.

Another important factor that drives the trend to replace conventional transmitters with power-DAC-like digital transmitters is technology scaling. At least down to the 28-nm production node, gate-length scaling has produced an ever-increasingly better switch, characterized by low on-resistance and small off-capacitance. To see this, note that the intinsic channel resistance per unit gate width improves in a similar manner to the intrinsic device transconductance per unit gate width $R'_{on} \propto 1/g'_m$, and likewise the device off-capacitance per unit gate width remains the same or slightly improves when the device gate length is scaled, $C'_{off} \propto C'_{gs} + C'_{gd}$. So, to first order, the switch figure of merit $\tau_{sw} = R_{on}C_{off} \propto (C_{gs} + C_{gd})/g_m = \omega_T^{-1}$. CMOS scaling continues to improve the device unity gain frequency ω_T, even considering the increasing negative impact of the source and drain resistances. This is not true for the device f_{max} (also affected by gate resistance, output conductance, and substrate parasitics), which is a good metric to estimate the amount of power gain available in a tuned amplifier operated close to activity limits $G \sim f_{max}/f$, implying that technology scaling is having a reduced impact, if any, on more traditional mm-wave circuits that rely on class-A/B biasing.

Given the prominence of digital modulation schemes in radio and fiber-optic communication, it is surprising that digital approaches have not been more prevalent solutions. One reason is that in a traditional transmitter much effort is expended to limit the bandwidth of the signal to maximize spectral utilization, curb spectral regrowth and interference to adjacent channels, and suppress ISI. This is why the baseband filters in a conventional transmitter play such an important role. If the carrier signal is modulated directly in the PA, it becomes increasingly difficult to filter the transmitted waveform, and great care is needed to avoid the aforementioned problems. Oversampling the data signal is an easy way to alleviate spectral emissions, and this is why the arrival of digital architectures for the transmitter has been partly enabled by the high switching speeds of modern CMOS and SiGe HBT devices.

9.2 Architectures for high efficiency/linearity

To break the aforementioned efficiency/linearity bottleneck, several transmitter architectures have been developed that utilize nonlinear PAs for the phase modulation and resort to other schemes to convey amplitude modulation. These architectures are briefly reviewed here for completeness since there are digital equivalents that have gained popularity, especially at lower carrier frequencies. For a more detailed treatment, the reader is directed to other chapters in this book. The polar architecture originated in the envelope elimination and restoration (EER) techniques of classic AM transmitters. While a nonlinear efficient switching PA processes the phase modulation, the amplitude modulation is imparted through modulation of the supply voltage. Since the amplifier operates in saturation mode, the output power is proportional to the supply voltage in a linear manner, thus re-creating both the amplitude and the phase of the signal. Owing to the different paths from modulation to the output, the phase and amplitude paths need to be delay matched to realize a given EVM specification. In a modern transmitter, the I and

338 Digital mm-wave silicon transmitters

Q baseband bits are easily converted to the amplitude and phase modulation through the well-known trigonometric identities

$$\phi(t) = -\tan^{-1}\left(\frac{Q(t)}{I(t)}\right),$$

and

$$A(t) = \sqrt{I(t)^2 + Q(t)^2}.$$

In the outphasing transmitter, also known by the acronym LINC (linear amplification using nonlinear components), the outputs of two separate nonlinear switching amplifiers are summed to produce the fully amplitude/phase-modulated waveform. Since each amplifier only amplifies a phase-modulated signal, they can be operated in nonlinear switching mode with high efficiency. Ideally

$$v_o(t) = A\cos(\omega_0 t + \phi_1(t)) + A\cos(\omega_0 t + \phi_2(t)) = A(t)\cos(\omega_0 t + \phi(t)).$$

In practice, the summation of two signals suffers from load-pulling-induced distortion, where the load seen by each PA is not only a function of R_L, but also of the phase of the other PA. If isolating summation is used, such as a Wilkinson combiner, then this problem is averted, but the output power is lower due to the combiner loss. Furthermore, the back-off characteristics are similar to those of a class-A PA, since the DC power consumption remains constant. On the other hand, if non-isolating combiners are employed, then the effect of outphasing causes a load-pull that increases the output impedance seen by the amplifiers, which improves the back-off characteristics. But, in addition to the distortion, the outphasing also produces a reactive load. The problem can be mitigated by using a Chireix combiner, which compensates the change of load reactance by a fixed susceptance [5].

9.2.1 Digital polar architecture

As shown in Fig. 9.4, in a digital polar architecture (DPA) [6, 7], the amplitude path is composed of a segmented power transistor, sometimes called an RF-DAC, that

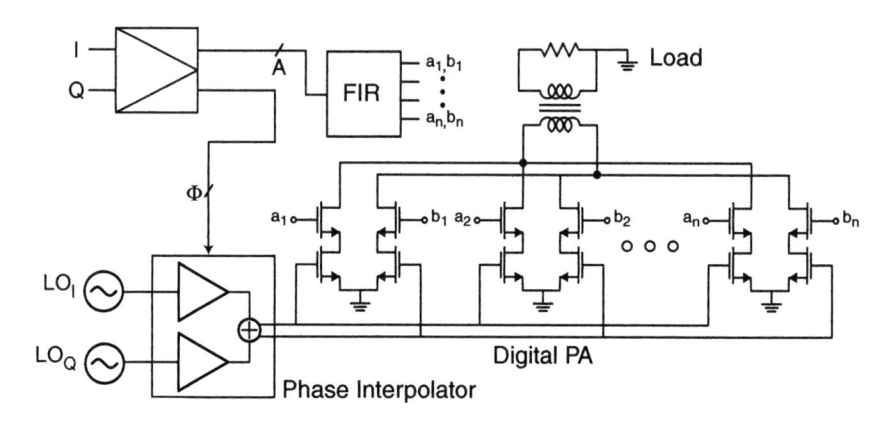

Fig. 9.4 The digital polar architecture (DPA).

reproduces the signal envelope by digitally turning on/off an appropriate number of elements. The segments can be unary weighted (unit element thermometer code) or binary weighted, or a combination of both. In essence, this architecture moves the DAC all the way to the antenna, but, since the DAC is designed to carry the carrier signal, only the envelope of the carrier needs to be processed by the DAC. The unit cell of the DAC is a switching amplifier hard switched by the RF carrier. When all the unit elements are on, the structure of the RF-DAC is identical to a switching PA and the switch is sized to deliver the required output power, trading-off efficiency with power gain. Larger devices have larger parasitics, and therefore are harder to tune out and drive, while smaller devices have more on-resistance. This unit cell is then divided into N_t thermometer unit cells or segments. A segmented structure allows unit elements to trace the coarse steps in the amplitude. To realize finer resolution, one cell is further divided into smaller elements, usually binary weighted, and it is used to generate the small quantization steps necessary to meet out-of-band emissions.

Each unit element can be controlled in several different ways. In the left-hand diagram of Fig. 9.5, the cascode device is conveniently used to control the state of a unit element, while in the right-hand part of Fig. 9.5, the entire element is gated by an AND gate. While more compact, the cascode solution suffers from voltage breakdown and poor back-off characteristics. Since the gate of the cascode device is grounded in the off-state, the drain-to-gate voltage experiences the full voltage swing of the output, and must be sized appropriately (usually using a thick-oxide device). Also, the power to drive the RF-DAC is relatively constant as a function of codeword (CV^2f for dynamic power), which implies that the efficiency at back-off is similar to that of a class-A stage. On the other hand, when a separate AND gate is used to control each unit element, the power consumption of the drivers is reduced dramatically in back-off mode since the LO waveform is cut off from the gate in off mode. Moreover, the gate of the cascode is biased either statically or dynamically to experience a substantially smaller fraction of the output swing. In either case, the AND function implicit in the cascode or the explicit AND gate effectively acts like a digital mixer, up-converting the baseband data to RF.

The number of RF-DAC bits and the clock rate are determined by the specifications of the EVM and out-of-band emissions, particularly quantization noise. Often, the baseband envelope of the carrier is quantized and mapped to the RF-DAC using a look-up

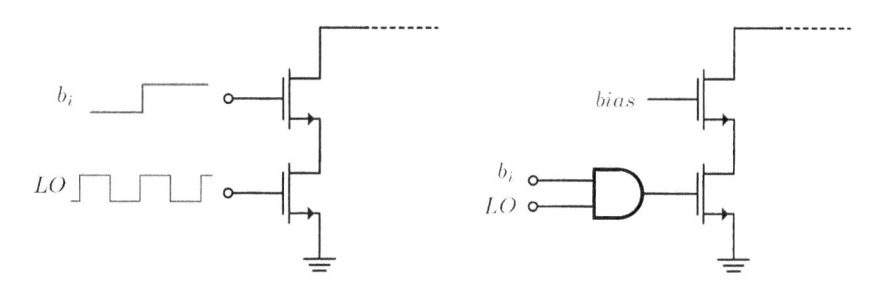

Fig. 9.5 Two realizations of the digital mixing operation using (left) a cascode device and (right) an explicit AND gate.

table (LUT) to minimize the spectral regrowth/EVM due to the DAC nonlinearity. While the bandwidth of the envelope signal is modestly higher than the I and Q baseband signals due to the nonlinear operation inherent in the transformation, the bandwidth of the phase signal is significantly larger than the baseband counterparts. The reason is that the arctangent nonlinearity causes significant bandwidth expansion, approximately 7× the baseband rate, which requires a very wideband phase-modulation path. In a polar transmitter, the carrier phase modulation is often produced using an offset PLL, or simply using an I/Q modulator. This bandwidth expansion is increasingly problematic as the modulation rates increase from 100 Mb/s to several gigabits per second. This predicament limits the viability of the DPA for mm-wave applications.

To circumvent the bandwidth limitation of a PLL, or a closed-loop approach, an open-loop digital modulation is proposed and demonstrated in [8]. I and Q LO signals are summed using two DACs to control the weights of the I and Q signals in the combination. In this way the phase of the LO is rotated in accordance with the modulation scheme. Sub-degree phase accuracy and gigahertz modulation bandwidth are easily achievable.

9.2.2 Digital outphasing architecture

A combination of dynamically variable PA size and a switched tuning capacitor bank is introduced in [9] to compensate the load modulation at multiple outphasing angles and eliminate the distortion. Recently, mm-wave outphasing transmitters have been demonstrated with both transformer combining and spatial combining [10, 11]. However, due to lack of load compensation, the power stays relatively constant and the back-off curve is close to class A. The major efficiency enhancement comes from the utilization of the power beyond P_{1dB}.

9.2.3 Digital Cartesian (I/Q) architecture

The digital Cartesian (or I/Q) architecture (DCA) (Fig. 9.3) is similar in spirit to the DPA, where the amplitude modulation is imparted in a quantized fashion using an array of unit cells biased in switching mode to realize high efficiency. The only difference is that the I and Q signals are quantized and summed directly at RF, instead of at baseband. Since the I and Q signals are never converted into amplitude and phase representations, there is no bandwidth expansion. The disadvantage of this architecture is the inherent power loss due to quadrature summation of signals. Consider an RF-DAC in a DPA sized at a width of W, such that the output current is proportional to this width. If we split this in half to realize two I/Q RF-DACs each of width $W/2$, since the currents of these DACs never sum in phase, there is an inherent $\sqrt{2}$ loss, or equivalently an increase in area to realize the same power levels.

The challenge in the DCA is in the signal summation of the I and Q paths. In the baseband Cartesian transmitter operating at sub-Gb/s data rates, these signals are current summed using high-impedance drivers and there is minimal loss or interaction between the drivers. When this is done at RF (or even in a multi-Gb/s baseband

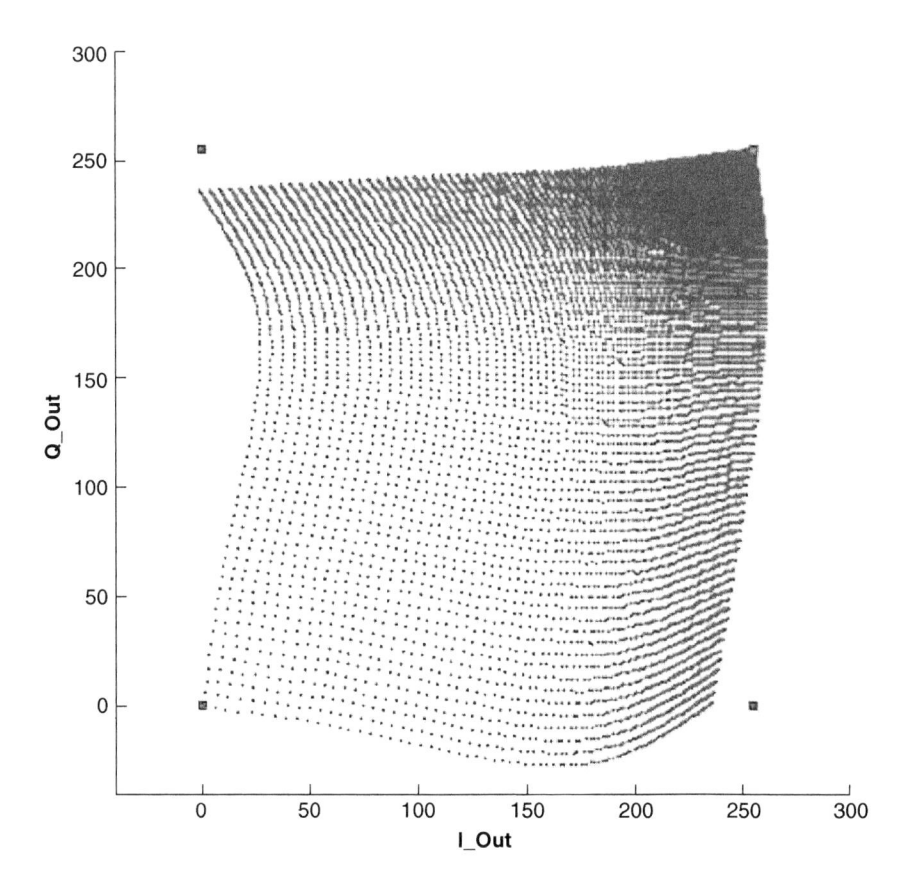

Fig. 9.6 Distorted constellation pattern of a digital Cartesian transmitter shows a 2D dependence on the codeword.

path) due to the lower impedance (especially at higher codewords when more elements are on), there is considerable interaction between the RF-DAC segments. This results in a two-dimensional (2D) distortion mechanism, which requires a quadratically larger look-up table (LUT) than for a DPA. For example, as a large I_n codeword is applied, the output impedance of the RF-DAC drops and tends to have a saturating compression effect on the output power. But the amount of compression is also a function of how many Q RF-DAC elements are on, so both I_n and Q_n are needed to determine the correct pre-distortion level. Example constellations for the Cartesian RF-DAC are shown in Fig. 9.6. In [12] a smaller 2D table is used to alleviate the memory and training requirements, and the codewords are interpolated for values not in the table.

9.2.4 Unique challenges for digital transmitters

In a conventional transmitter, the baseband resolution and oversampling are mainly set by EVM requirements, whereas the thermal noise of the transmitter determines

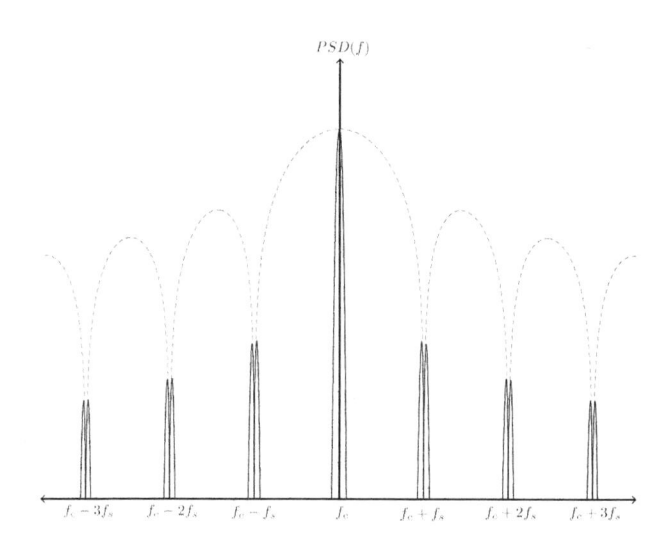

Fig. 9.7 Spectral aliases of a DAC are up-converted internally in the RF power DAC and only experience sinc ZOH filtering.

out-of-band emissions. In a digital transmitter, though, the out-of-band emissions are also defined by the quantization noise level, any noise shaping, and the oversampling ratio. This is because there is virtually no filtering of the digitally modulated data stream fed into the antenna. As sharp transitions are made from one amplitude level to the next, only the settling time of the RF-DAC and possible tuning networks determine the out-of-band emissions. By increasing the resolution, the quantization noise is reduced by 6 dB per bit, similar to a DAC. Oversampling helps to spread the quantization noise over a wider bandwidth and linearly improves the noise level.

Another problem unique to the digital transmitter also arises from the limited filtering, namely the fact that the RF-DAC produces the output spectrum at multiples of the baseband clock, due to the inherent aliasing in the DAC output. Without a filter, all the aliases of the signal are also modulated on the carrier, at multiples of the clock sampling frequencies f_s. For example, a zeroth-order-hold (ZOH) waveform produces a sinc spectral shape with the first nulls occurring at the clock rate, as shown in Fig. 9.7. The peaks of the sinc waveform die off very slowly, with the first peak only 13 dB down from the carrier. It is precisely the job of the filter to smooth the transitions between digital codewords, which also results in a faster roll-off of the alias spectra. Alternatively, some of this filtering can be done using the RF-DAC itself, by for example ramping the waveforms in a linear fashion. This so-called first-order hold has a faster roll-off, like $sinc^2$. In practice this linear ramping can be approximated by introducing a delay in the clock of elements [7]. In fact, generalizing this concept allows some of the filter functionality to be realized in the RF-DAC itself. If the RF-DAC elements are appropriately partitioned and delayed, an FIR filter can be realized. Since a polar RF-DAC only controls

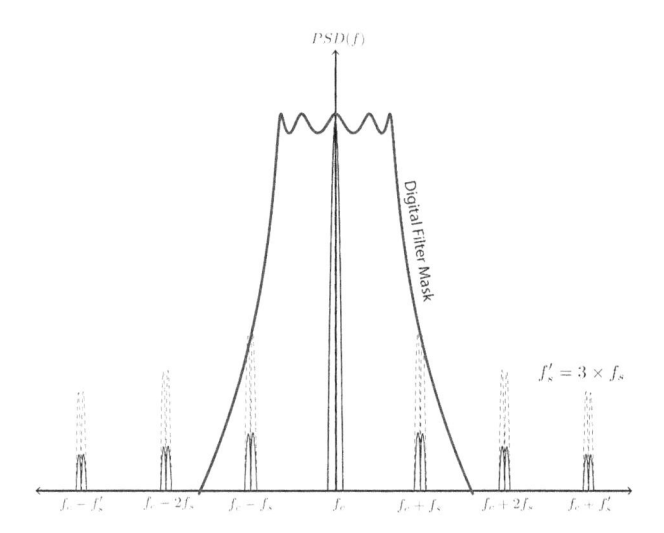

9.2 Architectures for high efficiency/linearity

Fig. 9.8 A digital approach to eliminate spectral images uses oversampling and filtering.

the envelope of the carrier, it cannot impact the phase, thus preventing the realization of negative coefficients. A Cartesian RF-DAC is therefore the preferred manner to realize the FIR filter.

On running the sampling clock at higher and higher rates, the images move further away, and some mild form of filtering can be used to knock these down to acceptable levels, often using the inherent filtering at the output of the RF-DAC (for instance in a tuned transformer). The downside is that this requires the digital signal processing to be run faster and faster, which incurs a power penalty. An alternate approach is run the sampling clock at the low Nyquist rate of the data, but then to oversample the output. An interpolation filter is required to smooth the transitions in time, which has the effect of knocking down the images. This scheme is shown in Fig. 9.8. The digital filter is thus an interpolating filter and either FIR or IIR topologies can be used. In [13] a 19-tap FIR filter is realized, running at 1 GHz, and it provides 30 dB of attenuation starting at the stop band of 180 MHz. The filter coefficients are optimized using a least-squares method to allow multiplier-less FIR filter implementation, which saves considerable power. With extensive optimization of the architecture, employing parallel pipelined stages running at 1/5 of the clock rate, and exploiting coefficient grouping, the entire FIR filter consumes only 1.8 mW. Coefficient grouping takes advantage of the repetitive data fed into the filter due to the ZOH operation followed by oversampling.

Another issue that must be carefully considered and controlled is the clock skew arising from different routing delays for the unit elements. The LO signal should be routed to the delay elements using a clock tree. If systematic delays exist on the clock tree, they can be compensated by using the delay at the output of the cells, similar to the gate/drain delay balance inherited in a distributed amplifier. The data fed to the DAC elements should also be retimed as close to the unit elements as possible to equalize delays. The clock should have minimal jitter in order to preserve the fidelity of the signal. System simulation should be performed to determine the acceptable clock jitter to meet a given specification.

9.3 Digital mm-wave transmitter architectures with on-chip power combining

9.3.1 OOK/ASK/BPSK

Simple modulation schemes lend themselves very well to digital RF modulation techniques. For example, in OOK/ASK transmitters all the information is contained in the amplitude of the carrier signal and the output amplitude can be modulated by turning on/off unit elements [14]. In BPSK modulation, the amplitude is constant but the phase is inverted, which can be done using a butterfly switch or by switching between two paths. In [15] a 135-GHz carrier is modulated on/off by simply shorting out the oscillator signal that is fed into the antenna or PA (Fig. 9.9). More bits can be added by partitioning the switch to control the level of attenuation. Measured scattering transmission parameters for a 6-bit modulator are shown in Fig. 9.10 versus frequency and codeword, indicating that the modulator is operational up to 90 GHz and fairly broadband. The kinks in the measured curves are a measurement artifact caused by the VNA switching from coaxial to waveguide signal sources in the 45–55-GHz range.

One problem with this direct ASK modulation approach is that the input and output impedances of the modulator become strongly codeword dependent, pulling the VCO frequency and distorting the output signal constellation. A more elegant solution is to use a Gilbert-cell quad biased at constant current and to partition it in unary- or binary-weighted segments. Each segment or binary-weighted grouping of segments is controlled by independent data streams, as shown later in Fig. 9.18 [16, 17], to form a

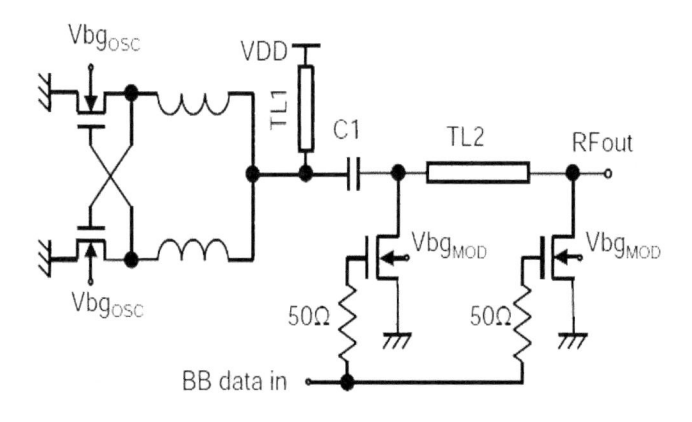

Fig. 9.9 An ASK/OOK digital RF modulator.

9.3 Digital mm-wave transmitter architectures

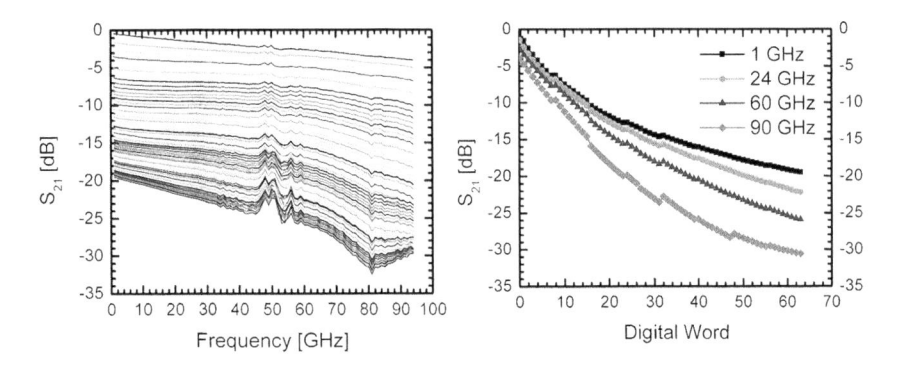

Fig. 9.10 The measured S-parameters of the 6-bit ASK modulator as a function of frequency (left) and codeword (right).

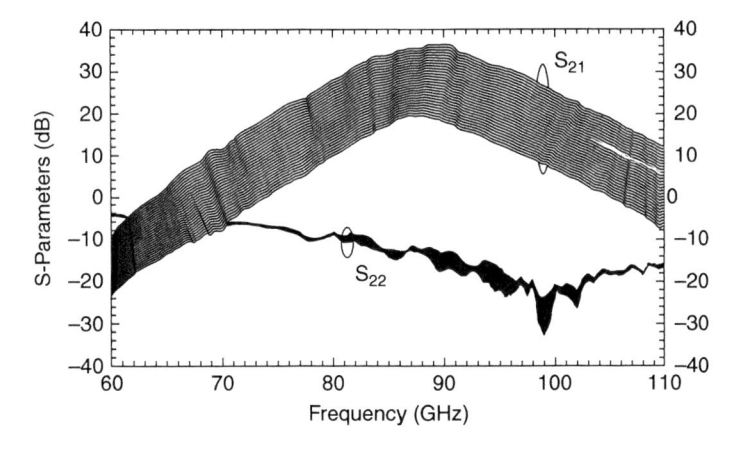

Fig. 9.11 Measured S_{21} and S_{22} of the 90-GHz RF power-DAC for all 256 codewords [18]. Reprinted with permission from the IEEE.

constant-current direct ASK modulator otherwise known as an RF-DAC. For example, if each MOSFET in the Gilbert-cell quad has 127 gate fingers, and each gate finger is 1 μm wide, binary groupings of Gilbert-cell quads with 1, 2, 4, 8, 16, 32, and 64 MOSFET gate fingers connected together will result in 6 bits of amplitude control. Owing to the double balanced topology of the Gilbert cell, the seventh bit performs sign inversion and could be used for BPSK modulation. Additionally, at the expense of a larger number of baseband bit streams and complexity, any or all of the higher-order Gilbert-cell groupings could be segmented and thermometer encoded to better handle transistor mismatch.

In [18] an 8-bit RF power-DAC based on a series-stacked segmented Gilbert-cell topology is used to modulate the amplitude of a 90-GHz carrier at a rate of at least 15 GBaud, to form a multi-level ASK constellation. The small-signal S-parameter measurements in Fig. 9.11 demonstrate that the output impedance remains codeword independent, unlike in [14, 15] or in Fig. 9.10.

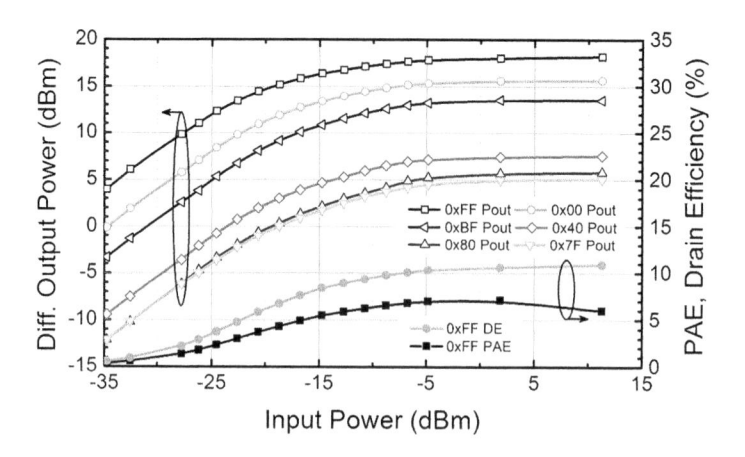

Fig. 9.12 Measured output power, drain efficiency, and *PAE* of a 90-GHz stacked Gilbert-cell RF power-DAC as a function of the 8-bit codeword applied to the segmented gates of the Gilbert cell [18]. Reprinted with permission from the IEEE.

As shown in Fig. 9.12, a dynamic range of 18 dB is measured in saturated power mode with 19 dBm maximum output power and 11% drain efficiency. This finite dynamic range is caused by the finite isolation and poor symmetry of the capacitive and inductive parasitics of the Gilbert-cell layout, which degrade with increasing carrier frequency. It indicates that the output power constellation can be controlled with three effective bits of precision. For higher precision and higher-order modulation at these frequencies, an on–off switch and antenna segmentation are required, as discussed later. The measured transmitter output spectrum is shown in Fig. 9.13, when one of the already-segmented most-significant bits switches at 15 Gb/s, for an effective aggregate data rate of 3×15 Gb/s = 45 Gb/s. As with the other multi-level ASK modulator, a weak point of the constant-current direct ASK modulator/RF-DAC topology is that its efficiency decreases from its peak value for output signal constellations with lower amplitude, since the DC power consumption remains constant. A solution to overcome this problem in m-ary QAM direct digital power-modulators is described later in this chapter.

In a BPSK transmitter [19], a single data bit is used to modulate the phase of the output by using a Gilbert quad stacked with the VCO, which results in a power-efficient modulator (Fig. 9.14). The SiGe prototype works at 65 GHz and directly drives 50-Ohm differential loads. A similar approach is described in [20] in CMOS technology, but the LO signal is generated separately. A transformer is used to couple the LO signal to the tail of the Gilbert quad, saving headroom on a 1.2-V supply (Fig. 9.14). One of the limitations of the direct ASK and BPSK modulation schemes is that, even with the Gilbert-cell approach, significant buffering is needed between the VCO and the modulator to alleviate data-dependent VCO pulling.

9.3.2 QPSK/QAM

To generate a complex modulation, both I and Q LO signals need to be modulated and combined. In [21] the baseband bits are applied to current DACs which modulate

9.3 Digital mm-wave transmitter architectures

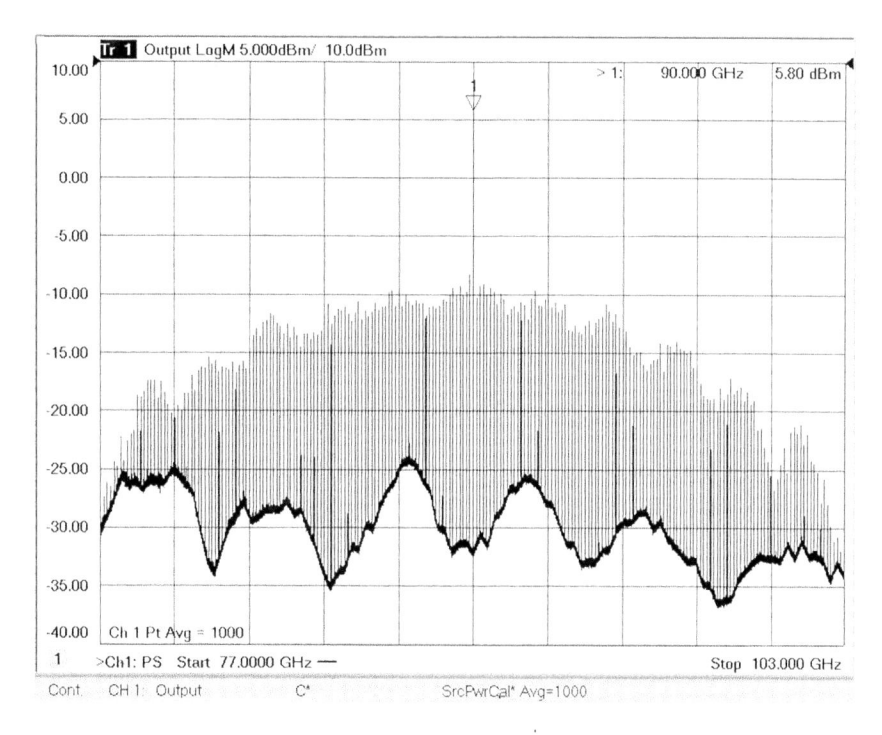

Fig. 9.13 Measured output spectrum of the 94-GHz power-DAC when the MSB in the Gilbert cell switches at 15 Gb/s.

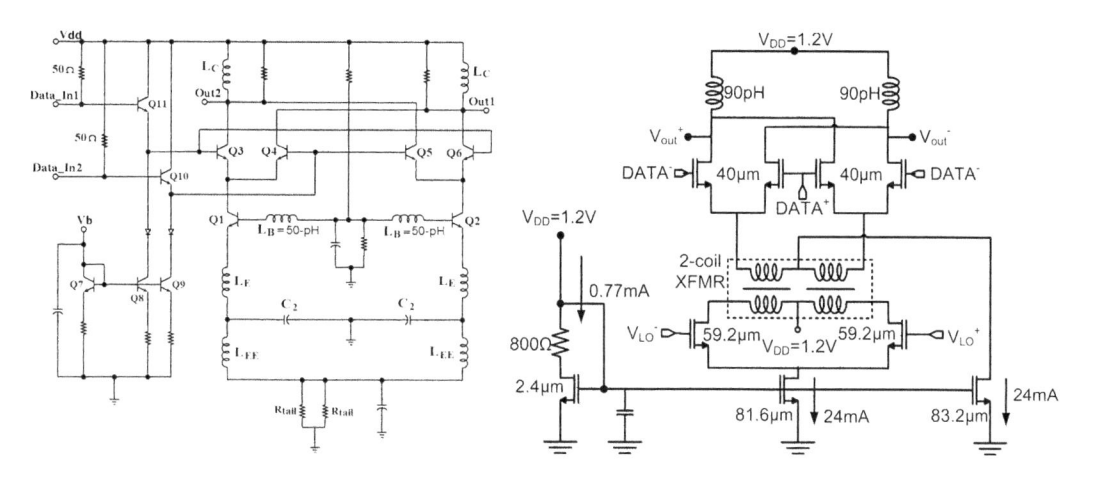

Fig. 9.14 (Left) A BPSK transmitter with stacked VCO and modulator presented in [19]. (Right) A BPSK modulator realization in CMOS [20]. Reprinted with permission from the IEEE.

the current of a Gilbert-cell mixer and are up-converted around the carrier frequency. The digital data from the on-chip PRBS generator are fed to the modulator, which consists of a fully differential, combined DAC-mixer structure (Fig. 9.15). The modulator uses a double-balanced Gilbert quad whose tail current sources are digitally switched by the input data. This structure re-uses the DAC current for the mixer and improves

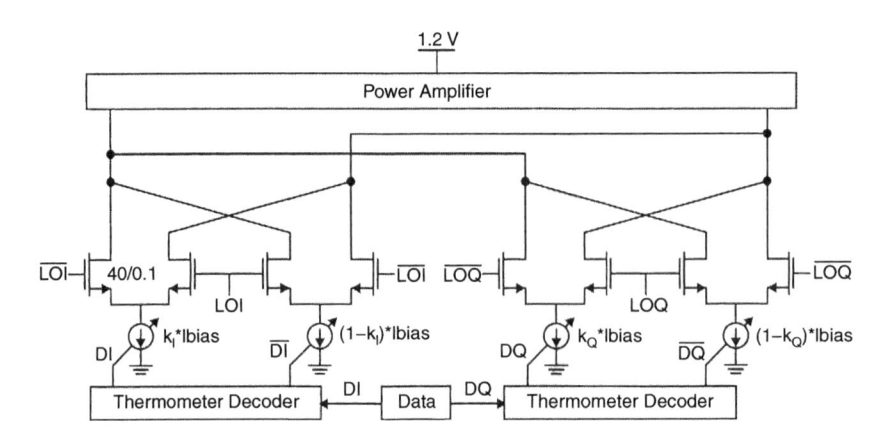

Fig. 9.15 A 60-GHz digital power mixer uses baseband I/Q DACs to directly modulate the LO signal. The modulator supports QPSK, 16QAM, and 64QAM.

linearity by omitting the transconductance (gm) stage of a conventional mixer. The LO signal derived from the VCO is converted to differential mode using a low-loss (1 dB) transformer and fed into the LO port of the Gilbert cell. Given the general structure of the modulator, any conceivable waveform can be synthesized, limited only by the DAC resolution and the switching speed (10 Gbps in this design). But since a separate PA is used after the RF-DAC, the PA efficiency would limit the application of this modulator to constant-envelope or weakly varying envelope signals, such as QPSK.

In [22] the current of the modulator is kept constant, and the data modulation is imparted after the LO signal, as shown in Fig. 9.16. We will call this topology, where the constellation is created directly at the carrier frequency at the output of the modulator, a direct QPSK modulator or direct 2-bit RF I/Q DAC, to distinguish it from the mixer-DAC described earlier in which the I and Q constellations are first separately created in current at baseband, and then up-converted and summed in quadrature at the carrier frequency. QPSK modulation at up to 18 Gb/s (9 GBaud) has been demonstrated at 70–80 GHz using this direct I/Q modulator approach [22]. Unlike the current-tail modulated DAC-mixer structure, where the output amplitude and the LO-port and output impedances of the DAC-mixer are codeword dependent, causing distortion, the input, output, and LO-port impedances, and the output signal amplitude of the constant-current direct modulator are codeword independent, allowing it to be operated in deep saturation with minimal distortion, large output power, and peak efficiency. Note that, to further maximize efficiency and output power, a current source need not be present, as illustrated in Fig. 9.17 [23]. The bias current and output power can be varied by appropriately setting the DC voltage on the gates of the bottom MOSFET differential pair with a current mirror. Transmitter output power levels exceeding 10 dBm were achieved throughout the 60-GHz band with 10% drain efficiency using this 2-bit RF I/Q power DAC without a power amplifier [23].

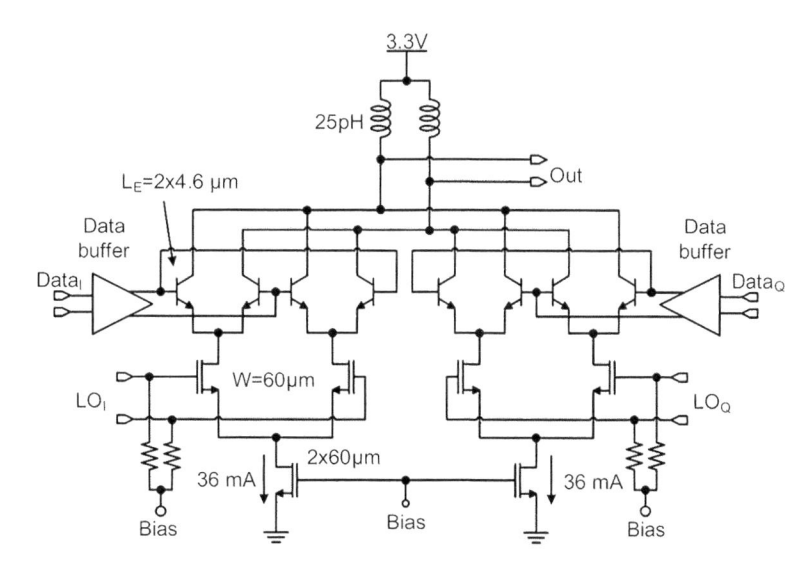

Fig. 9.16 An 80-GHz QPSK constant-current modulator [22]. Reprinted with permission from the IEEE.

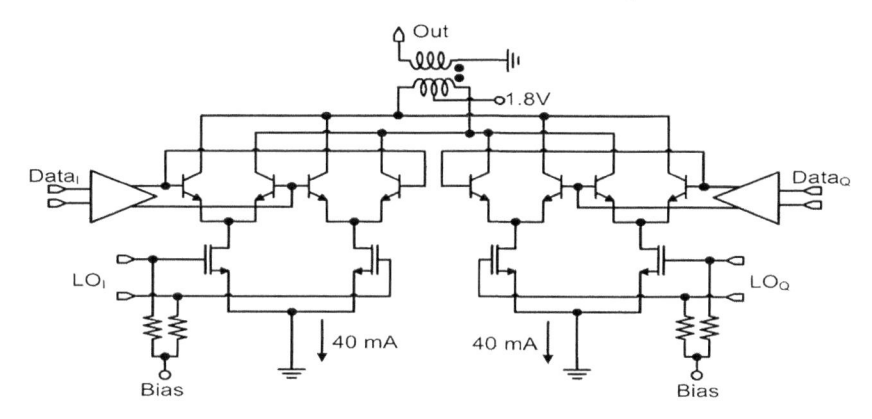

Fig. 9.17 A 60-GHz direct digitally modulated, 2-bit I/Q power-DAC [23]. Reprinted with permission from the IEEE.

9.3.3 RF I/Q Cartesian DAC

How can this constant-current I/Q modulator be modified for *m*-ary QAM constellations without compromising peak efficiency for any of the signal constellation points? The direct Cartesian I/Q modulator/DAC in Fig. 9.19 can operate either as a high-efficiency 16QAM power modulator using the external data bits *txd*[3 : 0] directly, or as a 12-bit I/Q DAC, using the internally generated data inputs $b_I[5:0]$ and $b_Q[5:0]$. Each of the I and Q sections consists of two 1.5-bit modulators (with output states 0, 1, and −1) in a 2:1 size ratio. The precision of this ratio is assured by segmentation, with the 2× modulator formed by two identical 1× modulator units connected in parallel. The output currents of all six modulator units are summed by a transformer which also acts as a differential-to-single-ended converter and provides output matching to 50 Ω. The schematic of the unit modulator is shown in the inset of Fig. 9.19. It is a modified version

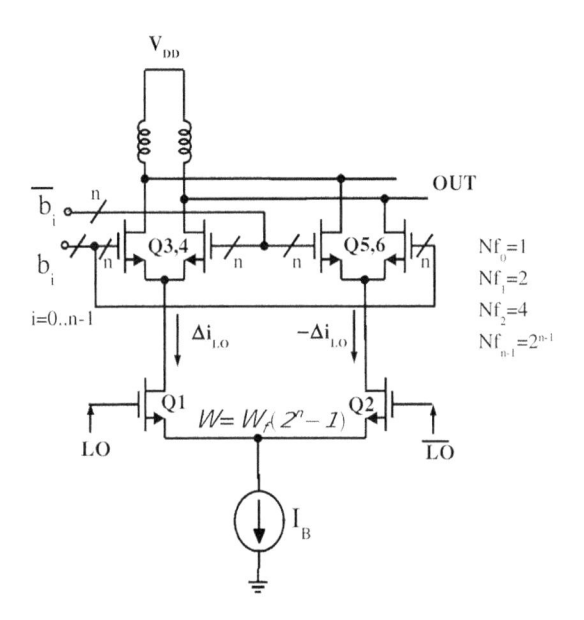

Fig. 9.18 A power RF-DAC employs constant-current and segmented-gate Gilbert quads driven by independent baseband bits [17].

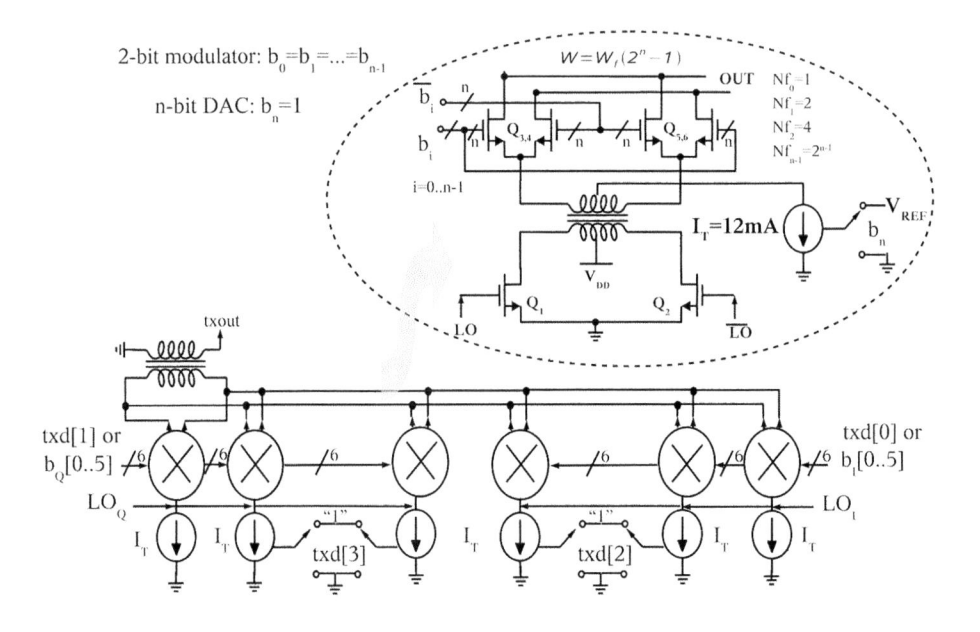

Fig. 9.19 Block diagram of a segmented 60-GHz power 16QAM modulator/RF I/Q power-DAC.

of the constant-current RF-DAC in Fig. 9.18 and consists of a differential gain stage with transformer load, driven into saturation by the large LO signal, and a segmented-gate Gilbert cell whose tail current can be turned on and off by bit b_n. When the transmitter operates as a 12-bit I/Q-DAC, the tail current of each unit modulator is always turned on (i.e. $b_n = 1$) and data bits $b_0 \ldots b_{n-1}$ are applied to binary weighted groupings of gate fingers of the Gilbert cells forming the I and Q sections of the DAC. A unit modulator features $N_f = 21$, $W_f = 0.91\,\mu\text{m}$ wide gate fingers, for a total of 63 gate fingers in each I and Q section of the DAC. The total bias current drawn by the I/Q-DAC/16QAM modulator is 72 mA, resulting in an output power of over +10 dBm. To derive the expression of the output voltage on the load impedance Z_L seen at the transformer primary, the Gilbert cells can be viewed as passive switches with $IP_{1dB} > 10$ dBm, which steer the tail current either to the positive or to the negative output of the I/Q-DAC. The tail current is modulated by the internal 2.5-Gb/s parallel data streams $b_I[5:0]$ and $b_Q[5:0]$, derived from $txd[0]$ and $txd[1]$, respectively,

$$V_{out} = g'_m W_f Z_L V_{LO} \left[\cos(\omega t) \sum_{i=0}^{5}(-1)^{b_I[i]} 2^i + j\sin(\omega t) \sum_{i=0}^{5}(-1)^{b_Q[i]} 2^i \right].$$

When the transmitter operates as a 16QAM modulator, the tail currents of two of the unit modulators in the I and Q sections are turned on and off (by external data bits $txd[2]$ and $txd[3]$, respectively) while all the gate fingers of the I-path segmented Gilbert cells are controlled by the external data stream $txd[0]$ and those of the Q-path by $txd[1]$:

$$V_{out} = 21 \times g'_m W_f Z_L V_{LO} \Big[(-1)^{txd[0]}(1 + 2txd[2])\cos(\omega t) \\ + j(-1)^{txd[1]}(1 + 2txd[3])\sin(\omega t) \Big],$$

where $g'_m = 0.8$ mS/μm is the n-MOSFET transconductance when biased at 0.3 mA/μm, and Z_L is the load impedance.

Since, when configured as a 16QAM modulator, 48 mA of the total 72-mA tail current is dynamically turned on and off by external bits $txd[2]$ and $txd[3]$, the transmitter behaves like a peak-efficiency-saturated PA for all 16QAM constellation points. QPSK modulation is obtained simply by setting $txd[2]$ and $txd[3]$ to logic "1." Note that if the 16QAM constellation is generated with the transmitter configured as a 12-bit I/Q-DAC, the efficiency degrades for the low-amplitude constellation points since the entire 72-mA tail current of the output stage is always on. The aggregate data rate will also be higher in 16QAM mode (four data streams $txd[3:0]$, each at 2.5 Gb/s, rather than two I and Q streams $txd[0]$ and $txd[1]$ at 2.5 Gb/s). One disadvantage of operation in the 16QAM mode is that the output impedance of the transmitter becomes slightly codeword dependent. This is not the case in I/Q-DAC mode. The measured QPSK and 16QAM constellation points at low data rates are illustrated in Fig. 9.20 after down-conversion with an external narrow-band mixer.

The peak-efficiency, direct 16QAM power modulator concept can be extended to higher-order schemes. Figure 9.21 sketches a 64QAM version with on–off switched binary-weighted RF-DAC segments and the antenna acting as the summing node.

9.3.4 Broadband digital transmitter with segmentation

Although the DAC-mixer and direct RF-I/Q-DAC topologies described earlier can operate with LO signals from several GHz to hundreds of GHz, their bandwidth and data

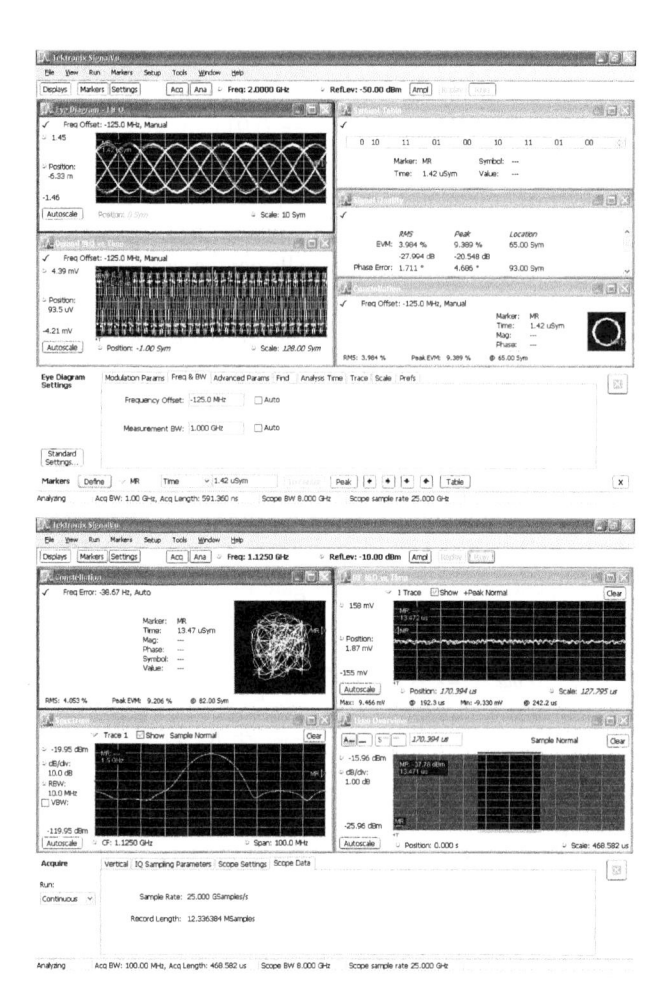

Fig. 9.20 Measured (top) QPSK and (bottom) 16QAM constellations points after down-conversion.

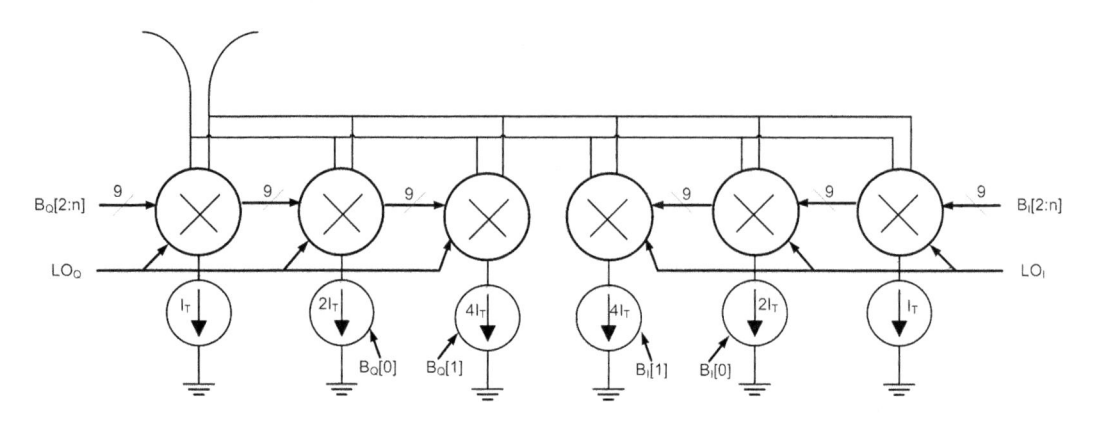

Fig. 9.21 A 64QAM power modulator with on–off switched binary-weighted RF-DAC segments and the antenna acting as the summing node.

9.3 Digital mm-wave transmitter architectures

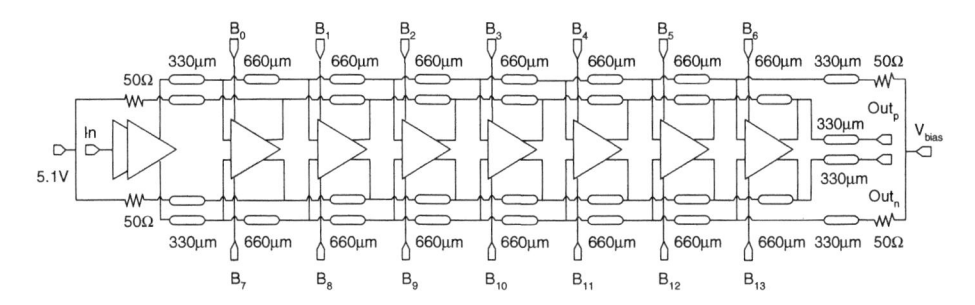

Fig. 9.22 Schematic of the distributed 7-stage DAC [24]. Reprinted with permission from the IEEE.

rate are limited by the bandwidth of the output transformer, which acts as a current summer and impedance transformer. Figure 9.22 illustrates a large-swing, broadband 6-bit RF-DAC topology with segmentation that can be used for wired and wireless communication links, as well as for instrumentation. Unlike the traditional baseband DACs used in high-data-rate fiber-optic communication links and which are limited to low-voltage swings because of the large capacitance accumulating at the output node, a distributed transmission line can be employed as a broadband summing node, where the output capacitance of each DAC segment is built into the transmission line. The data rate can be as high as the sampling clock and the maximum clock frequency in this particular realization is 50 GHz [24].

The DAC is designed for up to 3 V_{pp} output swing per side, 6 V_{pp} differential, and is realized as a 7-stage distributed amplifier (DA). The block diagram of the DA can be seen in Fig. 9.22. The 6-bit DAC features a segmented architecture with a total of 14 independently controlled thermometer-coded bits, seven of which represent the fully segmented most significant (MSB) 3 bits, while the other seven represent the least significant (LSB) 3 bits. Each DA cell consists of two independent data-retiming paths, each driving a BPSK modulator realized with a Gilbert cell, as seen in Fig. 9.23. This arrangement maximizes the number of bits, the swing, and bandwidth. The 2-mA cell employs $0.13 \times 1\,\mu m$ HBTs and $0.13 \times 6\,\mu m$ MOSFETs, while the 16-mA cell is scaled by a factor of 8 and hence uses $0.13 \times 8\,\mu m$ and $0.13 \times 48\,\mu m$ MOSFETs. Because the currents in each cell are constant, the output impedance is codeword independent. However, the efficiency of the DAC depends on the constellation point being formed at the output transmission line and decreases from its peak value for lower-amplitude constellation points.

The two data paths are scaled versions of each other, consisting of two buffers and a retimer followed by another two buffers, all operating from 2.5 V as seen in Fig. 9.23. The buffers are designed as CML differential pairs employing HBTs and inductive peaking. The retimer is designed as a BiCMOS latch using inductive peaking. The component values for each latch are summarized in Table 9.1. The retimer in the LSB data path uses two 2-mA latches while the retimer on the MSB data path uses a 6-mA latch followed by an 8-mA latch. Perfect scaling of the data paths was limited by the minimum tail current required to achieve 50 Gb/s switching operation in this production

Table 9.1 Summary of latch device sizing.

	2-mA latch	6-mA latch	8-mA latch
R	175 Ω	50 Ω	42 Ω
L	150 pH	113 pH	100 pH
L_{HBT}	1 μm	3 μm	4 μm
W_{MOS}	6 μm	18 μm	24 μm

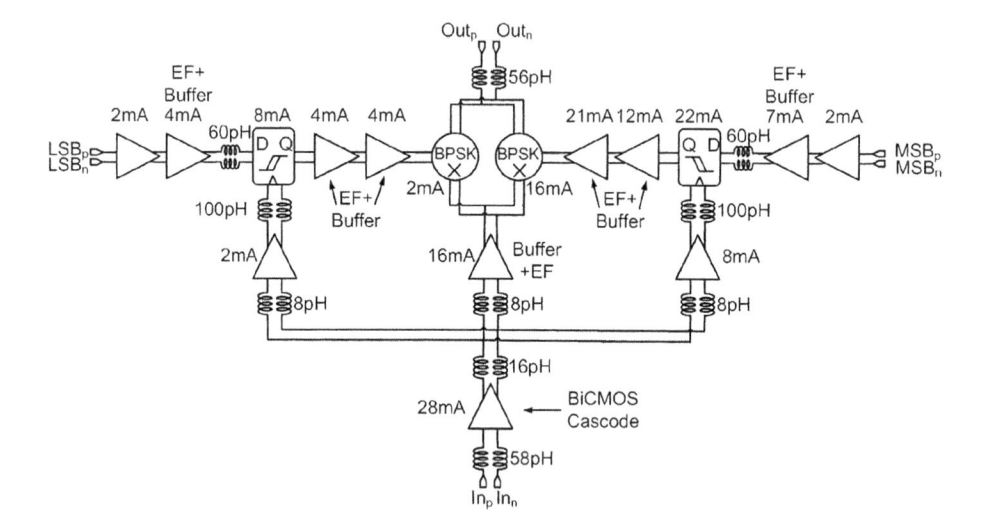

Fig. 9.23 DAC stage schematic illustrating the LSB and MSB output cells and data-retiming paths in a 1:8 ratio [24]. Reprinted with permission from the IEEE.

130-nm SiGe BiCMOS process. Power consumption, which is dominated by the 14 data-retiming paths, is contained by employing low-voltage BiCMOS CML topologies as in [25]. Simulations indicate that the power consumption of the data-retiming lanes will be reduced by more than a factor of 3 in the new 55-nm SiGe BiCMOS node [26]. However, in the future, for data rates up to 50 Gb/s or so, the CML logic can be replaced by inductively peaked CMOS inverter logic [27, 28], which further reduces power consumption in the data-retiming lanes and makes it adaptable with the data rate for a constant energy per bit, irrespective of the data rate. In order to distribute the clock signal in phase to each cell, a cascade of buffers was used, as seen in the left-hand part of Fig. 9.24. The first inverter and the first emitter follower (EF) stage operate from 3.3 V while the other inverters and EF stages operate from 2.5 V. The first buffer in each DA-DAC cell was designed as a BiCMOS cascode structure in order to decrease the loading presented to the transmission line. A schematic of this buffer can be seen in the right-hand part of Fig. 9.24. The other three buffers use HBT differential pairs with inductive peaking. Compact layout of each stage is of paramount importance in order to maintain the transmission line length at the desired value. The transmission lines between stages are 5 μm wide and 660 μm long (ℓ) formed in metal 6 over a metal 3 ground plane. The

9.3 Digital mm-wave transmitter architectures

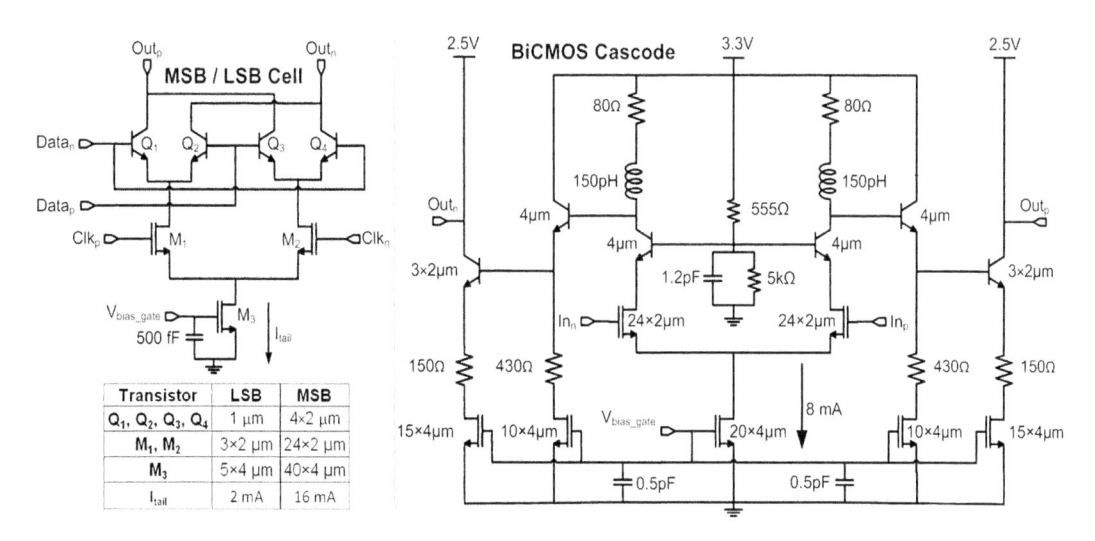

Fig. 9.24 (Left) MSB and LSB DAC cell output-stage schematics. (Right) Clock buffer schematics [24]. Reprinted with permission from the IEEE.

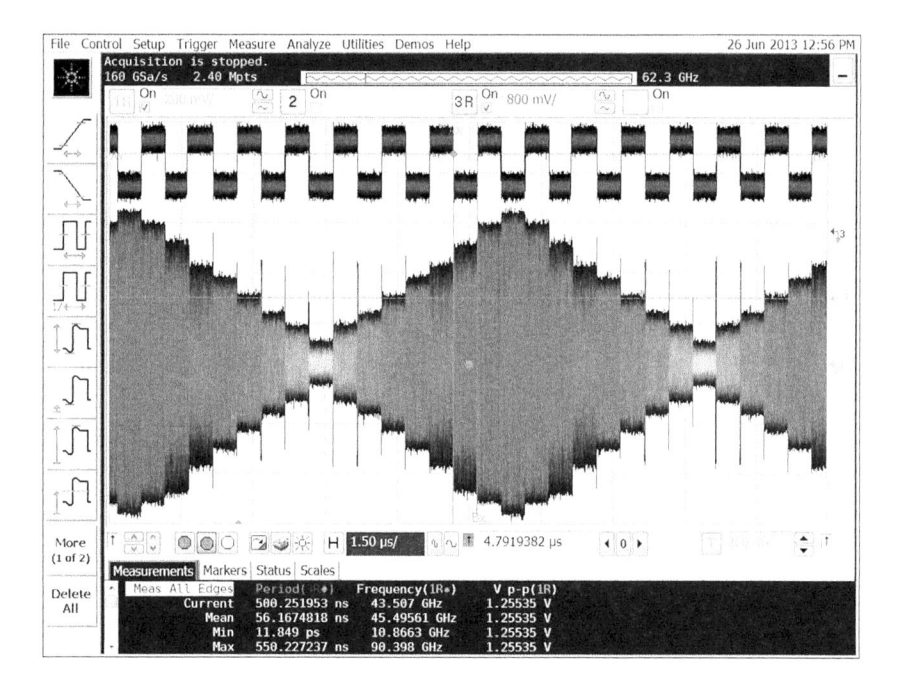

Fig. 9.25 Measured staircase response for a carrier frequency of 44 GHz with only the seven MSBs switching at low data rate.

end transmission lines are 330 μm long ($\ell/2$). They can be seen in the layout diagram of the chip in Fig. 9.26. The measured staircase response for a carrier frequency of 44 GHz is shown in Fig. 9.25 when the segmented seven MSBs are switched sequentially at a low data rate of 100 Mb/s without yet exercising the seven LSBs.

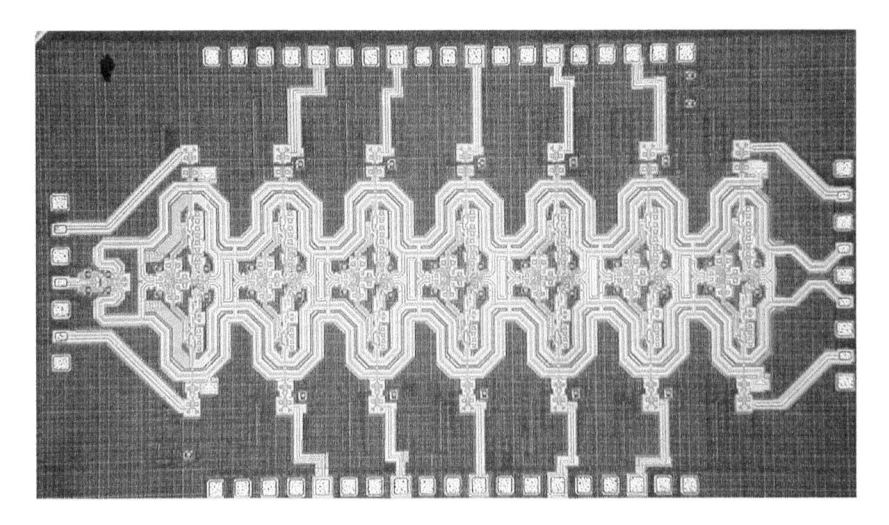

Fig. 9.26 Layout diagram of the distributed power-DAC [24]. Reprinted with permission from the IEEE.

9.4 Digital antenna modulation

While the digital techniques introduced in this chapter offer efficiency enhancements compared with linear PAs, there are several factors that limit the efficiency and linearity, especially for complex modulation schemes. In particular, as already mentioned, when unit elements are switched in and out of the RF-DAC, the output impedance varies and causes a load-pull, which results in both AM–AM and AM–PM distortion. The phase distortion arises from the output capacitance (and hence phase) variation with codeword. When both I and Q signals are combined, the effect of this load-pull is even more problematic due to the interaction between the I and Q elements. If an isolating combiner is used, there is a power penalty, which in a PA results in a large hit in efficiency. For example, a 40%-efficient PA using a Wilkinson combiner with 1.5 dB loss suffers an efficiency penalty of about 71%, with a net efficiency of 28%. One way to overcome this problem is to avoid combining the I and Q signals on chip. Instead, separate antenna elements are used for the I and Q paths, with the signals from each element being combined in space, as shown in Fig. 9.27 [29, 30]. This at least eliminates the need for a 2D LUT, and for simple modulation schemes with a small number of bits the LUT can be eliminated altogether.

In practice, a silicon mm-wave transmitter is often realized as a spatial combiner, often a phased array to reach sufficiently high output power level. Spatial combining allows low-power unit elements to combine in space and in the targeted direction, offering spatial multiplexing and efficient power, even in low-voltage nanoscale CMOS technologies. Given that the transmitter is already a phased array, an even more radical approach is to segment the antenna array itself, rather than the RF-DAC, as shown in Fig. 9.28 [29]. This approach has been named a digital quadrature (or I/Q) spatial combining transmitter and allows one to realize the highest levels of efficiency and linearity by taking advantage of the antenna array.

9.4 Digital antenna modulation

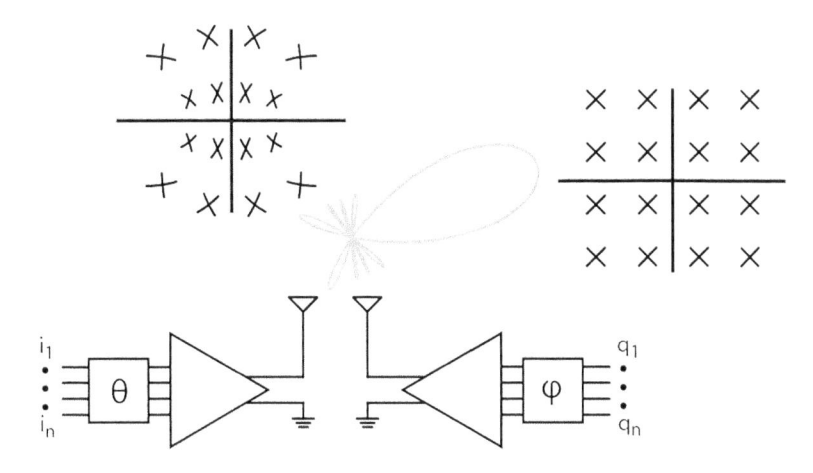

Fig. 9.27 A digital Cartesian (I/Q) spatial combiner uses an RF-DAC and two separate antennas for the I and Q paths and combines the signals in space.

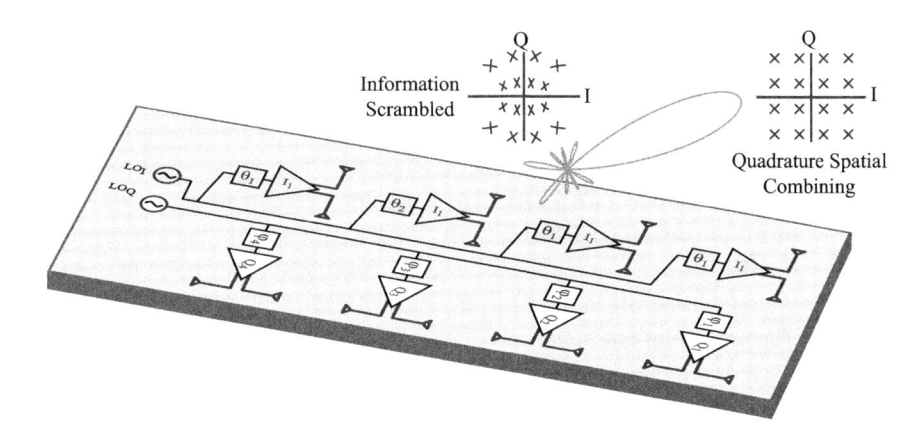

Fig. 9.28 A digital Cartesian (I/Q) spatial combiner uses an array of antennas as "bits" in the modulator to achieve high efficiency.

9.4.1 Antenna segmentation

Consider an N-bit spatial RF-DAC realized using the proposed antenna segmentation approach. These N bits correspond to 2^{N-1} discrete power levels. Unless N is small, it is impractical to use 2^{N-1} antenna elements, since the power per element would be very small. Moreover, such an array would become impractically large. These levels must therefore be partitioned between the antenna elements and further down within each antenna element driver. Except at frequencies above 100 GHz, for a mm-wave transmitter, using more than $M = 32$ antenna elements is impractical in a mobile profile, not only because of the space occupied by the array, but more because of the routing losses in going from a single chip to the elements. However, it should be noted that a square 64-element antenna array spaced $\lambda/2$ apart at 250 GHz occupies only 5 mm × 5 mm and is four times smaller than a 16-element square array at 60 GHz. Below 100 GHz, since typically $M < 2^{N-1}$, one cannot make each antenna element correspond to a single bit unless the modulation format is 16QAM or simpler.

Next consider a weighted scheme for the antenna segmentation. One simple approach would be to use binary weighting, so one element would be driven with the maximum power, and each subsequent element would be driven by a power level that is (2^i) smaller. However, such an approach would be space inefficient since most of the array elements would be driven with significantly less power than the largest element, sized at 2^{N-1}.

A more practical approach is to use a segmented structure with a coarse and a fine array. Similar to a segmented DAC, one can trace coarse steps from 0 power to $M \cdot P_u$ of power, where P_u is the power of each antenna element. This corresponds only to a resolution of $\log_2 M$ bits, so the remaining $P = N - \log_2 M$ bits of fine resolution should come from breaking one or more unit elements into an RF-DAC. If the power level becomes too small to realize, in other words $P_u/2^P$ corresponds to a device too small to realize or too small to provide reliable matching, then more than one non-unit element should be used to generate the fine steps.

In this approach, most of the antenna elements are driven in a binary fashion, either fully on or off, which means that the antenna element driver (or PA) can be realized by using a nonlinear switching amplifier. As long as the coupling between antenna elements (which depends on the antenna spacing and the material used to build the package) is smaller than −30 dB [30], there is minimal compression and distortion as more elements are turned on. The back-off characteristic is also linear (similar to class B), since the DC power consumption scales with the coarse codeword. Off elements can be completely turned off to save power and maintain peak efficiency for all unneeded cells. The remaining fine element(s) have a more traditional RF-DAC architecture, but, due to the reduced number of bits and lower power levels, the compression characteristics are manageable and the efficiency remains within $1/M\%$ of the peak value.

To make this discussion concrete, say that we wish to realize an array with 7 bits of resolution but we are limited to eight antenna elements. Since $8 = 2^3$, the antenna elements can only realize 3 bits of resolution. The remaining 4 bits can be realized by dividing the power of one element into 16 steps. In this manner, we have a total of 16*8 = 128 steps, exactly the desired resolution. This transmitter is shown schematically in Fig. 9.29.

Quadrature segmentation

From here we can see that, to make a complex modulator, all that is needed is to partition the array into an in-phase and quadrature-phase array. Such an example is illustrated in Fig. 9.30, where 90° hybrids inserted in the array replace Wilkinson power dividers at strategic locations to ensure that adjacent antenna elements are driven in quadrature by the LO signal. The desired signal constellation is formed by in-phase spatial power combining in the direction perpendicular to the antenna array, at its center of symmetry [30]. Alternatively, if no 90° hybrids are used but the antenna elements have phase shifters to realize a tunable phased array, the I and Q arrays should be designed to radiate into the same solid angle such that the spatial summation occurs in quadrature.

9.4 Digital antenna modulation

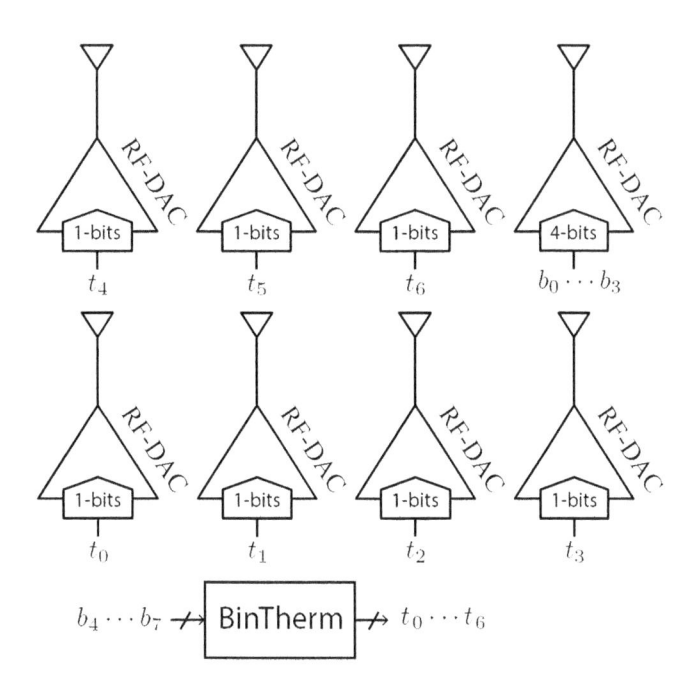

Fig. 9.29 A segmented antenna RF-DAC architecture.

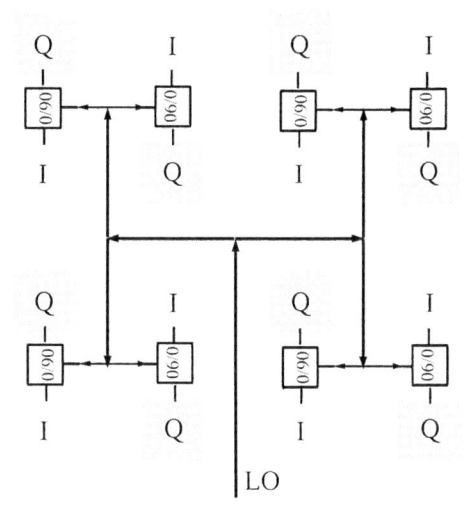

Fig. 9.30 A balanced arrangement of a segemented I/Q antenna array with eight I and eight Q elements.

In both approaches, if the receiver is not situated in the direction of the peak, the I and Q components do not add in phase and there is a distortion in the received constellation pattern. In a traditional transmitter, I and Q amplitude and phase mismatch is a common phenomenon that arises from a mismatch in components in the I and Q paths. Here the mismatch is due to the spatial directivity of the antenna elements. Equations relating EVM to amplitude and phase mismatch are well known in the literature and can be used

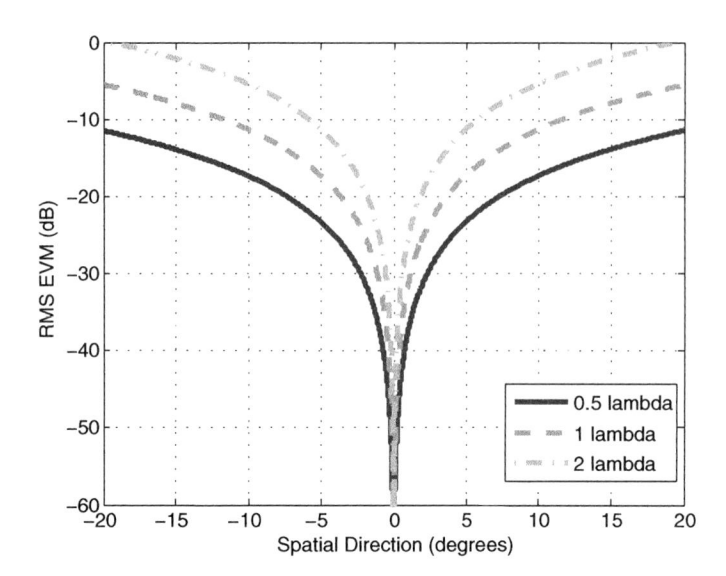

Fig. 9.31 The information beamwidth of an antenna spatial combiner compared with the pattern beamwidth.

in the context of the spatial I/Q combiner. This spatial directivity of EVM means that the antenna has an "information" beamwidth that corresponds to the solid angle over which information is received with sufficient EVM. This information beamwidth is unrelated to the pattern beamwidth of the array. For a small array, one finds that the directionality of such a modulator is dominated by the EVM distortion, whereas for large arrays the directionality of the array begins to dominate.

The information and beam width of an antenna array are plotted in Fig. 9.31. As is evident from the figure, the EVM limits imposed by the beam are only a limit for a small array. Moreover, if beam steering is used to maximize the link budget, this limitation can be easily overcome. The amount of induced I/Q phase mismatch is also a function of the spacing between the I and Q array. To minimize coupling between the I and Q arrays (which leads to load-pull distortion), it is desirable to space them far apart. But the farther apart the arrays are spaced, the more spatially directive the information beamwidth becomes. The spacing between the I and Q arrays is therefore a compromise between directivity and unwanted coupling.

To ensure good EVM as a function of position, the transmitter phased array should be designed with sufficiently fine phase resolution. Fortunately it is rather easy to design an accurate transmitter phase shifter, as baseband phase shifters can be realized essentially by varying DC currents in a DAC [31], while LO path phase shifters need not be broadband or linear, and precise phase control can be inserted in the 90° hybrid itself [18].

Digital transmitter beam pattern

Another interesting property of the digitally modulated antenna transmitter is the time-varying beam pattern. Since different elements are turned on and off, the antenna pattern changes as a function of time. For example, when a small-amplitude codeword is desired, only one or a few antenna elements are active. As more elements are activated, the beam narrows and widens as a function of the data pattern. But since each element has a static phase corresponding to the direction of radiation, and only the amplitude is modulated, the beam always points in the correct direction.

On the other hand, when the quadrature array is introduced with an in-phase and quadrature-phase component, then the antenna pattern begins to move depending on the data pattern. This is because the Q array has a 90° phase shift relative to the I array, and the addition of more Q elements relative to I tends to shift the pattern maximum. Note that the pattern maximum does not occur in the desired direction because at these spatial directions the I and Q arrays are adding in phase, rather than in phase quadrature. This means that, in the desired direction, the pattern gain will vary with codeword. Detailed analysis shows that while the pattern changes instantaneously with codeword, the average pattern remains constant for the array, as long as the baseband data are balanced [32].

9.4.2 A 60-GHz CMOS digital Cartesian quadrature spatial combiner

To prove the feasibility of a quadrature spatial combiner, a 60-GHz WiGig prototype has been designed and tested in a 65-nm technology node [33]. The block diagram, shown in Fig. 9.32, shows the architecture of the transmitter. The CMOS chip contains eight channels, divided into four I and four Q sub-channels. This means that each array can realize 2 bits of coarse resolution. One antenna element is further divided into four sub-elements, providing another 2 bits of fine resolution, for a total of 4 bits resolution for the I and Q array. Since the entire array is realized using thermometer coding, 7 bits are required to activate each element one-by-one. An on-chip PRBS generates baseband I and Q bitstreams and a fully digital FIR filter is realized using four

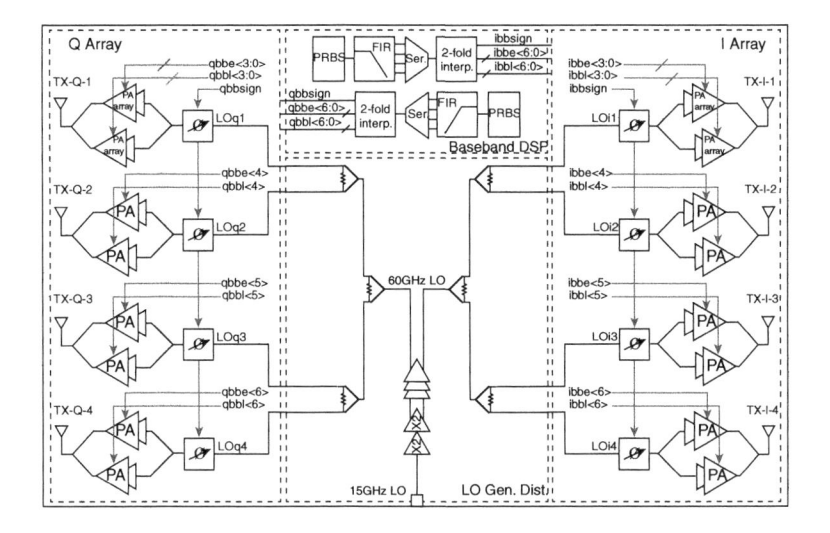

Fig. 9.32 Block diagram of 60-GHz WiGig prototype digital spatial quadrature combiner.

parallel sub-FIRs. A 4:1 serializer generates the I and Q filtered data streams. To further reduce spectral images, a 2-fold interpolation is implemented by splitting the arrays into two identical halves, where the first half is excited one half clock period earlier than the other. For example, for the Q channel, the bitstreams $qbbe[6:0]$ and $qbbl[6:0]$ correspond to the early and late signals. Since early and late signals are routed to each element, a total of $7 \times 2 \times 2 = 28$ baseband signals are routed. Each signal is oversampled by a factor of 4 to allow digital filtering, resulting in a power consumption of 30 mW.

Each channel is designed to drive an antenna element. An incoming LO signal at 15 GHz is quadrupled to 60 GHz and distributed to each element. On-chip transmission lines and Wilkinson dividers are used for signal distribution. The LO chain alone consumes only 35 mW of power. To realize the phase shifter, a baseband phase-rotator architecture similar to [8] is used. Each phase shifter takes the LO signal, splits it into I and Q components using a passive lumped hybrid, and takes a weighted sum of I and Q signals to generate a locally phase shifted LO. Each LO phase shifter consumes 6 mW of power.

As already noted, each PA element is divided into two, one activated with the early bit, and the next with the late bit, so that the output ramps, similar to a first-order hold. This 2-fold interpolation reduces the spectral image significantly compared with a zeroth-order hold. Without this scheme, an oversampling ratio of 8 would be required to meet the −30 dBc WiGig specification. Given that the baseband data is at 1.76 G, a 14-GHz baseband clock is impractical. By using this scheme, the oversampling ratio is reduced to 4×.

The schematic of the 2-bit PA RF-DAC unit element is shown in Fig. 9.33. The core realizes a class-E/F$_2$ switching amplifier, which has superior efficiency than linear PAs. To realize the 2 bits, the PA is divided into four sub-elements, non-uniformly sized in order to pre-distort the AM–AM curve. As a larger codeword is applied, the unit elements are made larger to combat the compression. To compensate for AM–PM

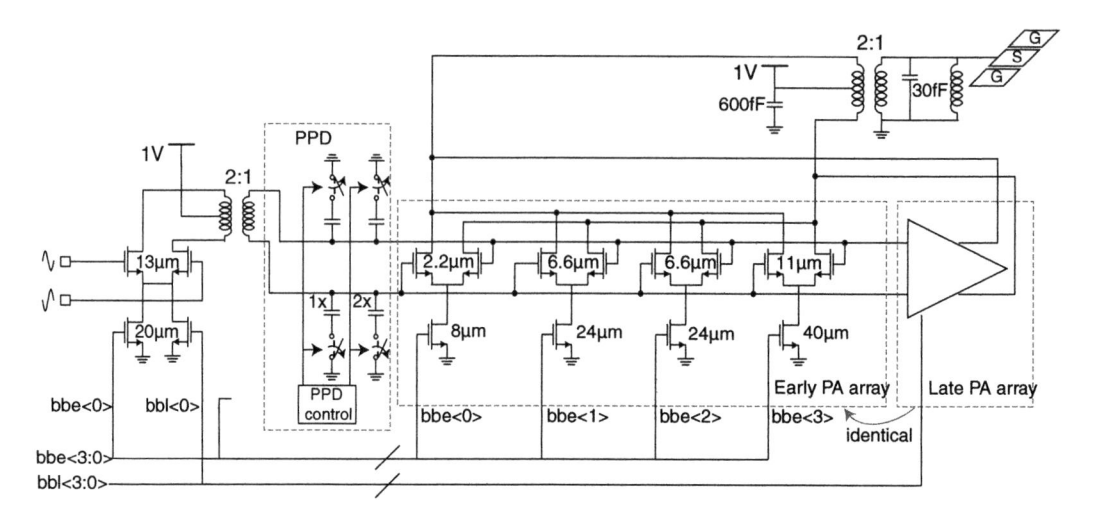

Fig. 9.33 Schematic of 60-GHz RF-DAC unit element.

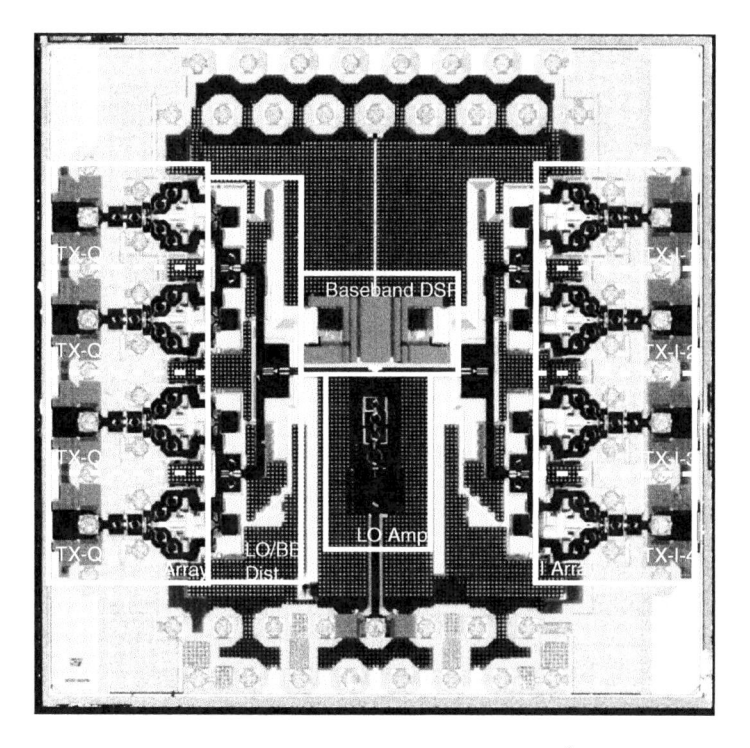

Fig. 9.34 Die photo of 60-GHz WiGig prototype. The die measured 3×3 mm^2.

distortion, dummy capacitors are used to load the output as a function of codeword. The capacitors are binary weighted with a unit element of 3.4 fF.

Class-E/F$_2$ waveform tuning is provided by carefully engineering the harmonics of the load impedance. Since controlling harmonics with multiple tanks is both lossy and impractical, only a single transformer is used to condition the first and second harmonic by exploiting the fact that the second harmonic excites the even-mode of the transformer, whereas the fundamental mode excites the odd-mode. In this manner, an inductive reactance of 1/4 of the fundamental mode reactance is synthesized at the second harmonic to generate the desired waveforms.

A die photo of the experimental prototype is shown in Fig. 9.34. The CW mode performance is measured by probing the output of each transmitter channel. The peak output power of each element is 9.6 dBm and the peak drain efficiency is 28.5%. The back-off characteristic of this transmitter can be observed in the PA efficiency versus the output power plot. The measured back-off curve closely matches a class-B PA behavior. At 6-dB CW power back-off, the PA still has an efficiency of 15%, which is twice the efficiency of a class-A PA in the same technology node. The static digital AM–AM and digital AM–PM curves of the first transmitter element are measured. Owing to non-uniform device sizing on the sub-PA thermometer cells, the PA output voltage amplitude is very linear with respect to the digital amplitude codeword, and the difference between I and Q PAs is negligible. When the phase-pre-distortion is off, the worst-case AM–PM distortion is more than 30°. By enabling the phase-pre-distortion, the AM–PM distortion

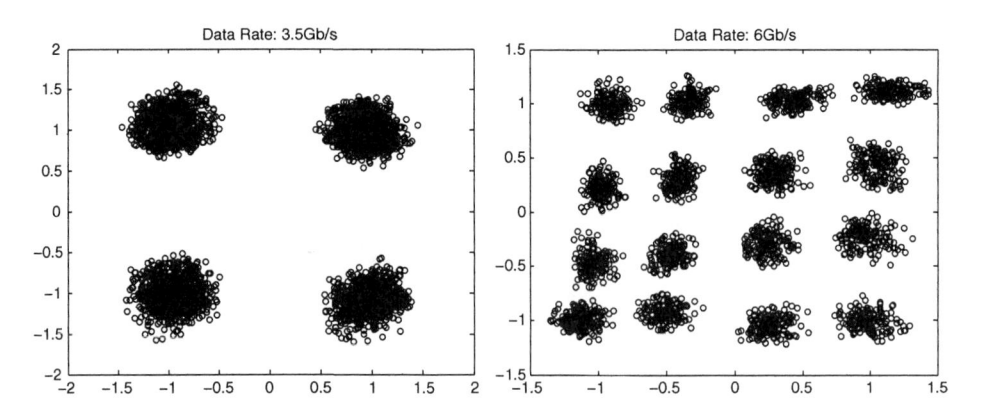

Fig. 9.35 Measured constellation points for transmission in QPKS (left) and 16QAM (right) modes.

can be reduced to 10°. The LO phase-shifter performance was measured by comparing the relative delay of two transmitter element output signals. The phase shifter has an RMS phase resolution of 2.2°, and the fluctuations in the phase steps are mainly due to the timing accuracy of the sub-sampling oscilloscope.

To verify the concept of quadrature spatial combining, a complete mm-wave module was also designed and fabricated. The package consists of a three-layer PCB made of Rogers 4350 material. The silicon die is attached to the front-side of the package in a flip-chip configuration and the patch antenna arrays are printed on the back-side. The patches are aperture coupled from the feed lines on the front-side. Unassembled PCBs are used for antenna characterization. The measured peak EIRP is 22 dBm at 60 GHz in the broadside direction. Modulated signal transmission was conducted by sending an I/Q pseudo-random bit sequence from the transmitter and recovering the data. Figure 9.35 (left) plots the received signal constellation for QPSK modulation at 3.5-Gbs data rate with a carrier frequency of 60 GHz. The corresponding EVM is −15 dB. Similarly, the constellation for 16QAM modulation is shown in Fig. 9.35 (right). The data rate for 16QAM modulation is 6 Gb/s and the EVM is −16.2 dB. The average EIRP is around 16.4 dBm. The transmitter EVM was also tested as a function of the receiving angle. Figure 9.36 shows the measured EVM pattern when the information direction is centered around 0°. QPSK modulation is used in this experiment with a carrier frequency of 60 GHz and a data rate of 2 Gb/s. The measured EVM spatial behavior matches closely with the theoretical prediction. Finally, the mixed-signal baseband signal processing function is verified by monitoring the transmitter output spectrum.

By comparing the down-converted output spectrum for QPSK and 16QAM modulation schemes respectively, one can demonstrate the importance of the FIR filter. The mixed-signal baseband signal processing can effectively suppress the image and lower the noise floor. With mixed-signal filtering on, the output spectrum is compliant with the WiGig mask. The entire circuit is powered from 1-V supply. The I/Q PA arrays consume a peak DC power of 229 mW and the entire transmitter consumes a peak power

9.4 Digital antenna modulation

Table 9.2 Chip performance summary.

CW mode	
Peak EIRP (dBm)	22
TX element P_{out} (dBm)	9.6
PA peak efficiency (%)	28.5
TX peak efficiency (%)	17.4
Total peak DC power (mW)	382
Data transmission at 6-dB back-off	
PA average efficiency (%)	16.5
TX average efficiency (%)	7
Peak data rate (Gb/s)	6
QPSK EVM	−15
16QAM EVM	−16.2
TX average DC power (mW)	260

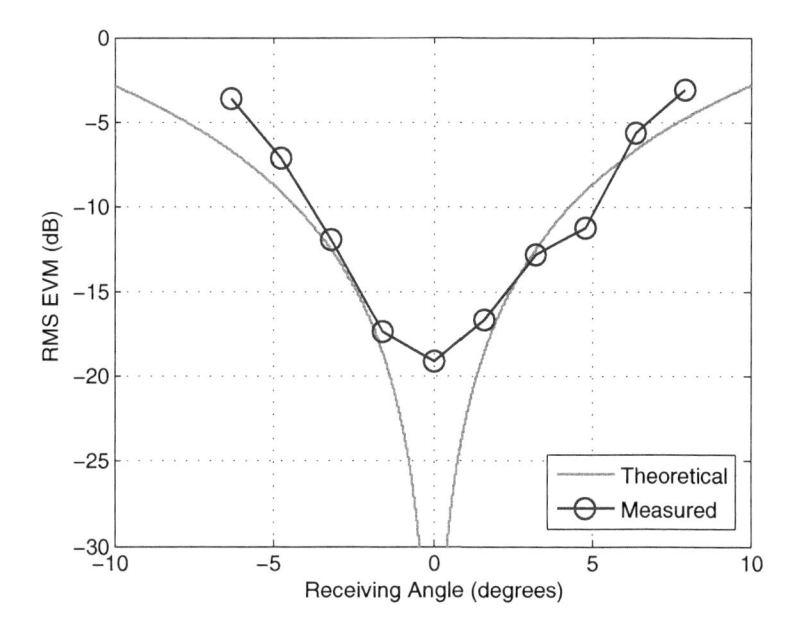

Fig. 9.36 Measured EVM as a function of the receiver angle shows the expected degradation due to the information beamwidth.

of 382 mW. The power consumption of other blocks is summarized in Table 9.2. The measured peak PA efficiency is 28.5% and the peak transmitter efficiency is 17.4%. At 6-dB back-off when transmitting QPSK or 16QAM data, the measured average PA efficiency is 16.5% and the average transmitter efficiency is 7%. Table 9.3 summarizes the chip performance.

Table 9.3 Comparison with efficiency-enhancing 60-GHz transmitters.

	[10]	[11]	[34]	This work
TX architecture	Outphasing	Outphasing	Class-AB	Cartesian digital-to-RF
Signal reconstruction	Spatial combining	Transformer combining	No need	Spatial combining
Single PA P_{sat} (dBm)	9.7	15.6	10	9.6
η_{PA} at P_{sat} (%)	11	25	10.8	28.5
η_{PA} at 6-dB back-off[a] (%)	N.A.	9	5.7[b]	15
TX beamforming?	Yes (2 PAs)	No	No	Yes (8 PAs)
Antenna package?	No	No	No	Yes (EIRP = 22 dBm)
Modulation	16QAM	16QAM	16QAM	16QAM
Peak data rate (Gb/s)	0.2	0.5	7	6
Technology	65 nm	40 nm	40 nm	65 nm

[a] CW mode.
[b] At 5-dB back-off.

9.4.3 Phase- and amplitude-modulated 45-GHz and W-band Cartesian (I/Q) power DACs for watt-level digital transmitters with antenna segmentation

To explore the potential of the digital I/Q architecture with antenna segmentation shown in Fig. 9.30, and to achieve both frequency and data-rate scalable watt-level high-efficiency mm-wave transmitters, prototypes were designed and fabricated at 45 GHz [30], 90 GHz [18], and 110 GHz [28] in a commercial 45-nm SOI CMOS process. In all cases, the transmitter is partitioned as in Fig. 9.37, with chiplets consisting of either one or two I/Q DAC cell pairs driving differential patch antennas. Each I/Q DAC cell pair features a broadband quadrature LO-tree based on lumped L-C 90° hybrids. Precision phase correction over ±5° is inserted after the 90° hybrid to compensate for systematic phase errors inherent in the design. The detailed block diagram of a 45-GHz chiplet is shown in Fig. 9.38 [30]. The aggregate output power generated by the four DAC cells is 30 dBm. With the exception of the antennas, and 90° hybrids, all circuit blocks operate from DC to 50 GHz.

In order to create an arbitrary m-ary QAM constellation, each I and Q antenna element features both BPSK and on–off amplitude modulation (Fig. 9.39). For the 16-element array in Fig. 9.30, this modulation scheme generates as many as 289 constellation points, all with peak transmitter efficiency [30]. For lower-order modulation formats such as 16QAM and QPSK, which require only 16 and 4 constellation points, respectively, the peak efficiency of the transmitter can be maintained for a range of output power levels by switching completely off the elements that are not needed. This permits the transmitter to instantaneously adjust its modulation format and output power to adapt to the link length and data rate with maximum efficiency. The on–off modulation is imparted in the output stage to maximize the bandwidth, while the BPSK modulator is inserted

9.4 Digital antenna modulation

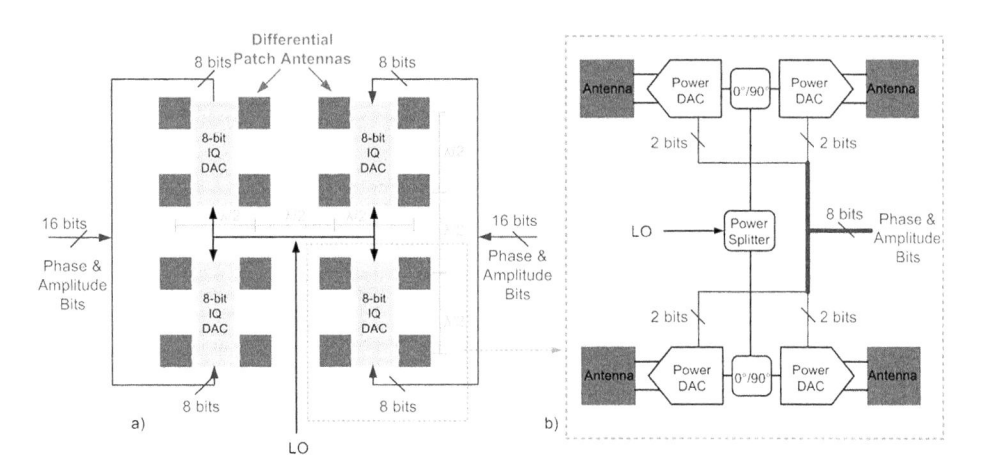

Fig. 9.37 Block diagram of a chiplet consisting of 2I and 2Q DAC cells.

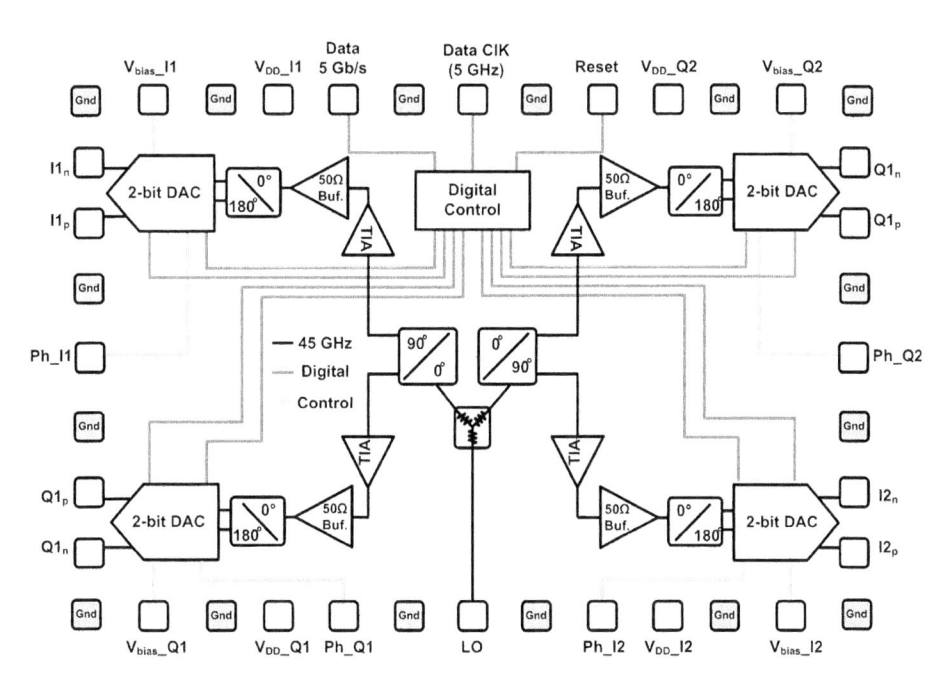

Fig. 9.38 Possible chiplet partition for frequency scalable watt-level I/Q antenna array with eight I and eight Q elements.

either immediately before the output stage [28, 30] or as part of the large-power 8-bit segmented Gilbert-cell output stage [18], as shown in Fig. 9.40. In either case, the output stage is designed as a high-efficiency class-AB [18, 28, 30] or class-D [35] PA. To overcome the low breakdown voltage of nanoscale MOSFETs and to maximize the optimal load impedance, R_{LOPT}, for a given output power, series stacking is used in all versions of the output stage. As many as eight n-MOSFETs and/or p-MOSFETs have

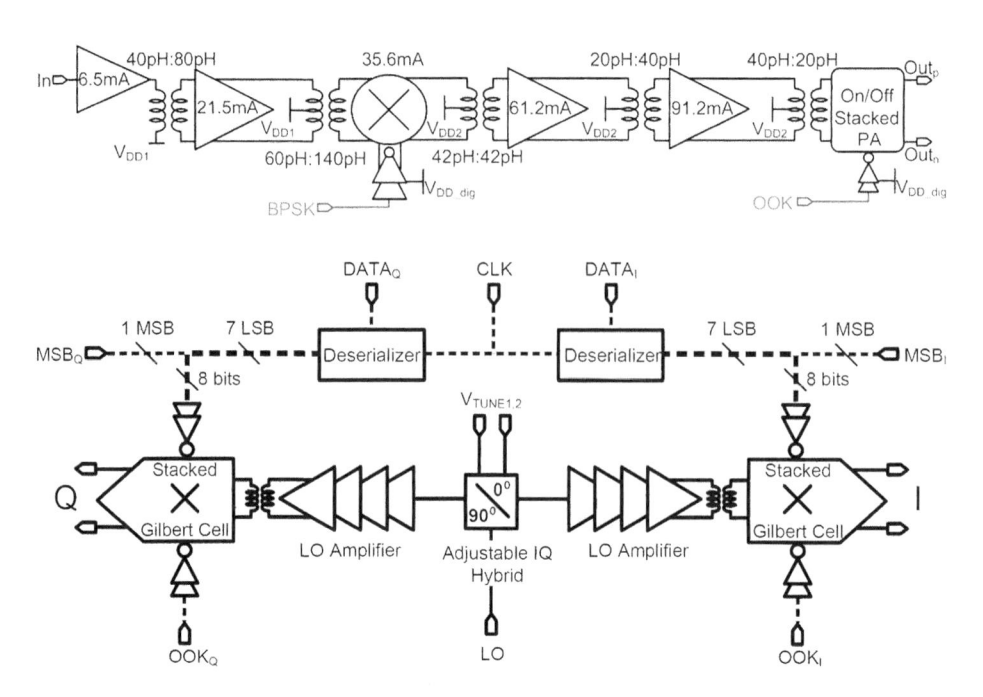

Fig. 9.39 Block diagram of tuned (top) 110-GHz 2-bit DAC lane [28], and (bottom) 90-GHz 18-bit I/Q DAC pair [18]. Both reprinted with permission from the IEEE.

been reliably stacked vertically to achieve 24 dBm of output power at 45 GHz with 24% drain efficiency [30], and 19 dBm with 11% drain efficiency at 90 GHz [18]. The 90-GHz chiplet has 8 bits of finer amplitude control in each antenna element which can be used for spectral shaping, quantization noise reduction, and transmitter array impairment correction. When the finer-resolution bits are exercised in one of the 16 antenna array elements, the transmitter efficiency degrades by less than 7% from the peak value. For example, if the peak efficiency of an antenna element is 40%, the overall transmitter array efficiency varies between 40% and 37%. Such an array has almost constant back-off efficiency, much better than any PA class.

The baseband data lanes are identical in all designs and employ inductively peaked CMOS inverters [35] capable of operation up to at least 44 Gb/s with an energy efficiency of less than 1.5 pJ/bit. The measured output eye diagram at 44 Gb/s of a baseband lane is reproduced in Fig. 9.41. The main advantage of the inductively peaked CMOS inverter baseband lane over a CML version is that its power consumption is scalable with the data rate and can be further reduced at low modulation rates by reducing the supply voltage.

The die photographs of the 45-GHz chiplet with two I/Q DAC pairs, and of the 90-GHz I/Q DAC pair are shown in Fig. 9.42 and Fig. 9.43, respectively. Figures 9.44 and 9.45 illustrate the measured output spectra of the 110-GHz DAC cell when 5-Gb/s and 44-Gb/s PRBS7 sequences, respectively, are applied to the BPSK bit. The maximum data rate is limited by the commercial PRBS generator available for measurements. At

9.4 Digital antenna modulation

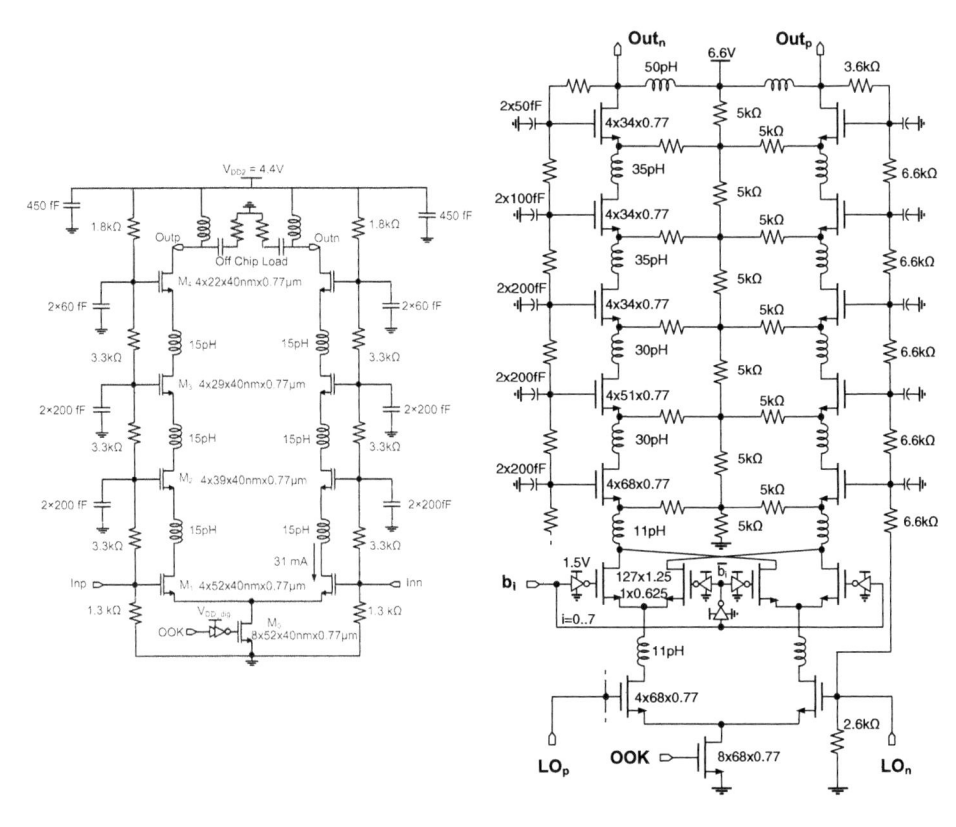

Fig. 9.40 Large-power output stage schematics: 110-GHz stacked n-MOSFET (left) [28], and 90-GHz stacked n-MOSFET Gilbert cell (right) [18]. Both reprinted with permission from the IEEE.

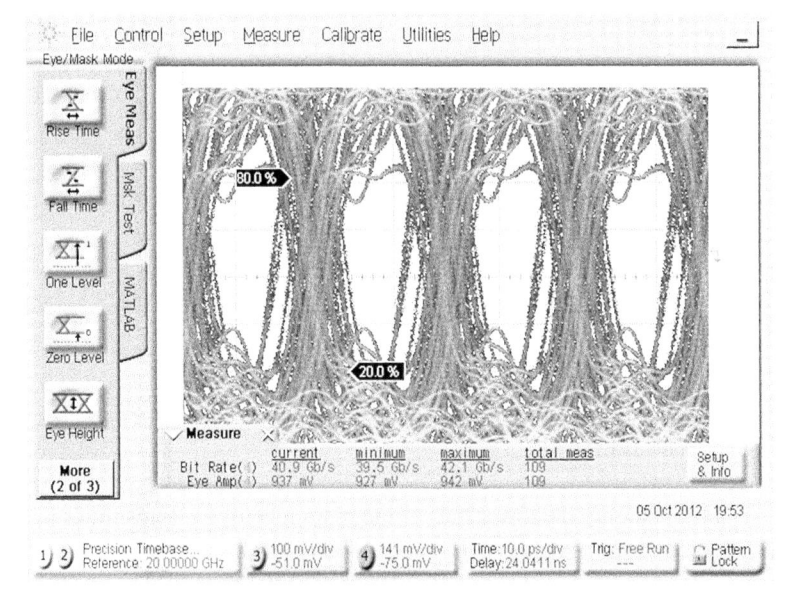

Fig. 9.41 Measured PRBS7 eye diagram of a CMOS baseband lane breakout driving 50 Ohm directly with over 1-V_{pp} swing at 40 Gb/s.

370 Digital mm-wave silicon transmitters

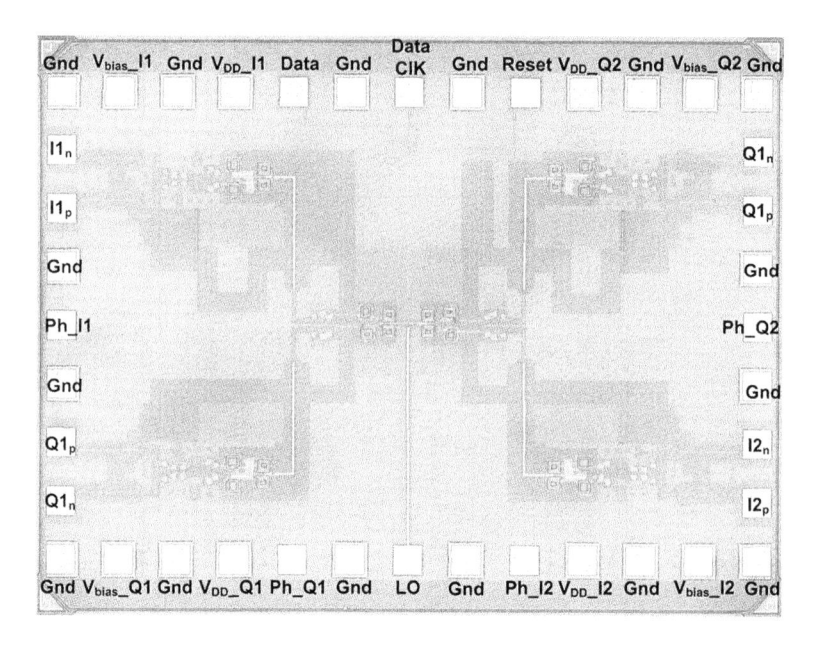

Fig. 9.42 Die photograph of the 45-GHz chiplet with two I/Q DAC pairs [30]. Reprinted with permission from the IEEE.

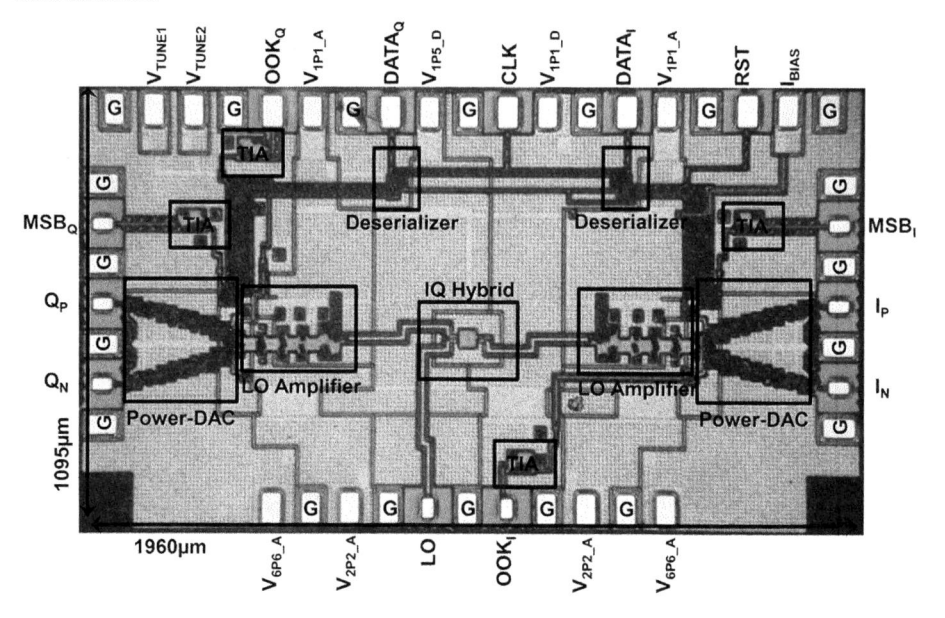

Fig. 9.43 Die photograph of the 90-GHz chiplet with one I/Q DAC pair [18]. Reprinted with permission from the IEEE.

44-Gb/s, only the lower half of the main lobe of the transmitter output spectrum can be captured in the bandwidth of the W-band VNA module.

It should be noted that, although 100-Gb/s baseband data rates can now be achieved using current-mode-logic topologies in SiGe BiCMOS or in some SOI CMOS

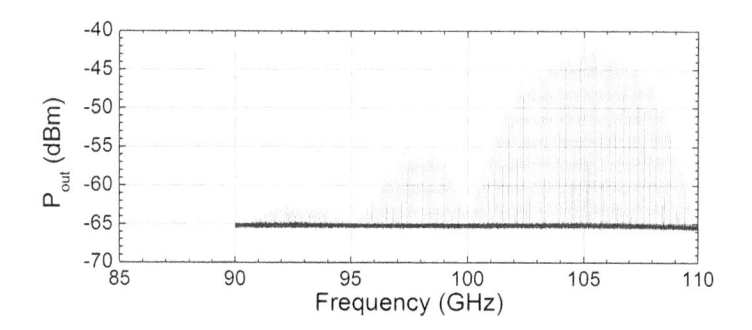

Fig. 9.44 A 5-Gb/s PRBS7 BPSK-modulated output spectrum of the 110-GHz DAC cell with a 105-GHz carrier signal [28]. Reprinted with permission from the IEEE.

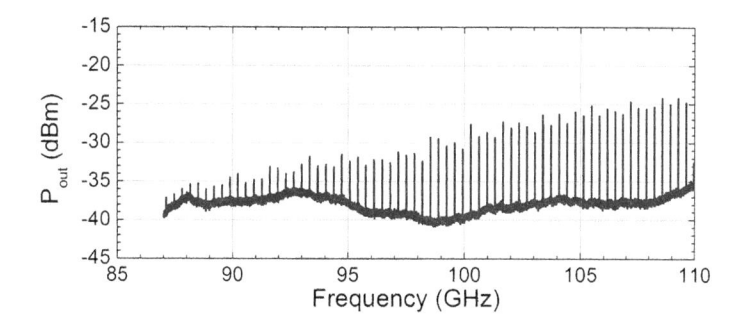

Fig. 9.45 A 44-Gb/s PRBS7 BPSK-modulated output spectrum (lower half of the main lobe only) of the 110-GHz DAC cell with a 110-GHz carrier signal [28]. Reprinted with permission from the IEEE.

technologies, the main lobe of the output spectrum of a carrier modulated at 100 Gbaud occupies 200 GHz of bandwidth (Table 9.4). Even for transmitters centered at 240 GHz, an antenna with 200-GHz bandwidth has yet to be demonstrated and no commercial instrument exists today that could display such a wide spectrum in one piece. Therefore, at least in the next few years, the symbol rate of wireless transmitters with aggregate data rates of hundreds of Gb/s will most likely be limited by the carrier frequency and by the antenna bandwidth to 50 Gbaud or less. CMOS logic will be sufficient for these systems and will allow the baseband power consumption to scale up or down with the data rate.

Finally, Fig. 9.46 shows the measured 9QAM constellation at 0.5 Gbaud (1.5 Gbs) obtained through free-space power combining from an IQ-DAC pair with on-die antennas at 100 GHz. The constellation was captured by a horn antenna 12 cm away from the die.

9.5 Conclusion

Throughout this chapter we have demonstrated digital/mixed-signal modulation and power-amplification techniques in mm-wave transmitters. In essence, the boundary between digital and analog is pushed as far forward as possible, allowing more

Table 9.4 Comparison with state-of-the-art mm-wave power-DAC/modulator results.

Ref.	Technology	Frequency (GHz)	P_{SAT} (dBm)	Peak PAE (%)	Peak data rate (Gb/s)	Number of bits
[18]	45-nm SOI CMOS	85–95	19	8.9	8×15	8 binary weighted + 1 OOK
[30]	45-nm SOI CMOS	0–50	24.3	19.6	5+2	1 BPSK + 1 OOK
[33]	65-nm CMOS	60	9.6	28.5	6	2 binary weighted + 1 BPSK
[22]	130-nm SiGe BiCMOS	70–80	9	—	18	2 BPSK (QPSK)

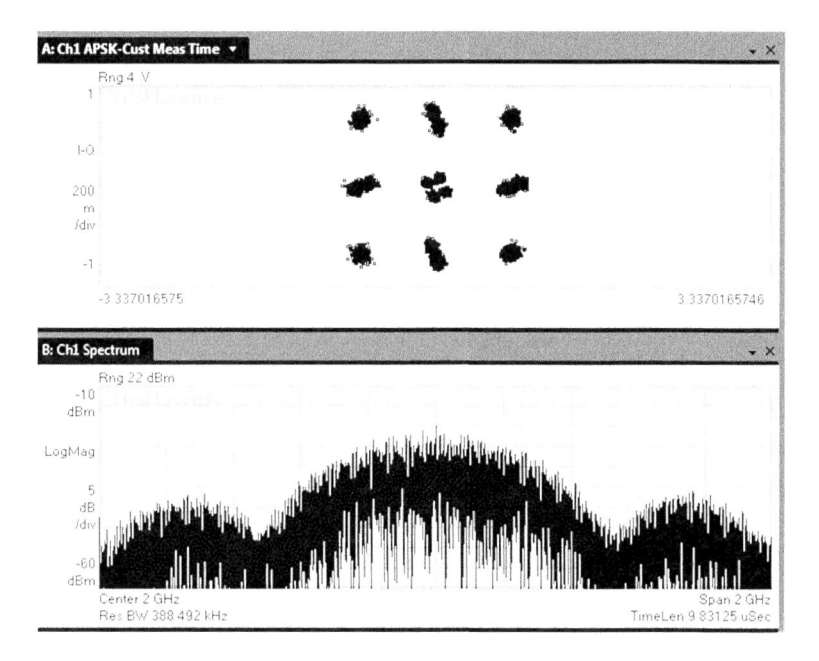

Fig. 9.46 Measured free-space generated 9QAM constellation obtained from a 100-GHz IQ-DAC pair with 16-dBm output power modulated at 0.5 Gbaud.

energy-efficient transmitter architectures. This is especially important in the mm-wave frequency bands, since power gain comes at a premium of DC power consumption. The techniques were demonstrated from relatively simple OOK and BPSK systems to complex I/Q modulators. The modulators are combined with the power amplifiers themselves in the Cartesian and polar digital architectures, allowing even more power efficiency but requiring careful management of spectral emissions at discrete clock aliases and the ever present quantization noise. These ideas are even pursued beyond the PA itself into the domain of the antenna array, where spatial combining can be done to avoid lossy on-chip power combining, and the modulation itself can be realized in space by utilizing switching amplifiers in the array transmitting a single bit of information. These systems involve the co-design of digital signal processing circuits, power-amplifier building blocks, high-speed clock and data distribution, and custom antenna array design. But the benefits to be gained far outweigh the design overhead, as the transmitter efficiencies can be boosted by an order of magnitude. This trend is likely to continue as CMOS and SiGe technology advances, since transistors become intrinsically better switches, offering lower on-resistance and lower off-capacitance. One may view this as a digitization of even the mm-wave frequency band, but in reality it's the other way around: mm-wave techniques and ideas are invading the digital world.

References

[1] D. S. McPherson, and S. Lucyszyn, "Vector modulator for W-band software radar techniques," *IEEE Trans. Microwave Theory and Techniques*, vol. **49**, no. 8, pp. 1451–1461, 2001.

[2] Y. M. Greshishchev, D. Pollex, S.-C. Wang, et al., "A 56 GS/s 6b DAC in 65 nm CMOS with 256 × 6b memory," *ISSCC Digest of Technical Papers*, pp. 194–195, Feb. 2011.

[3] M. Nagatani, H. Nosaka, and K. Sano, et al., "A 60-GS/s 6-bit DAC in 0.5-μm InP HBT technology for optical communications systems," *IEEE CSICS Digest*, Oct. 2011.

[4] S. D. Kee, I. Aoki, A. Hajimiri, and D. Rutledge, "The class-E/F family of ZVS switching amplifiers," *IEEE Transactions on Microwave Theory and Techniques*, vol. **51**, no. 6, pp. 1677–1690, 2003.

[5] H. Chireix, "High power outphasing modulation," *Proc. of the Institute of Radio Engineers*, vol. **23**, no. 11, pp. 1370–1392, 1935.

[6] A. M. Niknejad, "Integrated EER Proposal," Bell Labs, Aug. 1997.

[7] A. Kavousian et al., "A digitally modulated polar CMOS power amplifier with a 20-MHz channel bandwidth," *IEEE J. of Solid State Circuits*, vol. **43**, no. 10, pp. 2251–2258, Oct. 2008.

[8] Lu Ye, Jiashu Chen, Lingkai Kong, et al., "A digitally modulated 2.4 GHz WLAN transmitter with integrated phase path and dynamic load modulation in 65 nm CMOS," *ISSCC, Digest of Technical Papers*, pp. 330–331, Feb. 2013.

[9] W. Tai et al. "A transformer-combined 31.5 dBm outphasing power amplifier in 45 nm LP CMOS with dynamic power control for back-off power efficiency enhancement," *IEEE J. of Solid-State Circuits*, vol. **47**, no. 7, pp. 1646–1658, July 2012.

[10] C. Liang and B. Razavi, "Transmitter linearization by beam forming," *IEEE J. of Solid-State Circuits*, vol. **46**, no. 9, pp. 1956–1969, 2011.

[11] D. Zhao, S. Kulkarni, and P. Reynaert, "A 60 GHz outphasing transmitter in 40-nm CMOS," *IEEE J. of Solid-State Circuits*, vol. **47**, no. 12, pp. 3172–3183, 2012.

[12] Chao Lu, Hua Wang, C. H. Peng, et al., "A 24.7 dBm all-digital RF transmitter for multimode broadband applications in 40 nm CMOS," *ISSCC Digest of Technical Papers*, pp. 332–333, Feb. 2013.

[13] D. Chowdhury, Lu Ye, E. Alon, and A. M. Niknejad, "A 2.4 GHz mixed-signal polar power amplifier with low-power integrated filtering in 65 nm CMOS," *Custom Integrated Circuits Conference*, pp. 1–4, 2010.

[14] Jri Lee, Yenlin Huang, Yentso Chen, Hsinchia Lu, and Chiajung Chang, "A low-power fully integrated 60 GHz transceiver system with OOK modulation and on-board antenna assembly," *ISSCC Digest of Technical Papers*, pp. 316–317, Feb. 2009.

[15] N. Ono, M. Motoyoshi, K. Takano, et al., "135 GHz 98 mW 10 Gbps ASK transmitter and receiver chipset in 40 nm CMOS," *Symposium on VLSI Circuits*, pp. 50–51, June 2012.

[16] S. P. Voinigescu, A. Tomkins, M. O. Wiklund, and W. W. Walker, "Direct m-ary quadrature amplitude modulation (QAM) modulator operating in saturated power mode," United States Patent 8, 861, 627.

[17] E. Laskin, M. Khanpour, S. T. Nicolson, et al., "Nanoscale CMOS transceiver design in the 90–170 GHz range," *IEEE Trans. Microwave Theory and Techniques*, vol. **57**, pp. 3477–3490, Dec. 2009.

[18] S. Shopov, A. Balteanu, and S. P. Voinigescu, "A 19 dBm, 15 Gbaud/s, 9-bit SOI CMOS power-DAC cell for high-order QAM W-band transmitters," *IEEE ESSCIRC*, Sep., 2013.

[19] C. Lee, T. Yao, A. Mangan, et al., "SiGe BiCMOS 65-GHz BPSK transmitter and 30 to 122 GHz LC-varactor VCOs with up to 21% tuning range," *IEEE CSICS Technical Digest*, pp. 179–182, Oct. 2004.

[20] A. Tomkins, R. A. Aroca, T. Yamamoto, *et al.*, "A zero-IF 60 GHz 65 nm CMOS transceiver with direct BPSK modulation demonstrating up to 6 Gb/s data rates over a 2 m wireless link," *IEEE J. of Solid-State Circuits*, vol. **44**, no. 8, pp. 2085–2089, Aug. 2009.

[21] C. Marcu, D. Chowdhury, C. Thakkar, *et al.*, "A 90 nm CMOS low-power 60 GHz transceiver with integrated baseband circuitry," *ISSCC Digest of Technical Papers*, pp. 314–315, Feb. 2009.

[22] I. Sarkas, S. T. Nicolson, A. Tomkins, *et al.*, "An 18-Gb/s, direct QPSK modulation, SiGe BiCMOS transceiver for last mile links in the 70–80 GHz band," *IEEE J. of Solid-State Circuits*, vol. **45**, no. 10, pp. 1968–1980, Oct. 2010.

[23] E. Laskin, A. Tomkins, A. Balteanu, I. Sarkas, and S. P. Voinigescu, "A 60-GHz RF IQ DAC transceiver with on-die at-speed loopback," *IEEE RFIC Symposium, Digest of Papers*, pp. 57–60, June 2011.

[24] A. Balteanu, P. Schvan, and S. P. Voinigescu, "A 6-bit segmented RZ DAC architecture with up to 50-GHz sampling clock," *IEEE IMS*, Montreal, June 2012.

[25] T. O. Dickson, R. Beerkens, and S. P. Voinigescu, "A 2.5-V, 45-Gb/s decision circuit using SiGe BiCMOS logic," *IEEE J. of Solid-State Circuits*, vol. **40**, no. 4, pp. 994–1003, Apr. 2005.

[26] P. Chevalier, T. Lacave, E. Canderle, *et al.*, "Scaling of SiGe BiCMOS technologies for applications above 100 GHz," *IEEE CSICS Technical Digest*, Oct. 2012.

[27] I. Sarkas, A. Balteanu, E. Dacquay, A. Tomkins and S. P. Voinigescu, "A 45-nm CMOS class D mm-wave PA with > 10 Vpp differential swing," *ISSCC Technical Digest*, pp. 24–25, Feb. 2012.

[28] A. Balteanu, S. Shopov, and S. P. Voinigescu, "A 40 Gbaud/s wireless transmitter with direct amplitude and phase modulation in 45-nm SOI CMOS," *IEEE CSICS*, Oct. 2013.

[29] A. Niknejad, D. Chowdhury, J. Chen, J.-D. Park, and L. Ye, "mm-Wave quadrature spatial power combining: a proposal," Technical Report UCB/EECS-2010-110, U.C. Berkeley, http://www.eecs.berkeley.edu/Pubs/TechRpts/2010/EECS-2010-110.html.

[30] A. Balteanu, I. Sarkas, E. Dacquay, *et al.*, "A 2-bit, 24 dBm, millimeter-wave SOI CMOS power-DAC Cell for watt-level high-efficiency, fully digital *m*-ary QAM transmitters," *IEEE J. of Solid-State Circuits*, vol. **48**, no. 5, pp. 1126–1137, May 2013.

[31] M. Tabesh, J. Chen, C. Marcu, *et al.*, "A 65 nm CMOS 4-element sub-34 mW/element 60 GHz phased-array transceiver," *IEEE J. of Solid-State Circuits*, vol. **46**, pp. 3018–3032, Dec. 2011.

[32] J. Chen, *Advanced Architectures for Efficient mm-Wave CMOS Wireless Transmitters*, PhD Dissertation, U.C. Berkeley, 2013.

[33] J. Chen, L. Ye, D. Titz, *et al.*, "A digitally modulated mm-wave Cartesian beamforming transmitter with quadrature spatial combining," *ISSCC, Digest of Technical Papers*, pp. 232–233, Feb. 2013.

[34] V. Vidojkovic, G. Mangraviti, K. Khalaf, *et al.*, "A low-power 57-to-66 GHz transceiver in 40 nm LP CMOS with 17 dB EVM at 7 Gb/s," *IEEE ISSCC Technical Digest*, pp. 268–270, Feb. 2012.

[35] I. Sarkas, A. Balteanu, E. Dacquay, A. Tomkins and S.P. Voinigescu, "A 45-nm CMOS class D mm-wave PA with 10 Vpp differential swing," *IEEE ISSCC Technnical Digest*, pp. 24–25, Feb. 2012.

10 System-on-a-chip mm-wave silicon transmitters

Brian Floyd and Arun Natarajan

10.1 Introduction

Millimeter-wave (mm-wave) links feature large bandwidths which enable high-throughput, multi-gigabit-per-second (multi-Gb/s) wireless links. High-volume, low-cost applications for wireless communications require the transmitter to achieve a high integration level, avoiding both lossy off-chip interconnects at mm-wave frequencies and expensive packaging technologies. State-of-the-art CMOS [1] and SiGe BiCMOS [2–4] technologies achieve f_t and f_{max} in excess of 200 GHz, making integrated mm-wave circuits feasible (see Fig. 10.1); however, the relatively high operation frequency compared with f_{max} makes it challenging to achieve both high transmit output power and high efficiency. Earlier chapters have discussed high-efficiency power amplifiers and efficient spatial combining and modulation schemes. This chapter will discuss system-on-a-chip (SOC) approaches to achieve highly integrated mm-wave single-element and phased-array transmitters. It is important to note that mm-wave refers to frequencies from 30 GHz to 300 GHz and the feasibility of several complex transmitter architectures must be evaluated carefully in the context of operating frequency relative to the capabilities of the process technology. The broad range of frequencies also impacts integrated circuit topologies since these frequencies represent a natural yet ill-defined transition point between the use of on-chip lumped inductor/capacitor (LC) passives and on-chip distributed transmission-line (t-line)-based components.

10.2 Multi-Gb/s wireless links at mm-wave frequencies

Wireless standards for both short-range links (primarily at 60 GHz) and longer-range point-to-point links (at 40 GHz, 60 GHz, 70–86 GHz, and 94 GHz) have focused on achieving multi-Gb/s data rates. These data rates are substantially higher than those achieved in wireless links at frequencies less than 6 GHz (IEEE 802.11n, WiMax, LTE), an increase primarily achieved by enabling channel bandwidths exceeding 2 GHz (the European standard currently calls for 250-MHz channelization at 80 GHz but allows for channel bonding). The application space for mm-wave wireless links can be broadly divided into (a) short-range wireless local-area network (LAN) links and (b) point-to-point links for high-data-rate wireless backhaul, although proposals exist for the use of

10.2 Multi-Gb/s wireless links at mm-wave frequencies

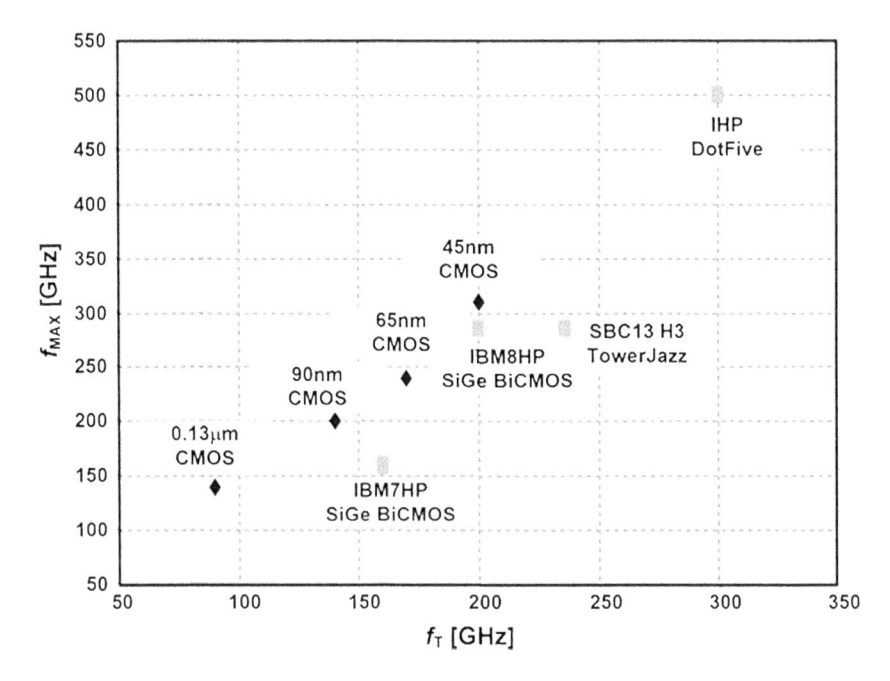

Fig. 10.1 Figures-of-merit for state-of-the-art silicon technology.

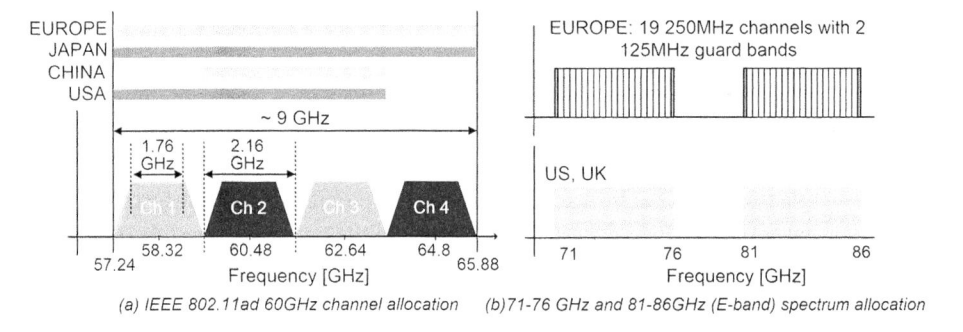

Fig. 10.2 Spectrum allocation at 60 GHz and in the E-band.

mm-wave for next-generation cellular systems [5]. Figure 10.2 summarizes standards for 60-GHz wireless links and licensed E-band (71–76 and 81–86 GHz) wireless links worldwide, demonstrating the large instantaneous bandwidth and wide frequency range that is required of mm-wave transmitter ICs for communications.

While the required transmitter output power is determined by link budgets and modulation formats specific to applications, maximum output power at the transmitter is limited by standards imposed by spectrum regulators such as the Federal Communications Commission (FCC) in the United States. The maximum power is typically defined in terms of the effective isotropic radiated power (EIRP), where the EIRP is given by

$$EIRP = P_{TX}G_{TX}. \tag{10.1}$$

Here, P_{TX} is the power supplied to an antenna and G_{TX} is the antenna gain in a particular direction relative to an isotropic antenna. In the 60-GHz license-free band the FCC allows a maximum EIRP of +43 dBm for an indoor wireless link and +82 dBm for outdoor 60-GHz links. The peak EIRP is lowered by 2 dB for every dB that the antenna gain is less than 51 dBi [6]. In the E-band and at 94 GHz, the output power is restricted to 150 mW every 100 MHz with a total maximum transmit power of +35 dBm. When the antenna gain is 50 dBi, the EIRP is +85 dBm (Table 101.113 in [7]). For a transmitter SOC on silicon, the required output power must be evaluated in the context of the maximum uncombined output power from an on-chip mm-wave PA, whose unit-cell output power depends upon technology and supply voltage, and is typically within the range of +5 to +17 dBm at saturation for CMOS and SiGe BiCMOS technologies.

While higher antenna gain can be used to achieve targeted EIRP (based on (10.1)), antenna gain is limited by applications as there is a trade-off between gain, G_{ant}, and the half-power beamwidths of the antenna in azimuth and elevation, BW_Θ and BW_Φ. These beamwidths determine the field of view of the transmitter, outside of which beam-steering would be required:

$$G_{ant}(dB) = 10\log\frac{41,253}{BW_\phi BW_\theta}. \quad (10.2)$$

For example, an antenna gain of 25 dBi corresponds to a half-power beamwidth of 11°. High antenna gain is suitable for point-to-point line-of-sight links at 60 GHz, E-band, or at 94 GHz where transmitter and receiver locations are static. Such point-to-point links operate over long distances and hence require high transmit output power, e.g. a transmitter output power of 3 W is required to achieve the maximum possible EIRP of +85 dBm at E-band assuming an antenna gain of 50 dBi (which corresponds to 0.6° half-power beamwidth). Clearly, the narrow beamwidth of high-gain antennas limits link flexibility and is therefore unsuitable for mobile applications, such as short-range links at 60 GHz where the relative positions of the transmitter and the receiver are dynamic.

10.2.1 Modulation schemes for mm-wave wireless links

Wide bandwidths enable multi-Gb/s data rates to be achieved using simpler modulation and channel access schemes. This potentially leads to lower baseband complexity than for Gb/s links operating at <10 GHz. For example, the 802.11ac wireless LAN standard calls for eight multi-input multi-output (MIMO) streams and 256 quadrature amplitude modulation (QAM) subcarriers to achieve Gb/s data rates; however, at 60 GHz, the IEEE 802.11ad achieves >2 Gb/s with a single transmitter and receiver using quaternary phase-shift keying (QPSK) modulation of a single carrier. In general, mm-wave standards attempt to span a wide range of data rates and applications and include various physical (PHY)-layer modes tailored to particular system configurations [8]. For example, a short-range (1–2 m) link capable of 1 Gb/s with simple BPSK or on–off keying (OOK) with non-coherent detection simplifies baseband and radio integration on-chip. To extend the range and data rate of a single-carrier system, QPSK and 16QAM constellations, frequency-domain equalization, and coherent detection are required.

10.2 Multi-Gb/s wireless links at mm-wave frequencies

Medium-complexity applications that would benefit from this improved single-carrier performance (e.g. video streaming between devices with fixed locations) would also require medium-access control (MAC) support for simple beamforming and more than two devices on the same network. At the top end of complexity and performance, multi-gigabit-per-second networking among multiple users with robustness to a wide range of usage environments and channel conditions may require multi-channel radios, OFDM modulation, and a MAC layer that can handle beamforming without noticeable latency. Figure 10.3 illustrates the required receiver sensitivity for different modulation formats at 60 GHz based on the IEEE 802.11ad standard.

The required transmitter output power can be found based on Friis' equation,

$$EIRP_{TX} = P_{RX,Sens} - G_{RX} + 20\log\left(\frac{4\pi d}{\lambda}\right) + LM_{dB}, \quad (10.3)$$

where $P_{RX,Sens}$ is the receiver sensitivity in dBm, which depends on receiver bandwidth and required signal-to-noise ratio, G_{RX} is the receiver antenna gain, d is the distance between transmitter and receiver, λ is the signal wavelength, and LM_{dB} is the link margin in dB, which would include reflection losses, implementation losses, polarization mismatch, etc.

As shown in Fig. 10.3 for a 2-Gb/s link with receiver sensitivity of −63 dBm and no loss due to any obstructions, transmitter EIRP of 22 dBm and 42 dBm are required for a 10-m and 100-m link, respectively (assuming NF of 8 dB, 0-dB link margin, and G_{RX} is 3 dB). Note the path loss is 88 dB and 108 dB for 10-m and 100-m links, respectively. The required $EIRP_{TX}$ influences transmitter architecture and technology – single silicon PA output powers are limited to ∼17 dBm, which can be adequate for short-range links, but longer links require high-gain antennas, sacrificing link flexibility, or multi-element arrays. From an SOC perspective, modulation formats such as OOK, BPSK, and minimum shift keying (MSK) are preferred as the lower modem complexity and allow the integration of some of the modulation and demodulation functions within the radio. SC modulation is also preferred for reduced power consumption due to the lower

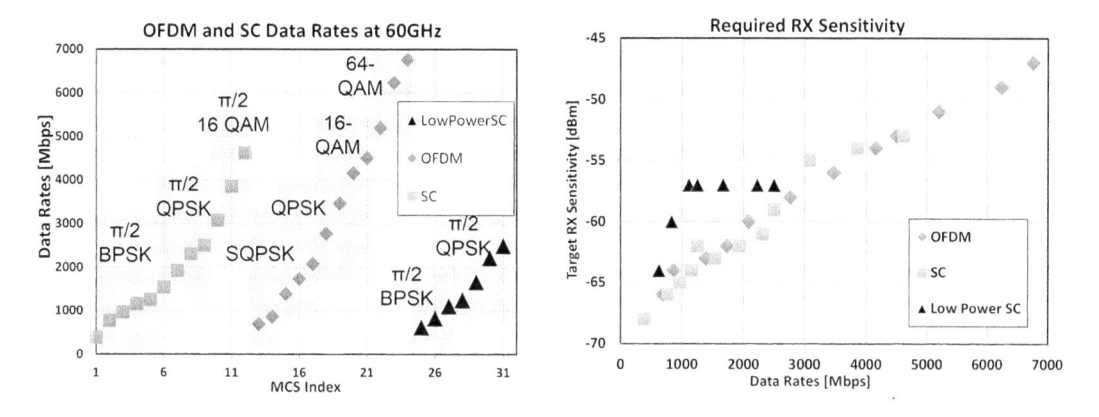

Fig. 10.3 Data rates and required sensitivity for 60-GHz wireless links based on the IEEE 802.11ad standard.

peak-to-average-power ratio (PAPR) than for OFDM modulation (∼3 dB for SC-QPSK as compared with ∼10 dB for OFDM-QPSK [9]). However, the SC modulation is sensitive to in-band amplitude variations, mainly made worse by the gain variations of analog circuits and multi-path delay spread (rms delay spread in indoor environments varies between 10 ns and 40 ns according to measurements in [10]). While the OFDM modulation scheme simplifies equalization in the presence of in-band amplitude variations and multi-path, it imposes higher transmitter linearity constraints leading to poorer transmitter efficiency. However, baseband system-level studies estimate comparable transmitter linearity requirements for SC and OFDM modulations in the presence of multi-path [11] and OFDM is required for achieving >6 Gb/s data rates in the 802.11ad standard (Fig. 10.2). In the following sections, architectural choices and system-level integration issues for single-element and multi-element transmitters are discussed in the context of applications and modulation formats.

10.3 On-chip mm-wave transmitter architectures

10.3.1 Direct up-conversion vs superheterodyne architectures

Figure 10.4a shows a generic direct up-onversion architecture in the left-hand diagram, comparing it with the super-heterodyne transmit architecture in Fig. 10.4b. Direct up-conversion potentially consumes less power, is more amenable to integration than super-heterodyne architectures, and has been extensively used at radio frequencies, particularly with increased digital calibration afforded by faster CMOS technologies. Integrated mm-wave transmitters at 60 GHz and E-band targeting high-data-rate wireless links adopt direct up-conversion to minimize silicon area and power consumption [12–15]. However, quadrature modulation with direct up-conversion requires in-phase (I) and quadrature-phase (Q) local-oscillator (LO) signals at mm-wave as well as a large tuning range (∼15%) in the frequency synthesizer to accommodate process, voltage, and temperature variations. Direct-conversion transmitters can also be impaired by LO pulling since the VCO oscillation frequency is the same as the PA output frequency [16]. Typically, integrated direct-conversion architectures at frequencies <6 GHz avoid LO pulling by using a VCO that operates at $2\omega_1$ followed by a frequency divider that generates quadrature outputs at ω_1. However, this option is less attractive for mm-wave transmitters since low-noise, wide-tuning-range, and low-power oscillators at $2\omega_1$ become challenging for $\omega_1 > 35$ GHz. Therefore, VCOs have to operate at

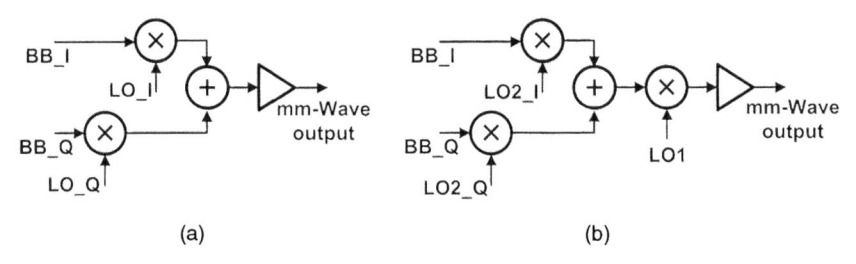

Fig. 10.4 On-chip direct and super-heterodyne up-conversion architectures.

$< \frac{\omega_1}{M}$ followed by frequency multiplication to generate the desired LO at ω_1, which complicates quadrature signal generation. Furthermore, direct conversion architectures require calibration to reduce IQ mismatch and carrier leakage (which can corrupt the desired mm-wave signal). The case study on a direct conversion architecture in Section 10.6.2 demonstrates calibration architectures to achieve LO I and Q phase and amplitude matching at mm-wave.

LO-pulling challenges are greatly reduced in super-heterodyne architectures where $\omega_0 \neq \omega_1$. Additionally, ω_1 is reduced in super-heterodyne architectures, shown on the right in Fig. 10.4. The image appears at the output at $\omega_1 - \omega_2$ (assuming $\omega_1 < \omega_0$) therefore, increasing ω_2 ensures that on-chip resonators in the transmitter mm-wave amplification chain attenuate the image sufficiently. This allows double sideband up-conversion in the second mixer that translates ω_2 to ω_0, avoiding quadrature LO signals at ω_1. Concerns related to LO pulling, challenges with building low-loss, accurate transmitter sensors for calibration at mm-wave, and the relatively low maturity of silicon design in mm-wave led to multiple early transmitters adopting the robust super-heterodyne architecture at 60 GHz ($\omega_1 = 53$ GHz, $\omega_2 = 8.83$ GHz) [17], 77 GHz ($\omega_1 = 51$ GHz, $\omega_2 = 25.5$ GHz) [18], and 94 GHz ($\omega_1 = 80.6$ GHz, $\omega_2 = 13.4$ GHz) [19].

10.3.2 Frequency synthesis at mm-wave

LO frequency generation plays a critical role in integrated mm-wave transmitter architectures. Wide LO tuning ranges are often required at mm-wave frequencies to take advantage of the large avavailable bandwidths. Low phase noise is desirable to reduce TX EVM and direct up-conversion architectures necessitate quadrature LO outputs. The phase noise of on-chip VCOs is constrained by the quality factor of the inductors and capacitors that form the VCO resonator. Suitable inductor design leads to inductor quality factors improving with frequency (~20 at 60 GHz); however, MOS varactor quality factors are significantly degraded at mm-wave with varactor Q of ~5–7 at 60 GHz for varactor channel lengths of ~130 nm. The effective quality factor of the resonator, Q_{RES}, with inductor quality factor, Q_L, and varactor quality factor, Q_C, can be expressed as [20]

$$\frac{1}{Q_{RES}} = \frac{1}{Q_L} + \frac{1}{Q_C}, \tag{10.4}$$

Wide-tuning-range requirements lead to smaller inductors and larger varactors, leading to higher power consumption and increased phase noise in the VCO. Techniques for LO signal generation at mm-wave for heterodyne and direct up-conversion transmitters are summarized in Fig. 10.5. Achieving fundamental oscillators with wide tuning range and quadrature outputs at mm-wave is challenging (Fig. 10.5a). The oscillator tuning-range requirements can be relaxed by dividing the required frequency range among multiple oscillators (two in [21]) and selecting the appropriate VCO for the required LO frequency. However, this leads to increased area and power consumption. A novel quadrature coupling technique is introduced in [22] that enables in-phase coupling (Fig. 10.6) between quadrature oscillators, leading to improved phase noise while achieving a wide tuning range.

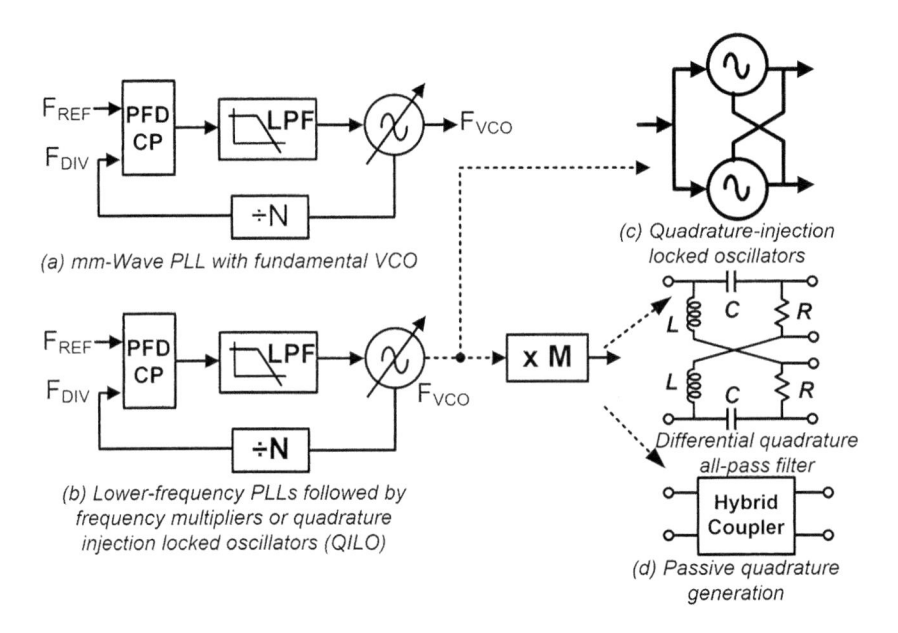

Fig. 10.5 On-chip LO generation for heterodyne and direct-conversion transmitters.

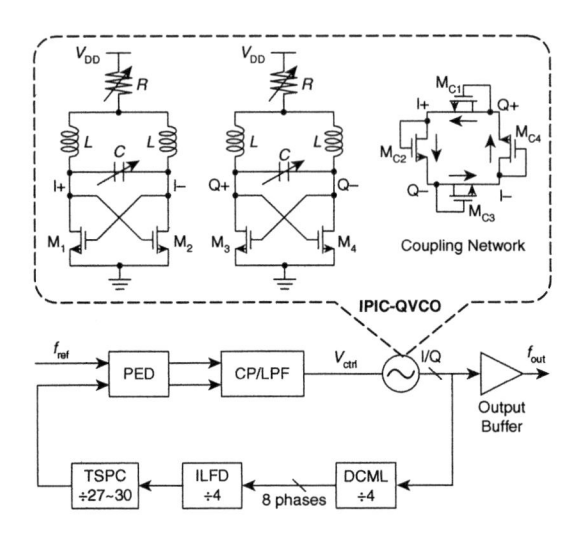

Fig. 10.6 In-phase injection-coupled 60-GHz QVCO for low-phase-noise quadrature signal generation (from [22]).

Push–push oscillators [13], frequency multiplication using active multipliers (×2 or ×3) [15], or quadrature injection-locked oscillators (QILO) [12] (Figs. 10.5b,c,d) enable robust VCO design with adequate tuning range at lower frequencies. In addition, such approaches enable one to use lower-frequency synthesizers, eliminating mm-wave frequency dividers that can be power hungry in current CMOS and BiCMOS technologies ($f_t < 300$ GHz and $f_{max} < 300$ GHz).

10.3 On-chip mm-wave transmitter architectures

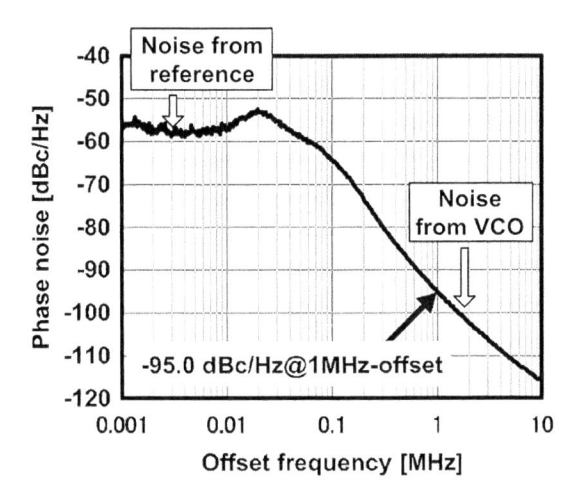

Fig. 10.7 Phase noise of state-of-the-art mm-wave synthesizer with $f_c = 60$ GHz (from [12]).

It must be noted that even a ×2 multiplier approach implies that PLLs operate at tens of GHz and leads to high division ratios (>1000) when operating from a low-frequency reference. Phase noise from the reference is multiplied by the division ratio, leading to high phase noise at low-offset frequencies within the PLL closed-loop bandwidth. Therefore, phase-error cancellation techniques are required for high-data-rate OFDM links. Given an OFDM modulated signal with OFDM symbol length of T_S, the phase noise at offset frequencies below $1/T_S$ can be corrected using OFDM pilot tones [23]. For example, the 60-GHz 802.11ad OFDM standard uses an FFT length of ~194 ns, and hence phase errors at offset frequencies below 5.15 MHz can be cancelled. As a result, synthesizer phase noise optimization for mm-wave transmitters must be focused at higher offset frequencies where the noise is dominated by the LC VCO.

Figure 10.7 shows the phase noise of a typical mm-wave LO in a transmitter [12]. The phase noise of the LO signal translates to distortion in the transmitter constellation. The error vector magnitude (EVM) is used to characterize this distortion and is defined as the ratio of the power of the error vector and the average power of the constellation (Fig. 10.8a) [24],

$$EVM_{dB} = 10 \log_{10} \left(\frac{P_{error}}{P_{avg,constellation}} \right), \quad (10.5)$$

and the EVM due to LO phase noise alone is given by [24]

$$EVM_{rms} = \sqrt{\frac{1}{SNR} + 2 - 2\exp\left(-\frac{\sigma^2}{2}\right)}, \quad (10.6)$$

where σ is the rms LO phase error. For example, achieving data rates > 3 Gb/s with the 802.11ad standard using SC and OFDM requires an EVM better than −20 dB (Fig. 10.8b). Assuming very high SNR, the rms phase error should be <5.5°, whereas 25 dB SNR requires rms phase error <3°.

(a) Error Vector Magnitude for Transmitter

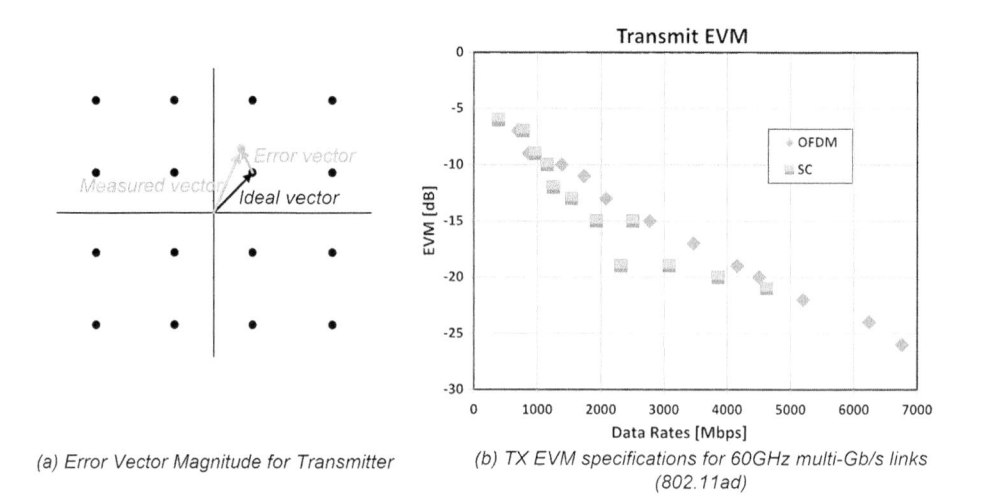

(b) TX EVM specifications for 60GHz multi-Gb/s links (802.11ad)

Fig. 10.8 Transmitter error vector magnitude (EVM) specifications for 802.11ad links at 60 GHz.

10.3.3 Quadrature generation

Quadrature LO signals can be generated from synthesized LO signals using passive couplers [14] or QILO [12]. High interconnect parasitics make it challenging to achieve wide tuning range in a QILO at nominal supply voltages. On the other hand, wideband quadrature signals can be achieved using passive techniques such as on-chip t-line-based quadrature couplers (hybrid coupler or the Lange coupler, if wider bandwidth is desired [25]). However, the dimension of t-line-based couplers is $\sim \lambda/4$. Lumped LC coupler implementations generally provide lower bandwidth than t-line based implementations but achieve smaller silicon area, particularly at lower mm-wave frequencies. The output signals for the differential all-pass filter shown in Figs. 10.5c,d are given by [26]

$$\begin{bmatrix} V_{OI,\pm} \\ V_{OQ,\pm} \end{bmatrix} = V_{in} \begin{bmatrix} \pm \frac{s^2 + \frac{2\omega_0}{Q}s - \omega_0^2}{s^2 + \frac{2\omega_0}{Q}s + \omega_0^2} \\ \mp \frac{s^2 - \frac{2\omega_0}{Q}s - \omega_0^2}{s^2 + \frac{2\omega_0}{Q}s + \omega_0^2} \end{bmatrix}, \qquad (10.7)$$

where $\omega_0 = \frac{1}{\sqrt{LC}}$ and $Q = \frac{1}{R}\sqrt{\frac{L}{C}}$. While the all-pass filter results in higher loss and smaller bandwidths, it occupies a much smaller area, which is critical for multiple-element RX arrays [26]. The transmitter EVM in the presence of quadrature phase error and amplitude mismatch is given by [24],

$$EVM_{rms} = \sqrt{\frac{1}{SNR} + 2 - \sqrt{1 + \frac{2g_t}{1+g_t^2} + \cos\phi_t + \frac{2g_t}{1+g_t^2}\cos\phi_t}}, \qquad (10.8)$$

where $g(t)$ and $\phi(t)$ are the transmitter gain and phase imbalance. Figure 10.9 plots the EVM across amplitude and phase mismatch, demonstrating that amplitude matching

10.3 On-chip mm-wave transmitter architectures

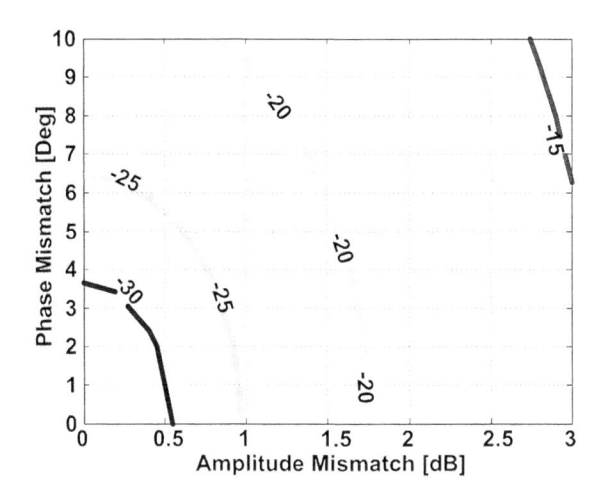

Fig. 10.9 Impact of quadrature error on transmitter EVM.

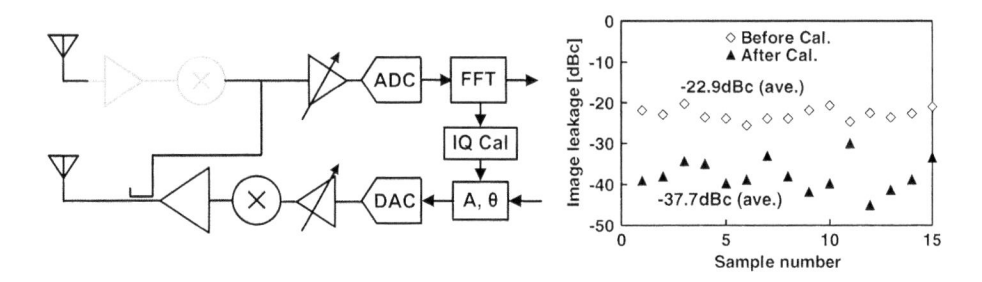

Fig. 10.10 IQ calibration to reduce IQ offset.

of ±0.7 dB is required in the presence of ±4° phase to achieve better than −25 dB EVM. Narrowband techniques to generate quadrature can lead to higher mismatches at the edge of the signal band. For example, the phase error in the all-pass filter is given by [26]

$$\Theta_{error}|_{\omega=\omega_0+\Delta\omega} = 90° - 2\tan^{-1}\left(\frac{\frac{1}{Q}\left(1+\frac{\Delta\omega}{\omega_0}\right)}{1+\frac{\Delta\omega}{\omega}+\frac{1}{2}\left(\frac{\Delta\omega}{\omega_0}\right)^2}\right). \tag{10.9}$$

The quadrature error can be reduced significantly by using calibration to eliminate static and slow-varying offsets. Fully integrated transceivers that include mm-wave receivers along with the transmitters can implement loop-back paths that sense the transmitter output signal and apply appropriate phase and amplitude compensation to the baseband I and Q signals (Fig. 10.10). Such IQ calibration techniques can improve image suppression by ∼15 dB, leading to the image-rejection of ∼ −38 dB, which translates to ∼ −28 dB EVM.

10.4 Single-element transmitters

Millimeter-wave applications that have a low-to-medium degree of complexity, such as short-distance (<3 m) "point-and-shoot" data transfer, require a radio with a small form factor, low power consumption, and low cost. It is important to note that on-chip PA output powers are ~17 dBm and have peak PAE of ~25%, with additional improvements anticipated. Therefore, low-power transmitters must rely on modulation schemes that support nonlinear PA operation to ensure high PA efficiency and low system power consumption.

Non-coherent transceiver architectures that rely upon energy detection sacrifice spectral efficiency, link robustness, and link range for circuit complexity and power consumption. Typically, such transmitters use the entire spectrum without channelization and hence require a single LO, relaxing tuning range requirements. A simple low-power OOK transmitter architecture that serves as an example of the high data rates that can be achieved by leveraging large available bandwidths is shown in Fig. 10.11 [27]. When combined with a non-coherent receiver, this architecture leads to simplified baseband processing enabling full system integration on a single IC. The modulator consists of a switch incorporated into the PA that modulates the doubler output. Such architectures achieve sub-50 mW transmitter power consumption with energy efficiencies of the order of 10–50 pJ/bit for 10 cm links.

The robustness of the link to multi-path and interference can be improved and higher SNR can be achieved by adopting coherent links. Constant envelope modulation schemes such as FSK or PSK ensure that nonlinear PAs with relatively higher efficiency can be used in the transmitter.

Heterodyne architectures adopted in silicon RF transmitters and III–V-based mm-wave transmitters can be adapted to achieve integrated CMOS and SiGe transmitters. However, direct-conversion architectures are advantageous for silicon implementations and are discussed in the context of a case study in Section 10.6.2.

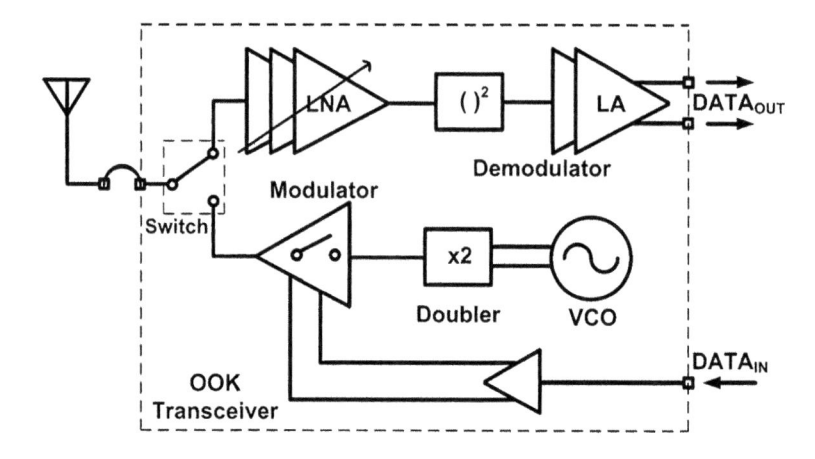

Fig. 10.11 Low-power OOK-based 60-GHz transceiver (from [27]).

10.5 Phased-array transmitters

Limitations on single PA output power in integrated mm-wave transmitters can be overcome by harnessing the output power from multiple elements (see Fig. 10.12). Multiple-element transmitters that focus transmitted energy in desired directions are essential to close the link budget for mm-wave mobile wireless links and for achieving high-resolution and desired SNR in mm-wave radar and active imaging. The short wavelengths at mm-wave make it feasible to achieve multiple antenna systems with small physical size – for example, a 4 × 4 antenna array at 60 GHz (with λ/2 spacing) occupies only 1 cm × 1 cm and a packaged 28 mm × 28 mm 16-element 60-GHz array and a 16 mm × 16 mm 64-element 94-GHz array have been demonstrated (Fig. 10.13) [19, 28]. Therefore, multiple-element arrays can be accommodated on handheld communication devices to provide increased range and data rates. Similarly,

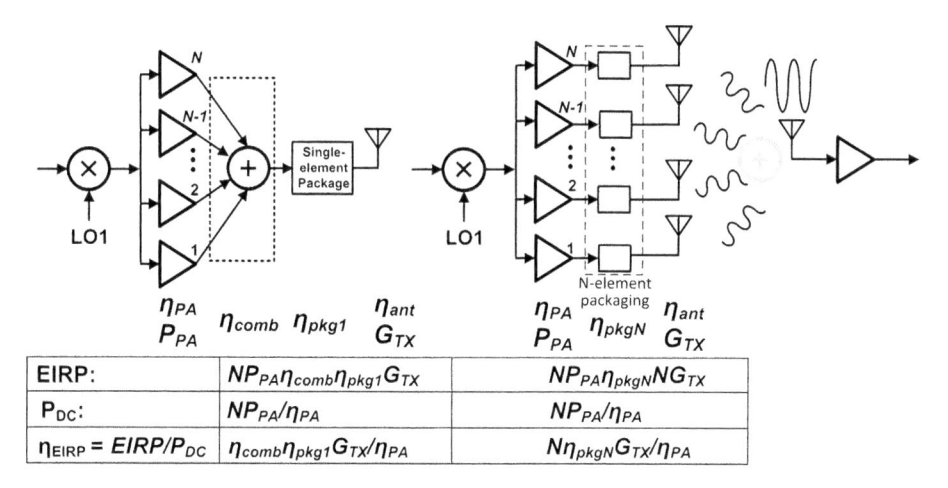

EIRP:	$NP_{PA}\eta_{comb}\eta_{pkg1}G_{TX}$	$NP_{PA}\eta_{pkgN}NG_{TX}$
P_{DC}:	NP_{PA}/η_{PA}	NP_{PA}/η_{PA}
$\eta_{EIRP} = EIRP/P_{DC}$	$\eta_{comb}\eta_{pkg1}G_{TX}/\eta_{PA}$	$N\eta_{pkgN}G_{TX}/\eta_{PA}$

Fig. 10.12 On-chip vs spatial power combining at mm-wave.

Fig. 10.13 Physically small, highly integrated mm-wave phased arrays at 60 GHz and 94 GHz with λ/2 antenna spacing.

large-aperture imaging arrays can be realized with small physical size, providing improved image resolution in azimuth and elevation.

The highest reported efficiency of on-chip power combining networks is ~80% [29, 30]. As shown in Fig. 10.12, combining power from on-chip amplifier cells using lossy power-combining networks degrades transmitter efficiency as well as maximum achievable EIRP. Spatial power combining promises higher efficiency at the cost of increased signal processing and packaging complexity in the transmitter. Note that spatial power combining results in a quadratic increase in EIRP for fixed unit PA output power as shown in Fig. 10.12 rather than a linear increase with on-chip power combining, since

$$EIRP = (N \cdot G_{ant}) \cdot (N \cdot P_{PA} \cdot \eta_{pkg,N}). \qquad (10.10)$$

Thus, spatial power combining translates to increased transmitter efficiency for constant EIRP. While CMOS and BiCMOS silicon technology have lossy passives, they are inherently well-suited for complex architectures that rely on a large number of transistors, making spatial power combining preferable. It must be noted that spatial combining can cause restrictions on bandwidth and antenna field-of-view in the transmitter that must be considered for specific applications. Such restrictions will be briefly mentioned in the following sections.

The multiple-element transmitter can be architected as a MIMO system in association with a multiple-element receiver. There is increasing interest in such MIMO systems at mm-wave for indoor and outdoor applications. However, high power consumption in the baseband and analog/RF ICs makes it challenging for such ICs to be on mobile devices that are power-constrained. Furthermore, the high losses in reflections at mm-wave make the line-of-sight (LoS) path dominant [10, 31]. This lowers the benefits of spatial diversity using multi-path reflections. However, MIMO benefits can be obtained by considering multiple LoS paths when TX–RX antenna pairs are spaced a few wavelengths apart [32]. Such MIMO systems may be attractive for future applications in advanced CMOS technologies.

Phased arrays are a special class of multiple-element transmitters with beam-forming and electronic beam-steering capabilities. The following section describes approaches for integrating mm-wave phased-array transmitters in silicon technologies.

10.5.1 Phased-array transmitter: principle of operation

A phased-array transmitter enables electronic beamforming and beam-steering and hence focuses the radiated energy in desired directions. As shown in Fig. 10.14, a beam is formed in the targeted direction by varying the relative delay in each element of a multi-element transmitter, compensating for the difference in propagation delays for signals from different elements. Electronic variation of the delay enables beam-steering without actual mechanical reorientation of the antennas. In the simplified n-element one-dimensional (linear) phased-array transmitter in Fig. 10.14, the input signal, $s(t)$, is distributed to elements that delay the signal by multiples of τ, the combined signal in a direction θ is given by

10.5 Phased-array transmitters

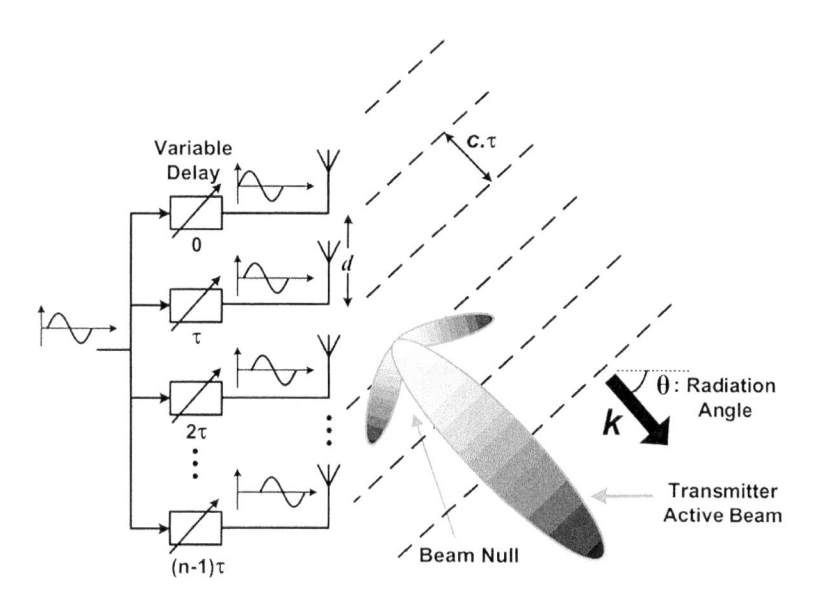

Fig. 10.14 Phased-array principle of operation.

$$S(t,\theta) = \sum_{k=0}^{n-1} s\left(t - k\tau - (n-1-k)\frac{d\sin\theta}{c}\right). \quad (10.11)$$

Therefore, the signals from all elements add up coherently in the direction θ_0

$$\theta_0 = \sin^{-1}\left(\frac{c\tau}{d}\right), \quad (10.12)$$

where d is the spacing between antennas and c is the velocity of light. This coherent addition increases the power radiated in the desired direction while incoherent addition of the signal in other directions ensures lower interference power at receivers that are not targeted. Varying the relative delay between adjacent elements, τ, enables beam-steering across different directions. For example, steering the beam to $\theta_0 = 30°$ requires $\tau = \frac{0.866d}{c} = \frac{0.433}{f}$ for $d = \lambda/2$.

Broadband phased-array operation in the transmitter requires a true-time delay in each transmitter element. In the architecture in Fig. 10.14, the required time delay is implemented in the RF path in each element. However, implementing a broadband low-loss true-time delay element at RF, which is capable of large variation, occupies a small area, and scales well with an increase in number of elements, is not feasible with current technology. The true-time delay can be implemented in the IF path or in the baseband/digital domain. An analog delay element at IF has the same problems as a delay at RF, and practical considerations of mixed-signal circuits and digital signal processors (DSPs) with multi-Gb/s performance limit digital array architectures.

If the bandwidth of interest is sufficiently narrow, the time delay (a linear phase shift in the frequency domain) can be approximated by a constant phase shift at the center frequency (see the left-hand diagram in Fig. 10.15), and can be implemented as an RF-path phase shift in each element (the right-hand diagram in Fig. 10.15), leading to a

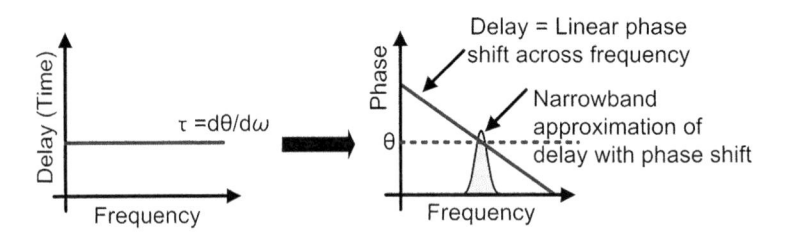

Fig. 10.15 Approximation of delay with phase-shift for narrow-band signals.

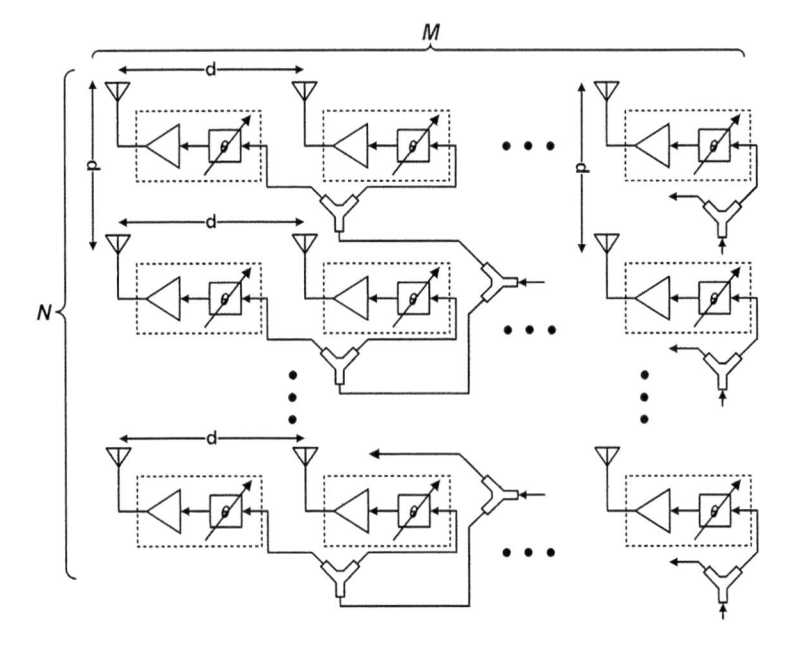

Fig. 10.16 Planar phased array with $N \times M$ elements and utilizing RF-path phase shifting architecture.

phased array as shown in Fig. 10.16. Given an $N \times M$-element two-dimensional array with input signal $s(t) = a(t)\cos(\omega_0 t)$, the total signal in a given direction (based on polar coordinates), (θ, ϕ), is

$$S(t) = AF(\theta, \phi) = \sum_{m=0}^{M-1}\sum_{n=0}^{N-1} a_{n,m} e^{j(k\sin\theta(nd_x\cos\phi + md_y\sin\phi) - \alpha_{n,m})}. \quad (10.13)$$

Figure 10.17 plots the array factor for a two-dimensional planar array where $AF(\theta, \phi)$, represents the normalized array gain in direction (θ, ϕ). This plot shows the array factor in dB scale for a 6×10 planar array plotted as a function of the direction cosines, u_x, u_y, where $u_x = \cos(\theta)\cos(\phi)$ and $u_y = \cos(\theta)\sin(\phi)$. Assuming that each element is connected to identical antennas with gain $G_{ant}(\theta, \phi)$,

$$EIRP(\theta, \phi) = AF(\theta, \phi)G(\theta, \phi)P_{TX} \quad (10.14)$$

$$EIRP_{max} = NM \cdot G(\theta, \phi)P_{TX}. \quad (10.15)$$

10.5 Phased-array transmitters

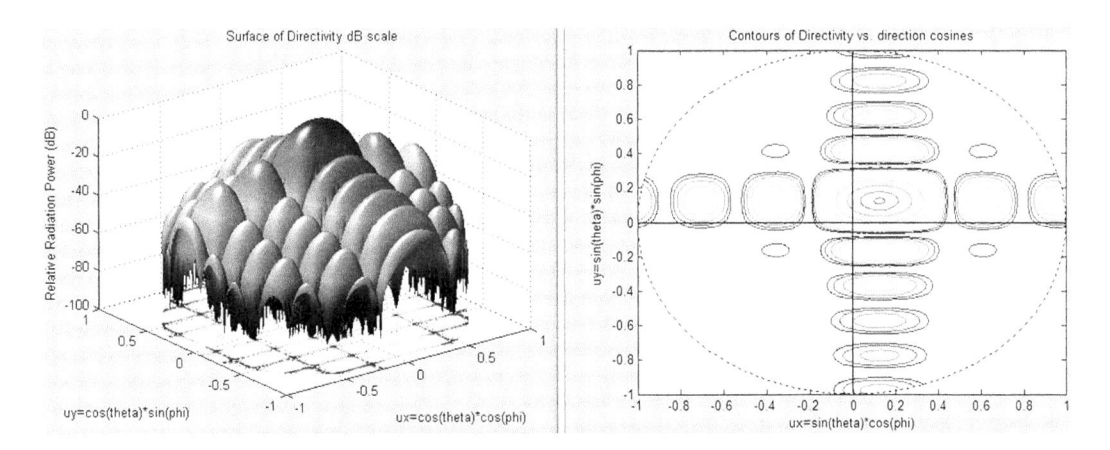

Fig. 10.17 (Left) Directivity of antenna array composed of 6 × 10 square lattice, steered to $\theta = 10°$, $\phi = 45°$, and (right) directivity contours for same situation.

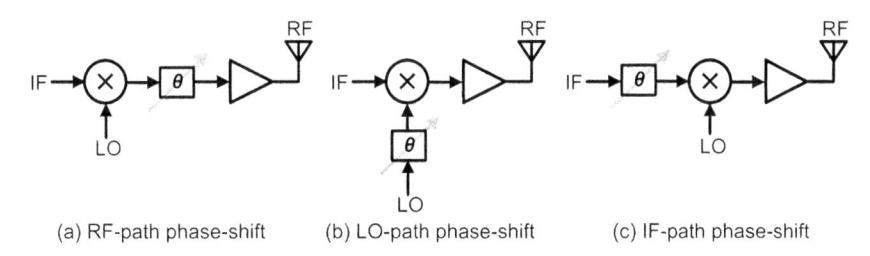

Fig. 10.18 Narrowband phased-array transmitter architectures.

Thus phased arrays focus transmit energy in directions determined by the relative phase and amplitude weights in each element, leading to a quadratic increase in EIRP with number of elements. Note that the narrowband approximation also translates to beam-pointing error for large-size arrays, which must be compensated for large-scale arrays [33].

10.5.2 Integrated narrowband phase-shift architectures

The required phase shift can be implemented in the RF-path, LO-path or in the IF-path (Figs. 10.18a,b,c). Implementing the phase shift in the RF-path simplifies the transmitter architecture with only one signal up-conversion chain from IF to RF. Hence, RF-path phase shifting promises smallest silicon IC area and lowest power dissipation, when compared with LO-path or IF-path phase-shifting architectures that require multiple transmitter up-conversion chains with multiple mixers and LO distribution. However, RF-path phase shifters operate across the entire signal bandwidth whereas the LO-path phase shifters operate on a single tone, and hence it is easier to implement the phase shift on an CMOS/BiCMOS IC in the LO-path than in the RF-path. In the following sections, we discuss the design of integrated phased arrays with RF-path and LO-path phase shifting, including circuits for phase shift and signal/LO distribution.

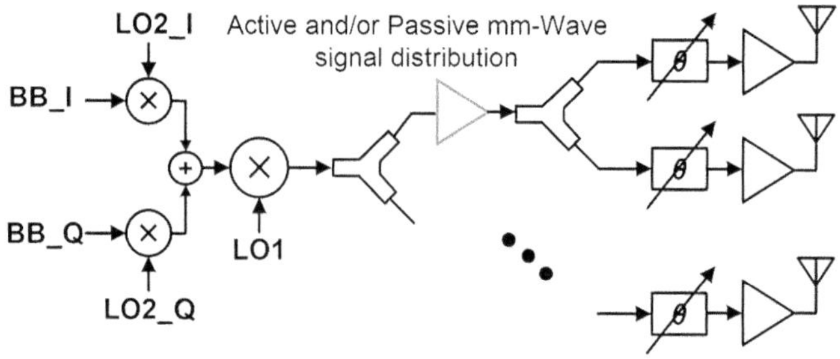

Fig. 10.19 Narrowband phased-array transmitter architectures.

10.5.3 Integrated phased-array transmitters with RF-path phase shifting

Figure 10.19 shows a RF-path phase-shifting architecture with an RF signal distributed to multiple front-ends that include RF phase shifters. The critical blocks that are added, when compared with single-element architectures, are the signal distribution circuits and the phase shifters.

On-chip mm-wave signal routing and power division

Single-element and phased-array architectures require extensive mm-wave signal distribution across an IC. Notably, dimensions of the IC are comparable to wavelength. Therefore, interstage and intrastage signal routing must be carefully modeled to ensure desired performance. Signal routing and impedance matching at mm-wave can be accomplished using on-chip transmission lines (t-lines) that provide lower loss and improved isolation.

On-chip t-lines can be implemented using microstrip structures with the top thick metal layer serving as the signal layer. However, better isolation between signal lines can be achieved by ground-shields that create a ground-shielded coplanar waveguide (GSCPW) t-line structure as shown in Fig. 10.20a. The presence of well-defined return currents also leads to good correlation between EM simulation and measurement. For typical metal stacks in silicon technologies, the highest achievable impedances using such t-line structures are $\sim$70 Ω. Differential GSCPW t-lines can also be implemented using similar structures (Fig. 10.20b). Figure 10.20c shows measured and EM simulations of losses of on-chip transmission lines. The ground layers isolate the signal from the lossy substrate, leading to t-line loss of $\sim$1–1.5 dB/mm. Figure 10.21 demonstrates the improved isolation using the GSCPW structure – simulations comparing isolation between adjacent 50 Ω ($W_s = 5\,\mu$m, $S = 7.5\,\mu$m, $W_g = 8\,\mu$m), 400-μm-long GSCPW lines with adjacent 400-μm, 50-Ω microstrip lines show more than 20-dB increase in isolation for spacing of 50 μm [35]. The increased isolation allows for aggressive meandering of electrically long t-lines required in broadband couplers or for power matching in amplifiers.

RF signal division can be accomplished using power dividers consisting of passive couplers, active amplifiers, or a combination of the two. Most integrated arrays are expected to operate under different transmitter output power requirements and therefore

10.5 Phased-array transmitters

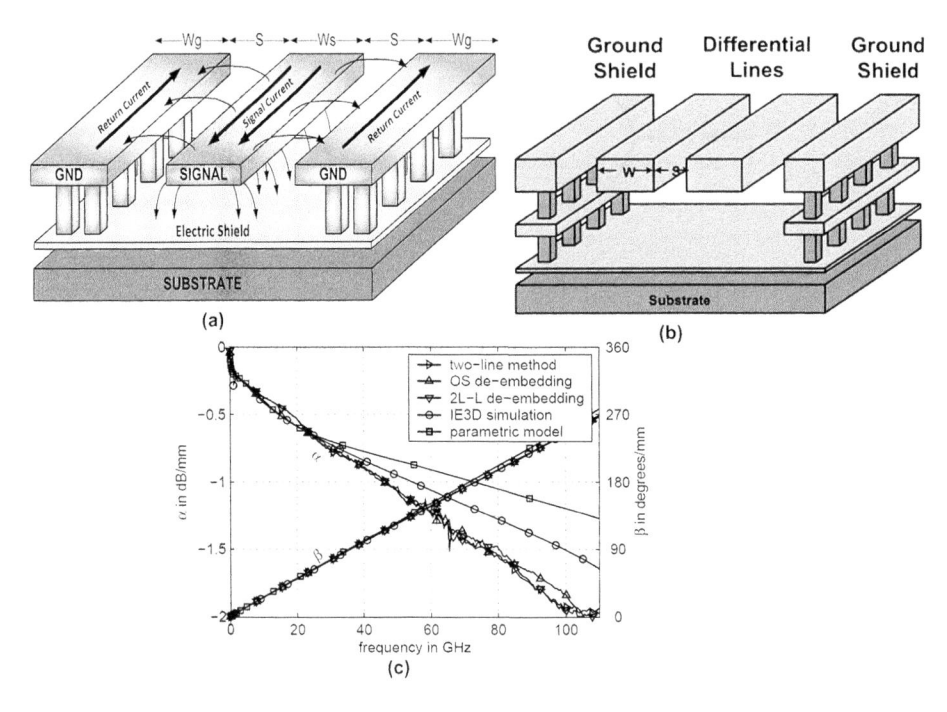

Fig. 10.20 On-chip transmission-lines for mm-wave signal distribution (from [34]). (a) Single-ended ground-shielded CPW (GSCPW), (b) differential GSCPW, and (c) comparison of measured and simulated α and β for on-chip microstrip lines.

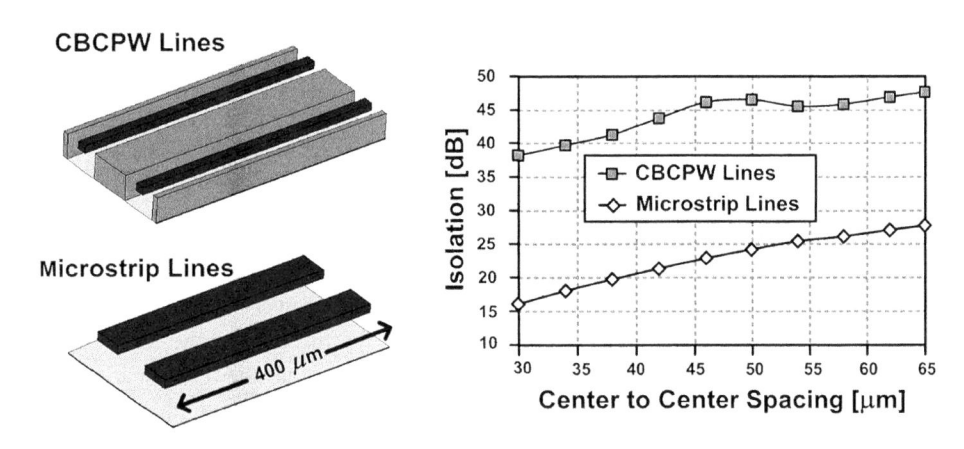

Fig. 10.21 Improved isolation with GSCPW t-lines compared with microstrip structures on-chip [35].

with different numbers of elements active. Additionally, the power divider should ensure isolation between different elements so that the phase shift in the *j*th element is not influenced by the phase setting in other elements. This requires the power division network to provide isolation between between elements.

The Wilkinson combiner (Fig. 10.22a) is a well-known power divider that provides isolation between outputs. The t-line based combiner needs $\lambda/4$ in each arm, which

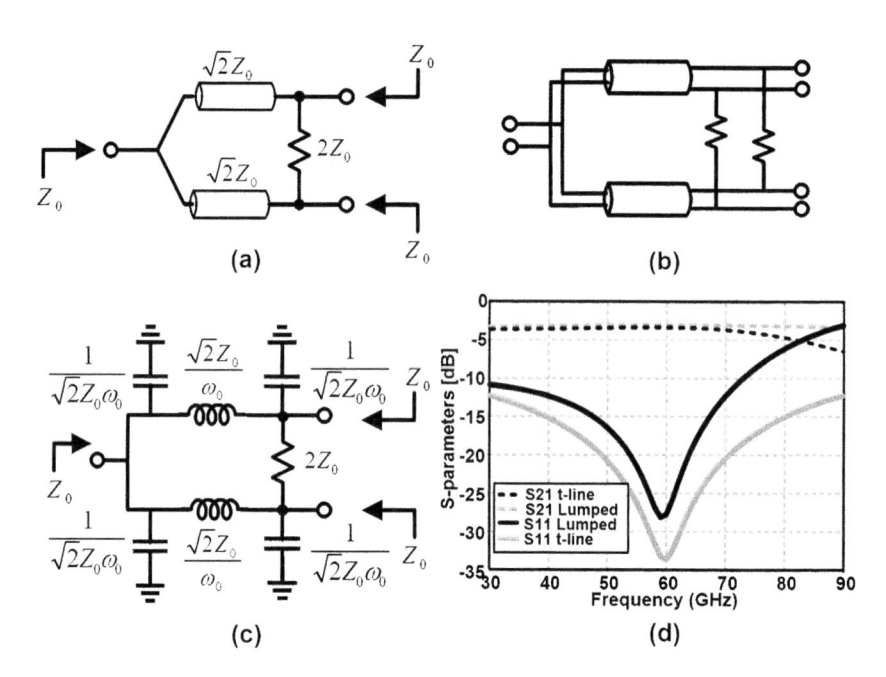

Fig. 10.22 (a) Single-ended Wilkinson power splitter, (b) differential Wilkinson power splitter, (c) Wilkinson divider implemented using lumped *L* and *C*, and (d) simulated performance of t-line based and lumped *LC* implementations at 60 GHz.

can lead to increased chip area, particularly at the lower end of mm-wave frequencies. Differential signal distribution is desirable on-chip to increase noise immunity – the Wilkinson combiner can be made differential as shown in Fig. 10.22b. The combiner can also be implemented using lumped *L* and *C* on-chip to save area (see Fig. 10.22c) leading to smaller chip area but lower bandwidth and higher loss. The bandwidth reduction due to the lumped approach can be seen in Fig 10.22d.

The effect of signal distribution on the gain across the array transmitter is shown in Fig. 10.23. Assuming 1-dB loss in the combiner as well as 1-dB loss in on-chip transmission-lines, it can be seen that passive power-division losses can degrade system linearity, which becomes limited by the linearity of the RF amplifier driving the signal distribution network. In order to ensure that the TX linearity is dominated by the linearity of the output power amplifier in each element, either driver linearity (amp A in Fig. 10.23) has to be increased or active power divider stages have to be introduced to ensure sufficient input power at each element. Since active stages consume power and can restrict distribution bandwidth, a hybrid approach with active and passive power dividers yields a balance between signal distribution output power, DC power consumption, and bandwidth [36].

RF-path phase shifting and variable gain

The phase shifters in each transmitter element must be capable of providing 360° of variable phase shift. On-chip narrow-band phase shifters can be implemented using active

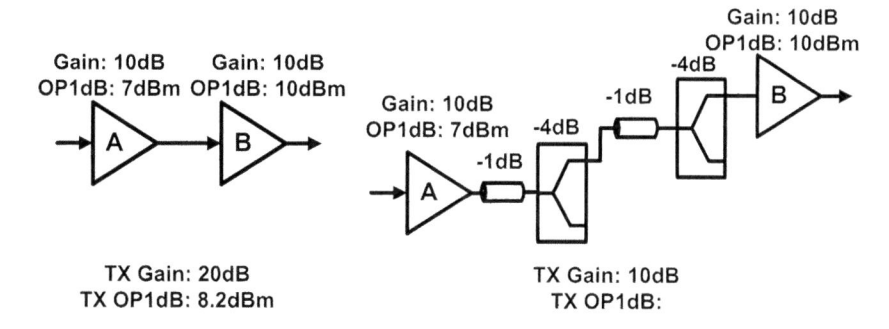

Fig. 10.23 Impact of combiner and interconnect t-line loss on overall TX element gain and linearity.

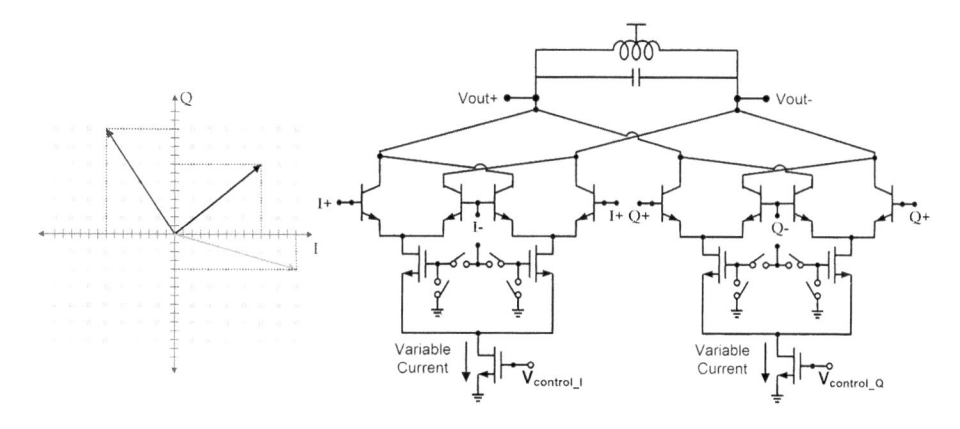

Fig. 10.24 RF-path phase shifter using a vector modulator.

and/or passive techniques at mm-wave frequencies. Figure 10.24 shows an active phase shifter that extends the multiplier to mm-wave frequencies. Such a phase shifter or vector interpolator requires the generation of differential quadrature signals. Techniques outlined in Section 10.3.3 can be used to generate quadrature at mm-wave – however, it is worth noting that the phase shifter at RF must operate over the entire signal bandwidth and therefore must have multi-GHz instantaneous bandwidth.

The output of the ideal interpolator can be expressed by using

$$v_I \cdot A \cos(\omega t) + v_Q \cdot A \sin(\omega t) = A_{out} \cos(\omega t - \theta) \tag{10.16}$$

$$A_{out} = A\sqrt{v_I^2 + v_Q^2} \tag{10.17}$$

$$\theta = \arctan\left(\frac{v_Q}{v_I}\right). \tag{10.18}$$

However, phase-shift dependence in the input matching and coupling between the IQ signals lead to the gain and phase shift shown in Fig. 10.24, as phase-shift settings are varied. The vector interpolator provides simultaneous variable gain and phase shift, and the phase-shift resolution is limited by the resolution with which the I and Q weights can

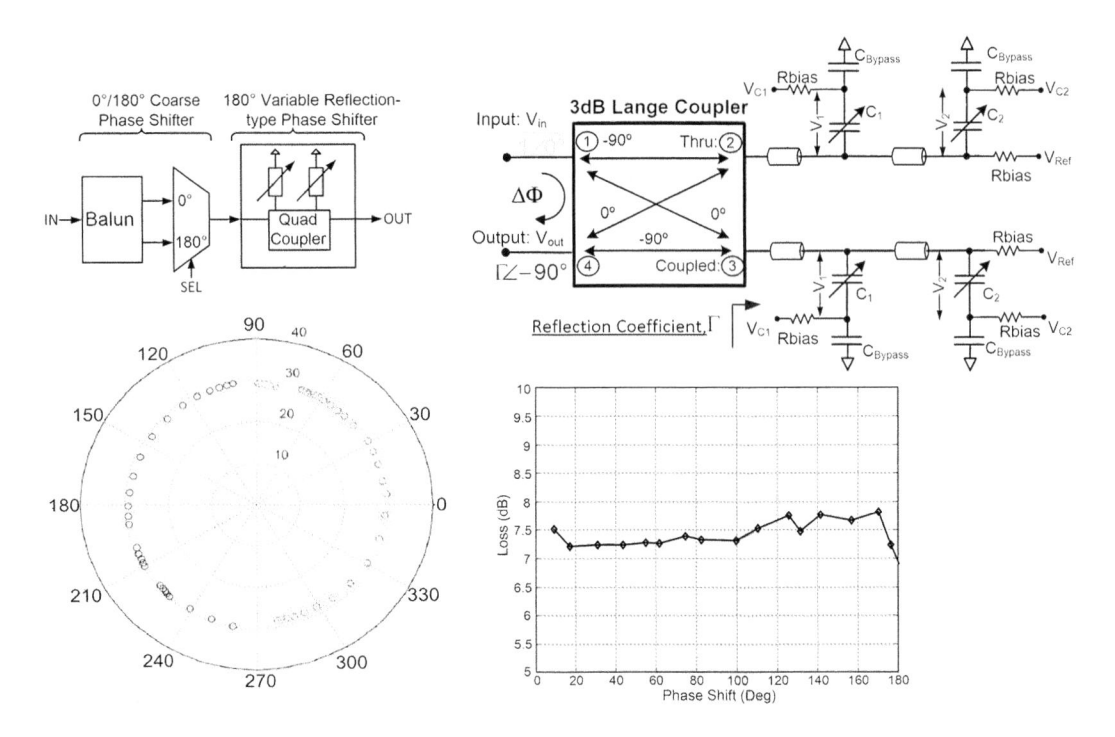

Fig. 10.25 RF-path phase shifter using a reflection-type phase shifter.

be generated. Phase shifters with 4-bit phase-shift resolution have been demonstrated across the range of mm-wave frequencies up to 100 GHz [15, 37, 38].

Passive phase shifters based on switches can be achieved at low mm-wave frequencies where switch losses are of the order of 1–2 dB. However, in such phase shifters, higher resolution requires an increased number of switches, leading to higher losses. Passive phase shifting can also be accomplished using reflective terminations on on-chip passive couplers. A reflection-type phase shifter using hybrid couplers is analyzed in Fig. 10.25. The input–output relationship is given by

$$\frac{V_{out}}{V_{in}} = \Gamma e^{\frac{-j\pi}{2}}. \qquad (10.19)$$

The variable termination can be visualized on a Smith chart. The phase of the reflection coefficient, Γ has to vary by 360° to provide a variable phase shift of 360° between input and output. Parasitic capacitances associated with on-chip passives make it unfeasible to get the entire phase variation from a single RTPS stage. Therefore, two-stage architectures with (a) an RTPS stage providing 180° of phase shift, followed by a discrete 0°/180° stage, or (b) two RTPS stages can be used to used to achieve the desired 360° phase shift.

The passive phase shifter achieves very high linearity, both in terms of amplitude and phase shift. Nonlinearities are introduced by varactors used to vary the terminating impedance. However, amplitude and phase linearity is maintained up to input powers approaching 10 dBm. In addition to linearity benefits, high phase-shift resolution can be achieved since the phase-shift resolution is limited by the resolution with which the control voltages across varactors are generated. In [19], a resolution of $11.25° \pm 4°$ is achieved with a control voltage step size of 25 mV.

10.5.4 Integrated phased-array transmitters with LO-path phase shifting

Implementing the phase shift in the LO path relaxes phase-shifter performance as the desired phase shift has to be achieved on a single tone as opposed to across the RF bandwidth (Fig. 10.26). The output of the phase shifter drives a mixer whose gain is insensitive to LO amplitude for large-enough LO amplitudes. Therefore, limiting amplifiers can be used after the phase shifter, eliminating requirements to maintain constant amplitude across phase shift. The phase shift in the LO-path can be implemented using the quadrature generators as described in Fig. 10.5 followed by vector modulators as described in Fig. 10.24.

LO-path phase shifting in each element requires the LO signal to be distributed across the entire IC, increasing layout complexity. At mm-wave frequencies, the size of the IC is comparable to the wavelength and therefore the distribution network consists of defined-impedance transmission lines and impedance-matched buffers. If each passive divider in a 16-element array has 1-dB loss in addition to the 3-dB division ratio, the LO power in each element is 16 dB lower than the input power. Therefore, buffer amplifiers are required to recover LO output power. Hence, distributing quadrature or multiple phases of the LO requires excessive area compared with single-phase LO distribution followed by local quadrature generation. When the quadrature signal has to be generated at a single frequency, a $\lambda/4$ line can be used to delay the signal by 90° at the target [18]. It is feasible to distribute a lower-frequency LO signal to each element, followed by frequency multiplication/frequency synthesis in each element using a PLL. For example,

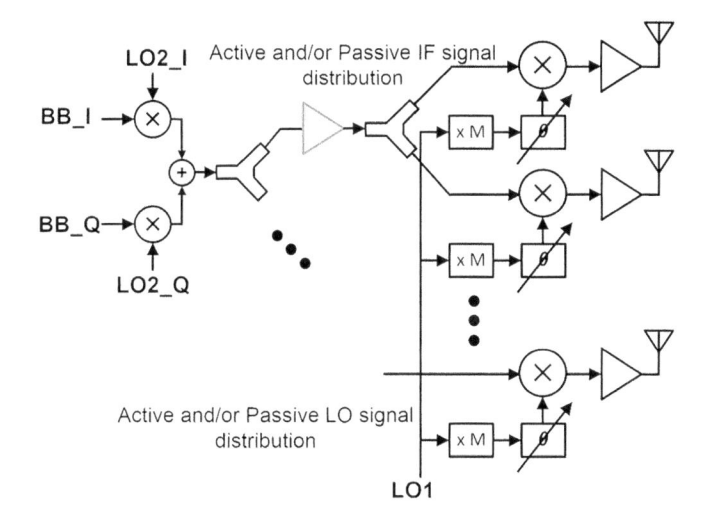

Fig. 10.26 Phased-array transmitter with LO-path phase shifting.

in the 280-GHz phased-array transmitter in [39], a 47-GHz LO signal is distributed to each element followed by ×6 multiplication ($M = 6$ in Fig. 10.26) to 282 GHz in each element.

10.6 Millimeter-wave transmitter examples

We now review three mm-wave transmitters to illustrate key trade-offs involved in the transmitter circuit and system architectures as well as to demonstrate the performance which is possible using advanced SiGe BiCMOS or CMOS technologies. The first example is a 16-element phased-array transmitter at 60 GHz, developed jointly by IBM and MediaTek using 0.13-μm SiGe BiCMOS technology [36]. The second example is a single-element transceiver at 60 GHz, developed by Panasonic in 90-nm CMOS technology, including built-in calibration with a companion baseband chip. These two examples will illustrate the trade-offs involved in designing for single-antenna fixed-beam operation or multi-antenna steered-beam operation. The third example is a radar transmitter at 76–81 GHz, developed by Freescale in SiGe BiCMOS technology, which will highlight the differences between highly integrated radar and communications transmitters.

10.6.1 Example 1: 60-GHz phased-array transmitter by IBM and MediaTek

Application requirements

The 60-GHz frequency band features a large available bandwidth which can support multi-Gb/s communications over a range of 1–10 m. Shorter-distance (1–3 m) line-of-sight links can be achieved using single-antenna transceiver solutions which feature simpler packaging and antenna solutions along with lower-power (RF and DC) and smaller-area transceivers. Longer-distance (10 m) non-line-of-sight links, however, require steered-beam multi-antenna solutions in the form of phased-array transceivers which are higher in power and larger in area. The IBM and MediaTek 60-GHz phased array [36] was developed for non-line-of-sight applications at >10 m range. A 16-element array architecture was chosen, using RF distribution and phase shifting for the beam former and a sliding-IF dual-conversion architecture for the transmitter core. The total measured RF power generated by the array at 1-dB compression can be adjusted between 21 and 25.5 dBm and the DC power consumption of the array is between 3.8 and 6.4 W depending on the RF power level. The transmitter was implemented using 0.13-μm IBM SiGe BiCMOS 8HP technology, featuring NPN transistors having $f_T/f_{MAX} = 200/265$ GHz. The array is standard-compliant to IEEE 802.15.3c.

The transmitter block diagram is shown in Fig. 10.27. The core of the transmitter uses a dual-conversion super-heterodyne architecture, where differential quadrature baseband signals are up-converted to a sliding intermediate frequency (IF) from 8.3 to 9.3 GHz and then converted to RF using a mm-wave mixer. A single phase-locked loop operating between 16.6 and 18.5 GHz generates the local oscillator (LO) signal, which is either divided by two to provide the LO for the IF mixer or multiplied by three to provide the LO for the mm-wave mixer. This LO solution results in a sliding-IF architecture,

10.6 Millimeter-wave transmitter examples

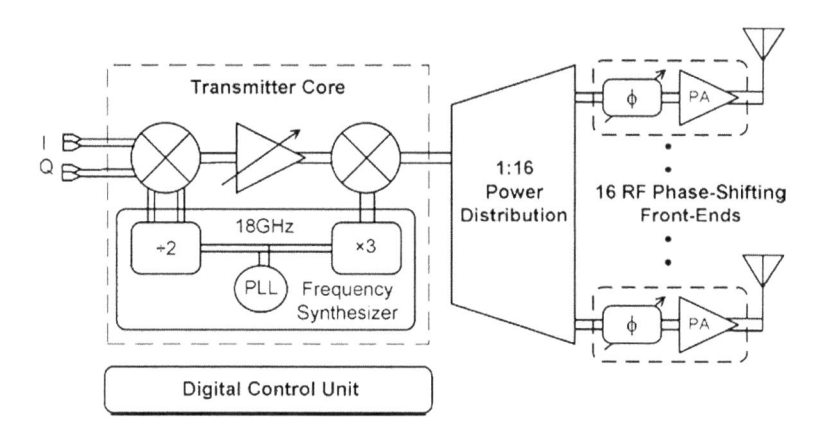

Fig. 10.27 Simplified block diagram of the 16-element phased-array 60-GHz transmitter.

where the IF is at 1/7th the desired RF frequency (IF = 8.33–9.26 GHz for RF = 58.32–64.80 GHz), the frequency tripler generates an output at 6/7th the desired RF frequency (3XLO = 49.99–55.54 GHz), and the opposite sideband or image generated by the mm-wave mixer occurs at 5/7th the desired RF frequency (IMG = 41.66–46.29 GHz). The modulated 60-GHz signals are then split through a 1:16 power distribution network and fed to 16 parallel RF front-ends, each of which includes RF phase shifting and power amplification.

Transmitter core and phase-locked loop

A simplified block diagram of the transmitter core is shown in Fig. 10.28. Baseband quadrature signals for the transmitter are first fed through a pair of programmable baseband attenuators having 6-dB coarse and 1-dB fine attenuation. These signals are then fed into a quadrature up-conversion mixer to convert to an 8–9 GHz intermediate frequency (IF). The I/Q mixer also can be configured as a direct modulator to generate minimum shift keying (MSK) at up to 2 Gb/s with minimum baseband overhead. The up-converted single-sideband signal is fed through an IF variable-gain amplifier (IFVGA) having 20 dB of programmable gain in steps of 1 dB. The cascaded filter response of the I/Q mixer and IFVGA forms a fourth-order bandpass characteristic with digitally controllable center frequency to accommodate the sliding IF frequency plan. The output of the IF VGA is fed to an RF mixer, which delivers 60-GHz modulated signals to the 1:16 RF power distribution network.

A single phase-locked loop generates a 16.6–18.5-GHz signal, from which is derived the LO for the IF mixers (through division by two) and the LO for the mm-wave mixers (through multiplication by three). Operating the VCO at a sub-harmonic of the desired output frequency allows for higher quality factors in the VCO tank (since varactor Q is higher at the lower frequency), wider tuning range, and smaller division ratios within the PLL, all of which result in improved phase-noise performance. Measured phase noise for the PLL is −90 dBc/Hz at 1 MHz offset for the 3XLO signal, suitable for 16QAM modulations.

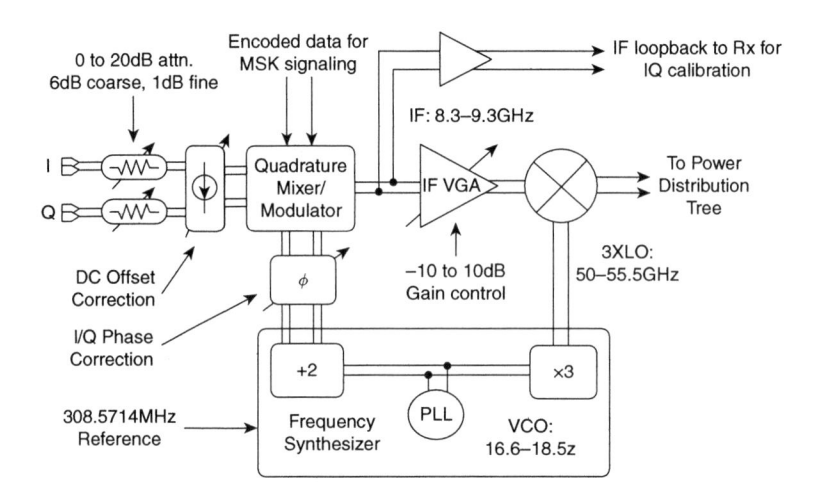

Fig. 10.28 Super-heterodyne transmitter core with sliding-IF frequency for 60-GHz array.

RF power distribution network

The splitting or distribution of RF signals to the individual transmit front-ends is difficult owing to the lossy cross-chip interconnects required as well as bandwidth, size, and linearity constraints. In this design, a hybrid architecture was used for the power distribution, where both active power splitters and passive power splitters were employed. Figure 10.29 shows a block diagram of the 1:16 power distribution tree, whereas Fig. 10.30 shows a circuit implementation of a 1:4 active/passive hybrid splitter. The hybrid architecture was chosen as a balance between the power consumption and linearity constraints.

The active splitter, shown in Fig. 10.30, is a fully differential amplifier that uses a split-cascode architecture, which relies on current-mode splitting at the intermediate node of the cascode. Current-mode splitting requires a relatively high impedance looking into the collector of the bottom transistor and relatively low impedance looking up into the emitter of the cascode device. The outputs of the active splitter are impedance-matched using transmission-line-based matching networks.

The passive splitter, also shown in Fig. 10.30, is a differential Gysel structure, which has been modified to include a phase inversion or twist in the isolation path. Similar to a Wilkinson splitter, the Gysel structure uses differential quarter-wave transmission lines to connect the input to each of the two outputs. The isolation network is composed of another pair of quarter-wave transmission lines which are then cross-connected at the center of the isolation arm together with a load resistor that will absorb any odd-mode signals arising from outputs 1 and 2. The cross-connection or twist can be imagined as a virtual half-wave-length line, which together with the two quarter-wave lines results in a full wave length of offset between the desired outputs. Note that this wave-length offset is also the case for a regular Gysel structure as well as a rat-race hybrid, and the overall function of the isolation arm for all three structures is similar. A key benefit of this modified Gysel structure is that there is no need to place the two output signals (OUT1

10.6 Millimeter-wave transmitter examples

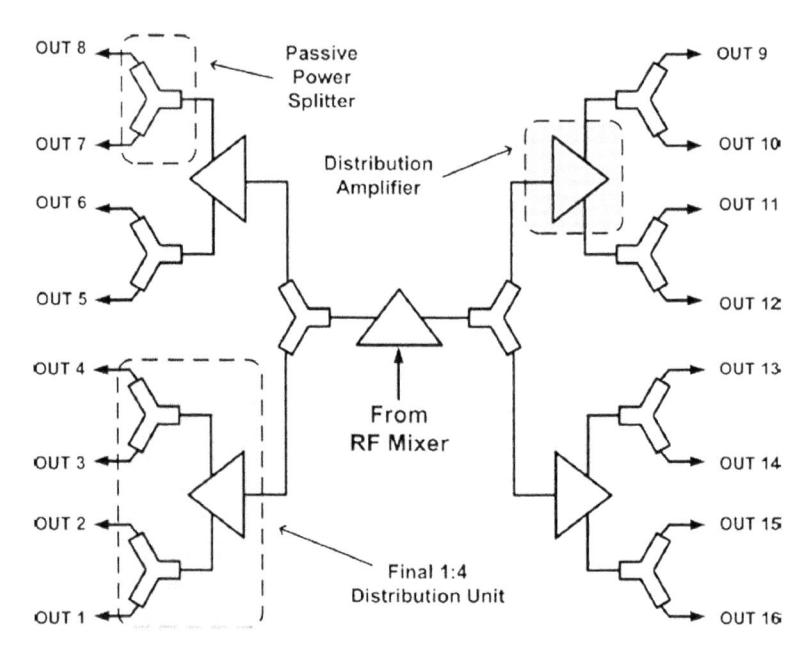

Fig. 10.29 Block diagram of the 1:16 power distribution tree for 60-GHz phased array.

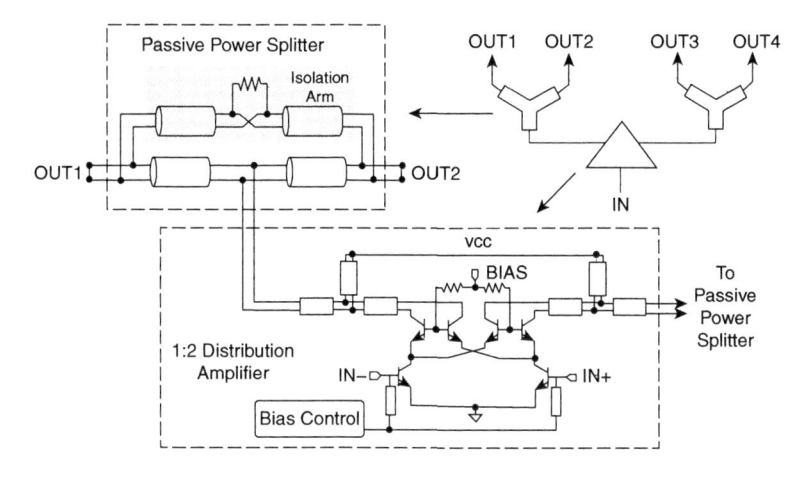

Fig. 10.30 Circuit-level implementation of each 1:4 distribution unit.

and OUT2) very close together, as in a Wilkinson splitter, as the isolation resistor is no longer immediately adjacent to the outputs. Instead, the routing of the signals that is necessary for the corporate distribution network can be absorbed into the splitter, resulting in an overall more-compact and lower-loss structure.

Phase-shifting front-end

Each phased-array element, shown in Fig. 10.31, consists of a balanced reflection-type phase shifter (RTPS), a fully differential variable-gain amplifier, which is used for gain

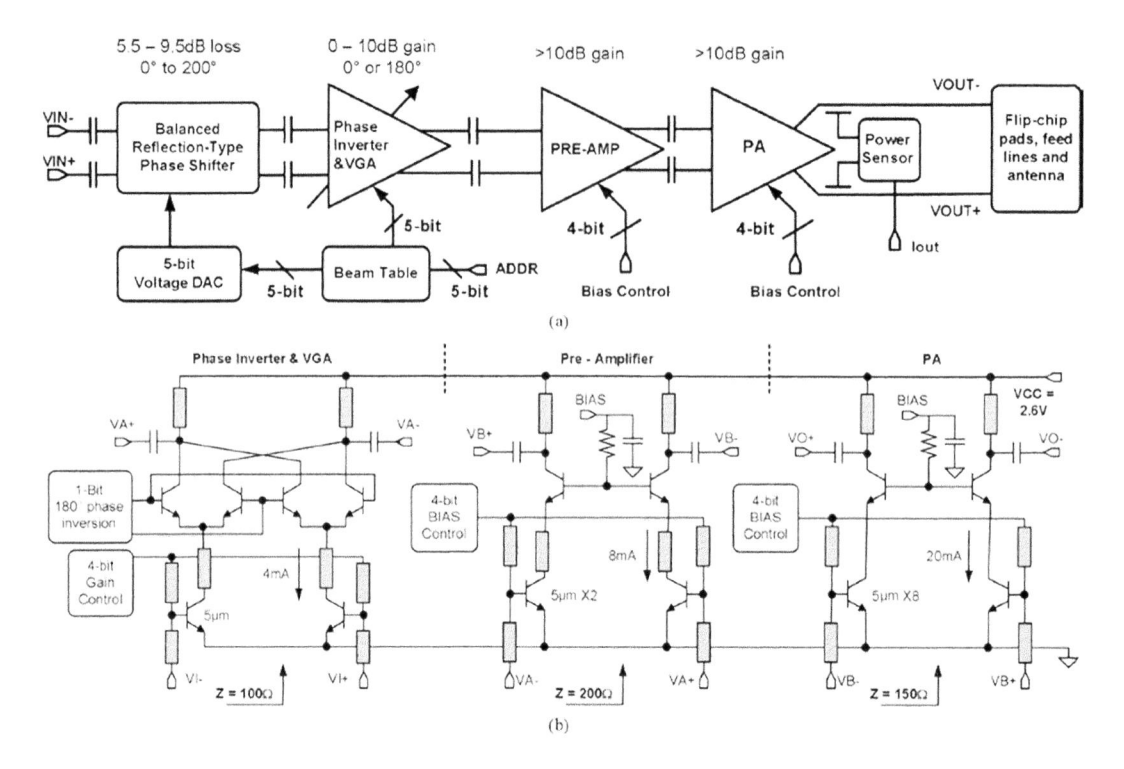

Fig. 10.31 (a) Block diagram of RF front-end and (b) circuit schematic of RF amplifiers.

compensation and 0/180° discrete phase switching, and a two-stage power-amplifier chain. A passive phase shifter is first used to generate 0–180° phase shift within a single stage. This phase shifter employs a reflection-type topology composed of a 90° hybrid coupler and reflective loads. A schematic of a single-ended RTPS is shown in Fig. 10.32. The coupler directs the input signal into two loads with 90° phase shift between the signals. These loads are designed to be highly reflective, with a tunable phase imparted through the reflection. Once reflected off the loads, the signals then recombine constructively at the RTPS output (port 4) and destructively at the RTPS input (port 1). The key to achieving a good performance in this design is having a reflective load with minimal absorption and a wide phase-shift tuning range, as indicated in Fig. 10.32, which shows plots of the reflection coefficient achievable for varying Q-factor. In this design, accumulation-mode MOS varactors are used together with transmission lines to realize the loads. The capacitance of the varactors is controlled through a 5-bit voltage DAC. The low quality factor of the varactors at 60 GHz results in a loss of 5–9 dB across a 200° phase-shift range. As a result, a variable-gain amplifier is used following the phase shifter to equalize the gain response.

The variable-gain amplifier employs cross-coupled Gilbert-cell topology. Additionally, a 1-bit 180° phase inversion can be realized together with the variable gain by simple commutation of the currents. The VGA has 10 dB of programmable gain, controlled through a 4-bit DAC. Altogether, the RTPS and the phase-inverting VGA

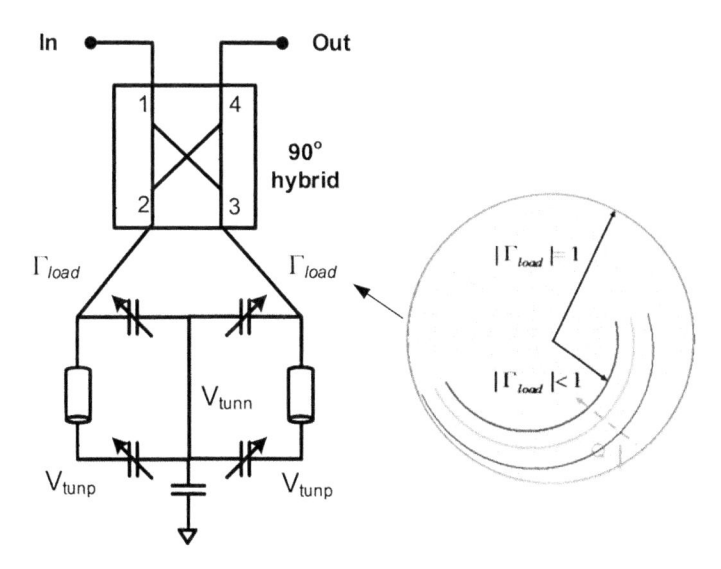

Fig. 10.32 RTPS schematic and plot of reflection coefficient vs varactor Q.

(PI-VGA) can provide for an equalized gain response with adjustable phase across a full 360° range, controlled through 10 digital bits (4-bits VGA gain, 1-bit VGA phase, 5-bit RTPS phase).

Following the RTPS and PI-VGA, there is a two-stage power amplifier (PA), shown in Fig. 10.31. Cascode circuit topologies are used to allow for high voltage swing approaching BV_{CBO} on the cascode device. Both cascodes are operated in class-AB mode to provide reasonable power gain. The complete PA is capable of providing +15 dBm output 1-dB compression point with 2.6-V power supply. A power detector is weakly coupled to each RF output to allow for built-in monitoring of the RF power of each front-end.

Chip layout

A chip micrograph of the phased-array transmitter is shown in Fig. 10.33. The die size is 6.5×6.75 mm^2. The core of the transmitter, including baseband, IF, and PLL, is located in the center bottom portion of the chip. The output of the mixer is located in roughly the center of the chip, which then drives an active 1:2 splitter and two 1:8 hybrid splitters, situated left and right. Banks of eight front-ends are placed along the right and left halves of the chip. The overall chip is designed for flip-chip packaging; hence, internal pads are used to maintain low-impedance power supplies throughout.

Measured results

The 60-GHz IBM/MediaTek phased array was characterized through a combination of wafer-probing and package-level experiments. The frequency response of the RF phase-shifting front-end is shown in Fig. 10.34. A gain of 25–30 dB is attained across 55–65 GHz. The large-signal performance is shown in Fig. 10.35, where the output 1-dB compression point is shown together with the complete front-end power-added

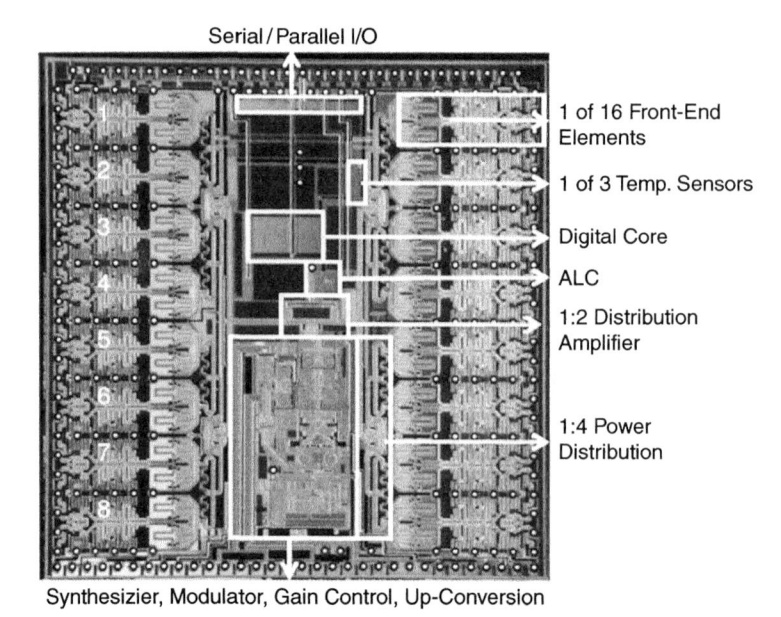

Fig. 10.33 Chip micrograph of 60-GHz phased array with 16 elements. Size 6.5 × 6.75 mm².

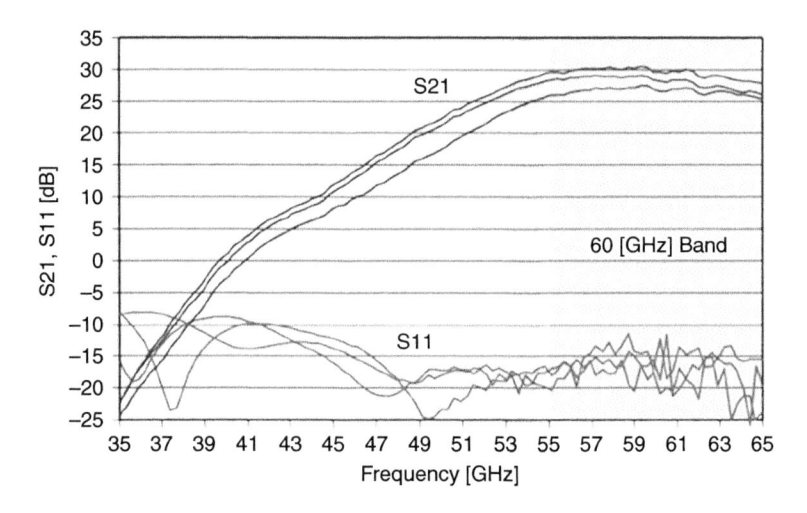

Fig. 10.34 Measured differential S-parameters of the RF phase-shifting front-end across three different phase settings where VGA gain is left unequalized.

efficiency across the four 60-GHz standardized channels. As can be seen, oP_{1dB} between +10 to +15 dBm is achieved, programmable through adjustment of PA bias current level. Also, the front-end peak PAE is between 6% and 8%. Note that this peak PAE is relatively low and represents an area for continued research improvement for mm-wave transmitters and power amplifiers.

10.6 Millimeter-wave transmitter examples

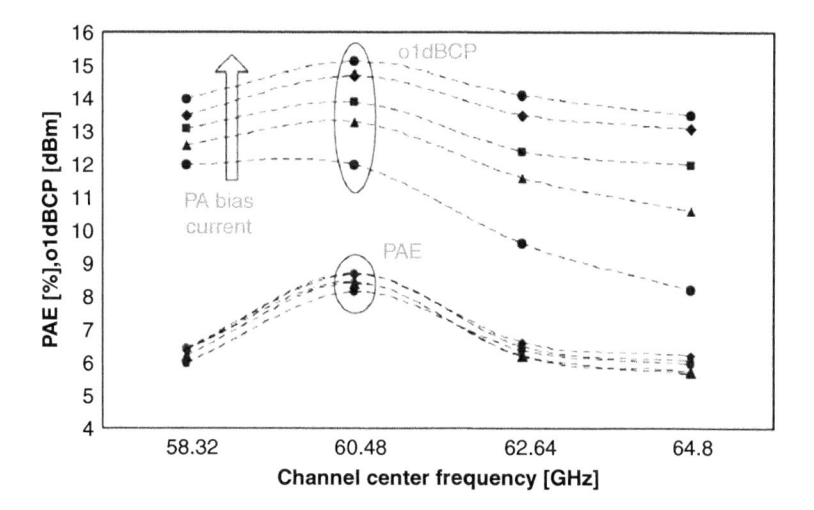

Fig. 10.35 Measured output 1-dB compression point and peak power-added efficiency (PAE) of the RF front-end for the four 60-GHz channels. Adjustable bias current in the PAs allows adjustable P_{1dB}.

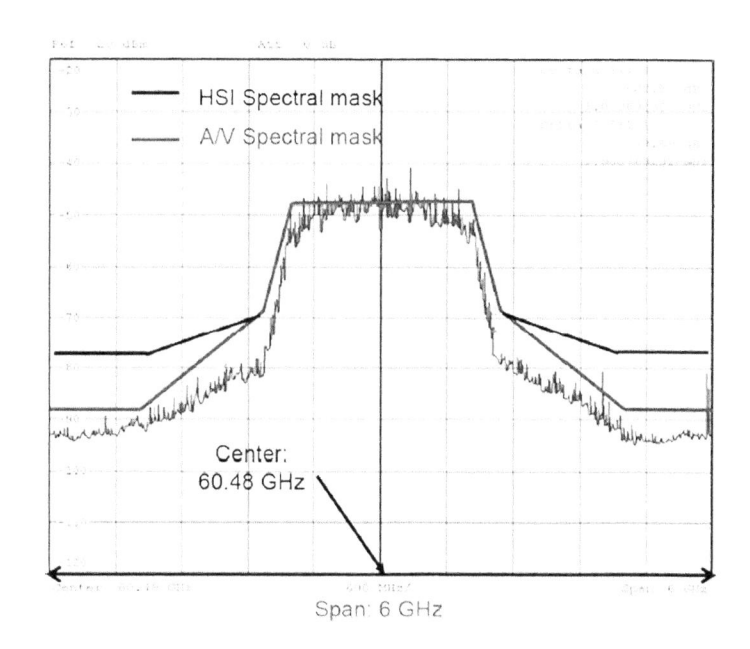

Fig. 10.36 Measured transmitter output spectrum in channel 2 (60.48 GHz) when applying a 1.6-Gb/s OFDM signal (MCS2 802.15.3c transmit mode).

The overall transmitter performance has been measured using modulated signals, and the transmitter output spectrum complies with 60-GHz spectral mask requirements. Figure 10.36 shows a measured spectrum for a 1.6-Gb/s OFDM signal. Beamforming performance was measured through antenna-pattern measurements for the transmitter

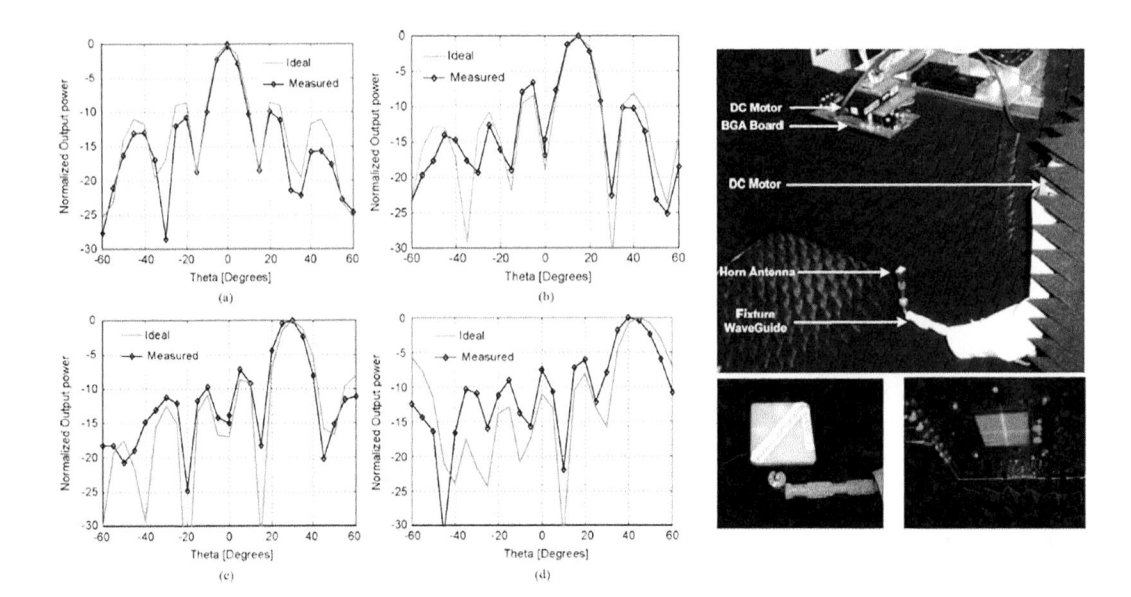

Fig. 10.37 Measured radiation patterns of the packaged 16-element phased-array for 0°, 15°, 30°, and 45° steering. Measurements obtained within the antenna chamber are shown on the right.

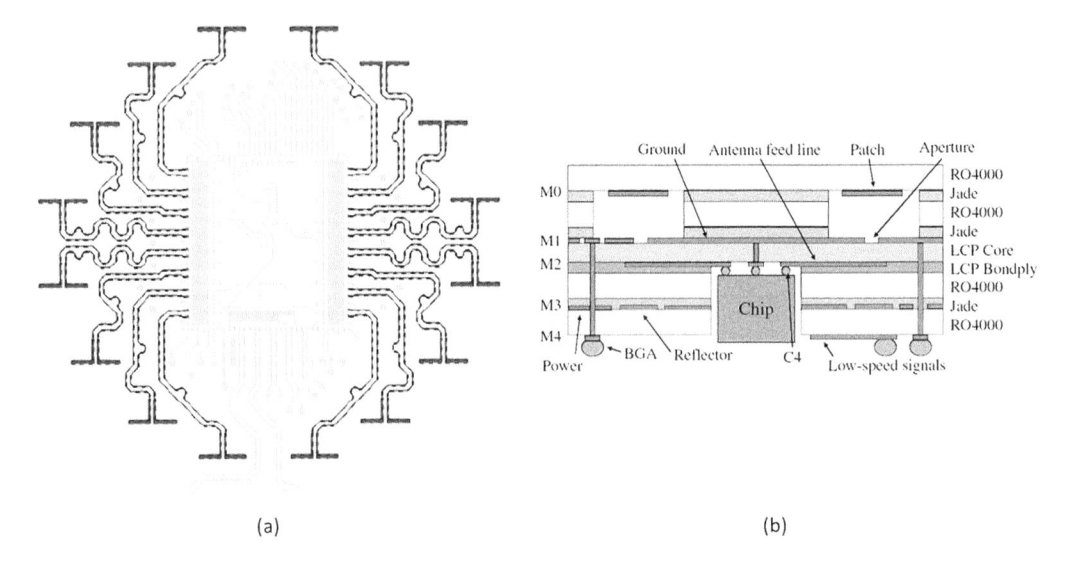

Fig. 10.38 (a) Layout of the 60-GHz interconnections within the package, showing how the antenna array is organized in a ring configuration. (b) Cross section of multi-layer organic package.

packaged together with 16 antennas in a multi-layer organic package, mounted within an antenna chamber. Figure 10.37 shows the measured patterns for the array when steered to 0°, 15°, 30°, and 45° elevation angles. These are compared with the ideal patterns which would be obtained using identical and isotropic antennas. Excellent agreement is attained between theory and measurements. Note that the relatively narrow beamwidth

10.6 Millimeter-wave transmitter examples

Table 10.1 Summary of the measured performance of IBM/MTK 16-element phased-array transmitter.

	Channel 1 58.32 GHz	Channel 2 60.48 GHz	Channel 3 62.64 GHz	Channel 4 64.80 GHz
Max. gain/element (dB)	35	35	34	32
oP_{1dB} 16 elements (dBm)	21.0	21.3	19.5	17.0
Phase noise at 1 MHz (dBc/Hz)	−90	−89	−89	−87
Phase tuning	\> 200° Continuous (5-bit), +180° discrete (1-bit)			
Amplitude mismatch	±1 dB (across elements)			
Carrier leakage	< −31 dBc			
I/Q gain and phase error	< 1 dB, < 1°			
−3-dB IF channel BW	±1 GHz			
Total power consumed	3.8 W at 22 °C			
Total area	6.5 × 6.75 mm²			

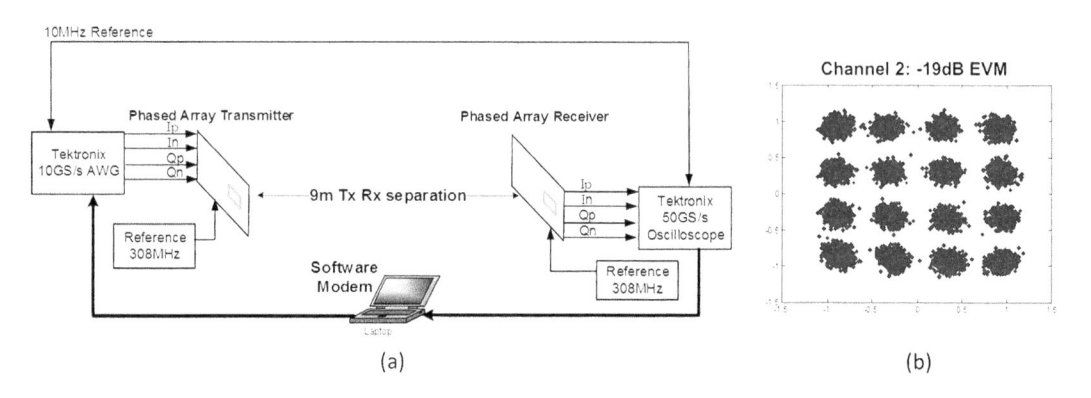

Fig. 10.39 Link measurement set-up and measured constellation of wireless link using 16-element phased-array transmitter and receiver at 9 m distance, using 16QAM OFDM modulation at 5.3 Gb/s.

and high sidelobe performance is expected and is due to the use of a 16-element antenna array arranged in a circular ring around the periphery of the chip. Figure 10.38a shows a top view of the package interconnect layer which contains the 60-GHz feed lines to each antenna. These feeds are aperture coupled to patch antennas located above the dipole-like structure, according to the cross section shown in Fig. 10.38b. The ring configuration was selected to simplify package-level interconnections and this sparse array will naturally result in narrow beamwidth and high sidelobes, neither of which is problematic for the 60-GHz application.

Table 10.1 summarizes the performance of the 16-element phased-array transmitter. This transmitter has been used together with a companion phased-array receiver to demonstrate up to 9-m non-line-of-sight links at 5.3 Gb/s using IEEE 802.15.3c standard modulation (both single-carrier and OFDM). The measurement set-up is shown in Fig. 10.39a, and Fig. 10.39b shows the measured constellation of the link at 5.3 Gb/s using 16QAM OFDM, indicating −19 dB EVM.

10.6.2 Example 2: 60-GHz single-element transceiver by panasonic

Application requirements

The previous 60-GHz phased array from IBM and MediaTek was designed to support 10-m non-line-of-sight links. As a result, a phased array employing multiple RF transmit chains connected to multiple antennas is required. In contrast, for shorter-distance line-of-sight links, less output power is needed and smaller-unit antennas can be used. The second example studied is a single-element (i.e. not a phased array) transceiver from Panasonic, implemented in 1.4-V overdrive 90-nm CMOS. The transmitter employs a direct up-conversion from quadrature baseband to 60 GHz. The total transmitter power consumption in the RF chip is 347 mW, used to generate an output power of +3.7 dBm at oP_{1dB} (+8 dBm P_{SAT}). The complete transceiver chipset (RF chip and baseband chip) includes built-in in-band calibration to improve EVM, and is compliant to WiGig/IEEE 802.11ad standards.

Transmitter description

The complete transceiver architecture is shown in Fig. 10.40. A direct-conversion scheme is used. Differential I/Q baseband signals drive the analog baseband composed

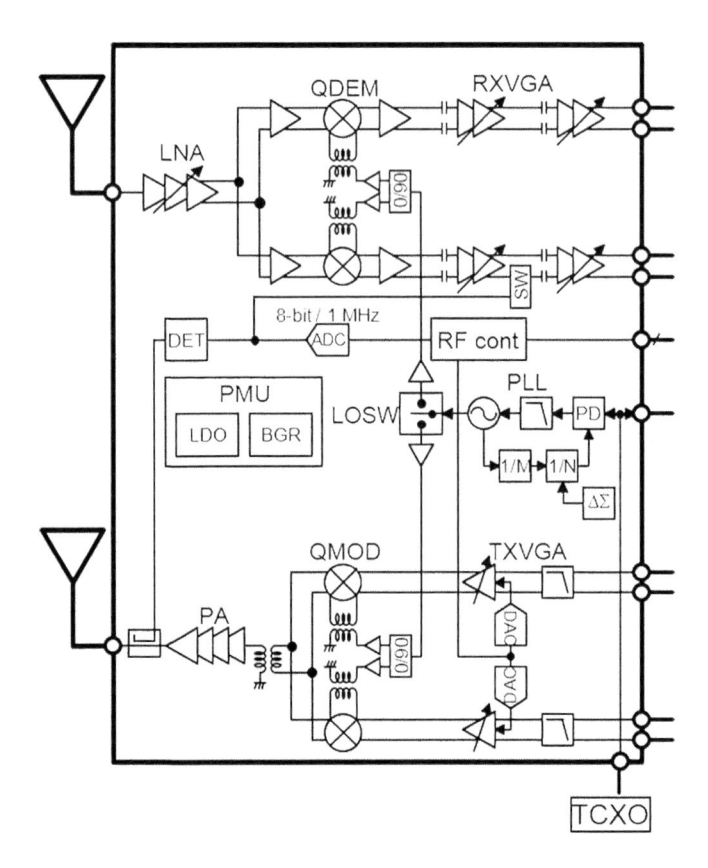

Fig. 10.40 Simplified block diagram of the Panasonic 60-GHz transceiver.

10.6 Millimeter-wave transmitter examples

of fourth-order passive anti-aliasing low-pass filters and baseband variable-gain amplifiers. A quadrature modulator up-converts the signals directly to the 60 GHz band. The output of the modulator drives a four-stage power-amplifier chain. The LO signal for the modulator is derived from an on-chip PLL operating at 30-GHz with a push–push output to generate the 60-GHz signal. Differential quadrature LO signals are generated at 60 GHz using a transformer-based hybrid coupler.

Frequency synthesizer

The frequency synthesizer for the Panasonic chip consists of a 30-GHz push–push VCO, a divide-by-three injection-locked frequency divider (ILFD), a multi-modulus counter, a phase-frequency detector, the charge pump, and control circuitry. This architecture can synthesize the required 60-GHz channel frequencies (58.32, 60.48, 62.64, and 64.8 GHz) with low phase noise. Measured results indicate a phase noise of -93 dBc/Hz or better at 1 MHz offset across all four channels, obtained with loop bandwidth of 50 kHz. In particular, a narrow loop bandwidth is used to minimize in-band phase-noise contributions which are amplified by the large division ratio when locking the 30-GHz signal to a 26–40-MHz reference. To keep the power consumption low for this PLL, a ring-oscillator-based injection-locked prescaler (divide-by-3) is used, and such a design requires a calibration technique to properly set both the VCO band and the ILFD free-running frequency.

Power amplifier

A simplified schematic of the power amplifier is shown in Fig. 10.41. A four-stage topology is used, where stage one is single-ended and stages two through four are balanced. Transformers are used for interstage matching and coupling. Stages three and four use a push–pull common-source configuration to improve linearity. The PA operates from a 1.25-V drain supply for improved lifetime, rather than using the 1.4-V supply used elsewhere on the chip. The PA is designed to achieve at least +3.7 dBm output 1-dB compression and +8 dBm saturated output power.

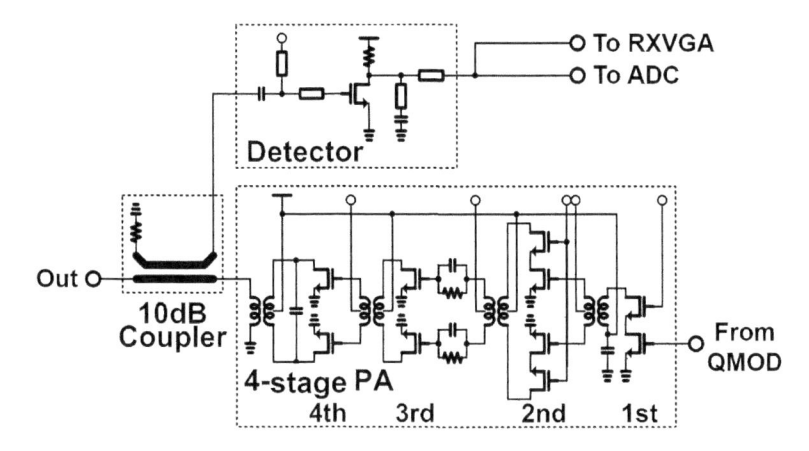

Fig. 10.41   Simplified schematic of the four-stage power amplifier together with built-in power detector.

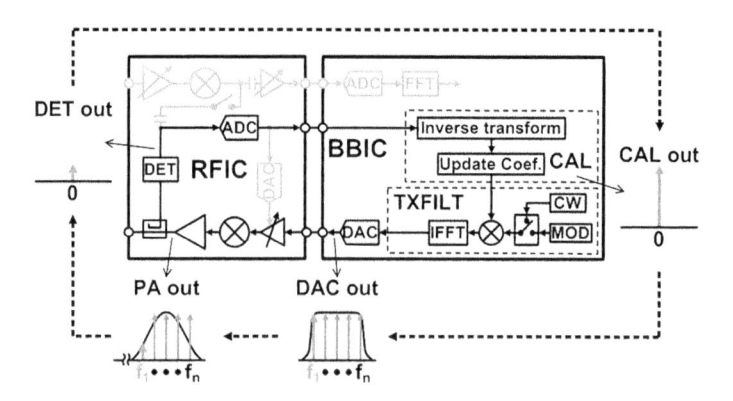

Fig. 10.42 Block diagram of the in-band amplitude calibration used in the transmitter.

Baseband and built-in test

The Panasonic chipset includes both the RF transceiver and a baseband chip. The baseband implementation is outside of the scope of this comparative study; however, we do point out a few of the key built-in test and calibration features of the chipset which are used to achieve lower EVM. The 60-GHz band features wide-bandwidth channels and, when used for single-carrier modulation (as opposed to OFDM modulation), it requires a flat frequency response over the entire channel. To obtain this, a built-in-test function has been integrated, where the DAC on the baseband chip generates tones stepped from −880 MHz to +880 MHz in 110-MHz steps. As shown in Fig. 10.42, these tones are modulated onto the 60-GHz carrier and then amplified by the PA. A 10-dB coupler at the output of the PA drives a power detector which convert the RF power to a DC signal, and this signal is then measured using an 8-bit 1-MHz ADC. Based on these measurements, filter calibration coefficients are calculated for the baseband IC. In addition to calibrating the in-band gain response, calibration is also used to suppress carrier and image leakage within the up-conversion mixer. The calibration loop involves the complete transmitter path, the power detector at the output of the PA, and then the analog baseband and digital baseband within the receiver. The inclusion of more sophisticated calibration and compensation represents a maturation of the 60-GHz application space, leveraging and expanding upon techniques successfully used for lower-frequency transceiver products.

Chip layout

A chip micrograph of the complete transceiver is shown in Fig. 10.43. The die size is 3.75×3.6 mm^2. The transmitter is located in the lower half of the die.

Performance results

Panasonic's 60-GHz transmitter was characterized through a combination of wafer-probing and package-level experiments. The measured oP_{1dB} and P_{SAT} in channels 2 and 3 are +3.7 dBm and +8 dBm, respectively, and conversion gain is 23 dB. Roughly 3-dB lower gain and 2–3-dB lower output power is observed in channels 1 and 4. The

10.6 Millimeter-wave transmitter examples

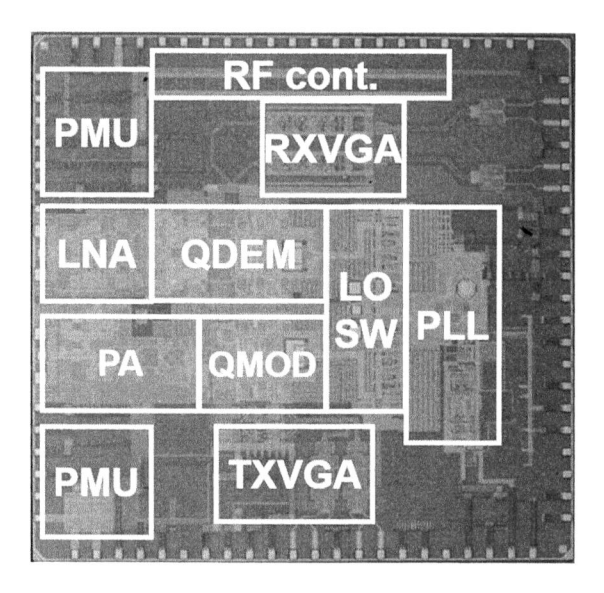

Fig. 10.43 Chip micrograph of Panasonic 60-GHz transceiver in 90-nm CMOS. Die size is 3.75 × 3.6 mm², approximately half of which is occupied by the transmitter.

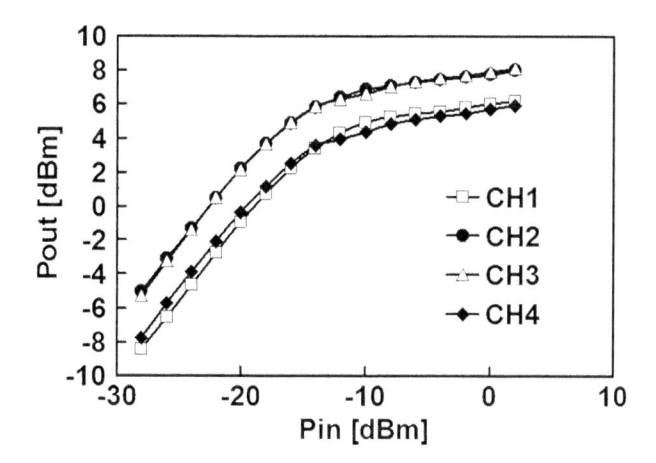

Fig. 10.44 Swept power measurement of the Panasonic 60-GHz transmitter.

measured PAE is 4.5% at oP_{1dB} and 16.4% at saturation. Using the built-in test and calibration, the carrier leakage and image leakage can be reduced from approximately −20 dBc to roughly −35 dBc or better.

Finally, the 60-GHz transceiver has been packaged and attached to a 2 × 2 array of patch antennas with a single feed, having 6.5-dBi antenna gain. The chipset achieves 1.8-Gb/s throughput (2.5-Gb/s PHY rate) at up to 0.4-m range and 1.5-Gb/s throughput (1.9-Gb/s PHY rate) at up to 1-m range. The measured EVM is −22 dB.

412 System-on-a-chip mm-wave silicon transmitters

Table 10.2 Comparison between IBM/MediaTek phased-array and Panasonic single-element transmitters.

	IBM/MTK transmitter	Panasonic transmitter
Frequency range	Channels 1–4 58.32–64.8 GHz	Channels 1–4 58.32–64.8 GHz
Number of RF elements	16	1
Antenna gain	16 dBi (array factor + single element)	6.5 dBi (single element)
Conversion gain	32–35 dB	20–23 dB
Total chip P_{1dB} (Channel 2)	+21 dBm (126 mW)	+3.7 dBm (2 mW)
Power consumption	3.8 W	0.35 W
Efficiency P_{1dB}/P_{DC}	3.3%	0.7%
Measured range	9 m	1 m
Die size (Tx only)	43.9 mm^2	6.75 mm^2

Comparison between phased-array and single-antenna transmitters at 60 GHz

Table 10.2 compares key performance results between the IBM/MediaTek phased array transmitter and the Panasonic single-element transmitter. This table illustrates the key differences that exist for designing a beam-steered non-line-of-sight link for 10-m range and a fixed-beam line-of-sight link for 1-m range. First, to support a 10× higher range (equal to 20 dB in link budget), higher RF power and higher-gain antennas are required. Second, the power consumption is about 10× larger in the phased array, and this power is primarily used to generate the larger RF power. Finally, the phased-array requires larger chips and larger packages, where the array is about 6.5× larger than the single-element design.

10.6.3 Example 3: 77-GHz FMCW radar transmitter by Freescale

Application requirements

Automotive radar sensors operating around 77 GHz are being used to create driver-assistance systems, such as adaptive cruise-control and collision-warning systems. These radars employ a frequency-modulated carrier-wave (FMCW) architecture, where the VCO frequency is swept linearly to allow both range and velocity measurements. Long-range radar (LRR) systems operate at 76–77 GHz for 10- to 250-m range, 0.5-m range resolution, and 0.1° angular resolution. These sensors can be used for high-speed forward-looking applications requiring narrow beamwidths, such as an adaptive cruise-control system. Short-range radar (SRR) systems operate at 77–81 GHz, for 0.15- to 30-m range, 0.1-m range resolution, and 1° angular resolution. These sensors can be used for low-speed obstacle detection and parking-aid systems. Two critical requirements for the radar transmitter include achieving and maintaining high RF output power at 77 GHz across the full automotive temperature range (−40 to +125 °C) and achieving a very low phase noise in the VCO. The Freescale chipset presented herein is designed for both LRR and SRR modes, and features a four-channel receiver and single-channel transmitter, implemented in 200-GHz f_T SiGe BiCMOS technology.

Transmitter architecture

A block diagram of the transmitter chip is shown in Fig. 10.45a. Two W-band VCOs are included, one optimized for LRR operation and centered at 76.5 GHz, the other optimized for SRR operation and centered at 79 GHz. The phase-noise requirements for LRR operation are significantly more stringent that those for SRR operation (i.e. LRR requires about −100 dBc/Hz at 1-MHz offset). As a result, a dual-VCO solution can offer additional freedom to optimize each individual circuit for its intended operation. A schematic for the VCO is shown in Fig. 10.46. As seen in the previous 60-GHz examples, the core of the oscillator operates at a sub-harmonic, in this case the half frequency, and a push–push architecture is used to double the frequency to the mm-wave range.

The VCOs are buffered and then drive both the PA as well as an LO distribution network. The PA is a balanced three-stage cascode design. The LO distribution path is used to provide the LO signal to the receiver chip(s) and the external PLL. A dynamic frequency divider generates a 38–40.5-GHz signal which is then amplified and sent off-chip to the receiver chips. An additional divide-by-768 is used to generate a 50-MHz range signal for the external PLL.

Though this chapter is not focused on receiver design, a brief summary is included to help explain the key system operation. As shown in Fig. 10.45b, the 38-GHz LO signals are buffered, multiplied by two, and then split to drive the four independent

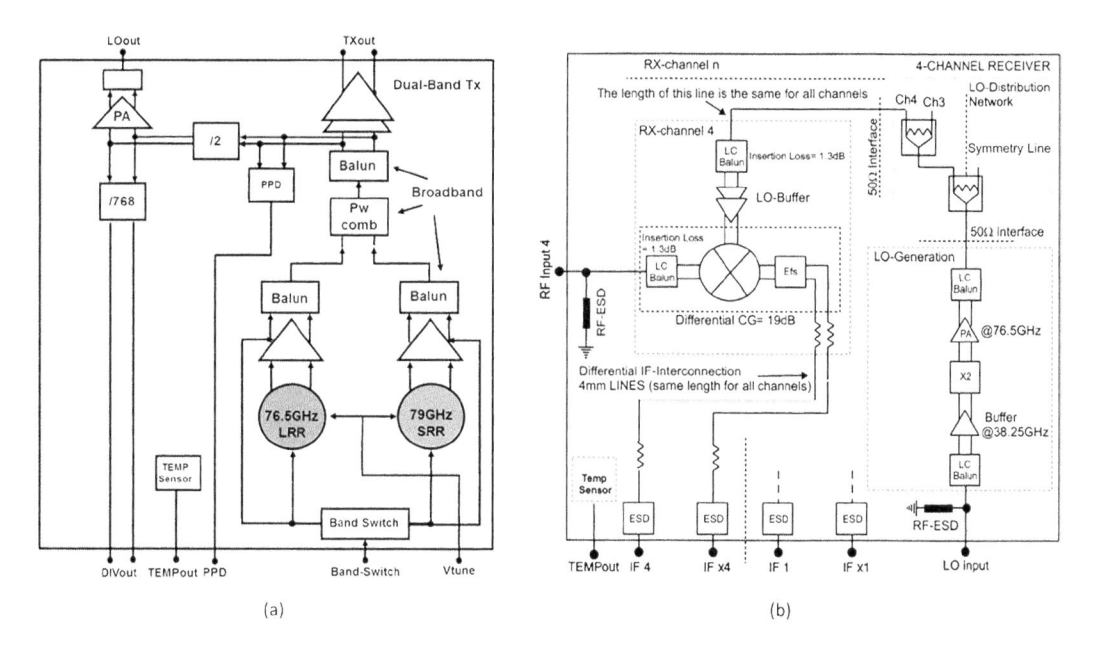

Fig. 10.45 Block diagrams of Freescale (a) transmitter and (b) receiver chips.

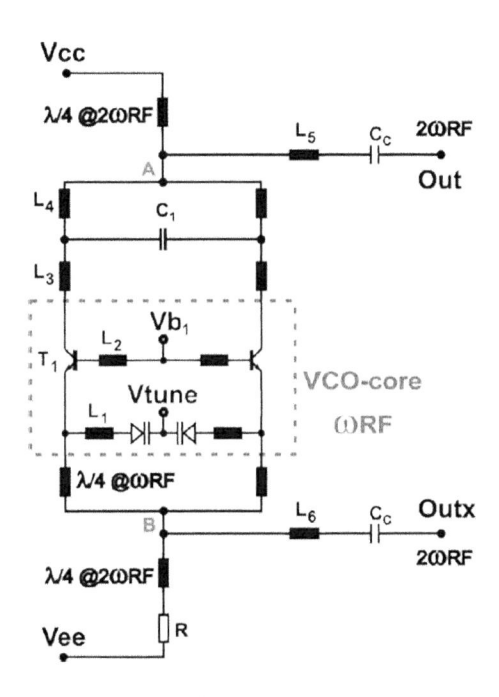

Fig. 10.46 Schematic of the Freescale push–push VCO.

Fig. 10.47 Chip micrograph of Freescale radar transmitter. Total chip size is 6.5 mm².

receiver channels. Each receiver channel includes a mixer with LO buffer and IF buffers. Mixer-noise figure and linearity are both important for the overall system performance. A die photograph of the transceiver is shown in Fig. 10.47. The total chip size is 6.5 mm².

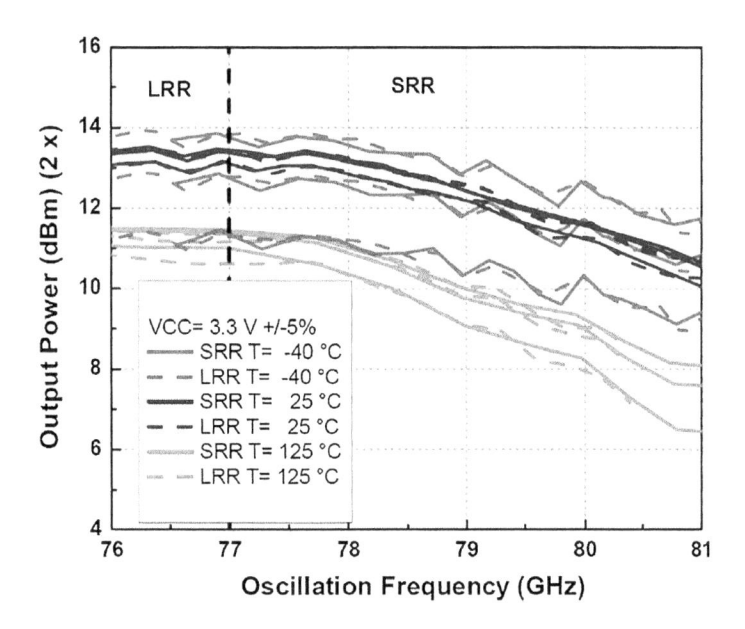

Fig. 10.48 Measured (single-ended) output power of the transmitter across frequency and temperature.

Performance results

The Freescale transmitter has been characterized through both on-wafer and package-level measurements. Figure 10.48 shows the measured output power of the transmitter across frequency and temperature. Each half of the PA generates +13 dBm under the typical case. At 125 °C, the output power drops by 2 dB in LRR mode (from 76–77 GHz) and 5 dB in SRR mode (at 81 GHz). The two VCOs cover the required frequency range and the measured phase noise is better than −96 dBc/Hz at 1-MHz offset in the LRR mode and better than −95 dBc/Hz at 1-MHz offset in the SRR mode. Figure 10.49 shows the measured phase noise. Total power consumption of the transmitter is 1.7 W. When compared with the 60-GHz transmitters, some key differences are as follows: (a) the radar transmitter has more stringent output power and temperature requirements; (b) the radar transmitter operates in saturation and therefore the linearity requirements are relaxed; and (c) the VCO requires both a wide tuning range and very low phase noise and, as a result, band-switched architectures cannot be used to keep K_{VCO} low.

10.7 Conclusion

This chapter has discussed the implementation of mm-wave system-on-a-chip solutions. First, we have highlighted architectural choices for the transmitter, including direct-conversion versus dual-conversion, frequency synthesis, and quadrature generation. Second, we have discussed both single-element and phased-array transmitter architectures. We have focused in particular on the phased array, as multi-antenna solutions are prevalent at mm-wave frequencies to provide both sufficiently high EIRP

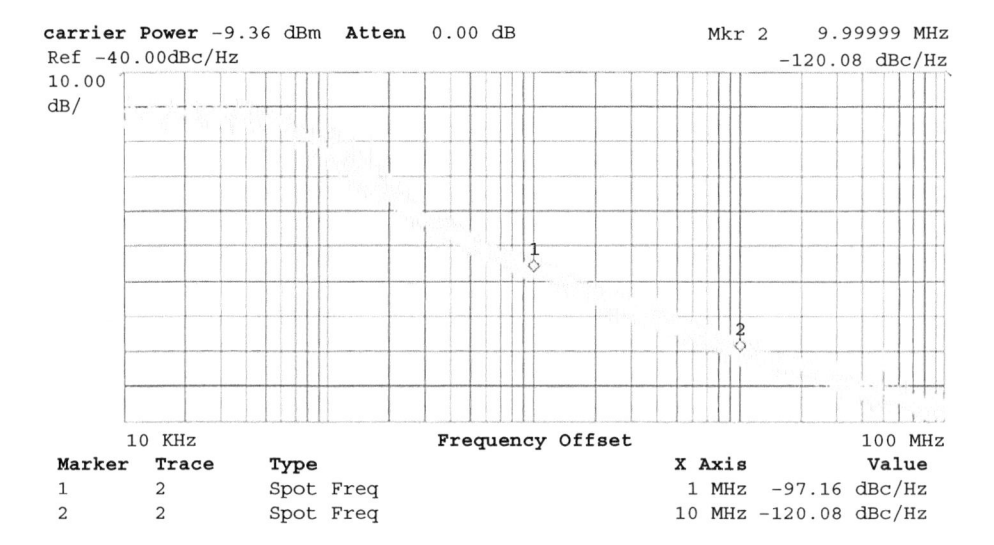

```
carrier Power -9.36 dBm   Atten  0.00 dB              Mkr 2      9.99999 MHz
Ref -40.00dBc/Hz                                                 -120.08 dBc/Hz
10.00
dB/
```

```
        10 KHz              Frequency Offset              100 MHz
Marker  Trace   Type                        X Axis         Value
1       2       Spot Freq                   1 MHz          -97.16 dBc/Hz
2       2       Spot Freq                   10 MHz         -120.08 dBc/Hz
```

Fig. 10.49 Measured phase noise of the transmitter output signal within a package.

and beam-steering capability. Finally, we have highlighted the performance of three recent mm-wave transmitter examples, including a 60-GHz phased array, a 60-GHz single-element transmitter, and a 77-GHz frequency-modulated radar transmitter. These examples show that both very high levels of performance and high levels of integration can be achieved in mm-wave system-on-a-chip transmitters.

References

[1] A. Cathelin, B. Martineau, N. Seller, et al., "Deep-submicron digital CMOS potentialities for millimeter-wave applications," in *2008 IEEE Radio Frequency Integrated Circuits Symp. (RFIC) Dig.*, June 2008, pp. 53–56.

[2] B. Jagannathan, M. Khater, F. Pagette, et al., "Self-aligned SiGe NPN transistors with 285 GHz f_{MAX} and 207 GHz f_t in a manufacturable technology," *IEEE Electron Device Lett.*, vol. **23**, no. 5, pp. 258–260, May 2002.

[3] J. Pekarik, J. Adkisson, R. Camillo-Castillo, et al., "A 90 nm SiGe BiCMOS technology for mm-wave applications," *Proc. of the GOMAC-Tech*, pp. 1–4, 2012.

[4] A. Kar-Roy, E. J. Preisler, G. Talor, et al., "270 GHz SiGe BiCMOS manufacturing process platform for mm wave applications," in *Proc. SPIE 8188, Millimetre Wave and Terahertz Sensors and Technology IV*, Oct. 2011, p. 81 880F.

[5] T. Rappaport, S. Sun, R. Mayzus, et al., "Millimeter wave mobile communications for 5G cellular: It will work!" *IEEE Access*, vol. **1**, pp. 335–349, May 2013.

[6] ——, "Revision of Part 15 of the Commission's Rules Regarding Operation in the 57–64 GHz Band," FCC, Tech. Rep. Report and Order FCC 13-112, Aug. 2013.

[7] FCC, "Allocations and Service Rules for the 71–76 GHz, 81–86 GHz, and 92–95 GHz bands," FCC, Tech. Rep. Memorandum Opinion and Order 05-45, Mar. 2005.

[8] IEEE, "IEEE Standard for Information Technology – Telecommunications and information exchange between systems – Local and metropolitan area networks – Specific requirements – Part 11: Wireless LAN Medium Access Control (MAC) and Physical

Layer (PHY) Specifications Amendment 3: Enhancements for Very High Throughput in the 60 GHz Band," *IEEE Std 802.11ad-2012 (Amendment to IEEE Std 802.11-2012, as amended by IEEE Std 802.11ae-2012 and IEEE Std 802.11aa-2012)*, pp. 1–628, 2012.

[9] D. Sobel, "Opportunities and challenges in 60 GHz wideband wireless system design," *BWRC Retreat*, 2004.

[10] H. Xu, V. Kukshya, and T. Rappaport, "Spatial and temporal characteristics of 60-GHz indoor channels," *IEEE J. Select. Areas Commun.*, vol. **20**, no. 3, pp. 620–630, Mar. 2002.

[11] S.-K. Yong, P. Xia, and A. Valdes-Garcia, *60 GHz Technology for Gbps WLAN and WPAN: From Theory to Practice*. Wiley, 2011.

[12] K. Okada, K. Kondou, M. Miyahara, et al., "Full four-channel 6.3-Gb/s 60-GHz CMOS transceiver with low-power analog and digital baseband circuitry," *IEEE J. Solid-State Circuits*, vol. **48**, no. 1, pp. 46–65, Jan. 2013.

[13] C. Marcu, D. Chowdhury, C. Thakkar, et al., "A 90 nm CMOS low-power 60 GHz transceiver with integrated baseband circuitry," *IEEE J. Solid-State Circuits*, vol. **44**, no. 12, pp. 3434–3447, Dec. 2009.

[14] A. Tomkins, R. Aroca, T. Yamamoto, et al., "A zero-IF 60 GHz 65 nm CMOS transceiver with direct BPSK modulation demonstrating up to 6 Gb/s data rates over a 2 m wireless link," *IEEE J. Solid-State Circuits*, vol. **44**, no. 8, pp. 2085–2099, Aug. 2009.

[15] S. Shahramian, Y. Baeyens, N. Kaneda, and Y.-K. Chen, "A 70–100 GHz direct-conversion transmitter and receiver phased array chipset demonstrating 10 Gb/s wireless link," *IEEE J. Solid-State Circuits*, vol. **48**, no. 5, pp. 1113–1125, May 2013.

[16] B. Razavi and R. Behzad, *RF Microelectronics*, 2nd edn. Prentice Hall, New Jersey, 2012.

[17] S. K. Reynolds, B. A. Floyd, U. R. Pfeiffer, et al., "A silicon 60-GHz receiver and transmitter chipset for broadband communications," *IEEE J. Solid-State Circuits*, vol. **41**, no. 12, pp. 2820–2831, Dec. 2006.

[18] A. Natarajan, A. Komijani, X. Guan, A. Babakhani, and A. Hajimiri, "A 77-GHz phased-array transceiver with on-chip antennas in silicon: transmitter and local LO-path phase shifting," *IEEE J. Solid-State Circuits*, vol. **41**, no. 12, pp. 2807–2819, Dec. 2006.

[19] A. Valdes-Garcia, A. Natarajan, D. Liu, et al., "A fully-integrated dual-polarization 16-element W-band phased-array transceiver in SiGe BiCMOS," in *2013 IEEE Radio Frequency Integrated Circuits Symp. (RFIC) Dig.*, June 2013, pp. 375–378.

[20] A. Van Der Wel, S. L. J. Gierkink, R. Frye, V. Boccuzzi, and B. Nauta, "A robust 43-GHz VCO in CMOS for OC-768 SONET applications," *IEEE J. Solid-State Circuits*, vol. **39**, no. 7, pp. 1159–1163, July 2004.

[21] K. Scheir, G. Vandersteen, Y. Rolain, and P. Wambacq, "A 57-to-66 GHz quadrature PLL in 45nm digital CMOS," in *2009 IEEE Int. Solid-State Circuits Conf. Dig. of Tech. Papers (ISSCC)*, Feb. 2009, pp. 494–495.

[22] X. Yi, C. C. Boon, H. Liu, et al., "A 57.9-to-68.3 GHz 24.6 mW frequency synthesizer with in-phase injection-coupled QVCO in 65 nm CMOS," in *2013 IEEE Int. Solid-State Circuits Conf. Dig. of Tech. Papers (ISSCC)*, Feb. 2013, pp. 354–355.

[23] F. Herzel, M. Piz, and E. Grass, "Frequency synthesis for 60 GHz OFDM systems," in *Proc. 10th International OFDM Workshop*, 2005.

[24] A. Georgiadis, "Gain, phase imbalance, and phase noise effects on error vector magnitude," *IEEE Trans. Veh. Technol.*, vol. **53**, no. 2, pp. 443–449, Feb. 2004.

[25] D. M. Pozar, *Microwave Engineering*. Wiley, 2009.

[26] K.-J. Koh and G. Rebeiz, "0.13-μm CMOS phase shifters for X-, Ku-, and K-band phased arrays," *IEEE J. Solid-State Circuits*, vol. **42**, no. 11, pp. 2535–2546, Nov. 2007.

[27] C. W. Byeon, C. H. Yoon, and C. S. Park, "A 67-mW 10.7-Gb/s 60-GHz OOK CMOS transceiver for short-range wireless communications," *IEEE Trans. Microw. Theory Techn.*, vol. **61**, no. 9, pp. 3391–3401, Sep. 2013.

[28] D. G. Kam, D. Liu, A. Natarajan, et al., "LTCC packages with embedded phased-array antennas for 60 GHz communications," *IEEE Microw. Wireless Compon. Lett.*, vol. **21**, no. 3, pp. 142–144, Mar. 2011.

[29] R. Bhat, A. Chakrabarti, and H. Krishnaswamy, "Large-scale power-combining and linearization in watt-class mm wave CMOS power amplifiers," in *2013 IEEE Radio Frequency Integrated Circuits Symp. (RFIC) Dig.*, June 2013, pp. 283–286.

[30] A. Niknejad, D. Chowdhury, and J. Chen, "Design of CMOS power amplifiers," *IEEE Trans. Microw. Theory Techn.*, vol. **60**, no. 6, pp. 1784–1796, June 2012.

[31] E. Torkildson, H. Zhang, and U. Madhow, "Channel modeling for millimeter wave MIMO," in *2010 IEEE Information Theory and Applications Workshop (ITA)*, 2010, pp. 1–8.

[32] E. Torkildson, U. Madhow, and M. Rodwell, "Indoor millimeter wave MIMO: feasibility and performance," *IEEE Trans. Wireless Commun.*, vol. **10**, no. 12, pp. 4150–4160, Dec. 2011.

[33] R. J. Mailloux, *Phased Array Antenna Handbook*. Artech House Boston, 2005.

[34] T. Zwick, Y. Tretiakov, and D. Goren, "On-chip SiGe transmission line measurements and model verification up to 110 GHz," *IEEE Microw. Wireless Compon. Lett.*, vol. **15**, no. 2, pp. 65–67, Feb. 2005.

[35] A. Komijani and A. Hajimiri, "A wideband 77-GHz, 17.5-dBm fully integrated power amplifier in silicon," *IEEE J. Solid-State Circuits*, vol. **41**, no. 8, pp. 1749–1756, Aug. 2006.

[36] A. Valdes-Garcia, S. Nicolson, J. Lai, et al., "A fully integrated 16-element phased-array transmitter in SiGe BiCMOS for 60-GHz communications," *IEEE J. Solid-State Circuits*, vol. **45**, no. 12, pp. 2757–2773, Dec. 2010.

[37] M.-D. Tsai and A. Natarajan, "60 GHz passive and active RF-path phase shifters in silicon," in *2009 IEEE Radio Frequency Integrated Circuits Symp. (RFIC) Dig.*, June 2009, pp. 223–226.

[38] F. Golcuk, T. Kanar, and G. Rebeiz, "A 90–100-GHz 4×4 SiGe BiCMOS polarimetric transmit/receive phased array with simultaneous receive-beams capabilities," *IEEE Trans. Microw. Theory Techn.*, vol. **61**, no. 8, pp. 3099–3114, Aug. 2013.

[39] K. Sengupta and A. Hajimiri, "A 0.28 THz 4×4 power-generation and beam-steering array," in *2012 IEEE Int. Solid-State Circuits Conf. (ISSCC) Dig. of Tech. Papers*, Feb. 2012, pp. 256–258.

11 Self-healing for silicon-based mm-wave power amplifiers

Steven M. Bowers, Kaushik Sengupta, Kaushik Dasgupta, and Ali Hajimiri

11.1 Background

11.1.1 Motivation for self-healing

The rise of digital computation and personal computing has led to continual advances in semiconductor technologies at an exponential pace, following Moore's Law. In each successive processing node, the minimum feature size decreases, improving performance, but also bringing some trade-offs in terms of variation between chips as well as between transistors on the same chip [1–4]. One major source of this variation is random dopant fluctuations (RDFs) in the channel of a transistor [5, 6]. A typical 130-nm complementary metal–oxide–silicon (CMOS) process will have several hundreds of dopant atoms in the channel region. In contrast, in a 32-nm process, only a few tens of dopants control important transistor characteristics like threshold voltage, etc. A second source of variation is line-width control in these advanced processes. Line-edge roughness (LER) caused by lithographic and etching steps directly impacts the overlap capacitances as well as other device parameters like drain-induced barrier lowering (DIBL) and threshold voltage [7]. Figure 11.1 shows how threshold voltage variations scale with process technology node. As can be seen, the variation is much more manageable at larger nodes, and the variation is expected to continue to increase at smaller nodes as the total number of dopant atoms as well as the channel length reduces even further. If the variation can be dealt with, however, the smaller transistors can enable new applications for mm-wave power generation, enabling transmitters and amplifiers at higher frequencies, powers, and efficiencies. Another issue that analog designers face is that, due to the digital processing market being the driving force pushing the scaling, the models provided by the foundries early in the node's development stage are primarily designed for digital use, and are often not reliable at mm-wave frequencies.

In addition to these static sources of variation, dynamic temperature variations across the same die can give rise to varying sub-threshold leakage, supply voltage variations thereby directly affecting overall system performance. Variability in operating environment of power generation systems can adversely affect their performance. This comes in the form of temperature variation, degradation due to aging [8], and, in the case of power amplifiers that are driving antennas, load impedance mismatch caused by voltage standing wave ratio (VSWR) events [9] that occur when objects in the environment interact

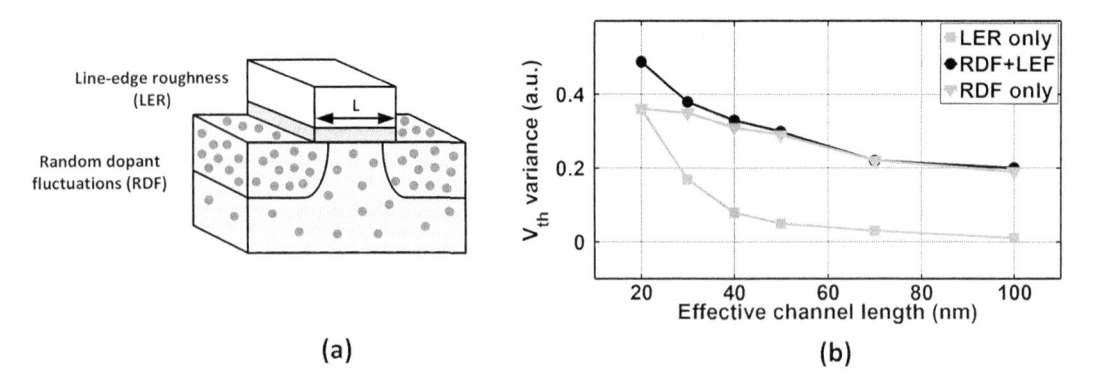

Fig. 11.1 (a) Two main sources of variation in a MOSFET. (b) Threshold voltage variation over technology node [1].

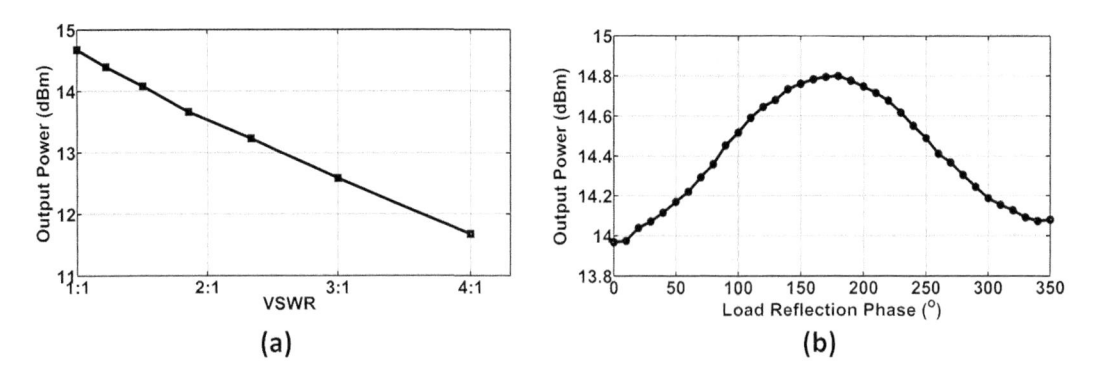

Fig. 11.2 (a) Output power variation with VSWR magnitude for a fixed phase, and (b) variation with phase variation for a fixed VSWR magnitude for one design of a mm-wave PA.

in the near field of the antenna, as can be seen in Fig. 11.2. Dealing with this issue is critical, especially when the amplifiers are in a phased array, as interactions between antennas can allow signals from other elements in the array to couple back through the antenna [10–15]. Power amplifiers are generally tuned to provide the optimal output power at maximum efficiency for a designed load, so when that load changes, the performance drops, and in extreme cases, can damage the chip if care is not taken to ensure breakdown voltages are not exceeded for any expected VSWR events.

This chapter will present self-healing as a method to reduce the adverse affects of process and environmental variation for mm-wave power amplifier (PA) design. Along the way, the design and measurement of a proof of concept 28-GHz self-healing power amplifier from [16] will be used as an example to explain the various self-healing concepts. Section 11.2 gives an introduction to self-healing and other reconfigurable circuit techniques and presents the design goals and architecture of the example PA. Section 11.3 then looks at various sensors that can be used to detect the performance metrics of mm-wave PAs, followed by an examination of some of the ways these circuits can be actuated in Section 11.4. The data converters are covered in Section 11.5 and design of the digital algorithm in Section 11.6. Finally, a case study of system level measurements of the example PA is then presented in section 11.7, with concluding remarks in Section 11.8.

11.2 Introduction to self-healing

There are two different approaches to solve these issues of performance degradation due to variation. The first approach is to design more and more variation-tolerant systems by adopting architectures and circuit topologies that are less sensitive to process and mismatch variations. These techniques have been widely adopted in CMOS digital as well as analog designs over the years. Some of the commonly used strategies include supply voltage optimization [17], optimum device sizing [18], PVT insensitive biasing methods [19], etc. The main limitation of the first approach is that with increasing variability in nanometer CMOS, these techniques become harder to implement. In addition, for high-frequency designs, a lot of them cannot be implemented due to severe degradation in circuit performance.

The second, more-scalable approach is to sense the performance degradation once it occurs and then adjust the system performance by using various knobs. The ability to dynamically sense and actuate critical blocks of the system eliminates the additional design complexity of variation insensitive circuits. Self-healing is a design methodology that takes advantage of the vast digital processing power that is available on modern CMOS processes to reduce the effects of process and environmental variation on the analog circuits in the system. It uses a feedback loop through a digital processing core to heal the chip back to its optimum performance levels when facing performance degradation caused by these variations. A block diagram of the general self-healing concept is shown in Fig. 11.3. The self-healing loop starts with integrated sensors that detect the performance of the mm-wave circuit. These sensors need careful consideration as they are being implemented on the same chip as the mm-wave circuit, and thus are subject to the same variations. They must be designed to be robust to these variations, so that their outputs are a true measure of the mm-wave circuit's performance, and not dependent on

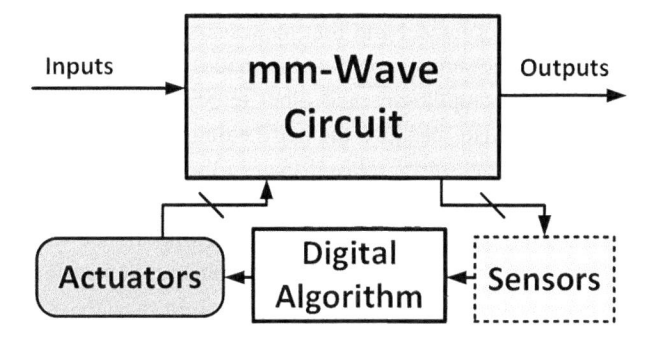

Fig. 11.3 Block diagram of generic self-healing system for mm-wave circuits [16].

the variation within the sensor. The sensor's outputs are converted to digital bits with an analog to digital converter (ADC) that sends that data to an integrated digital core. This core takes that data, runs an optimization on it, and controls digital to analog converters that set actuation points within the mm-wave circuit. The actuation space of the actuators has to be sufficiently large to cover all of the expected variation the chip may experience, while inducing a minimal amount of performance loss due to their presence. This loop can then be iterated until an optimum actuation state is found.

While this methodology is an excellent fit for healing high-performance mm-wave circuits, it has also been used as a resiliency technique to improve the yields of digital design. One such system is found in [20], which presents a system which utilizes several adaptive processing units in addition to regular processing units and periodically turns off the system and performs a functional check and replaces the regular units with their adaptive counterparts as necessary. An example of an analog-based circuit that uses a reconfigurable test scheme is reported in [21], where performance of an LNA is detected by reconfiguring it in a feedback configuration and subsequently healed by bias control as well as passive network switching. A healable mm-wave power amplifier is reported in [22] wherein, based on output power levels, the bias of the core amplifier is dynamically adjusted to provide constant gain, thereby improving the 1-dB compression point. In addition, built-in phase compensation provides ability to correct for constellation nonidealities. An indirect method of sensing phase noise is reported in [23], which utilizes correlation between phase noise and integrated sensor outputs to optimize the system for best phase noise. On-chip PVT compensation has been demonstrated in a 2.4-GHz LNA [24] where the input of the LNA switches between off-chip and an on-chip VCO during measurement and calibration phases respectively. A closed-loop method for healing PLL reference spurs is implemented in [25], where both spur estimation and correction are performed at the VCO control voltage. The work presented in [26] demonstrates integrated phase error detection and self-healing for use in phased-array-based receivers and transmitters.

11.2.1 Self-healing blocks

- **Sensors** The most vital aspect of sensor design for self-healing circuits is robustness. The mm-wave circuit is designed to be cutting edge, to push the envelope of possible performance, and that means that variations can significantly degrade performance. The sensors, on the other hand, can be designed more conservatively, with robust design topologies and other techniques such as using non-minimum-length transistors. Also, because the self-healing loop can be duty cycled, the DC power requirements are more relaxed than in the mm-wave circuit, robust designs that are more power hungry can still be acceptable in some circumstances. In order to design the sensors, the performance metrics of importance must be determined. It is also important to determine if the absolute value of the metric is important, or if it is a metric that must be maximized or minimized. In the case of mm-wave power amplifiers, the output RF power is a metric that often is bounded, but, on the other hand, efficiency is one that should always be maximized. For metrics that are not bounded, the most important aspect of the sensor is that it is monotonic, so that the optimizer does not get stuck on an artificial local maximum or minimum. The exact value reported in this case does not matter as much, as the optimization will always try to maximize or minimize the metric. For bounded metrics, the sensors need to be much more robust, as the optimizer will attempt to set the chip to a specific operating point that is based upon the absolute value that the sensor is reporting.

- **Actuators** To enable the mm-wave circuit to adapt to process and environmental variations, actuators need to be built into the design. Ideally these actuators will not affect the performance of the mm-wave circuit, but in practice there will be some cost associated with the actuator, in the form of performance degradation and/or power and area consumption. This means that the actuators must be placed sparingly and be designed with the expected variation in mind. In the case of power amplifiers, the expected variation from process variation comes in the form of threshold-voltage variation and parasitic-capacitance variation, while the environmental variation can include temperature, aging, and load mismatch. Thus, the aspects of the design that should be controlled by the actuator are the operating points of the amplifying transistors and the matching networks.

  Possible types of actuators can be separated into four broad categories, that cover most of the parameters that a designer has control over when initially designing the amplifier. The first is control of gate-bias voltages. By controlling the DC voltage on the gates of the amplifying transistors, the operating point of the transistor can be controlled. This will affect DC current, transconductance, f_{max} of the transistor, gain, RF saturation power, linearity, parasitic capacitances, and stability. In the case of cascode or stacked amplifiers, a combination of these parameters can be controlled by independently controlling the various gate-bias points. The second type of actuator is tuning of the passive power-combining and matching elements. Tuning of the matching elements can ensure that the input and output impedances seen by the amplifier are optimal, and, in the case of power-combining PAs, can adjust the power combiner in the case of mismatch between the output stages.

  The third type of actuator actuates the supply voltage. This can enable higher efficiency in low-power back-off, while still enabling a high-power mode with a higher supply voltage, but requires an efficient tunable DC–DC converter.

  The final type of actuator involves changing the sizes of the transistors themselves. This can be done by placing many transistors in parallel and switching them in and out of the circuit. This again can help in efficiency in low-power back-off by reducing the size of the amplifying transistors, but the impedance mismatch and the loss associated with the switches can make this type of actuator cost prohibitive. In the example design of the self-healing power amplifier, gate-bias control and matching network tuning are employed as actuators.

- **Data converters** In order to bridge the analog and digital domains, data converters must be implemented on-chip. The sensor data must pass through an analog-to-digital converter (ADC) to be converted to digital signals that can be read by the digital algorithm. These ADCs must be robust, as the digital algorithm only has access to the ADC's digital output bits, so the aggregate variation of the sensors combined with the

ADC determines the accuracy of the performance metric measurement. The analog actuators also need digital-to-analog converters (DACs) to enable the digital core to control analog voltages. To keep overhead down, low-power current-mode DACs are used to control the analog actuators. Some actuators such as the tunable transmission lines, however, directly take digital inputs and thus do not require the use of DACs.

- **Digital algorithm** In order to reach the optimum actuation state, the digital algorithm core needs first to determine the performance of the mm-wave circuit from the sensors, and then decide how to adjust the actuators to improve that performance. There are several components to the digital core that can be separated and implemented independently. There will be components for sensor reading, actuator writing, optimization, and global control.

The global control component will first tell the sensor-reading component to deliver the sensor data by querying the ADCs to return the state of all of the sensors. The global-control component then sends this information to the optimization component. The optimization component calculates the performance from these sensor readings and determines the next state to set the actuators to. This is where the algorithm itself is implemented. The global-control component then delivers this actuation-state information to the actuator-writing component, which sends out the bits to the DACs (for the analog actuators) or directly to the digitally controlled actuators. By grouping the digital core into these separate module components, more of the code can be reused between different designs, and changes can be made to update the optimization component's algorithm without recoding the other supporting components.

11.2.2 Block level vs global healing

It is important also to consider block level vs global healing. While the scope of this chapter deals with block-level healing of just the power-amplifier block, global healing of the entire transmitter can be useful as well. For example, if the transmitter is operating in a low-power back-off mode, it may be advantageous to adjust the gains of the preamplifiers and mixers, not just the bias points of the power amplifier to lower the overall DC power consumption and improve efficiency, or to trade off between linearity and power consumption. The entire healing digital core for the transmitter can be contained in a single centralized core, or it can be broken up into smaller block-level cores that are controlled by a transmitter-level global healing core. By incorporating self-healing both at the block level and at transmitter level, maximum performance can be achieved.

11.2.3 Design considerations and architecture for example self-healing power amplifier

The example PA will be a fully integrated self-healing PA at 28 GHz implemented in a standard 45 nm SOI CMOS process (Fig. 11.4) [16]. It is a two-stage, 2-to-1 power-combining class-AB PA matched to 50 Ω at the input and output. The interstage matching network and output power-combining matching network are designed to provide the optimum impedance for maximum saturated output power. The first stage is

11.2 Introduction to self-healing

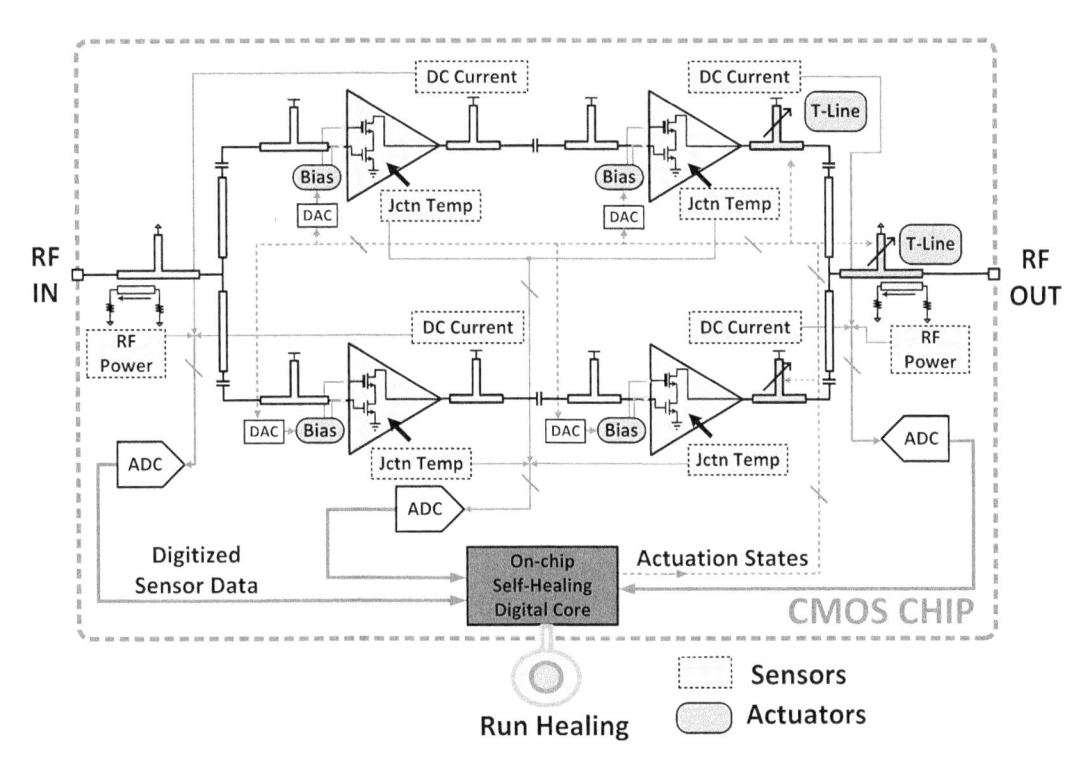

Fig. 11.4 Block-level architecture of the example integrated self-healing PA. Data from three types of sensors are fed through ADCs to an integrated digital core. During self-healing, the digital core closes the self-healing loop by setting two different types of actuators to improve the performance of the power amplifier [16].

half the transistor size of the output stage, to ensure that the output stage can be fully driven into saturation. Class-AB design was chosen in order to enable linear operation and to allow for non-constant envelope modulation schemes to be implemented.

To increase the gain of each amplifying stage, each stage is a cascode amplifier, and two stages are used to further increase the gain. The common source transistors are 56-nm analog transistors, while the cascode transistor is a 112-nm-thick gate-oxide transistor to increase the voltage breakdown of the amplifier. A schematic of one of the output amplifying stages is shown in Fig. 11.5, displaying the connections through the matching networks, as well as several of the sensors and actuators that will be discussed in the following sections. The three matching networks use a two-stub matching technique, and biasing is done through the AC short circuits at the end of the stubs. A metal AC coupling capacitor is used to allow for independent biasing of the inputs and outputs of the amplifying stages. Full 3D (three-dimensional) electromagnetic simulations of the matching networks including the capacitors and pads were performed to ensure proper functionality.

There are three main metrics of interest that will define the performance of the example mm-wave power amplifier: output power, power gain, and DC power. This

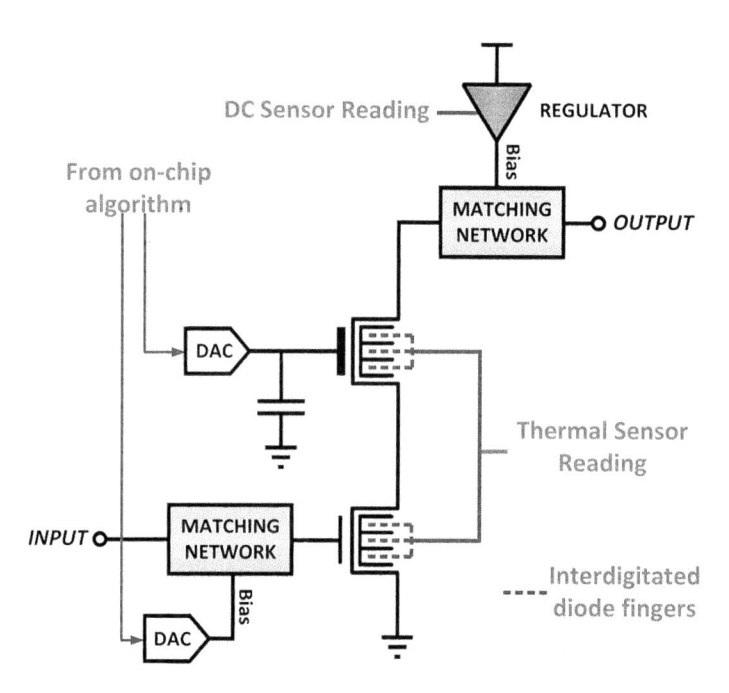

Fig. 11.5 Schematic of a single cascode amplifying stage showing connections to matching networks, gate-bias actuators, DC sensor, and temperature sensor [16].

means that, to be able to calculate the performance of the PA, the self-healing sensors need to be able to detect input and output RF power, and DC power. RF power sensors are thus placed at the input and output ports of the amplifier, and DC power is sensed both electronically through DC sensors and thermally through temperature sensors. On the actuator side, gate-bias actuators are implemented on all amplifying stages, and the stubs of the output power-combining matching network are tunable to enable tuning of the output network. These sensors and actuators will be discussed in more detail in subsequent sections.

11.3 Sensing: detecting critical performance metrics

As discussed in the previous section, on-chip closed-loop autonomous healing requires regular monitoring of the relevant system parameters, and applying self-correcting measures until the desired performance level is achieved [25–27]. In the case of a power amplifier, the relevant parameters may include input and output power (and therefore gain), DC power consumption, junction temperature of the core power devices, and linearity measurements such as P_{1dB}, output inter-modulation products, spectral regrowth, amplitude, and phase distortions or output signal constellation.[1] The sensor system

---

[1] Low-power on-chip linearity measurements are very challenging and it is an active field of research. A proof-of-concept self-healing embedded system was demonstrated in [28] for transmitter image and OIM3 healing for 60-GHz radio.

needs to measure these parameters on-chip "reliably" and send the collected data to the central healing unit for performance optimization. Evidently, the on-chip sensors will be affected by the same process variations, mismatches, and other factors which degrade the performance system under consideration which needs to be healed. Therefore, the design of the sensors needs to ensure that their performance is less affected by the same variations than the system block or sub-block that is to be healed. In this section, we will discuss the overview of sensor design, design considerations, and tradeoffs, for some design examples of low-overhead sensor implementation for a mm-wave PA amplifier. Specifically, we will discuss sensor design related to measurement of true RF and mm-wave power (which takes into consideration load mismatches) and low-overhead DC current sensors. The applications of such sensors are very broad in a transceiver design, and could be applicable to various other circuit blocks such as LNA, mixers, oscillators, etc.

11.3.1 Overview of sensor design

There are several design considerations for reliable on-chip sensor design, which often directly trade off with each other. The important characteristics are listed below.

- **Responsivity** Responsivity of a sensor is defined by the change in the sensor response corresponding to unit change in the sensed parameter. Evidently, a standalone measurement of the responsivity of the sensor system without the noise contribution does not correlate with the achievable sensor sensitivity. Higher responsivity could be obtained by adding successive gain stages, but this also amplifies the noise contributed by those stages. It is, therefore, important to optimize the physical mechanism that is exploited for sensing the desired parameter. In the case of RF power and device temperature measurement (as described later in this section), it could be the nonlinearity of the transistor transconductance, which is a function of its operating condition or the placement and layout of thermal diodes relative to the PA core whose rise in temperature needs to be sensed.
- **Noise and sensitivity** The sensitivity of the sensor is limited by the output noise spectral density due to the active and passive elements constituting the sensor circuit and the quantization noise of the following digitizer. The power spectral density of the noise at the sensor output can necessitate techniques to avoid being swamped by the $1/f$ noise such as chopping, correlated double sampling, etc. In the power amplifier example described in this chapter, the sensor outputs such as the input and output RF power, DC power, and device temperature are converted to a DC output. The sensor resolution is primarily limited by the resolution of the subsequent ADCs, since the noise is low-pass filtered at the sensor output.
- **Dynamic range** Dynamic range often trades off directly with responsivity and this may necessitate a nonlinear responsivity profile to cover a large dynamic range. A logarithmic amplifier was implemented in [29] to facilitate gain detection in decibels. In the presented case of the power amplifier, the entire range of the RF input power is expected to lie within 0–10 mW, while the output power varies within 0–100 mW,

consuming 0–40 mA and 0–80 mA per input and output stage respectively. All the sensors are designed to be linear within this operating range.

- **Monotonicity, linearity, and offset** The linearity requirements of the sensors depend on the self-healing algorithm, the performance metrics which are needed to be optimized and if they are pre-calibrated. Assuming that the sensors are not calibrated and if we need to actuate the PA into its maximum output power state, then monotonicity of the output RF sensor can guarantee successful convergence. However, if the on-chip algorithm needs to evaluate the output power corresponding to the maximum PAE, then the corresponding sensors need to be linear. The linear response can also help in a two-point on-chip calibration method that can measure the offset and responsivity. The offset is measured and canceled at the start of the algorithm for the amplifier presented.
- **Response time** The settling time of the sensor determines the speed of the healing process, since the sensors need to be measured every time an actuation signal is commanded from the central healing core. However, faster sensor response also means larger bandwidth at the sensor output, leading to higher output noise power. Thus, the response time can trade off directly with sensitivity. In this design process, the settling time is designed to be less than 1 μs so that even the exhaustive search healing algorithm can run in less than 1 s.
- **Power and performance overhead** The power-area overhead for the sensor needs to be carefully optimized so that the performance of the system is not adversely affected by the addition of the sensors. It could be prudent to integrate the sensors with the other existing ancillary circuits. In the case of the PA for example, the DC current sensor is integrated with the on-chip regulator. The digital core is designed to fit within the area between two PA paths, which would have been vacant otherwise. The thermal diodes are closely packed with the PA core, so as to reduce any stray parasitic effects, and the RF sensors are kept short. The total sensor overhead for the power amplifier is less than 6%.

It is to be noted that all the relevant parameters need not be measured directly. In the example shown later, the DC power dissipation in the PA is measured in the non-electrical domain by the resultant rise in temperature of the device core [30].

11.3.2 Measurement of RF power

Input and output RF power sensors are necessary to estimate the gain and power delivered to the load. Ordinarily voltage sensors placed at the output port can sense the true RF power delivered as long as the output load remains constant. However, due to environmental changes or load-impedance mismatch events, the output load is rarely ever 50 Ω. In this section, we present RF sensors which detect both the forward and the reflected power to estimate RF parameters in face of VSWR events.

At mm-wave frequencies, coupled transmission lines can be used to sense power. It is to be remembered that in a self-healing setting, the coupling strength is desired to be

11.3 Sensing: detecting critical performance metrics

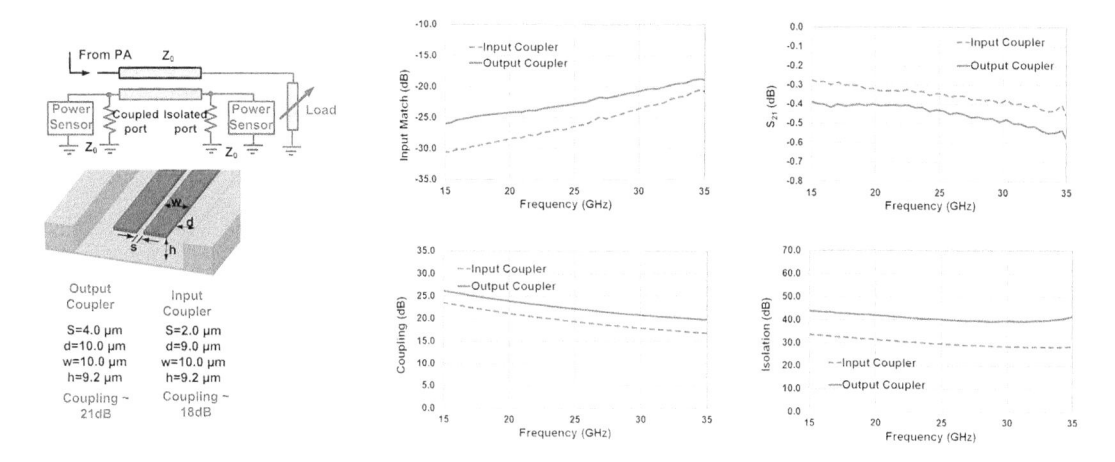

Fig. 11.6 True RF power measurement through coupled and isolated port power detection. The figure also shows the measured S-parameter of the input and output couplers with the integrated power sensors. The coupling coefficients of the input and output couplers are designed to be near 18 dB and 21 dB respectively, while the insertion loss is 0.3–0.5 dB at 28 GHz [16].

small, to avoid significantly perturbing the power flow through the main circuit. Therefore, unlike traditional couplers, the sensors do not need $\lambda/4$-long transmission lines, as explained in detail below. At lower RF frequencies, transformer-based coupling may be the more efficient method of coupling a fraction of the power to be sensed. In either case, the sensor designs must be made both area efficient and power efficient. In the following design example for a PA operating near 28 GHz, two coupled transmission lines are implemented for the input and output port of the PA as shown in Fig. 11.6. The couplers are kept compact and short at 220 μm for low insertion loss of 0.5 dB at 28 GHz. This also minimizes the coupling coefficients, which are designed to be at 18 dB and 21 dB at the input and output respectively so that only a small fraction of the input and output power is sensed by the couplers for estimation of gain and the power delivered. The transmission lines are implemented with 1.2-μm-thick copper line in a metal ground tub as shown in Fig. 11.6. The figure also shows the measured S-parameters of the input and output coupler. The input matching is better than 18 dB in the range 25–45 GHz and the isolation is more than 28 dB. The coupling coefficients vary by less than 3 dB in the range 25–35 GHz.

Once the true RF power has been converted into voltage swings at the matched coupled and isolated ports, it is measured by the power RMS voltage detector circuit as shown in Fig. 11.7. Integrated RMS voltage detectors have been demonstrated in bipolar devices based on trans-linear principles [31] and in CMOS [32, 33] at RF frequencies. The sensor is biased at cut-off and it relies on the nonlinearities of the transistor to generate a DC current proportional to the input power which is given by [34]

$$i_d(\theta) = \frac{i_{max}}{1 - \cos(\alpha/2)}(\cos\theta - \cos(\alpha/2)) \tag{11.1}$$

$$i_{dc} = \frac{1}{2\pi}\int_{-\alpha}^{\alpha} i_d(\theta)d\theta = \frac{i_{max}}{\pi}, \tag{11.2}$$

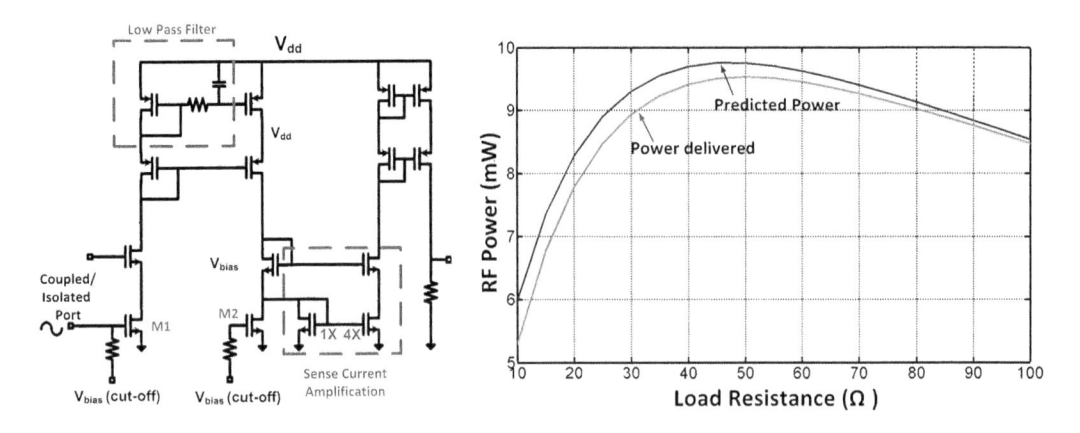

Fig. 11.7 RF power sensor schematic showing rectification, low-pass filtering, and detection current amplification. The figure also shows the predicted and simulated true RF power delivered to the load as the impedance is varied in the range 10–100 Ω [16].

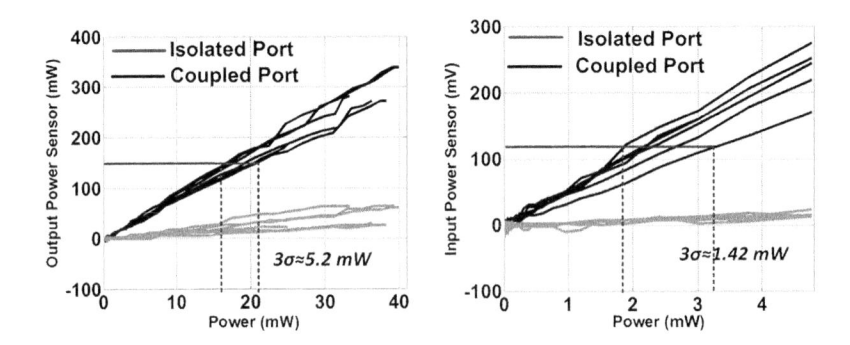

Fig. 11.8 Measured RF power detector response of the output and input power sensors at 28 GHz over six chips (coupled and isolated ports) [16].

where i_{max} is the peak current and α is the conduction angle. The rectified signal is then low-pass filtered and amplified in the current domain, and measured across a resistor as shown in Fig. 11.7. A matching transistor M2 subtracts the standby current of M1, increasing the dynamic range. Figure 11.7 also demonstrates the ability to evaluate the true RF power delivered to the load in case of a load-mismatch event. The predicted power is given by $P_{del} = \frac{P_c - P_I}{k}$, where P_c and P_I are the powers delivered at the coupled and isolated ports respectively, k is the coupling coefficient, and P_{del} is the estimated power delivered to the load (through-port). The difference between the two plots is the effect of neglecting the insertion loss. Figure 11.8 demonstrates the output and input power detectors where couplers achieve responsivity of 8.3 mV/mW and 54 mV/mW respectively and consume 1.2 mW of DC power each. The 3σ spreads of the true RF power for a measured sensor output are approximately 1 dB and 2 dB for output and input sensors respectively over six chips. The variation of the power sensor output across chips primarily comes from the active circuitry, which exploits the nonlinearity of the sensor transistor to convert an input RF signal into a DC signal. For many self-healing algorithms such as minimization of output power, the monotonicity of the sensor performance is enough to guarantee convergence. In algorithms where an absolute power requirement needs to be met, a conservative estimate based on measurement of multiple such sensors can be used. It is to be remembered that since the coupler and the accompanying sensor circuits are broadband (unlike the PA), their performances are less sensitive to process variations and mismatches than is the tuned circuitry whose on-chip performance is being measured.

11.3.3 Measurement of DC power consumption and efficiency

It is challenging to measure the DC power drawn by a high-power, high-frequency PA during operation. One possible method is to sense the DC current through a small resistor in series with the amplifier [35]. However, due to large currents in the PA, this can lead to excess power loss and parasitics. The method adopted in this example centers around mirroring the current accurately from the PA through another path and sensing it across a load. However, in order to mirror the current accurately without sacrificing efficiency, overhead transistors are kept in deep triode under voltages from 10 mV to 30 mV as shown in Fig. 11.9 under a 2.12 V supply. The current though a PA is scaled down by a factor of 100 and mirrored through matched transistors M1 and M2 as shown in Fig. 11.9. An accurate mirroring of the current will, therefore, require both the source nodes of M1 and M2 to be held at the constant potential which can be set externally by V_{ref}, as shown in Fig. 11.9. This allows us to accomplish supply regulation and current-sensing at the same time. As is obvious, the op-amp A1 forces the source nodes of M1 and M2 to be at the same voltage for accurate mirroring of the current, while the sensor current through M2 is converted into the sensor DC voltage as shown in Fig. 11.9 [36]. The effective voltage regulation is shown in Fig. 11.10, which illustrates

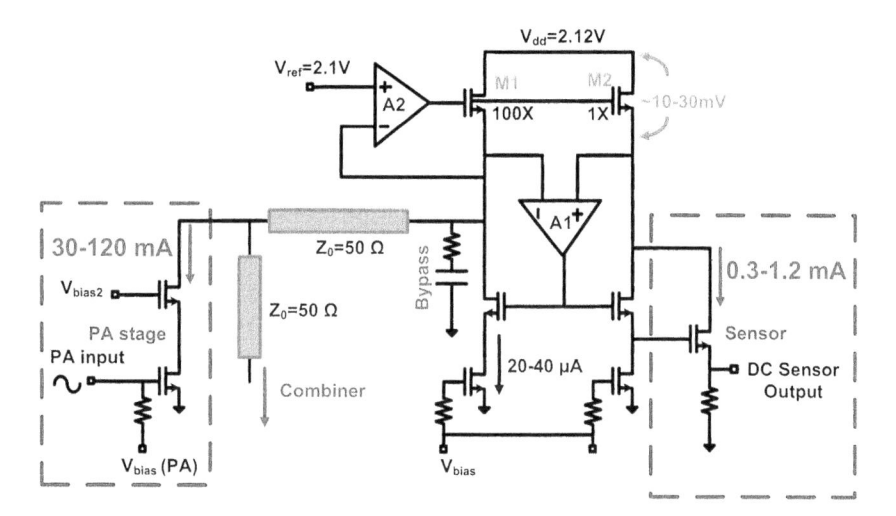

Fig. 11.9 Schematic of the DC sensor of the PA. The current through the PA is mirrored by a factor of 100 and sensed, also accomplishing voltage regulation with a headroom of only 10–30 mV [16].

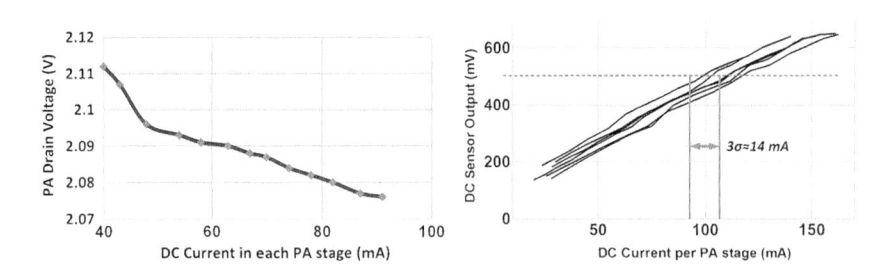

Fig. 11.10 The figure shows that the measured headroom for the sensor is maintained below 50 mV over the 2.12 V supply as the PA current is varied in the range 0–90 mA. The DC sensors, therefore, are maintained in extreme linear region, thereby not affecting the PA efficiency. The figure also shows the measured DC sensor output with PA output current, showing the spread over five chips [16].

that the headroom is held below 50 mV for the entire range of current actuation within 0–90 mA, thereby not affecting the PA efficiency. Figure 11.10 also shows a measured responsivity of 4.2 mV/mA current drawn based on five chips. The 3σ spread of the PA current for a measured sense voltage is less than 14% over all the chips. The total power consumption of the sensor is less than 1.7 mW.

11.3.4 Local temperature sensing to measure PA efficiency

In order to monitor the PAE (PA efficiency) on-chip, the DC power drawn can either be measured directly by sensing the current as described previously or it could be measured indirectly in a non-electrical domain. The concept relies on the key idea that, due to power dissipation, the core device and an area around it will experience a rise in temperature proportional to the power dissipated during the PA operation.

In order to predict accurately the temperature change during PA operation, thermal simulations were carried out using Ansoft ePhysics 2.0. The silicon substrate is an excellent thermal pathway; however, metal connections and vias also affect the temperature change and therefore need to be incorporated in the simulation [37]. Simulations and theoretical prediction indicated that the temperature profile falls sharply beyond a radius of 20–30 μm of the PA transistors during operation, with the core rising by 10 °C [30]. Hence, p-n junction sensor diodes were laid within the input and output PA transistors (thermally active region), both at the common source and the cascade transistor as shown in Fig. 11.11. In order to increase the responsivity, the ×1 diode is placed outside the local thermally sensitive region (∼ 40 μm away) as shown in Figure 11.11, leading to a factor of 10 increase in responsivity. It can be proved that if diodes of sizes ×1 and ×N are kept at two different temperatures T and $T + \Delta T$, and if the same currents are forced through them, the difference between their V_{be} is given by

$$\Delta V_{be} = V_T(m+4)\ln\left(1 + \frac{\Delta T}{T}\right) + V_T \ln N + \left(\frac{E_g}{q} - V_{be,n}\right)\frac{\Delta T}{T + \Delta T}, \quad (11.3)$$

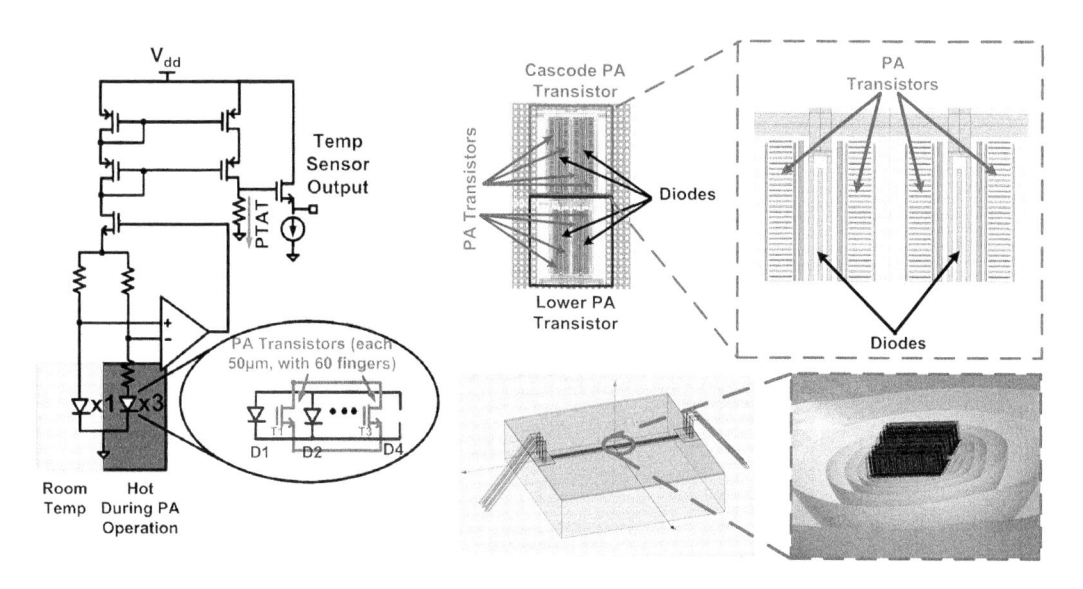

Fig. 11.11 Schematic of temperature sensor showing "hot" sensor diodes interspersed within PA transistor and the simulated thermal profile for 80 mW power dissipation in the process [16].

where $V_T \approx 26$ mV, $V_{be,n} \approx 700$ mV, $E_g \approx 1.12$ eV (bandgap of silicon), and $m \approx -\frac{3}{2}$ [38]. A regular temperature sensor, where both the diodes experience the same global temperature change, takes advantage of the term $V_T \ln N$ to establish $\frac{\partial V_{be}}{\partial T} \approx \frac{k}{q} \approx 8.67 \times 10^{-5}$ V/K. By placing the ×1 diode in a constant-temperature region, the designed local temperature sensor takes advantage of the last term $\left(\frac{E_g}{q} - V_{be,n}\right) \frac{\Delta T}{T+\Delta T}$ to establish $\frac{\partial V_{be}}{\partial T} \approx \frac{\frac{E_g}{q} - V_{be,n}}{T} \approx 1.33 \times 10^{-3}$ V/K. Since the range of temperature change is limited, we sacrifice linearity over a large range with sensitivity.

Figure 11.12 shows the measured responsivity of 2 mV/mW of dissipated power under a 2.12 V supply. The thermal and the DC sensor show similar monotonic variation with DC power drawn by the PA. The response of the thermal sensor is, however, limited by the thermal time constant. The measured results are in close agreement with predicted results from thermal and circuit simulations.

The sensor performances are summarized in Table 11.1.

11.4 Actuation: countering performance degradation

In order for the self-healing system to be able to improve the performance of the mm-wave circuit, actuators that are controlled by the digital algorithm are required. It is important for the actuation space cover the expected space of variation, so that the optimum point for any expected variation is still within the actuation space.

The expected process variations that are considered include chip-to-chip process variation, and mismatch between devices on the same chip. The main sources of this

Table 11.1 On-chip sensors implemented for the self-healing PA.

Sensors	Measured entities	Responsivity	Range	Sensor 1-bit resolution
True RF power	In. power	54 mV/mW	0–10 mW	55 μW
	Op. power	8.3 mV/mW	0–100 mW	300 μW
DC sensor	DC drawn by			
	Ip. stage	8.5 mV/mA	0–60 mA	280 μA
	Op. stage	4.2 mV/mA	0–120 mA	560 μA
Thermal sensor	Power dissipated			
	Ip. stage	4.0 mV/mA	0–130 mW	0.75 mW
	Op. stage	2.0 mV/mA	0–260 mW	1.5 mW

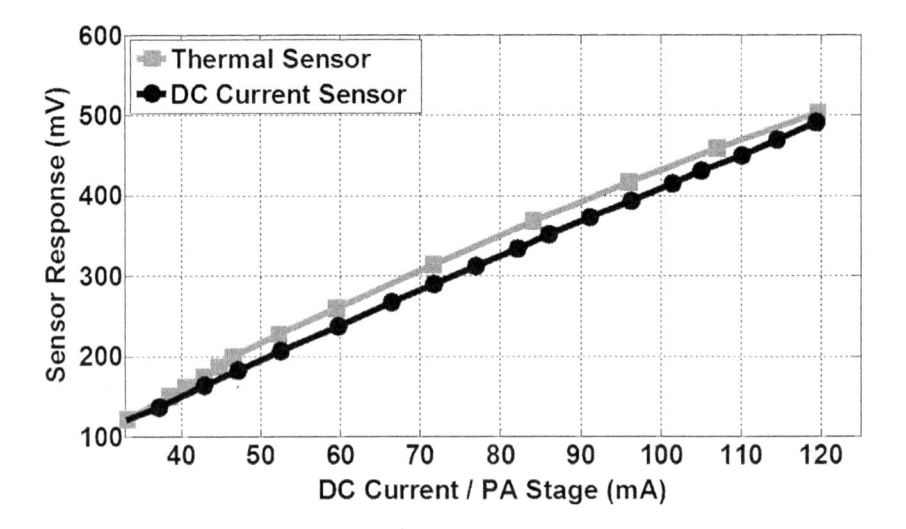

Fig. 11.12 Measured temperature sensor output and DC sensor output with increasing DC power dissipation in PA [16].

variation are variation in dopant levels and variation in gate thicknesses that affect threshold voltage and parasitic capacitance. Within the context of power amplifier design, these types of variations will change the operating points of the transistors, can change the maximum gain and f_{max} of the transistors, and will affect the desired input and output impedances for maximum power generation. The environmental variations that will be considered include load-impedance mismatch, temperature variation, and transistor degradation, and failure due to effects such as aging. Load-impedance mismatch will affect the output matching network, and will detune the system, lowering the output RF power supplied by the output amplifying transistors as well as possibly increasing the loss due to the output power combiner matching network. Temperature variation will mainly show up in transistor performance, and will affect the matching, bias, and saturated power of the amplifying transistors. To a lesser extent, it will affect

11.4 Actuation: countering performance degradation

the passives, and will change the matching networks, but this effect is much less dominant. Transistor degradation due to aging will reduce the gain and saturated output power of the transistors, and will affect the output matching, to the point where complete failure of some of the output transistors can severely degrade the matching network's ability to provide the optimum load impedance to the other functional transistors. There will thus be degradation to the output signal due both to less power being generated by the output transistors and to degradation due to mismatch in the output power-combining matching network.

How then do we select appropriate actuators to cover the effects of all of these variations? A good actuator should cover a large actuation space while having a minimal impact on the performance of the amplifier in the nominal case. Every actuator will cause some degradation, whether it be decreased RF power, increased DC power, or even occupying area on the chip, but keeping this degradation to a minimum will make the system have a net positive effect on the performance once the healing loop is turned on.

There are several possible actuators for mm-wave PAs that generally fall into four categories: gate-bias voltage actuators, passive matching network tuning actuators, supply voltage actuators, and transistor architecture actuators.

11.4.1 Gate-bias actuators

Control of the gate voltages of the transistors is one of the most-effective, lowest-cost actuators available to the designer. By adjusting the gate of the common-source transistors and any stacked or cascode transistors, the operating points of the transistors can be changed. These actuators can be very low cost, due to the high DC gate resistance of CMOS transistors, so adjusting the voltage of the gate does not inherently draw any extra DC power. In general, for class-AB PA operation, the transistors should be biased such that the f_{max} of the transistor is maximized. This biasing point is heavily dependent on the threshold voltage of the transistors and can vary greatly over process variation. This can enable maximum performance at class-AB saturation situations, but control of the gate bias also enables a lowering of the DC power consumption of the power amplifier in power back-off scenarios. There is a trade-off on gain vs DC power in back-off.

One way to reduce the output power and achieve back-off power levels is by using a variable gain amplifier before the PA and lowering the input power of the PA, but, because the PA is the dominant source of DC power consumption, this leads to low efficiencies in back-off because the PA is drawing relatively the same amount of DC power as in saturation, but is producing much less RF power. The other way to achieve back-off power levels is to trade some of the gain in the power amplifier for lower DC power consumption. By lowering the gate bias of the common-source transistors, the PA will saturate at lower power levels, while simultaneously reducing the DC power consumption, significantly raising the efficiency of the amplifier in back-off conditions. These actuators require analog inputs and thus will need DACs that will be discussed in the next section to convert the digital signal sent by the algorithm to the analog voltage for the actuator. Gate-bias actuators are employed on the example PA for the gates of the common-source and cascode transistors of both paths of the input and output amplifying stages. Looking back to Fig. 11.5, the gate-bias actuators for the common-source transistor are input through a stub in the matching network, while the gate-bias control for the cascode stage is biased at the AC short circuit created by the AC coupling capacitor.

11.4.2 Passive matching-network tuning actuators

The second type of actuators are passive matching-network tuning actuators. The goal of these actuators is to be able to dynamically retune the matching networks on the chip after fabrication to account for the variations in parasitic capacitances and optimal output loads of the amplifying transistors. Actuators can be added to any type of passive element, and, as an example, two types of actuators to create tunable transmission lines for transmission line matching networks will be considered in more depth. For the output power-combining matching network, keeping the loss to a minimum is critical, because the output network deals with the highest power levels, and, since the loss will show up in the form of attenuation in dBs, this means that loss in the output network will result in the largest actual loss in terms of watts. Also, for maximum performance, the amplifying transistors will be operating near their voltage breakdown and the matching network will transform the impedance to 50 Ω from a much smaller impedance, which means that the voltage swing within the output network and at the output will be much larger than the transistor breakdown, and any actuator transistors must not be placed such that they break down.

Using transistors as switches on various parts of the matching network also can lead to trade-offs in transistor size. To have low-loss switches in the on state, the size of the transistor is increased, but this increases capacitance in the off state. This can initially lead to a compromised design that both loads the circuit with capacitance in the off state, and causes loss due to resistance in the on state.

One way to overcome both of these challenges is to tune just the transmission-line stubs of the matching network that are terminated with short circuits. The effective length of the transmission line can be changed by shorting out the stub at various points near the end of the line, as shown in Fig. 11.13a.

Large actuation ranges can be achieved simply by placing actuation switches toward the end of the line, which, due to their proximity to the short circuit at the end of the line, keeps the voltage swing low when the switches are off. On top of that, because the switches are being placed at regular intervals within the transmission line, the off capacitance can be considered distributed, and can be absorbed into the transmission-line model and included in the match. This means that much larger transistors can be used and low on resistances can be achieved. These stubs can be considered as shunt inductors at their connection to the rest of the output network, and measurements of one of these tunable transmission-line stubs at 28 GHz shown in Fig. 11.13b show a tuning range from 25 pH to 71 pH with just eight switches, showing the large actuation space that these actuators can achieve. Because these actuators are already

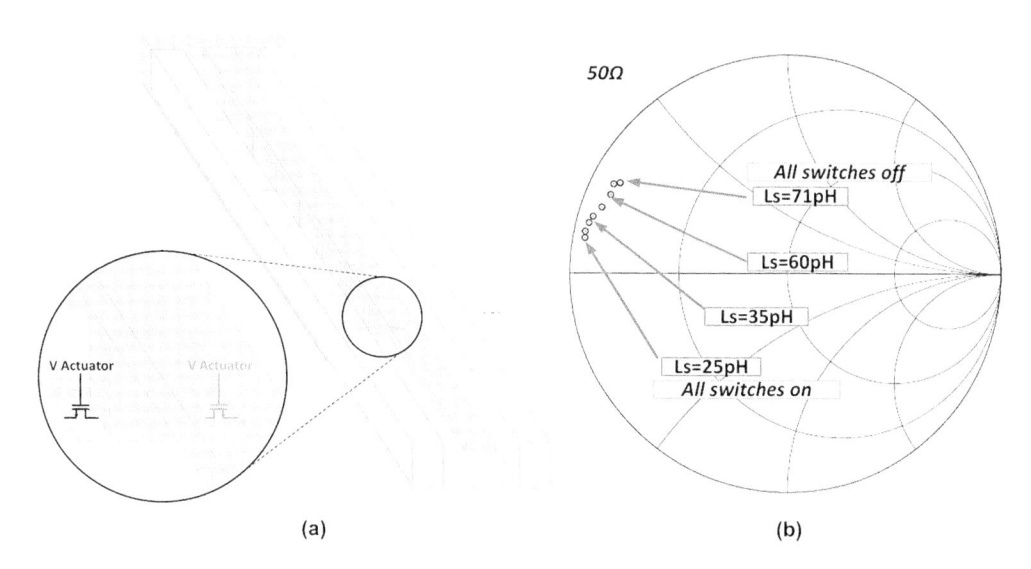

11.4 Actuation: countering performance degradation

Fig. 11.13 (a) Tunable transmission-line stub has switches placed at various points along the line to short out the signal line to the ground, changing the effective length of the stub; and (b) measured effective inductance seen looking into the stub for the different actuation states at 28 GHz [16].

digital switches, they can be set directly from the algorithm block without the need for any DACs.

The example self-healing PA utilizes these tunable transmission-line stubs on all three of the output power-combining matching-network stubs. They were not used on the input matching network or the interstage matching networks because the expected variation in impedances of the loads for those networks (gates of the first and second amplifying stages) was small enough that the advantage of the matching networks did not warrant the additional complexity and loss the actuators come with. The output power-combining matching network is shown in Fig. 11.14, with the nominal impedance mapping from the 50 Ω load to the optimal impedance of the output stage with typical transistors, which for this example is $(7 + 7j)$ Ω at 28 GHz. One additional benefit of the matching network being tunable is that the two output amplifiers did not need to be isolated. If there is mismatch between them, it can be minimized by tuning the output matching network.

Using this network, the overall actuation space impedance looking into an input port of the combiner with a nominal load of 50 Ω is shown in Fig. 11.15a, which is larger than the simulated variation of optimal loads that the transistors will experience. Looking from the other way, the actuation space of load impedances that can still be transformed back to the typical optimum impedance of $(7 + 7j)$ Ω at 28 GHz is shown in Fig. 11.15b. This means that most loads within the 4-1 VSWR circle can be transformed back to $(7 + 7j)$ Ω, and the ones that cannot, can be transformed back to something much closer to the optimal load and the performance can still be significantly increased compared with a static matching network.

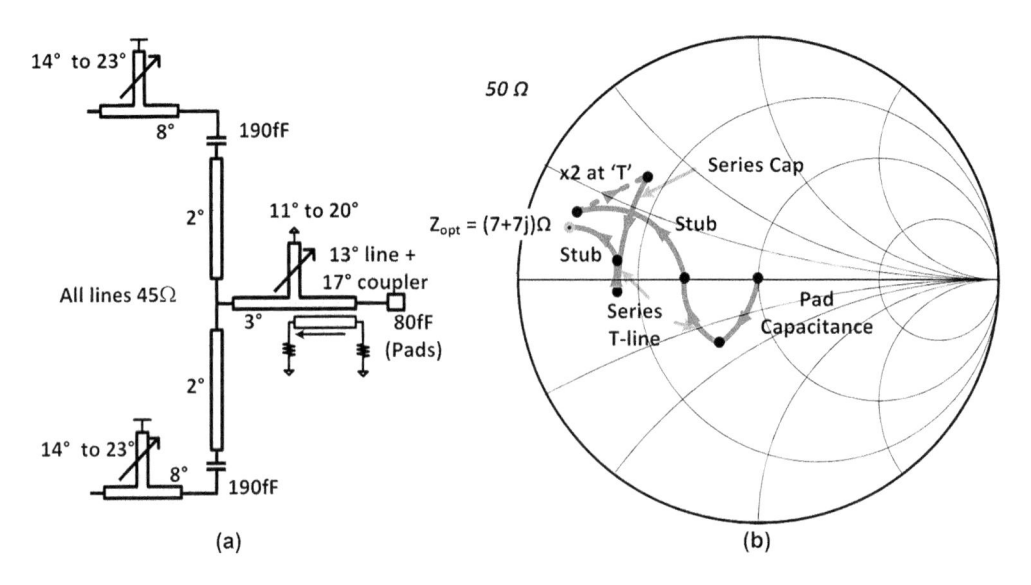

Fig. 11.14 (a) Schematic of the output power-combining matching network showing the three tunable stubs and the output RF power sensor coupler; and (b) mapping of the 50 Ω load through each element of the output power-combining matching network to the nominal optimum impedance at the output amplifying stage of $(7 + 7j)$ Ω at 28 GHz [16].

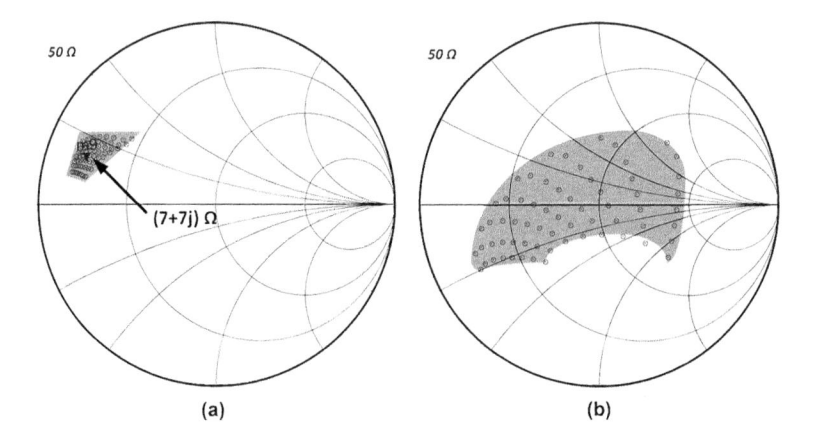

Fig. 11.15 (a) Simulated actuation space of the output power-combining matching network from a 50 Ω nominal load; and (b) actuation space of loads that can be matched to the nominal optimal impedance of $(7 + 7j)$ Ω at 28 GHz [16].

11.4.3 Supply voltage actuators

Another way to reduce the DC power consumption of the PA is to reduce the supply voltage to the amplifiers. Simply lowering the voltage on-chip by dumping the power into a transistor or resistor does not actually decrease the DC power, so efficient tunable DC–DC conversion is required for this type of actuator to be effective. Since these types of converters fall beyond the scope of this book, this type of actuator is not considered further, but should be kept in the designer's mind, especially if such a block is already being implemented on-chip for other reasons and can be taken advantage of by the self-healing system.

11.4.4 Transistor architecture actuators

Changing the size or architecture of the transistors themselves is also a possibility for actuators for mm-wave PAs. One can imagine that an easy way to deal with saturated output RF power degradation is simply to increase the size of the amplifying transistors. Switching fingers of transistors in and out of the circuit, however, can cause significant losses in performance from two dominant sources. The first is that the switches used to switch the transistors in or out will either have higher on resistance or off capacitance, and both will cause signal degradation. Second, even if perfect switches were available, changing the size of the transistors will significantly change the desired matching network, and put much more demand on the tunable matching networks, resulting either in more loss in the matching networks or in mismatch for some transistor sizes. For the example PA, it was determined that the best performance could be achieved by making the transistors large enough to cover the desired saturated output power and using the gate-bias actuators to achieve higher efficiencies while in back-off, so transistor architecture actuators were not used.

11.5 Data converters: interfacing with the digital core

An integral part of any self-healing system is data conversion. Analog-to-digital converters (ADCs) are required for converting analog sensor output to digital form for use with the digital healing algorithm. In addition, some of the actuators, for example bias actuators, can be controlled directly by the digital ASIC through digital-to-analog converters (DACs).

Because self-healing is a low-overhead technique for improving yield, the requirements on these data converters become somewhat stringent.

11.5.1 Analog-to-digital converters

As with all other self-healing enabling blocks, the on-chip ADCs need to be extremely low-power as well as area-efficient designs. However, lower power generally implies lower speeds, which will lead to a slower computation time for the digital healing algorithm, which in turn affects the total healing time. In addition to the overhead vs speed trade-off, the resolution of the ADC also directly affects the accuracy of the healing algorithm. This resolution is chosen based on the sensor responsivity as well as the sensitivity of the system to the different actuators. For example, an RF power sensor responsivity of 10 mV/mW on a 5-bit ADC implies a 1-bit resolution of around 2 mW, which may be greater than the sensitivity of the system to one particular actuator. Depending on the nature of the on-chip algorithm, this may lead to convergence issues, thereby rendering the self-healing loop unusable.

The fastest ADCs are flash ADCs, which operate in a parallel fashion leading to simultaneous generation of output bits [39, 40]. However, because of this parallel computation, these converters usually are among the most power-hungry ADCs. In addition, matching requirements for comparing elements become extremely stringent for higher resolutions. Both power as well as design complexity thus limit the use of flash ADCs in self-healing systems. A good compromise between resolution, design complexity, and speed is a pipelined ADC [41]. However, these ADCs usually occupy larger area as well as consume significant amounts of power. The added complexity in timing synchronization of such ADCs also makes them less suitable for low-overhead self-healing applications. The successive approximation register (SAR) based ADCs are serial ADCs, which offer several desirable features such as low power consumption, high resolution, reduced design complexity as well as low area overhead [42]. In addition to being suitable for medium-data-rate applications due to the inherent serial nature of the data output, such an ADC is an ideal candidate for self-healing systems where routing complexity for multiple high-speed digital signals across the chip also needs to be minimized. An 8-bit SAR ADC was chosen for the current self-healing system. Data from various sensors (four DC sensors, four thermal sensors, and two RF sensors) were multiplexed and fed to three ADCs placed throughout the chip. The sensor choice is governed by the digital ASIC. The block diagram of the implemented SAR ADC is shown in Fig. 11.16a. Global clock and the digitization initialization signals are the only inputs to the ADC in addition to the sensor select bits. Significant attention must be paid to the SAR DAC since it directly affects the linearity as well as the monotonicity of the ADC. In addition, specifically for SAR ADCs where comparison and setting of the output register both happen in a synchronous fashion, care must be

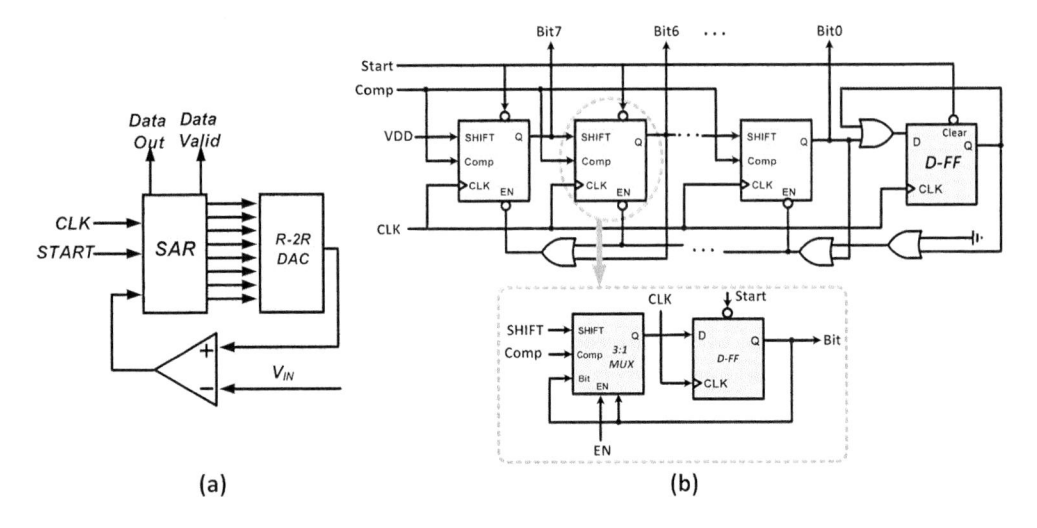

Fig. 11.16 (a) ADC block diagram; (b) fully synchronous SAR.

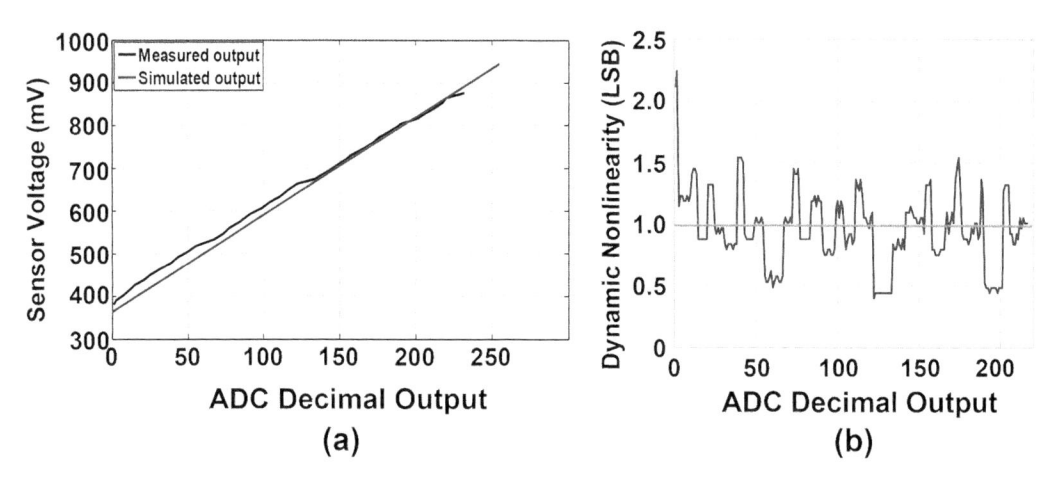

11.5 Data converters: interfacing with the digital core   441

Fig. 11.17 (a) ADC measured and simulated characteristics ($V_{refn} = 350$ mV and $V_{refp} = 950$ mV); (b) measured dynamic nonlinearity (DNL).

taken to ensure that the comparator does not change its output over the approximation window itself. Synchronous SAR architectures (Fig. 11.16b) are also popular choices to alleviate timing issues associated with asynchronous registers.

The ADC measurements were performed as part of the full self-healing system. Sensor voltages were DC probed and the corresponding readouts were obtained through the digital ASIC. The ADC was verified to operate at 25 MHz clock frequency, which translates to 2.5 Msps for an 8-bit SAR with initialization and data ready bits. Figure 11.17a shows measurement vs simulation results of the implemented ADC. The average DNL was -0.04 LSB and the worst-case DNL was measured to be -0.605 LSB, as shown in Fig. 11.17b. These ensure that the ADC is monotonic, which ensures proper operation of the digital ASIC. The SAR ADC draws 1.6 mW from a 1-V supply.

11.5.2 Digital-to-analog converters

A significant portion of the variations associated with a mm-wave power-generation system are due to quiescent point fluctuations due to process and temperature changes. In fact almost all the performance metrics of a mm-wave PA directly correlate to the bias current, for example, efficiency, saturated output power, etc. Most PAs are designed to operate at or near their maximum saturated power or maximum efficiency point. However, in a typical communication system, the PA operates <10% of the time near its peak output power. A dynamic biasing scheme addresses both these issues; it optimizes the operation of the PA for maximum performance when required and it can also reduce the DC power consumption of the system at back-off, leading to significant improvement in efficiency.

Owing to the cascode nature of the power stage, both the common-source (CS) as well as the cascode (CG) transistor biasing points have significant effects on the overall

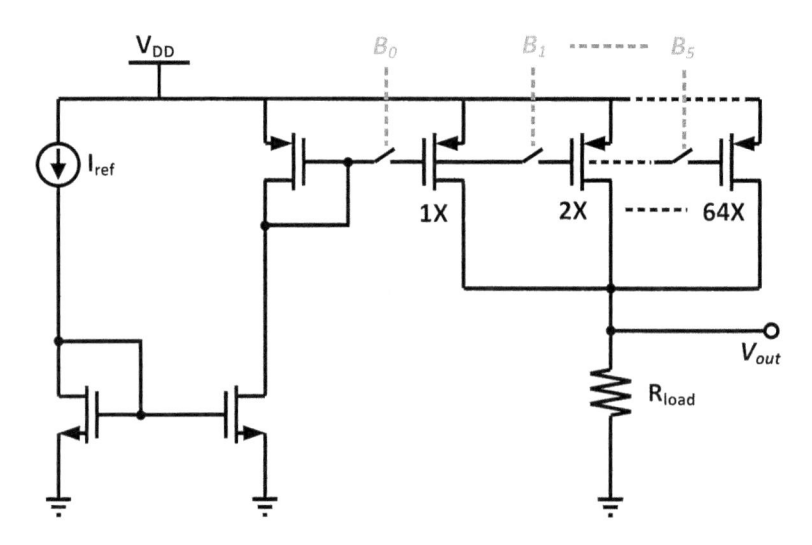

Fig. 11.18 A 6-bit binary weighted current-mode DAC [16].

system performance. Thus, bias control in the form of DACs has been implemented for both these transistors. The requirements for these DACs are the following: they need to be extremely low-power blocks so as to reduce overall self-healing overhead; however, they also need to drive the output stage transistors at relatively high speeds, which keeps the healing time to a minimum. In the present design, current-source based DACs are implemented. These binary weighted current sources are laid out in a common-centroid fashion to ensure good matching. To minimize variations due to process, the transistors in the DAC are long-channel devices which are less susceptible to variations such as line-edge-roughness. The DAC for the CS transistor provides output voltages in the range 450 mV–1.05 V, whereas the CG DAC provides 1.1–1.95 V. A 6-bit control ensures adequate resolution in the healing search space. For calibration purposes, the CS DAC has one extra OFF bit which sets the bias voltage to 0 V, thereby turning the PA off. Figure 11.18 depicts schematics of the implemented current-mode DAC. The two DACs were verified to operate at 25 MHz, which was limited by the test setup. Measurements from the CG and CS DACs are shown in Fig. 11.19.

11.6 Algorithms: setting the actuators based on sensor data

The integrated self-healing core of the system takes sensor data as inputs, and based on those data it sends instructions to the actuation mechanisms to optimize performance. This procedure can occur in an iterative fashion and the entire communication link between the sensors and actuators can be controlled by an on-chip digital core. Independent control of the various actuators and capability of monitoring the resultant effects through the various sensors allows us to minimize a defined cost-function, which could be represented as

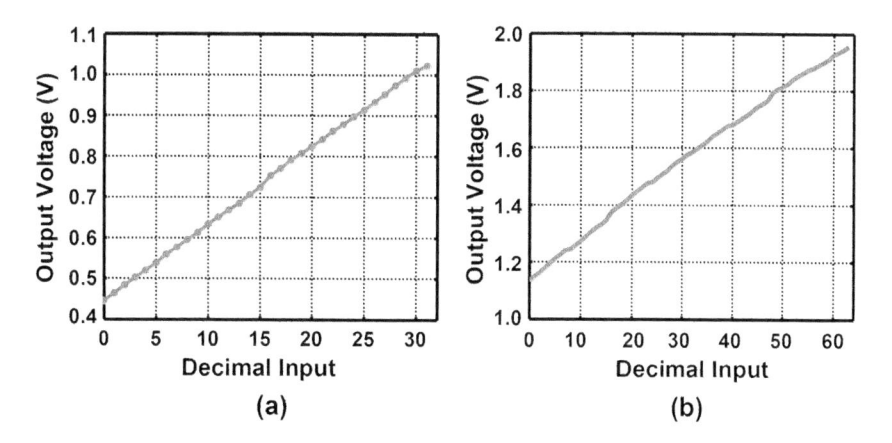

Fig. 11.19 Measured response of (a) common-source and (b) common-gate DAC [16].

$$C = \sum_i w_i(p_i - p_{i0})^2, \; p_i = f(s_1, s_2, \ldots), \tag{11.4}$$

where p_i is a performance parameter of interest that is desired to be set at a value of p_{i0} and can be expressed as a function of multiple sensor readings as $p_i = f(s_1, s_2, \ldots)$, and w_i is the weight of the corresponding error function. By manipulating the different weights, the system may be geared towards optimizing certain parameters across multiple frequency bands, trading off with the other less important ones. The mapping $f : \vec{s} \rightarrow p$, which computes the desired performance parameter from the sensory response data, can be a direct function such as measuring output power from the power sensor or PAE from a combination of measurement from input and output RF sensors and DC current sensors. The mapping can also be indirect as described in [43, 44], where an optimized multi-tone input was applied to a mixer under test, and, from the envelope information extracted from the output power sensor, a nonlinear regression model was built off-chip to extract the desired parameters such as gain, IIP3, etc. In [45], the use of external stimulus was avoided by connecting the device-under-test (LNA) in a feedback network to produce oscillation that was analyzed on-chip to extract the parameters of interest. In the case of the power amplifier presented here, we will focus on the PA characterization by the input and output RF power, gain, and PAE at the mm-wave frequency of interest. These metrics can be directly computed on-chip using the sensor data.

The rationale behind choosing a particular algorithm will depend on the particular application in hand, the performance metrics of interest, and the nature of actuation space. If the space is convex or linear, then various fast-converging algorithms exist that can guarantee convergence to the optimum set of performance parameters within the actuation space. Depending on the system under test, the nature of the algorithm could either be geared towards individually optimizing blocks or sub-blocks. However, the possibility exists for a global self-healing process, where information of multiple components of the system may go to a centralized healing core which can direct multiple sub-healing cores responsible for healing the various components. In this case, we will focus our attention on the healing process of the power amplifier related to its input and output power and efficiency. In references [22, 46], an iterative procedure was carried out to extend the linear range of the V-band power amplifier (55.5–62.5 GHz) by a method of adaptive biasing. In the case of the example PA, two methods of automated healing were implemented, as explained in the later sections.

The algorithm was implemented as a custom digital core, which was coded in VHDL and synthesized. The self-healing digital core was built with a set of instruction sets corresponding to several different modes of operation such as fully automated self-healing, reading sensor data without actuation, or step-by-step healing with off-chip control, etc. Once the desired instruction set was chosen, the global-state machine controlled all the necessary communication between the various component blocks for actuation loading, sensor reading, and optimization. Owing to this modular code setup, many different types of complex optimization algorithms can be incorporated into this general fully integrated self-healing framework. Two modes of fully automated healing algorithms were implemented within the digital core. This is illustrated in Fig. 11.20. All

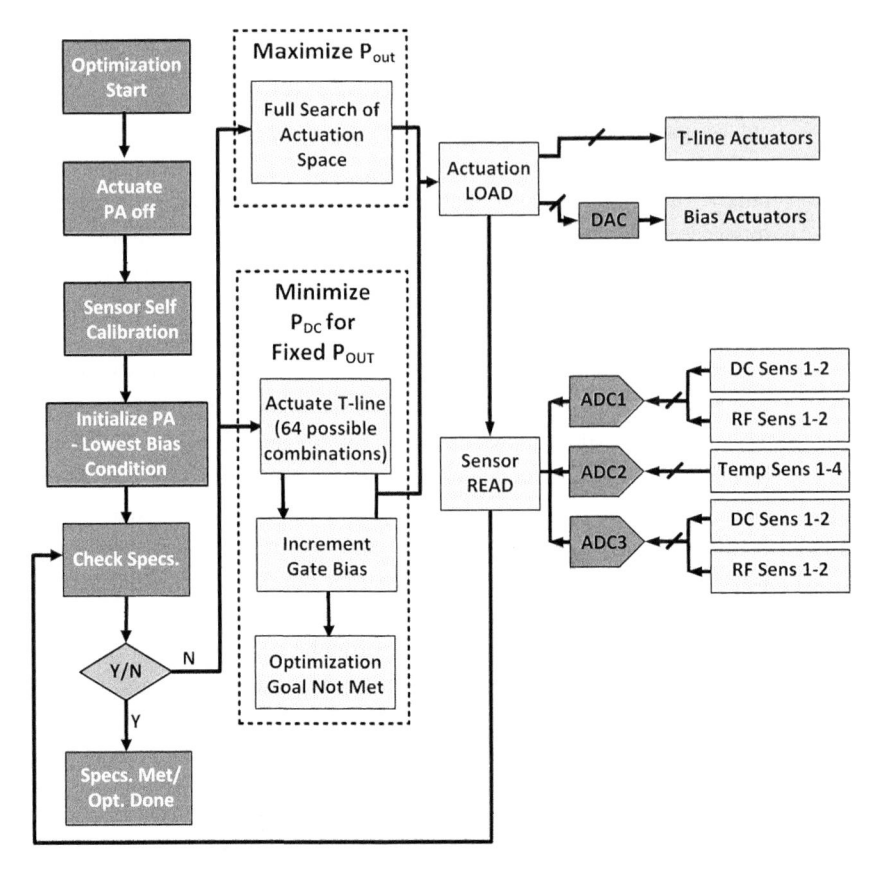

Fig. 11.20 Flowchart showing details of self-healing digital core and the possible modes of fully automated self-healing [16].

the possible modes of operation start with an automated offset calibration step, which measures the DC offset setting of the sensors when the PA is turned off and subtracts it from all future measurements.

The first self-healing mode optimized the actuation settings (both bias and t-line combiner settings) for the highest output power. This was an exhaustive search among all the possible 262 144 states. The algorithm starts with the lowest bias settings (lowest DC current) and then continues to increase the bias actuation one bit at a time, iterating through all possible combiner settings for each DC current setting. The settings of the driver and output stage are varied independently. The second mode of automated healing tries to find the most efficient state of the PA that can deliver at least a given amount of output RF power. As shown in Fig. 11.20, this mode also starts with the lowest bias setting, reads the sensor data through the shared ADC in a time-multiplied manner, and checks for the desired output power condition for all combinations of the t-line combiner setting. If the output power requirement is not reached, the bias current settings are incremented until the performance goal is reached. As before, the bias settings for the driver stage and the output stage are varied independently. If the desired output power has still not achieved after all actuation states have been searched, the state with the highest output power is returned.

In the current implementation, the digital core uses a test-setup-limited clock of 25 MHz (though the on-chip core is verified to operate without timing errors until 500 MHz) and requires 3 µs per optimization iteration (set actuators, read all sensors, decide on next actuation state). This results in a maximum healing time of 0.8 s when the algorithm is an exhaustive search visiting all possible actuation states.

11.7 System measurements of a fully integrated self-healing PA

The example PA was fabricated using the self-healing blocks presented in the previous sections, and the system-level measurements will be presented in this section. The measurement setup of the PA is shown in Fig. 11.21. The PA was mounted on a PCB and probed at 28 GHz. It was driven by an Agilent 83650B signal generator, and probed

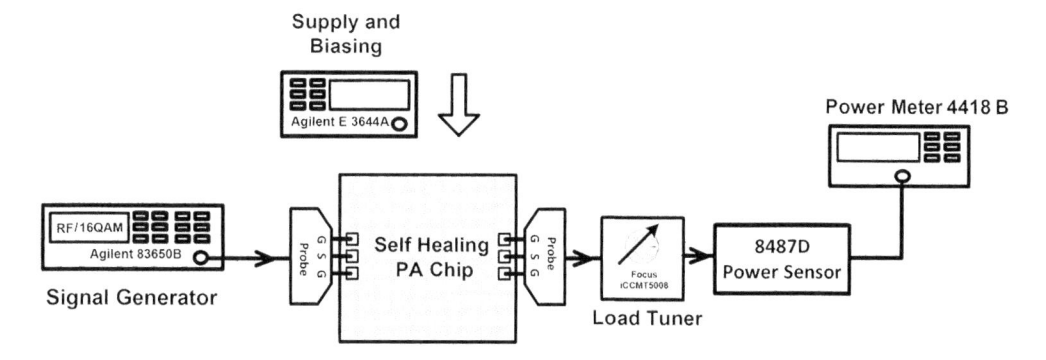

Fig. 11.21 Measurement setup for the fully integrated self-healing power amplifier [16].

with Cascade Z-probes. The output went through a calibrated network that included a mm-wave load tuner to an Agilent 8487D power sensor.

A comparison with an unhealed PA is useful to look at the benefits of the self-healing system, so a default actuation state must be selected as the "default" state. The default state is chosen to be the state that had the best saturated performance in simulation and represents the PA that would have been taped out if the self-healing system was not being used. For all of the measurements without self-healing, the DC power consumption of the self-healing blocks was omitted when looking at the performance of the default state.

The entire self-healing system was integrated on a single chip, which means that the only external input given to the chip was the mode of operation, which algorithm to run, the desired output power (if needed), and then a go command, and no external calibration or external performance information was used during healing.

11.7.1 Healing process variation with a nominal 50-Ω load

The healing ability of the amplifier for 50-Ω loads is presented first. Figure 11.22a shows the output power vs input power for an amplifier in its default state, as well as one that has been healed for maximum output power at low input-power levels, and one that has been healed for maximum output power at the 1-dB compression point, the point where the gain has been compressed by 1 dB compared with the small-signal gain. This plot

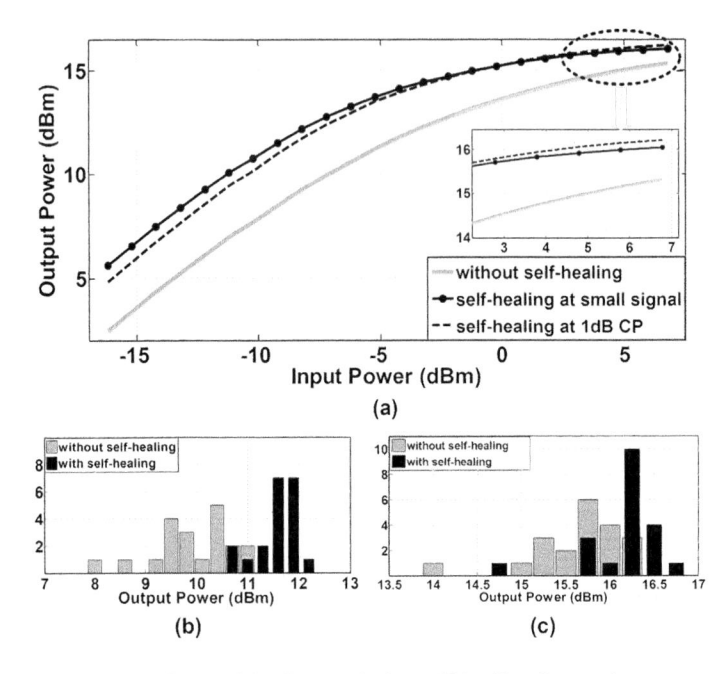

Fig. 11.22 Measured output power before self-healing, and after self-healing for maximum output power, both for healing done at small signal and at the 1-dB compression point (a), and histograms of 20 measured chips before and after self-healing at small signal (b), and at the 1-dB compression point (c) [16].

shows the improvement in output power that can be achieved using self-healing, but also shows that there is no one optimal state for all input powers. The optimum load impedance for the maximum output power from the output amplifying transistors varies based upon the input power levels, and thus, by tuning the matching network for the current power level, the corresponding optimal matching network can be found. This is an added benefit of the self-healing system, as it can be healed for the desired power level, and if that desired power level changes, it can be healed again. Near saturation where the default state was designed, it is close to the optimum, and thus there is not as much room for improvement. However, the default match at small signal is farther from the optimum, and thus there self-healing can provide larger improvements to the performance.

To show improvement for process variation, 20 chips were measured and histograms showing results with and without self-healing for maximum output power are shown in Figs. 11.22b, c, respectively. The small-signal gain after healing is 21.5 dB, with a saturated output power of 16 dBm and a 1-dB compression point of 12.5 dBm while consuming 520 mW of DC power at 28 GHz.

The second algorithm to minimize the DC power while maintaining a desired output power is shown next. Figure 11.23a shows the DC power consumption of 20 chips for various output power levels, with a histogram cross section of that plot for 12.5 dBm output power, near the 1-dB compression point, shown in Fig. 11.23b. Because the state is not changing, the DC power levels without self-healing for each chip are relatively flat until the transistors really start to saturate and the power increases slightly. Once self-healing is turned on, there is still high DC power required to achieve very high output powers, but once the desired output power becomes even a couple of dB below the saturated power, significant reduction in DC power consumption is observed. The DC power required to produce 12.5 dBm output power sees a 47% reduction in average power level over 20 chips, with a 78% decrease in the standard deviation between chips. This means that self-healing is improving the performance, but also making it much more consistent across chips than in the default case. Once the power is near small-signal levels, reductions of greater than 50% for every single chip measured are achieved.

11.7.2 Healing VSWR environmental variation with load mismatch

The load impedance was varied using a focus microwave mm-wave load tuner that produces loads within the 12-1 VSWR circle at the tuner. When calibrating for the loss of the cable and probes, it becomes a load variation within the 4-1 VSWR circle at the probe tips, which corresponds to a resistance from 25 Ω to 200 Ω on the real impedance axis. The ability of the RF power sensors to detect actual power going to the load, not just the voltage, enables the self-healing system to know the power delivered to the load even under load-impedance mismatch. This means that the algorithm can still heal and doesn't require knowledge of the load impedance to run the optimization, as the metric of interest, the output RF power, is already known.

The results of healing for maximum output power when the load impedance is swept within the 4-1 VSWR circle are shown as contour plots on the Smith chart in Fig. 11.24,

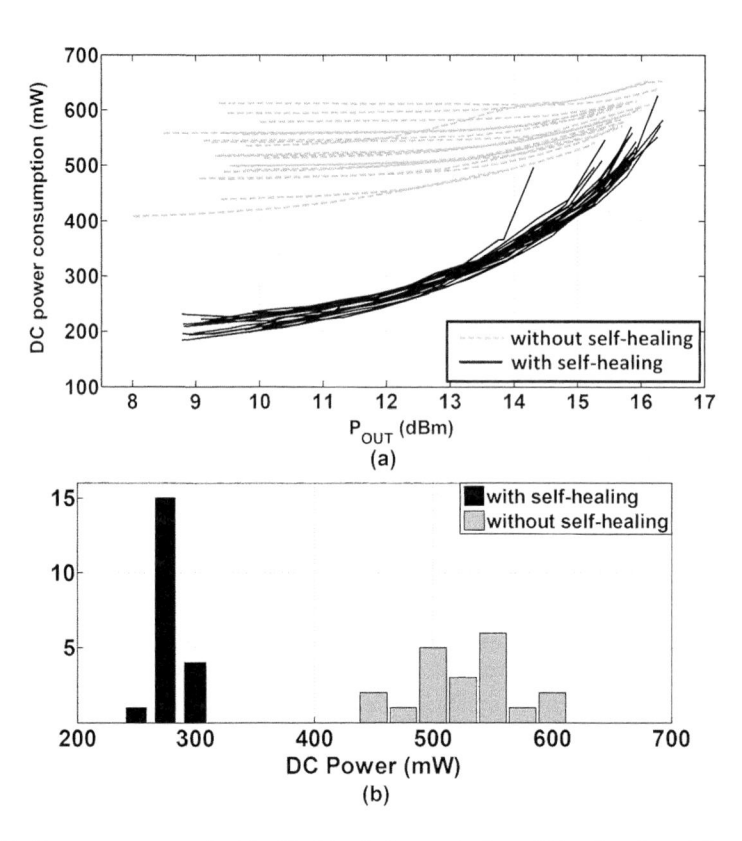

448 Self-healing for silicon-based mm-wave power amplifiers

Fig. 11.23 Measured DC power consumption for 20 chips before and after self-healing for minimum DC power while maintaining a desired RF power level is used (a), and a histogram cross section of 20 chips (b) of the DC power consumption before and after self-healing to maintain an output power of 12.5 dBm, near the 1-dB compression point [16].

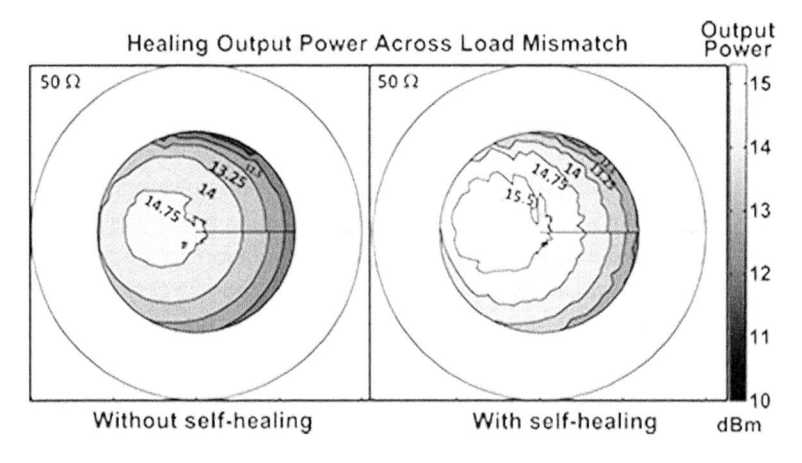

Fig. 11.24 Contour plots before and after self-healing for maximum output power for load impedance mismatch show improvement in output power over the entire 4-1 VSWR impedance circle [16].

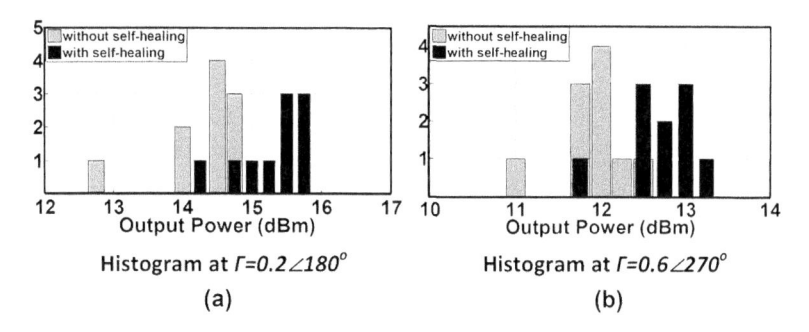

11.7 System measurements of a fully integrated self-healing PA

Fig. 11.25 Histograms of 10 measured chips showing output power before and after self-healing at two representative load impedance points, one near the maximum output power (a), and the other on the edge of the 4-1 VSWR impedance circle (b)[16].

and show an improvement in output power across the entire 4-1 VSWR circle. Ten chips were measured, and the results of self-healing for maximum output power for two representative load impedances are shown in Fig. 11.25. The first is near the maximum output power, and the second on the edge of the 4-1 VSWR circle. Again both show improvement overall in the output power, as well as a reduction in the variation between chips.

The second algorithm to minimize DC power for a desired output power was also tested under load-impedance mismatch, and the results for a desired output power of 12.5 dBm are shown as contour plots in Fig. 11.26. The outermost contour represents the loads where 12.5 dBm was achieved, with the shading and all subsequent contours representing the DC power consumed. For the default state, the DC power consumption remains constant regardless of the load-impedance mismatch. With self-healing, however, the power can be substantially reduced by up to 35% at impedances near 50 Ω, where the PA was designed, while still maintaining the desired output powers at the more extreme impedance mismatches.

11.7.3 Healing for linearity

The PA is designed as a linear amplifier to enable the use of non-constant envelope modulation schemes. The linearity of the PA has been verified using a 100 ksps 16-quadrature amplitude modulation (QAM) signal to measure the error-vector magnitude (EVM) for 10 chips, shown for 12.5 dBm output power as a histogram in Fig. 11.27. While the self-healing system does not specifically attempt to improve linearity, a reduction in average EVM from 5.9% to 4.2% is observed when self-healing for maximum output power was applied. The healed chips are able to provide higher output powers while not being pushed as far into saturation, and thus the linearity for a given output power is improved.

11.7.4 Healing for partial and total transistor failure

Partial or total transistor failure can be caused by aging, transistor stress such as from voltage spikes, or other phenomena. To show the self-healing system's ability to heal for

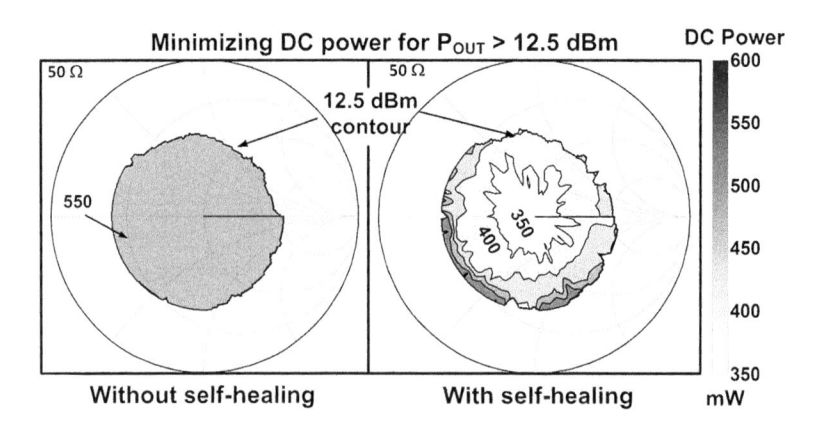

Fig. 11.26 Contour plots before and after self-healing for minimum DC power consumption while maintaining 12.5 dBm desired output RF power for load-impedance mismatch show improvement in output power over the entire 4-1 VSWR impedance circle [16].

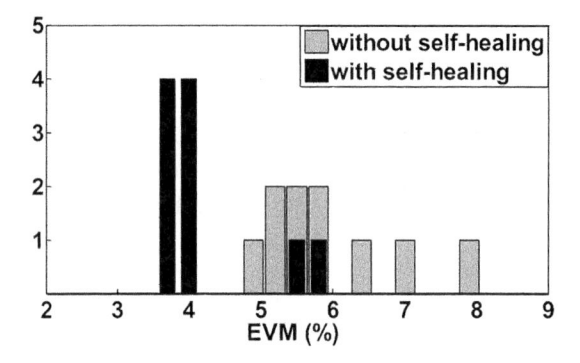

Fig. 11.27 Error-vector magnitude of 10 chips before and after self-healing for maximum output power show an improvement in linearity after self-healing [16].

partial and total transistor failure, a laser trimmer was used to blast away various parts of one of the output-stage transistors. The output stage was chosen as the transistors to cut as they are the ones that are pushed closest to breakdown and are likely to be the first ones to fail. Only one of the two output stages was cut to cause a worst-case scenario from a mismatch standpoint. As a reference, the amplifier before any laser blasting is shown in Fig. 11.28a. Measurements taken in the default state and after healing for maximum output power with half of the common-source transistor cut out are plotted in Fig. 11.28b. Figure 11.28c shows the results with and without self-healing when half of the cascode stage was additionally cut out. Finally, Fig. 11.28d shows the results once the entire output stage has been blasted away. This means that the matching network that was expecting to have two similar drives at both inputs now has only a single input, and then a large stub where the other input used to be, destroying the original match that was shown in Fig. 11.14.

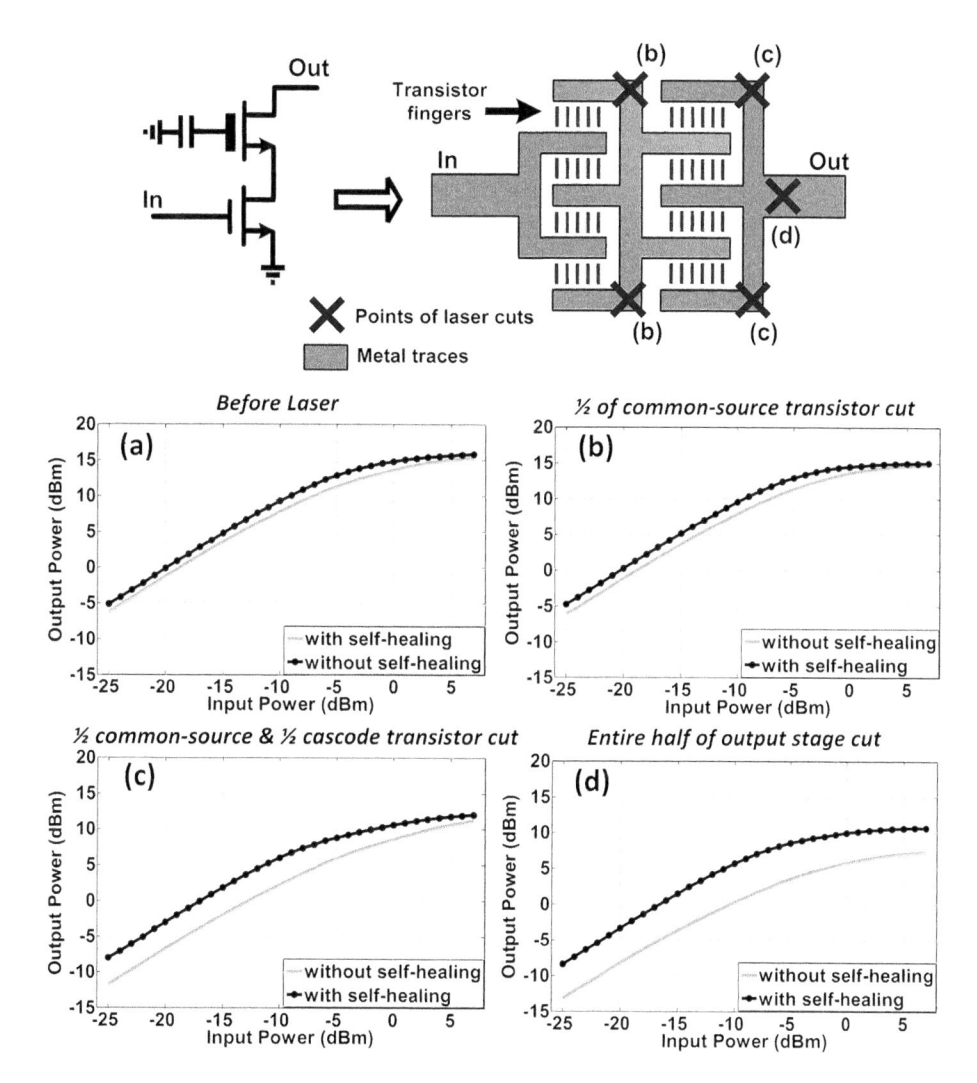

11.7 System measurements of a fully integrated self-healing PA

Fig. 11.28 Schematic and layout location of laser trim points, and measurements before and after self-healing for maximum output power at various stages of transistor failure due to laser blasting show more than 5 dB improvement when self-healing is used in the worst-case scenario of an entire output stage failing [16].

The default case at small signal loses 7.2 dB in output power from when the output stage is whole to when it is completely cut out: 3 dB of that is due to having only one of the two output stages providing power, but another 4.2 dB is caused by mismatch of the matching network. Once healing is applied, the loss due to cutting out the output stage is only 3.3 dB, which when taking into account the 3 dB loss from only having a single output stage means that the tunable matching network was able to heal back to the point where there was only 0.3 dB additional loss from this catastrophic event. This is one of the very strong points of self-healing, as under nominal conditions, where the default

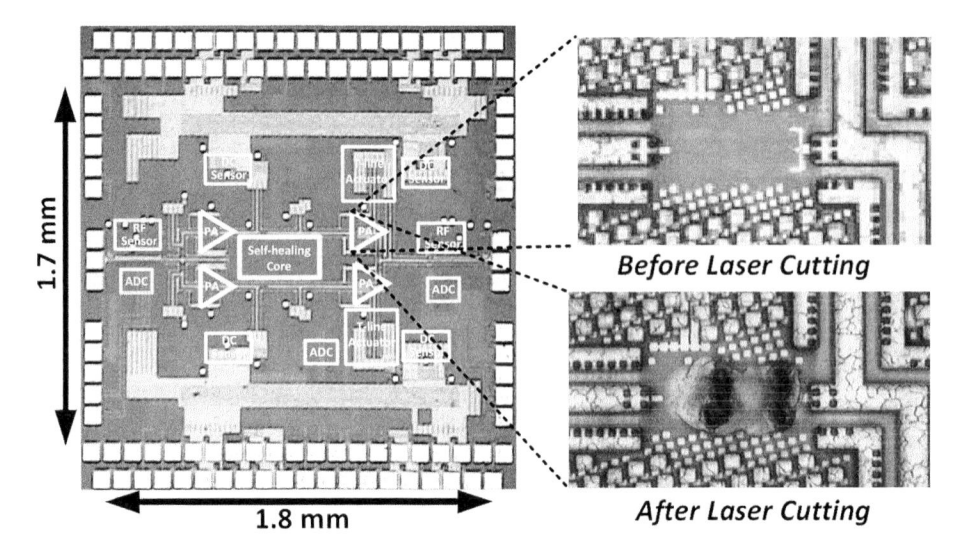

Fig. 11.29 Die photo of the self-healing PA with close-up views of one output stage before and after laser blasting [16].

design is close to optimum, self-healing can only possibly improve the circuit by at most the deviation from the optimum, but in cases such as this where the default falls far from the optimum point, and would normally register as a total failure of the entire circuit, self-healing can provide very significant gains and keep the circuit operational even under these types of extreme conditions. A die photo of the entire chip with close-up images before and after the laser blasting is shown in Fig. 11.29.

11.7.5 Yield improvement

An effective self-healing system should improve the yield of the design, that is, to improve the percentage of chips that achieve a minimum performance specification for acceptable use. It is important also to consider any additional area requirements that the self-healing system requires, as it can reduce the number of chips that can fit on a given wafer size. However, in this design, the self-healing circuitry is placed within the confines of the PA itself in area that otherwise would have gone to waste, and thus the post-healing yield can be compared chip-to-chip with the pre-healing yield. For this design, the specifications included a saturation output power >15.5 dBm, gain >20 dB, and a power-added efficiency >6%, and a 4-1 VSWR tolerance <3 dB, defined in dB as the worst-case output power falloff within the 4-1 VSWR circle compared with a nominal 50-Ω load. The PA was able to achieve best-case metrics of 16.5 dBm saturated output power, 23.7 dB gain, 7.2% efficiency, and 2.28 dB 4-1 VSWR tolerance. Across 20 chips, the yield of the saturated output power improved from 20% to 90%, the gain improved from 20% to 100%, and the efficiency improved from 5% to 100%. For the 10 chips measured under load-impedance mismatch, the yield improved from 0% to 80%, with an overall aggregate yield for all performance specifications improving from 0% to 80%.

11.8 Conclusions

Self-healing can enable significant improvements in performance if a mm-wave design has become susceptible to performance degradation caused by process or environmental variation. The effectiveness of the self-healing system is dependent on how far from optimal a non-healing design becomes. By programming different algorithms into the digital core, the amplifier can become dynamic, and trade-offs that traditionally had to be made at the design level before fabrication can be adjusted based upon the current environment and use case of the mm-wave circuit.

Self-healing enables aggressive mm-wave designs even in process nodes where process variation and mismatch can cause significant variation in transistor operation. By using robust design to sense the performance of the mm-wave circuit during the self-healing optimization process, the core mm-wave circuit can take advantage of that aggressive design and push performance levels while maintaining yield. It is expected that, as the minimum feature size continues to decrease, process variation will continue to increase, and the need for self-healing or other reconfigurable circuit techniques will increase.

Environmental variation can also be mitigated using self-healing. Variation in PA performance due to load-impedance mismatch caused by VSWR events can be handled by making the output matching network tunable, so that the output amplifier stages still see their optimum impedance for maximum output power. Transistor degradation and failure due to effects such as aging can also be counteracted by adjustment of the DC operation point of the transistors and by adjusting the matching networks as the optimal impedances change.

A 28-GHz power amplifier was presented as a case study of how such an integrated self-healing PA could be implemented. Measurements of multiple chips demonstrate the viability of an integrated self-healing system that requires no external calibration of any kind. Integrating the sensors, actuators, digital algorithm, and data converters on a single chip allows for a completely automated healing system that improves aggregate yield from several performance specifications from 0% up to 80%, while healing for process variation, mismatch, load impedance mismatch, and partial and total transistor failure.

Acknowledgment

The authors would like to thank Professor Aydin Babakhani of Rice University (formerly of Caltech) and Arthur Chang for their valuable technical discussions and contributions. They would also like to thank Professor Sanjay Raman, Tony Quach, Christopher Maxey, the Defence Advanced Research Projects Agency (DARPA), and the Air Force Research Laboratory (AFRL) for support.

References

[1] S. Nassif, N. Mehta, and Y. Cao, "A resilience roadmap," in *Design Automation Test in Europe Conference Exhibition*, 2010, pp. 1011–1016.

[2] K. Bernstein, D. J. Frank, A. E. Gattiker, *et al.*, "High-performance CMOS variability in the 65-nm regime and beyond," *IBM J. of Research and Development*, vol. **50**, no. 4.5, pp. 433–449, July 2006.

[3] T. Mizuno, J. Okumtura, and A. Toriumi, "Experimental study of threshold voltage fluctuation due to statistical variation of channel dopant number in MOSFETs," *IEEE Trans. on Electron Devices*, vol. **41**, no. 11, pp. 2216–2221, Nov. 1994.

[4] P. Stolk and D. Klaassen, "The effect of statistical dopant fluctuations on MOS device performance," in *International Electron Devices Meeting, IEDM*, Dec. 1996, pp. 627–630.

[5] S. Borkar, "Designing reliable systems from unreliable components: the challenges of transistor variability and degradation," *IEEE Micro*, vol. **25**, no. 6, pp. 10–16, 2005.

[6] K. Kuhn, "CMOS transistor scaling past 32 nm and implications on variation," in *IEEE/SEMI Advanced Semiconductor Manufacturing Conference*, 2010, pp. 241–246.

[7] A. Asenov, "Simulation of statistical variability in nano MOSFETs," in *IEEE Symposium on VLSI Technology*, 2007, pp. 86–87.

[8] M. Ruberto, O. Degani, S. Wail, *et al.*, "A reliability-aware RF power amplifier design for CMOS radio chip integration," in *IEEE International Reliability Physics Symposium*, 2008, pp. 536–540.

[9] O. Hammi, J. Sirois, S. Boumaiza, and F. Ghannouchi, "Study of the output load mismatch effects on the load modulation of Doherty power amplifiers," in *IEEE Radio and Wireless Symposium*, 2007, pp. 393–394a.

[10] A. Natarajan, A. Komijani, X. Guan, A. Babakhani, and A. Hajimiri, "A 77-GHz phased-array transceiver with on-chip antennas in silicon: Transmitter and local LO-path phase shifting," *IEEE J. of Solid-State Circuits*, vol. **41**, no. 12, pp. 2807–2819, Dec. 2006.

[11] A. Babakhani, X. Guan, A. Komijani, A. Natarajan, and A. Hajimiri, "A 77-GHz phased-array transceiver with on-chip antennas in silicon: Receiver and antennas," *IEEE J. of Solid-State Circuits*, vol. **41**, no. 12, pp. 2795–2806, Dec. 2006.

[12] H. Hashemi, X. Guan, A. Komijani, and A. Hajimiri, "A 24-GHz SiGe phased-array receiver-LO phase-shifting approach," *IEEE Trans. on Microwave Theory and Techniques*, vol. **53**, no. 2, pp. 614–626, Feb. 2005.

[13] A. Natarajan, A. Komijani, and A. Hajimiri, "A 24 GHz phased-array transmitter in 0.18 μm CMOS," in *IEEE Int. Solid-State Circuits Conference Digest of Technical Papers (ISSCC)*, Feb. 2005, pp. 212–594, Vol. 1.

[14] S. Alalusi and R. Brodersen, "A 60 GHz phased array in CMOS," in *IEEE Custom Integrated Circuits Conference (CICC)*, Sept. 2006, pp. 393–396.

[15] A. Valdes-Garcia, S. T. Nicolson, J.-W. Lai, *et al.*, "A fully integrated 16-element phased-array transmitter in SiGe BiCMOS for 60-GHz communications," *IEEE J. of Solid-State Circuits*, vol. **45**, no. 12, pp. 2757–2773, Dec. 2010.

[16] S. M. Bowers, K. Sengupta, K. Dasgupta, B. D. Parker, and A. Hajimiri, "Integrated self-healing for mm-wave power amplifiers," *IEEE Trans. on Microwave Theory and Techniques*, vol. **61**, no. 3, pp. 1301–1315, 2013.

[17] R. Kumar and V. Kursun, "Voltage optimization for temperature variation insensitive CMOS circuits," in *48th Midwest Symposium on Circuits and Systems*, 2005, pp. 476–479, Vol. 1.

[18] S. Sakurai and M. Ismail, "Robust design of rail-to-rail CMOS operational amplifiers for a low power supply voltage," *IEEE Journal of Solid-State Circuits*, vol. **31**, no. 2, pp. 146–156, 1996.

[19] K. Siwiec, T. Borejko, and W. Pleskacz, "PVT tolerant LC-VCO in 90 nm CMOS technology for GPS/Galileo applications," in *IEEE International Symposium on Design and Diagnostics of Electronic Circuits Systems*, 2011, pp. 29–34.

[20] D. Sylvester, D. Blaauw, and E. Karl, "ElastIC: an adaptive self-healing architecture for unpredictable silicon," *IEEE Design Test of Computers*, vol. **23**, no. 6, pp. 484–490, 2006.

[21] A. Goyal, M. Swaminathan, A. Chatterjee, D. Howard, and J. Cressler, "A new self-healing methodology for RF amplifier circuits based on oscillation principles," *IEEE Transactions on Very Large Scale Integration (VLSI) Systems*, vol. **20**, no. 10, pp. 1835–1848, 2012.

[22] J.-C. Liu, A. Tang, N. Wang, et al., "A V-band self-healing power amplifier with adaptive feedback bias control in 65 nm CMOS," in *IEEE Radio Frequency Integrated Circuits Symposium (RFIC)*, 2011, pp. 1–4.

[23] S. Yaldiz, V. Calayir, X. Li, et al., "Indirect phase noise sensing for self-healing voltage controlled oscillators," in *IEEE Custom Integrated Circuits Conference (CICC)*, 2011, pp. 1–4.

[24] K. Jayaraman, Q. Khan, B. Chi, et al., "A self-healing 2.4 GHz LNA with on-chip S11/S21 measurement/calibration for in-situ PVT compensation," in *IEEE Radio Frequency Integrated Circuits Symposium*, 2010, pp. 311–314.

[25] F. Bohn, K. Dasgupta, and A. Hajimiri, "Closed-loop spurious tone reduction for self-healing frequency synthesizers," in *IEEE Radio Frequency Integrated Circuits Symposium (RFIC)*, June 2011, pp. 1–4.

[26] H. Wang, K. Dasgupta, and A. Hajimiri, "A broadband self-healing phase synthesis scheme," in *IEEE Radio Frequency Integrated Circuits Symposium (RFIC)*, June 2011, pp. 1–4.

[27] S. M. Bowers, K. Sengupta, K. Dasgupta, and A. Hajimiri, "A fully-integrated self-healing power amplifier," in *IEEE Radio Frequency Integrated Circuits Symposium (RFIC)*, June 2012, pp. 221–224.

[28] A. Tang, F. Hsiao, D. Murphy, et al., "A low-overhead self-healing embedded system for ensuring high yield and long-term sustainability of 60 GHz 4 Gb/s radio-on-a-chip," in *IEEE Int. Solid-State Circuits Conference Digest of Technical Papers (ISSCC)*, Feb. 2012, pp. 316–318.

[29] Y. Huang, H. Hsieh, and L. Lu, "A build-in self-test technique for RF low-noise amplifiers," *IEEE Trans. on Microwave Theory and Techniques*, vol. **56**, no. 5, pp. 1035–1042, 2008.

[30] K. Sengupta, K. Dasgupta, S. M. Bowers, and A. Hajimiri, "On-chip sensing and actuation methods for integrated self-healing mm-wave CMOS power amplifier," in *IEEE MTT-S International Microwave Symposium Digest (MTT)*, June 2012, pp. 1–3.

[31] Q. Yin, W. Eisenstadt, R. Fox, and T. Zhang, "A translinear RMS detector for embedded test of RF ICs," *IEEE Trans. on Instrumentation and Measurement*, vol. **54**, no. 5, pp. 1708–1714, 2005.

[32] C. de La Cruz-Blas, A. Lopez-Martin, A. Carlosena, and J. Ramirez-Angulo, "1.5-V current-mode CMOS true RMS-DC converter based on class-AB transconductors," *IEEE Trans. on Circuits and Systems II : Express Briefs*, vol. **52**, no. 7, pp. 376–379, 2005.

[33] A. Valdes-Garcia, R. Venkatasubramanian, J. Silva-Martinez, and E. Sanchez-Sinencio, "A broadband CMOS amplitude detector for on-chip RF measurements," *IEEE Transactions on Instrumentation and Measurement*, vol. **57**, no. 7, pp. 1470–1477, 2008.

[34] S. Cripps, *RF Power Amplifiers for Wireless Communications*. Artech House Microwave Library, 2006.

[35] Y. Huang, H. Hsieh, and L. Lu, "A low-noise amplifier with integrated current and power sensors for RF BIST applications," in *25th IEEE VLSI Test Symposium*, 2007, pp. 401–408.

[36] F. Cheung and P. Mok, "A monolithic current-mode CMOS DC–DC converter with on-chip current-sensing technique," *IEEE J. of Solid-State Circuits*, vol. **39**, no. 1, pp. 3–14, 2004.

[37] S. Lee and D. Allstot, "Electrothermal simulation of integrated circuits," *IEEE J. of Solid-State Circuits*, vol. **28**, no. 12, pp. 1283–1293, 1993.

[38] B. Razavi, *Design of Analog CMOS Integrated Circuits*. McGraw-Hill, 2001.

[39] S. Park, Y. Palaskas, A. Ravi, R. Bishop, and M. Flynn, "A 3.5 GS/s 5-b flash ADC in 90 nm CMOS," in *IEEE Custom Integrated Circuits Conference (CICC)*, Sept. 2006, pp. 489–492.

[40] Y.-Z. Lin, Y.-T. Liu, and S.-J. Chang, "A 5-bit 4.2-GS/s flash ADC in 0.13-μm CMOS," in *IEEE Custom Integrated Circuits Conference (CICC)*, Sept. 2007, pp. 213–216.

[41] J. Li and U.-K. Moon, "A 1.8-V 67-mW 10-bit 100-MS/s pipelined ADC using time-shifted CDS technique," *IEEE J. of Solid-State Circuits*, vol. **39**, no. 9, pp. 1468–1476, Sept. 2004.

[42] S. Mortezapour and E. Lee, "A 1-V, 8-bit successive approximation ADC in standard CMOS process," *IEEE J. of Solid-State Circuits*, vol. **35**, no. 4, pp. 642–646, Apr. 2000.

[43] S. Akbay and A. Chatterjee, "Built-in test of RF components using mapped feature extraction sensors," in *Proceedings of IEEE VLSI Test Symposium*, 2005, pp. 243–248.

[44] S. Devarakond, V. Natarajan, S. Sen, and A. Chatterjee, "BIST-assisted power aware self healing RF circuits," in *IEEE International Mixed-Signals, Sensors, and Systems Test Workshop*, 2009, pp. 1–4.

[45] A. Goyal, M. Swaminathan, A. Chatterjee, D. Howard, and J. Cressler, "A new self-healing methodology for RF amplifier circuits based on oscillation principles," *IEEE Trans. on Very Large Scale Integration (VLSI) Systems*, vol. **20**, no. 10, pp. 1835–1848, 2012.

[46] J.-C. Liu, R. Berenguer, and M. Chang, "Millimeter-wave self-healing power amplifier with adaptive amplitude and phase linearization in 65-nm CMOS," *IEEE Trans. on Microwave Theory and Techniques*, vol. **60**, no. 5, pp. 1342–1352, 2012.

Index

1/f noise, 39, 101

Active load modulation, 171
Adjacent channel power ratio (ACPR), 5
Advanced Extremely High Frequency (AEHF), 9
AM/AM distortion, 98, 124
Analog-to-digital Converters (ADC), 439
Asymmetric Multilevel Outphasing (AMO), 312
Automotive radar, 9, 412

Backhaul, 2, 68, 257, 376
Backoff, 171, 177, 314
Balun, 280
Baseband, 307, 310, 316, 331, 334, 338, 341, 345, 353, 378, 385
Base Current Reversal, 41, 190
Base Resistance, 19, 25, 34, 37, 43, 59
Bipolar junction transistors (BJT), 18, 218
Bit error rate (BER), 5
Breakdown voltage, 41, 63, 103, 143, 187
BSIM model, 85, 119, 131
Built-in self-test (BIST), 3, 8

Cartesian, 4, 340, 349, 361, 366
Cascode, 107, 142, 219, 230, 280, 339, 400, 403, 413, 423, 425, 435, 441
Channel capacity, 5, 316
Chireix, 202, 314, 338
Co-channel interference, 6
Collector-base breakdown voltage for an open emitter (BVCBO), 41, 60, 63, 143, 167, 193, 195, 403
Collector-emitter breakdown voltage for open base (BVCEO), 41, 63, 143, 167, 188
Conduction Angle, 156, 162
Coplanar Waveguide (CPW), 148
Coupler, 218, 260, 267, 384, 392, 402, 429

Deep N-well (DNW), 279
Deep trench (DT), 29, 30, 45
Defense Advanced Research Projects Agency (DARPA), 10
Digital to Analog Converter (DAC), 201, 318, 335, 346, 360, 366, 410, 439

Doherty, 170, 174
DOTFIVE, 64
Drain Induced Barrier Lowering (DIBL), 82, 131, 150, 419

Early effect, 53
Effective isotropic radiated power (EIRP), 377, 388
Efficient Linearized All-Silicon Transmitter ICs (ELASTx), 10
Electromigration, 61, 63
Envelope Elimination and Restoration (EER), 200, 337
Envelope Tracking (ET), 8
Error vector magnitude (EVM), 5, 201, 320, 334, 337, 339, 341, 359, 381, 383, 384
ESD, 131

f_{max}, 25, 86, 149, 152, 195, 226
Fifth-generation commercial wireless standard (5G), 2, 9, 257
Forward-bias current stress, 60

Gate resistance, 87, 99, 149, 151, 230, 241
Ground-shielded Coplanar Waveguide (GSCPW), 392
Gysel, 400

Harmonic, 92, 108, 121, 129, 142, 155, 168, 180, 188, 215, 240, 334, 363
High Current Model (HICUM), 52
Homodyne, 6

IEEE 802.11ad, 8, 378, 380, 383, 408
IIP3 (Third Order Intercept Point), 95
IM3 (Third Order Intermodulation), 91
Impact ionization, 40, 124, 188
Impedance Transformation, 7, 146, 176, 210, 288
Industrial, scientific, and medical (ISM), 2, 8
Intermodulation distortion (IMD), 91, 96

Johnson Figure of Merit (JFoM), 132

Kahn Transmitter, 200
Kirk effect, 22, 53
Knee Voltage, 78, 141, 189, 215

Index

Linearity, 1, 7, 78, 83, 91, 139, 188, 200, 223, 304, 322, 329, 334, 337, 348, 356, 380, 394, 424, 449
Loadline resistance, 145, 161, 168, 173
Load Pull, 108, 119, 182, 338, 356, 360
Long-range radars (LRR), 9, 412
Loss factor, 142, 146, 165

Maximum Available Gain (MAG), 149, 153
Memory effect, 98, 102, 104
Metal-Insulator-Metal (MIM) capacitor, 51, 115
Metal-Oxide-Metal (MOM) capacitor, 115, 323
Mixed-mode stress, 60
Mixer, 322, 339, 347, 381, 397
Most Exquisite Transistor Model (MEXTRAM), 52
Multi-Input Multi-Output (MIMO), 378

Noise, 5, 37, 53, 98, 342, 427
Noise figure, 37
Nonlinearity, 5, 91, 97

Orthogonal frequency division multiplexing (OFDM), 8, 200, 308, 335, 379, 380, 383, 405
Outphasing transmitter, 202, 302, 340

Patch antenna, 364, 366, 407, 411
Patterned Ground Shield (PGS), 117
Peak-to-average power ratio (PAPR), 139, 317, 380
Phased array, 2, 5, 388, 389, 391, 398
Phase locked loop (PLL), 6, 307, 383, 397–399, 409, 413, 422
Phase Noise, 381, 413
p-i-n diode, 49
Polar, 4, 200, 337
Power added efficiency (PAE), 7, 129, 140, 163, 182, 191, 235, 302
Process, voltage, and temperature (PVT), 6, 421, 422
PSP model, 131

Quadrature Amplitude Modulation (QAM), 139, 171, 200, 335, 346, 366, 378
Quality Factor, 48, 77, 116, 142, 146, 163, 183, 288, 381

Quantization, 339, 342, 373

Receiver, 5, 359, 379, 385, 413
Reliability, 58, 105, 216
Reverse bias emitter-base stress, 59
RF DAC, 338, 340, 350, 357

Safe Operating Area (SOA), 63
SATCOM, 10
Schottky Barrier Diodes, 47
Self-heating, 43, 103, 123, 232
Shape factor, 141, 159, 163
Short-range radars (SRR), 9, 412
Silicon-on-insulator (SOI), 77, 98, 121, 150, 175, 218, 224
Spectral regrowth, 5, 201, 320, 325, 329, 334, 337, 340, 426
Spice Gummel–Poon Model, 52
Stability, 151, 221, 230, 264, 267, 302
System-on-a-chip (SOC), 2, 3, 5, 10, 11, 290, 378, 379

Thermal effect, 44, 102, 119
Thermal resistance, 43, 63, 103, 232
   Mutual thermal resistance, 46
Thin-film microstrip (TFMS), 263, 266, 279
Through-silicon-via (TSV), 51
Transformer, 117, 146, 171, 218, 238, 260, 268, 270, 273, 315

Vertical Bipolar Inter Company Model (VBIC), 52
Voltage Standing Wave Ratio (VSWR), 420, 453

Wilkinson Power Combiner/Divider, 238, 250, 260, 265, 312, 314, 338, 356, 358, 362, 393, 400
Wireless Gigabit Alliance (WiGig), 8, 361, 362, 364, 408

Zero-derivative of voltage switching (ZdVS), 181, 182, 235, 237
Zero-voltage Switching (ZVS), 142, 180–182, 235, 237